The Mollusca

VOLUME 4

Physiology, Part 1

The Mollusca

Editor-in-Chief

KARL M. WILBUR

Department of Zoology
Duke University
Durham, North Carolina

The Mollusca

VOLUME 4
Physiology, Part 1

Edited by

A. S. M. SALEUDDIN

Department of Biology
York University
Toronto, Ontario, Canada

KARL M. WILBUR

Department of Zoology
Duke University
Durham, North Carolina

1983

ACADEMIC PRESS
A Subsidiary of Harcourt Brace Jovanovich, Publishers
New York   London
Paris   San Diego   San Francisco   São Paulo   Sydney   Tokyo   Toronto

ACADEMIC PRESS, INC.
111 Fifth Avenue, New York, New York 10003

United Kingdom Edition published by
ACADEMIC PRESS, INC. (LONDON) LTD.
24/28 Oval Road, London NW1 7DX

Library of Congress Cataloging in Publication Data

Main entry under title:

The Mollusca.

Includes index.
Contents: v. 1. Metabolic biochemistry and
molecular biomechanics / edited by Peter W.
Hochachka -- v. 2. Environmental biochemistry
and physiology / edited by Peter W. Hochachka --
v. 3. Development / edited by N.H. Verdonk &
J.A.M. van den Biggelaar & A.S. Tompa -- v. 4-5.
Physiology / edited by A.S.M. Saleuddin & Karl M.
Wilbur.
    1. Mollusks--Collected works. I. Wilbur, Karl M.
QL402.M6 1983   594   82-24442
ISBN 0-12-751404-X (v. 4)

PRINTED IN THE UNITED STATES OF AMERICA

83 84 85 86   9 8 7 6 5 4 3 2 1

Contents

1. Cytology of Muscle and Neuromuscular Junction

GHISLAIN NICAISE AND JACQUELINE AMSELLEM

2. Neuromuscular Transmission and Excitation–Contraction Coupling in Molluscan Muscle

YOJIRO MUNEOKA AND BETTY M. TWAROG

v

3. Biochemical and Structural Aspects of Molluscan Muscle

P. D. CHANTLER

4. Locomotion in Molluscs

E. R. TRUEMAN

5. The Mode of Formation and the Structure of the Periostracum

A. S. M. SALEUDDIN AND HENRI P. PETIT

6. Shell Formation

KARL M. WILBUR AND A. S. M. SALEUDDIN

7. Shell Repair

NORIMITSU WATABE

8. Endocrinology

J. JOOSSE AND W. P. M. GERAERTS

9. Physiological Energetics of Marine Molluscs

B. L. BAYNE AND R. C. NEWELL

Contributors

Numbers in parentheses indicate the pages on which the authors' contributions begin.

Jacqueline Amsellem (1), Laboratoire d'Histologie et Biologie Tissulaire, Université Claude Bernard, 69622 Villeurbanne, France

B. L. Bayne (407), Natural Environment Research Council, Institute for Marine Environmental Research, Plymouth PL1 3DH, England

P. D. Chantler[1] (77), Department of Biology, Brandeis University, Waltham, Massachusetts 02254

W. P. M. Geraerts (317), Department of Biology, Free University, 1007 MC Amsterdam, The Netherlands

J. Joosse (317), Department of Biology, Free University, 1007 MC Amsterdam, The Netherlands

Yojiro Muneoka (35), Faculty of Integrated Arts and Sciences, Hiroshima University, Higashi-senda-machi, Hiroshima 730, Japan

R. C. Newell (407), Natural Environment Research Council, Institute for Marine Environmental Research, Plymouth PL1 3DH, England

Ghislain Nicaise (1), Laboratoire d'Histologie et Biologie Tissulaire, Université Claude Bernard, 69622 Villeurbanne, France

Henri P. Petit (199), Baylor College of Dentistry, and Dallas Museum of Natural History, Dallas, Texas 75246

A. S. M. Saleuddin (199, 235), Department of Biology, York University, Toronto, Ontario M3J 1P3, Canada

E. R. Trueman (155), Department of Zoology, University of Manchester, Manchester M13 9PL, England

Betty M. Twarog (35), Department of Biology, Bryn Mawr College, Bryn Mawr, Pennsylvania 19010

---

[1]Present address: Department of Anatomy, Eastern Pennsylvania Psychiatric Research Institute Division, Medical College of Pennsylvania, Philadelphia, Pennsylvania 19129.

Norimitsu Watabe (289), Department of Biology, Electron Microscopy Center, and Belle W. Baruch Institute for Marine Biology and Coastal Research, University of South Carolina, Columbia, South Carolina 29208

Karl M. Wilbur (235), Department of Zoology, Duke University, Durham, North Carolina 27706

General Preface

This multivolume treatise, *The Mollusca,* had its origins in the mid 1960s with the publication of *Physiology of Mollusca,* a two-volume work edited by Wilbur and Yonge. In those volumes, 27 authors collaborated to summarize the status of the conventional topics of physiology as well as the related areas of biochemistry, reproduction and development, and ecology. Within the past two decades, there has been a remarkable expansion of molluscan research and a burgeoning of fields of investigation. During the same period several excellent books on molluscs have been published. However, those volumes do not individually or collectively provide an adequate perspective of our current knowledge of the phylum in all its phases. Clearly, there is need for a comprehensive treatise broader in concept and scope than had been previously produced, one that gives full treatment to all major fields of recent research. *The Mollusca* fulfills this objective.

The major fields covered are biochemistry, physiology, neurobiology, reproduction and development, evolution, ecology, medical aspects, and structure. In addition to these long-established subject areas, others that have emerged recently and expanded rapidly within the past decade are included.

The Mollusca is intended to serve a range of disciplines: biological, biochemical, paleontological, and medical. As a source of information on the current status of molluscan research, it should prove useful to researchers of the Mollusca and other phyla, as well as to teachers and qualified graduate students.

Karl M. Wilbur

Preface

Physiology of Mollusca, published as two volumes in 1964 and 1966, interpreted physiology broadly and included chapters on biochemistry, embryology, and ecology. Because of the recent expansion and diversification of molluscan studies, nine volumes of the present treatise were required for the same disciplines. Biochemistry and physiology, fields that are so closely integrated, can scarcely be clearly separated, nor has such a separation been attempted in *The Mollusca.* Rather, we consider the four volumes that treat the disciplines of physiology and biochemistry (Volumes 1, 2, 4, and 5) to be a single major unit in the study of molluscs. The subject areas covered in each of the four volumes are given in the pages that follow. We call special attention to R. Seed's chapter, opening Volume 1. The chapter is a helpful introduction to the phylum Mollusca and provides information about structural organization, classification, and evolutionary changes.

Hormone action (endocrinology) and cellular defense mechanisms (immunobiology), both subjects that have been significantly developed by molluscan physiologists during the past decade, are covered in Volumes 4 and 5, respectively. Advances in the various areas of neurobiology are discussed in separate volumes. The treatment of each field is intended to be comprehensive, but because of constraint upon length, emphasis has been placed upon recent research and current status.

We are indebted to our many friends and colleagues who have worked with us on these volumes. Throughout the months of writing and editing, our authors have been unfailingly cooperative and understanding of all we have asked of them. We are also grateful to Mrs. L. W. Caldwell, Mrs. Dorothy Gunning, and A. M. Bernhardt for their secretarial assistance. The editorial staff of Academic Press has provided invaluable guidance from the early planning stages of *The Mollusca.*

A. S. M. Saleuddin
Karl M. Wilbur

Dedication

Sir C. M. Yonge, C.B.E., F.R.S.

This volume is dedicated to Sir Maurice Yonge in recognition of his great contribution to the study of molluscs. His research spans some six decades and includes most of the molluscan groups in respect of their form, biology, and evolution. Yonge was one of the earliest students of molluscan physiology with his investigations in the 1920s into their feeding mechanisms and digestion. More recently he served as coeditor of two well-known volumes on the Physiology of Mollusca.

Contents of Other Volumes

1

Cytology of Muscle and Neuromuscular Junction

GHISLAIN NICAISE
JACQUELINE AMSELLEM

Laboratoire d'Histologie
et Biologie Tissulaire
Université Claude Bernard
Villeurbanne, France

I. Tentative Multidimensional Classification of Muscle Cells

Most previous classifications of molluscan muscle cells into cross-striated, obliquely striated, and smooth muscles, with subclasses, are based on their appearance in living cells under light microscopy (e.g., Hanson and Lowy, 1960;

1

Hoyle, 1964; Millman, 1967). The striation (i.e., the organization of Z discs and dense bodies at the ultrastructural level) is a crucial morphological feature. Although its meaning is not perfectly clear, this organization readily gives a general indication of the speed of shortening, an important physiological property. However, the exceptional diversity of molluscan muscles cannot be expressed by a single gradient in which structure and function are seen to evolve from the genuine fast, cross-striated cell (Fig. 1) to the slow, smooth type (Fig. 2). In

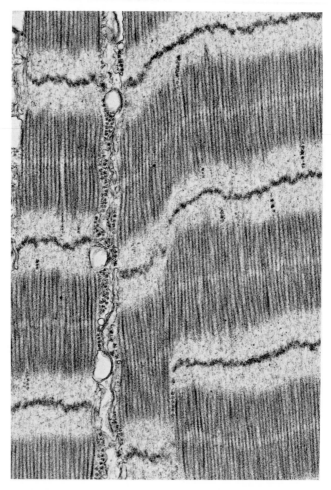

Fig. 1. Longitudinal section of the translucent adductor muscle of *Aequipecten irradians,* a Z1 L1 F1 T3 R2 M1 type. The structure is typical of a cross-striated muscle, with A and I bands and Z discs. (27,000X.) (From Nunzi and Franzini-Armstrong, 1981.)

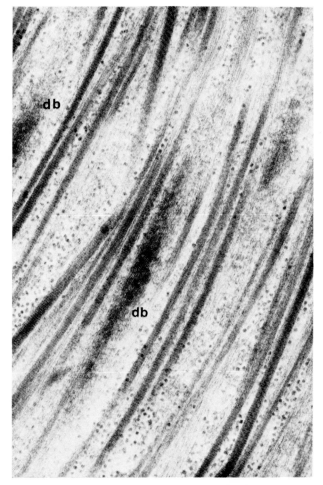

Fig. 2. Longitudinal section of the ABRM of *Mytilus,* a Z5 L4 F3 T3 R2 M1 type. Note the longitudinal periodicity of the thick filaments and the relationship between the thin filaments and the dense bodies (db). (45,000X.)

contrast, a multiplicity of parameters gives a unique opportunity for factorial analysis of structure–function relationships.

Table I represents a selection of molluscan muscles in which every cell type differs from the others in at least one of the following parameters: organization of the Z elements (Z), length of the thick filaments (L), main diameter of the thick filaments (F), plasma membrane invaginations (T), sarcoplasmic reticulum (R), and relative abundance of mitochondria (M).

TABLE I

Selected Molluscan Muscle Types Exemplifying Cytological Differentiation by Six Morphological Parameters

Genus	Muscle	Z^a	L^b	F^c	T^d	R^e	M^f	Reference
Placopecten	Translucent adductor	1	$(1)^g$	1	3	2	1	Morrison and Odense (1974)
Achatina	Heart	2/4	1	1	(2?)	1	2	Nisbet and Plummer (1966, 1968)
Ferrissia	Buccal bulb	2	1	1	3	1	2/3	Richardot and Wautier (1971)
Rossia	Heart	2	2	2	1	1	3	Jensen and Tjønneland (1977)
Loligo	Chromatophore	3	(1)	2	(3)	1	3	Florey (1969); Florey and Kriebel (1969)
Symplectotheutis	Mantle	3	(2)	1/2	3	1	2	Moon and Hulbert (1975)
Sepia	Buccal mass	3	2	3	3	1	2	Amsellem and Nicaise (1980)
Achatina	Radular muscle	4	(2)	1	2	1	2	Nisbet and Plummer (1966, 1968)
Busycon	Radula protractor	(4)	2	2	2	1	2	Sanger and Hill (1972, 1973)
Crassostrea	Translucent adductor	4	2	3	(3)	2	1	Hanson and Lowy (1961)
Sepia	Stomach (plexiform layer)	4	3	4	3	1	2	Amsellem and Nicaise (1980) (and unpublished data)
Venus	Heart	5	1	2	(3)	1	3	Kelly and Hayes (1969)

TABLE I (Continued)

Genus	Muscle	Z[a]	L[b]	F[c]	T[d]	R[e]	M[f]	Reference
Lymnaea	Head retractor	5	3	1	2	2	1	Plesch (1977a)
Philine	Buccal mass	5	3	2/3	1	2	1	Dorsett and Roberts (1980)
Lymnaea	Shell	5	4	2	2	2	1	Plesch (1977a)
Lymnaea	Diagonal	5	4	2	2	1	2+	Plesch (1977a)
Mytilus	ABRM	5	4	3	3	2	1	Sobieszek (1973); Twarog (1967)
Buccinum	Hypobranchial gland	5	?	4	2/3	2	1	Hunt (1972)

[a] Z1, presence of continuous Z discs; Z2, discrete dense bodies arranged in planes perpendicular to the myofibril axis; Z3, dense bodies arranged in planes forming an oblique angle with the main axis of the cell; Z4, the same disposition as Z3, but less regular; Z5, apparently random pattern of organization of the dense bodies.

[b] L1, length of the thick filaments less than 2 μm; L2, length between 2 and 5 μm; L3, length between 5 and 10 μm; L4, length greater than 10 μm.

[c] F1, main diameter of the thick filaments less than or equal to 20 nm; F2, main diameter between 20 and 40 nm; F3, main diameter between 40 and 100 nm; F4, main diameter greater than 100 nm.

[d] T1, presence of T tubules in the contractile myoplasm; T2, presence of pear-shaped caveolae or finger-like invaginations of the sarcolemma; T3, no obvious invaginations of the plasma membrane.

[e] R1, presence of subsarcolemmal cisternae connected to a net of sarcoplasmic reticulum tubules which penetrate the contractile myoplasm; R2, sarcoplasmic reticulum confined to the noncontractile periphery of the cell.

[f] M1, mitochondria rare; M2, mitochondria almost always present in the plane of section; M3, numerous mitochondria often forming large packets.

[g] The numbers in parentheses are not deduced from written indications of the authors themselves but rather from our observation of their electron micrographs.

A. Z Discs and Dense Bodies

Our Z gradient is not entirely divisible into distinct categories. There are intermediaries between Z1 and Z2 (Hawkins et al., 1980) and between Z2 and Z4 (see Table I), and the boundary between Z4 and Z5 is not clear. The cross-striated (Z1) type has been found thus far only in the fast part of the adductor muscle of Pectinidae (Fig. 1) and the regular, obliquely striated type (Z3) only in cephalopods. Cephalopods seem to have essentially obliquely striated muscles

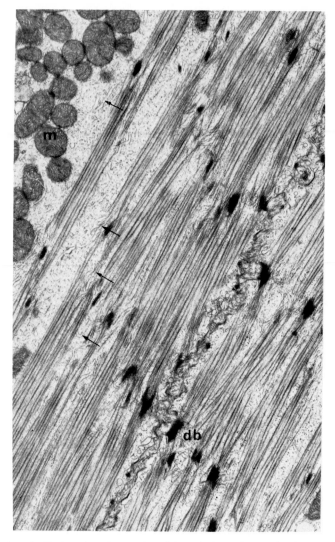

Fig. 3. Longitudinal section of a muscle cell from the buccal mass of *Aplysia rosea*, a Z4 L2 F2 T2 R1 M3 type. The dense bodies (db) are approximately aligned. The arrows point to the abundant tubular reticulum; m, mitochondria. (10,500X.) (From Amsellem and Nicaise, 1976, by permission of Springer Verlag.)

(Z2 and Z3) (Amsellem and Nicaise, 1980). The advantages of oblique striation have been discussed in detail by several authors (Lanzavecchia, 1977; Rosenbluth, 1972). Some particularly thin muscle cells (e.g., epineural muscle cells in *Aplysia;* Rosenbluth, 1963) do not have dense bodies in the myoplasm, but there are always attachment plates which are probably the equivalent of dense bodies abutting on the plasma membrane (see Figs. 6, 8, 10, and 14). Therefore, there is no known molluscan muscle devoid of dense bodies or their functional equivalent.

There is little doubt that the dense bodies in molluscs represent, as they probably do in other phyla, the functional equivalent of the Z line or Z discs in striated muscle (Hanson and Lowy, 1961; Sanger and Hill, 1973; Sobieszek, 1973). Numerous authors have described the insertion of the thin filaments on these structures (Fig. 2) (Hanson and Lowy, 1961; Hayes and Kelly, 1969). The thin filaments themselves are essentially actin (Szent-Györgyi et al., 1971), and the variety of thicknesses reported in different publications, from 5 to 8.5 nm, is probably due to the difficulty of measuring accurately such a small diameter on a structure that is not perfectly cylindrical.

It is obvious that in the distinctly striated muscles (Z1, Z2, Z3), the space between two consecutive dense bodies is organized into A and I bands like the sarcomere of vertebrate cross-striated muscle. It is probable that the same basic organization prevails in smooth muscles (Z5), as shown by the study of Sobieszek on the anterior byssus retractor muscle (ABRM) of *Mytilus* (1973). Groups of thick filaments with tapering free ends are situated between two dense bodies, at least in the resting muscle cell. It is likely that during shortening the extremities of the thick filaments can pass the dense bodies, as in certain cross-striated muscles (Hoyle *et al.*, 1965). However, it has been proposed that some smooth (Z5) muscles are even less organized than the ABRM and that the thick filaments do not belong to a fixed pair of dense bodies (Plesch, 1977a).

B. Thick Filament Length

It is generally agreed that when passing from striated (Z1) to smooth (Z5) types, the speed of shortening diminishes but the ability to generate tension increases (Ruegg, 1971). The speed of shortening is certainly linked to the number of contractile units arranged in series, which, in turn, is related to the length of the filaments (Josephson, 1975). Indeed, the gradient L1–L4 of Table I roughly follows that of Z1–Z5. In molluscs, there are no striated muscles with long sarcomeres, although they exist in other invertebrate phyla (see Dewey et al., 1977; Hoyle et al., 1973; Josephson, 1975).

There are, however, some smooth muscles with long and some with short, thick filaments (Table I). If one regards the smooth (Z5) muscles as having a hidden sarcomere structure, then the speed of contraction ought to decrease with increasing filament length. If thick filaments can pass from one dense body to the next, then the variation of lengths from L1 to L4 in Z5 muscles might be related to another functional characteristic. Comparative physiological data which might permit further speculation along these lines are lacking.

C. Thick Filament Diameter

Most authors who discuss molluscan muscle cytology give thick filament length as a constant, and they sometimes consider the sliding filament theory as being contradictory to a shortening of the thick filaments during contraction (see references in Lowy et al., 1964). Even though thick filaments have tapering ends (Hanson and Lowy, 1961; Sobieszek, 1973), thick filament diameter is easier to measure than length, and the measure of this diameter in a given muscle is considered to reflect the length (Dewey et al., 1977; Hanson and Lowy, 1961). There are at least two molluscan muscles in which thick filament diameter has been reported to increase during contraction or as a consequence of repetitive contraction: the penis retractor muscle of *Helix pomatia* (Foh, 1969; Wabnitz, 1975) and the white (tonic) part of the posterior adductor muscle of *Anodonta cygnea* (Zs.-Nagy et al., 1971). In the latter, however, the authors stress indirect evidence for an increase in length at the same time as the increase in width; they conclude that their data do not favor filament contraction. A study of isolated filaments is needed to elucidate this point (Dewey et al., 1977). The bimodal distribution of thick filament diameters observed in the opaque adductors of *Arctica islandica* (F2 + F3) and *Astarte undata* (F3 + F4) is interpreted by Morrison and Odense (1974) as evidence of two distinct populations of thick filaments. In other muscles, only one average or maximal diameter is given (Table I).

Figures 4–7 illustrate the four categories of thick filament diameters that we have arbitrarily distinguished in molluscan muscles. The smaller category (F1) (Fig. 4) generally presents an electron-lucent core which is thought to represent paramyosin (Millman and Bennett, 1976), the myosin being located at the periphery (see Szent-Györgyi et al., 1971); thicker filaments appear homogeneous in cross section; in longitudinal sections, they present a banding, considered typical of paramyosin (formerly termed *tropomyosin A*), with a 14.5-nm periodicity and (sometimes) a 7.2-nm subperiod. Thicker myofilaments contain more paramyosin (see Margulis et al., 1979). In Table I, it appears that the shorter filaments (L1) are often also thinner (F1), but there are long and thin (L3 F1) or short and thick (L2 F3) filaments as well. To our knowledge, the fact that filament length and diameter can vary independently has not been discussed.

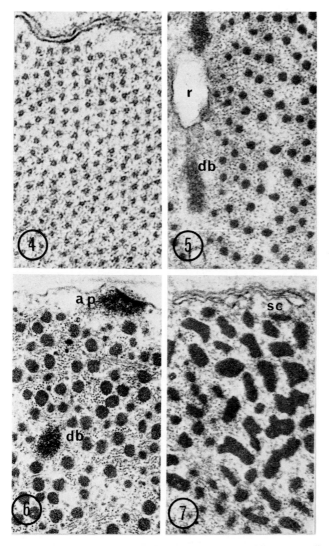

Fig. 4. Cross section of a muscle cell from the arm of *Sepia officinalis* to illustrate an F1 type. Note the electron-lucent core of the thick filaments. (60,000X.)

Fig. 5. Cross section of an F2 type from the buccal mass of *Sepia officinalis*. The dense bodies (db) are roughly organized into a line, interrupted by tubules of sarcoplasmic reticulum (r). (60,000X.)

Fig. 6. Cross section of an F3 type, the ABRM of *Mytilus*; ap, attachment plate; db, dense body. (60,000X.)

Fig. 7. Cross section of an F4 type in the plexiform layer of the stomach of *Sepia officinalis*; sc, subsarcolemmal cisterna. (60,000X.)

D. Plasma Membrane Invaginations

We consider the penetration of tubular infoldings of the plasma membrane into the contractile myoplasm, which is seen in a few molluscan muscles (T1, Table I), as equivalent to a T system. Subsarcolemmal cisternae are seen to come into close apposition with the plasma membrane of these tubules, forming dyads (Dorsett and Roberts, 1980). The essential difference between this and the T system of the vertebrate cross-striated muscle is that in molluscs, the basal lamina penetrates the T system, which may even contain collagen fibrils (Dorsett and Roberts, 1980). Another type of tubular sarcolemmal invagination, more often seen in molluscan muscle (see the review by Hunt, 1981), is morphologically closer to the T system of vertebrates. Like the latter, it contains no basal lamina or visible extracellular component; however, unlike the latter, it does not penetrate into the contractile part of the cell and does not form specialized appositions with the cisternae of the sarcoplasmic reticulum. It is likely that these invaginations are of the same nature as the *caveolae intracellulares* which are seen in other molluscan muscles (Figs. 8, 9) and many muscles in various other phyla (Hunt, 1981; Prescott and Brightman, 1976).

We listed these two types of invaginations under the label T2. It could be argued that the tubules provide a larger increase in membrane surface (more than double in the case studied by Sanger and Hill, 1972), but the caveolae may form chains with a comparable result. Another similarity between tubular invaginations and the caveolae is the decoration of membranes by ridges in conventional electron microscopy or by rows of particles in freeze-fracture (Fig. 9). This intrinsic marker of the tubular or caveolar membrane may not be present in all molluscan muscles investigated (Hunt, 1981). It was used by Prescott and Brightman (1976) to establish that caveolae can flatten reversibly when the plasma membrane is stretched; the authors even suggested that the caveolae could be stretch receptors for the muscle cell. Some molluscan muscles, such as the cross-striated adductor of Pectinidae, have practically no plasma membrane invaginations. In this case, the cells are small or flattened and the contractile filaments are less than 0.5 μ from the plasma membrane. We have not been able to collect enough reliable data to allow us to correlate systematically the development of membrane invaginations (from T1 to T3) with the distance between the more deeply situated filaments and the cell surface, as has been done in other phyla (Bone and Ryan, 1973).

Fig. 8. Longitudinal section of a smooth (Z5) muscle cell from esophagus of *Glossodoris valenciennesi*. The contrast is due exclusively to ruthenium red-osmium postfixation. The stain marks the plasma membrane, including the wall of the caveolae. Chains of caveolae are indicated by arrows. ap, Attachment plate; cbs, "calcium"-binding site on the inner side on the plasma membrane of the muscle cell; db, dense body; n, nerve ending, containing large, dense-cored vesicles; t, teloglia with typical glio-interstitial granules. (18,400X.)

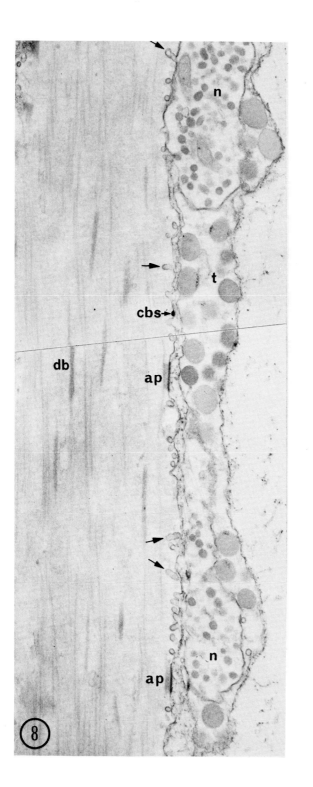

Fig. 9. Replica of a fracture across the cytoplasma of an *Aplysia rosea* perineurial muscle cell. L, Cytoplasmic leaflet of the sarcolemma; O, ostium of a caveola; X and Y represent the two complementary faces of the plane of fracture inside the caveolar membrane. One leaflet, X, bears rows of particles; the other leaflet, Y, has matching grooves. (30,300X.) (From Prescott and Brightman, 1976, by permission of Longman Group Ltd.)

E. Sarcoplasmic Reticulum

Although there are great differences in sarcoplasmic reticulum development in different molluscan muscles, we could not ascertain an R3 type: reticulum which is very rare or absent. In all the well-known molluscan muscles, there are subsarcolemmal cisternae (Figs. 7, 8, 10, and 14) connected to thin reticular tubules. The tubular net may either penetrate the contractile myoplasm (type R1) or be restricted to the noncontractile periphery (type R2). The difference between R1 and R2 does not rest upon the abundance of subsarcolemmal cisternae. At least, quantitative data to substantiate a classification based on this criterion are lacking.

The cisternae form dyads with the sarcolemmal membrane; electron-dense bridges span the gap between the reticular and plasma membranes (Fig. 10) (Amsellem and Nicaise, 1976, 1980; Nunzi and Franzini-Armstrong, 1981; Richardot and Wautier, 1971).

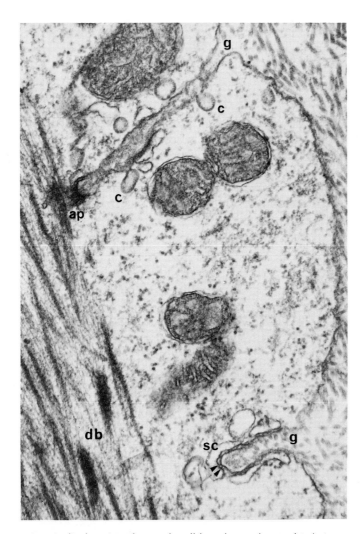

Fig. 10. Longitudinal section of a muscle cell from the esophagus of *Aplysia rosea*, a Z5 F2 T2 R1/2 M2 muscle. The periphery of the cell presents membrane invaginations. These invaginations are sometimes interpreted as rudimentary T systems (Richardot and Wautier, 1971). Here, the membrane does not form a tubule but rather a groove (g), which is probably temporary. During contraction, it is pulled in by an attachment plate (ap); c, caveolae; db, dense bodies; sc, subsarcolemmal cisternae. Triangles point to electron-dense bridges between the reticular and plasma membranes. (30,000X.)

It has been proposed that the subsarcolemmal cisternae could be a source of intracellular calcium (see Section II) and that the tubular reticulum ensures relaxation by calcium uptake. If this is true, then the R1-type muscles may relax faster than those of the R2, provided their contractile apparatus is not too different.

F. Mitochondria

Here again, quantitative data are badly needed, and the classification into M1, M2, and M3 given in Table I is less than satisfactory. As in vertebrates, molluscan heart muscle is very rich in mitochondria, which is consistent with the suggestion that the abundance of that organelle is linked essentially to endurance (see Hoyle and McNeill, 1968; Josephson, 1975; Plesch, 1977a).

G. Conclusion

We have selected six different and relatively independent parameters of cytological differentiation in molluscan muscles which were the best documented. Relevant information could be obtained from the thin/thick filament ratio, for example. It can be as low as 6 in *Crassostrea* translucent adductor muscle (Hanson and Lowy, 1961) and as high as 17 in the ABRM of *Mytilus* (Sobieszek, 1973). This ratio appears to be proportional to the thick filament diameter, at least in the nine examples of *Lymnaea* muscles studied by Plesch (1977a) and in the ABRM of *Mytilus*. This proportionality is confirmed by the biochemical study of Margulis et al. (1979). It can be speculated that this ratio is directly related to the number of cross-bridges, and consequently to the maximum force of the muscle (see Hoyle and McNeill, 1968; Josephson, 1975; Plesch, 1977a). It is also likely that wider thick filaments can establish cross-bridges with more thin filaments, but this remains to be confirmed. Similar information is gained from a consideration of total myofilament density, which is known for only a few muscles (Plesch, 1977a).

Catch contraction may not have a morphological correlate (Plesch, 1977a); however, it is impossible to decide because too few of the listed muscles in Table I were studied for their catch properties.

Other morphological attributes, such as the development of extracellular spaces and glio-interstitial tissue (see Section III), are still less well documented.

One is forced to conclude that despite the large number of ultrastructural studies on molluscan muscles, only some of which are covered by this chapter, the authors did not take full advantage of the extreme diversity of their subject. It is hoped that the few considerations listed above will encourage workers in this field to collect more quantitative and comparative data.

II. Cytochemistry of Calcium Compartments

Physiological experiments have shown that in several molluscan muscles the action potential is characterized by an inward current of calcium ions (Ca^{2+}). However, this current may not always be the direct activator of contraction (see Chapter 2, this volume). A contraction-inducing Ca^{2+} release from intracellular stores has been demonstrated in the ABRM of *Mytilus* (see references in Twarog, 1976), the radula protractor muscle of *Busycon* (Hill et al., 1970), the penis retractor muscle of *Helix* (Wabnitz, 1976), and the longitudinal body wall muscle of *Dolabella* (Sugi and Suzuki, 1978).

In the ABRM of *Mytilus* and in the longitudinal muscle of *Dolabella*, Atsumi and Sugi (1976) and Suzuki and Sugi (1978) attempted to visualize Ca^{2+} movement through the ultrastructural localization of pyroantimonate deposits after fixation with potassium pyroantimonate-containing osmium tetroxide. In the resting state, the calcium-containing pyroantimonate precipitates are restricted to peripheral sites: subsarcolemmal vesicles, mitochondria, and the inner face of the plasma membrane, whereas in the contracted fiber, the precipitate is scattered in the myoplasm. These results are interpreted by the authors as a demonstration of Ca^{2+} movement from peripheral intracellular stores to the contractile apparatus (see also Fig. 11). They further speculate that in the muscle studied, three intracellular sources of Ca^{2+} are mobilized during contraction.

The role of intracellular compartments in provoking relaxation through removal of Ca^{2+} from the cytosol is less well documented. Stössel and Zebe (1968, in Twarog, 1976) isolated a microsomal fraction from the ABRM of *Mytilus,* which accumulated Ca^{2+}; Ca^{2+} accumulation was accompanied by ATP breakdown. Heumann (1969) demonstrated that the subsarcolemmal vesicles of glycerinated ABRM were able to accumulate Ca^{2+} *in situ* when provided with oxalate and ATP; there was no mitochondrial or other accumulation. Huddart et al. (1977) suggested that the subsarcolemmal vesicles and the mitochondria of whelk columellar muscle were "both able to bind Ca^{2+} in a capacity and manner similar to that of the sarcoplasmic reticulum of skeletal muscle" of vertebrates. The observation of pyroantimonate precipitates suggests that after peak tension, the calcium ions "are gradually reaccumulated to their original sites of localization" (Atsumi and Sugi, 1976), but the subsarcolemmal vesicles are the last of the three sites in which the precipitate was relocalized during relaxation (Atsumi, 1978). The author considered this as an indication that the role of the sarcoplasmic reticulum was "important when concentration of intracellular Ca^{2+} becomes low" (Atsumi, 1978). When the subsarcolemmal cisternae are connected to a well-developed intramyoplasmic reticulum (type R1 in Table I), it is possible that the Ca^{2+} is removed from the sarcoplasma by the tubular reticulum and afterward accumulated in the cisternae to be used for

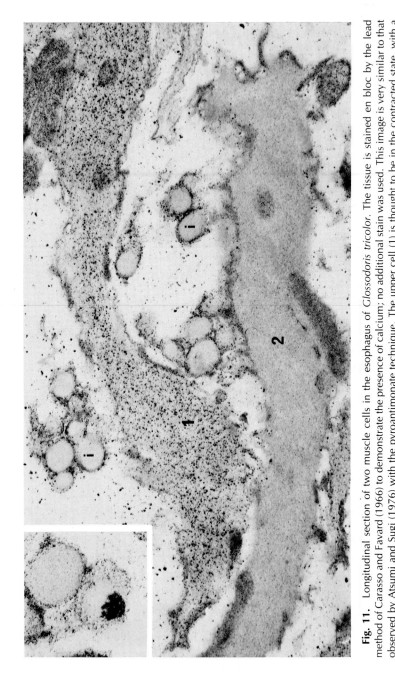

Fig. 11. Longitudinal section of two muscle cells in the esophagus of *Glossodoris tricolor*. The tissue is stained en bloc by the lead method of Carasso and Favard (1966) to demonstrate the presence of calcium; no additional stain was used. This image is very similar to that observed by Atsumi and Sugi (1976) with the pyroantimonate technique. The upper cell (1) is thought to be in the contracted state, with a diffuse precipitate in the myoplasm; the lower cell (2) is in the relaxed state, with the precipitate localized at the periphery. The glio-interstitial granules of the surrounding interstitial processes (i) are generally free of precipitate except at the level of occasional geometrical structures (insert). Such intragranular periodic structures are also stained by ruthenium red (Nicaise, 1973). (30,000X.)

another contraction (Sanger and Hill, 1972). This could explain why Suzuki and Sugi (1978) found a faster Ca^{2+} translocation cycle in the longitudinal body wall muscle of *Dolabella* (type R1) than Atsumi and Sugi (1976) in the ABRM of *Mytilus* (type R2).

The discrete calcium-binding sites on the inner side of the plasma membrane which are often described after aldehyde fixation of a variety of cells may be seen in certain molluscan muscles (see Fig. 8) (Twarog, 1977). Although they are hardly studied in muscles in general (Nicaise et al., 1982), they deserve further attention because they may be related to the calcium store described by Atsumi and Sugi (1976) at the inner side of the plasma membrane in *Mytilus* ABRM.

Ruthenium red cytochemistry and calcium-precipitating or substitution techniques also reveal strong anionic sites capable of binding Ca^{2+} in the granules of the glio-interstitial cells present in certain molluscan muscles (Nicaise, 1973) (See Fig. 11, insert). The glio-interstitial granules of freeze-substituted *Mytilus* ABRM may contain more than 0.1 M calcium, as measured by X-ray microanalysis (Hemming, 1981).

III. Intercellular Connections

The various types of connections among nerve, muscle, and glio-interstitial tissue are schematized in Fig. 12.

A. Myomuscular Junctions

Physiologically, the most significant connections between muscle cells are nexus or gap-junctions; in practically all the cases so far examined, they represent the morphological counterpart of electrotonic coupling (see Bennett and Goodenough, 1978). They have been reported in a variety of molluscan muscles, including ABRM of *Mytilus* (Brink et al., 1979; McKenna and Rosenbluth, 1973; Twarog et al., 1973, Fig 13), chromatophore muscle of squid (Cloney and Florey, 1968; Florey and Kriebel, 1969), oyster and mussel hearts (Irisawa et al., 1973), and longitudinal body wall muscle of *Dolabella* (Sugi and Suzuki, 1978). Gap junctions between molluscan myocytes are similar to those described in other tissues of various phyla.

Another kind of myomuscular junction can be seen in molluscs; it has a wider (6–16 nm) gap, and it occasionally extends over several micra; the cleft does not contain a basal lamina (Figs. 13, 14); (Amsellem and Nicaise, 1980; Hunt, 1981; McKenna and Rosenbluth, 1973; Price and McAdoo, 1979; Sugi and Suzuki, 1978; Zs.-Nagy and Salanki, 1970). Although these junctions were termed *intermediate junctions* by Suzuki and Sugi (1978), they lack the characteristic densities of adhering zonules. They may represent uncoupled gap-junction sites

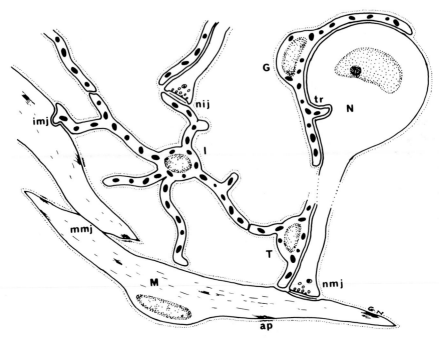

Fig. 12. Highly speculative schematic drawing summarizing the various kinds of junctions encountered in molluscan muscles. ap, Attachment plates; G, glial cell; I, interstitial cell; imj, interstitio-muscular junction; M, muscle cell; mmj, myomuscular junction; N, neuron; nij, neuro-interstitial junction; nmj, neuromuscular junction; T, teloglia; tr, trophospongium.

(Gilloteaux, 1976), but they are not necessarily a result of poor fixation, as they often appear close to well-preserved gap junctions.

The third type of myomuscular connection is the desmosome (Amsellem and Nicaise, 1980; Nunzi and Franzini-Armstrong, 1981; Rogers, 1969; Sugi and Suzuki, 1978; Zs.-Nagy and Salanki, 1970). These desmosomes of molluscan muscles can be considered as paired attachment plates facing each other; a mechanical function is generally attributed to them.

B. Interstitiomuscular Connections

When a nerve terminal is accompanied by glia, the glial processes are often seen in close apposition with the muscle membrane, without interposition of a basal lamina (see references in Nicaise, 1973). In certain molluscan muscles, this teloglia is continuous with a net of interstitial cells spreading through the connective tissue spaces between the muscle cells (Figs. 11, 12, and 18). This glio-interstitial tissue is easily characterized by its content of large, oval, membrane-

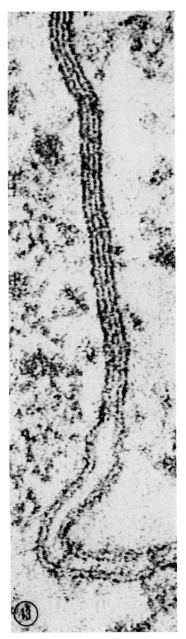

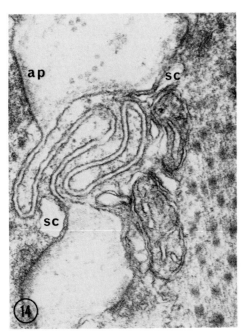

Fig. 13.   (left) High magnification of a gap-junction or nexus between two muscle cells of *Mytilus* ABRM. In the right-hand part of the micrograph, the cells are separated by a wider gap of approximately 10 nm, without a basal lamina. (405,000X.) (From Twarog et al., 1973, by permission of Rockefeller University Press.)

Fig. 14.   (top) Junctional apposition between two muscle cells in the plexiform layer of the stomach of *Sepia officinalis*. This may represent an uncoupled junctional site, as parts of these appositions occasionally form a gap-junction. ap, Attachment plate; sc, subsaredemmal cisternae. (60,000X.) (From Amsellem and Nicaise, 1980, by permission of *Journal of Submicroscopic Cytology*.)

bound granules (Figs. 8, 11, 12, 15, 17, 18, and 19). Since the review by Nicaise (1973), wherein interstitial elements in nine gastropod species were listed, their presence has been reported in *Aplysia* esophagus (Amsellem and Nicaise, 1976) and radular muscle of *Busycon* (Hill and Sanger, 1974), and has been confirmed in *Mytilus* ABRM (Gilloteaux, 1975) as well as in the digestive tract of *Sepia* (G. Nicaise and J. Amsellem, unpublished observations).

Interstitial cells and processes have been seen only in muscles that had well-developed connective tissue spaces (Amsellem and Nicaise, 1976). They seem to be lacking in fast muscles. It has been shown in *Mytilus* ABRM that the glio-interstitial processes were more numerous when the animals had been adapted to hypersaline seawater; conversely, their number was reduced after acclimatization to dilute seawater (Hemming and Nicaise, 1982). In general, the presence of a glio-interstitial tissue, as characterized by its granule content, is more conspicuous in marine than in freshwater species.

Figure 15 represents an extreme example of the connection between interstitial processes and muscle: The relationship between these two cell types is similar to the trophospongium which exists between glial cells and neuronal perikarya. Other outstanding examples are seen in regenerating muscle tissue (Nicaise, 1973). Interpenetration and narrow apposition between interstitial and muscle cells are also known to exist in vertebrates (see the references in Nicaise, 1973). The significance of these contacts is still obscure. It has been proposed that the glio-interstitial cells and their granules may play a role in the ionic regulation of the nerve and/or the muscle (Nicaise, 1973; Treherne et al., 1969).

C. Neuromuscular Junctions

The finer peripheral nerves, usually reduced to a single axon, establish differentiated contacts with the myocytes. The nerve terminal, often accompanied by a teloglial process, usually lies in a groove of the muscle cell. Sometimes the neurite is ensheathed by the muscle cell, which forms a mesaxon; in that case, there is no teloglia (Bogusch, 1972a; Ducros, 1979; Graziadei, 1966; Plesch, 1977; Rogers, 1969). The nerve and muscle membranes are closely apposed, with no collagen fibrils or basement lamina interposed between them. The intercellular cleft is of constant width in the nerve–muscle junctional region proper; its dimensions vary from 10 to 30 nm (see Heyer et al., 1973), depending on the species and perhaps also on the neurotransmitter (two different nerve endings on the same muscle cell may have different cleft widths; Fig. 18 in Amsellem and Nicaise, 1976). The junctional region is further characterized by an electron-dense, amorphous lamina in the middle of the cleft (see early references in Nicaise, 1973; see also McKenna and Rosenbluth, 1973; Plesch, 1977b) and by the presence of synaptic vesicles in the nerve ending (Fig. 16). Postsynaptic

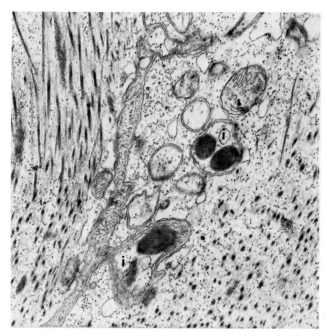

Fig. 15. Interstitio-muscular junction in the ABRM of *Mytilus*. Interstitial processes (i) pene-trate the muscle cell periphery, in the manner of glial cells forming a trophospongium with neurons. (24,750X.) (From Nicaise, 1973.)

differentiations are absent or reduced to a fine fibrillar coat, but presynaptic, dense projections have occasionally been noted (Amsellem and Nicaise, 1976, 1980; Foh and Bogusch, 1969; Graziadei, 1966; McKenna and Rosenbluth, 1973).

In summary, these contacts present most of the morphological features of a chemical synapse as it is known in other phyla and in the ganglionic neuropil of the molluscs themselves (e.g., Chalozonitis, 1969; Elekes, 1978; Pentreath et al., 1975). In the absence of constant differentiations, apart from vesicles, the dense lamina in the middle of the synaptic cleft (with or without dense bridges between the pre- and postsynaptic membranes) is important, as it probably represents a basic element in synaptic connectivity (Pfenninger, 1971). The existence of morphologically well-characterized neuromuscular junctions does not pre-clude the possibility that neuromuscular command occurs at other sites (Burn-stock and Iwayama, 1971; Heyer et al., 1973), but those differentiated junctions are likely to be privileged loci, at least.

Synaptic vesicles have been the object of much attention: Their morphology

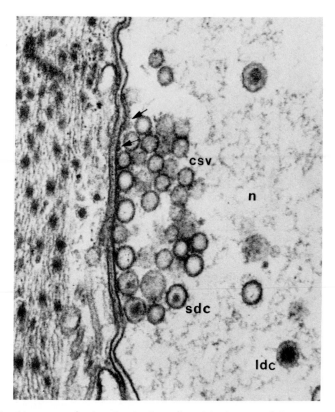

Fig. 16. Neuromuscular junction in the yellow (phasic) part of the posterior adductor muscle of *Anodonta*. Note the electron-dense lamina in the middle of the synaptic cleft and the dense projections (arrows). csv, Clear synaptic vesicles; ldc, large, dense-core vesicles; n, nerve ending; sdc, small, dense-cored vesicles. (80,000X.) (From Amsellem et al., 1973, by permission of Birkhäuser Verlag.)

could give information on their neurotransmitter content. Nerve endings have been classified by their vesicular content, taking into account the shape, diameter, electron opacity, and relative abundance of vesicles. In the tentacle of the slug *Vaginula*, Barrantes (1970) thus demonstrated the presence of two types of nerve endings, both containing clear synaptic vesicles accompanied either by small or larger granulated vesicles. In an exemplary study on *Mytilus* ABRM, McKenna and Rosenbluth (1973) tried to identify the cholinergic and tryptaminergic nerve endings, which are presumed to be present on the basis of firm physiological evidence. They suggested that the endings containing mostly clear synaptic vesicles (50 nm in diameter) were cholinergic and those containing mainly dense-cored vesicles (100 nm in diameter) were tryptaminergic, but they were not able to establish a statistical separation between the two types. The identification of neurotransmitters at the ultrastructural level is further discussed in Section IV).

D. Neurointerstitial Junctions

Connections morphologically similar to neuromuscular junctions have been reported between neurites and glio-interstitial elements in the foot and rhinophore of two species of *Glossodoris* (see Nicaise, 1973), the heart of *Aplysia*

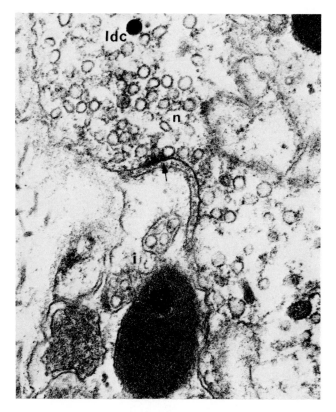

Fig. 17. Junction between a neurite (n) and a glio-interstitial process (i) in the esophagus of *Aplysia rosea*. Note the dense projections (arrow). The nerve ending contains clear synaptic vesicles and a large, dense-core vesicle (ldc). (60,000X.) (From Amsellem and Nicaise, 1976, by permission of Springer Verlag.)

californica (Taxi and Gautron, 1969), and the esophagus of *A. rosea* (Amsellem and Nicaise, 1976). As shown by Fig. 17 and contrary to what was said previously (Nicaise, 1973), those junctions may present a dense lamina in the middle of the intercellular cleft, which we regard as the best morphological criterion to locate the active zone of a molluscan synapse. Synaptic-like junctions have been observed between neurites and interstitial cells of the gut in various vertebrates (early references in Nicaise, 1973; Yamamoto, 1977), as well as between neurites and glial cells of the myenteric plexus (Gabella, 1972).

The extensive glio-interstitial tissue described in certain molluscan muscles is probably not so much a peculiarity of this phylum as a requirement of a given type of muscle or neuromuscular physiology.

IV. Cytochemistry and Neuromuscular Transmission

A. Cholinesterases

Attempts to visualize the postsynaptic part of neuromuscular junctions by cholinesterase (ChE) histochemistry, as in vertebrate motor end plates, have led to inconclusive or negative results (Bowden, 1958; Cambridge et al., 1959; Deane and Twarog, 1957; Nicaise, 1969; Zs.-Nagy 1964). The precipitate, indicating enzyme activity, is often uniformly distributed on the muscle membrane (Bogusch, 1972b; Watts and Pierce, 1978) and in mitochondria or, at least, in organelles having both the position and size of the mitochondria (Germino et al., 1968; Gilloteaux, 1978).

ChE activity was also localized in the glio-interstitial tissue, in the mesentery and esophagus of *Glossodoris* (Nicaise, 1973), in the ABRM of *Mytilus* (Gilloteaux, 1978; Hemming et al., 1983) (Fig. 18), and possibly in the lateral region of the snail foot (Chiang et al., 1972).

Some authors have demonstrated the presence of ChE activity in peripheral nerves. In the reports by Bogusch (1972b) and Gilloteaux (1978), all the neurites were marked by the histochemical reaction. In contrast, in those of Chiang et al. (1972) and Peretz and Estes (1974), there was a clear distinction between ChE-positive and ChE-negative neurites. Such an irregular distribution may permit the identification of cholinergic nerve endings, although this is debatable (McCaman and McCaman, 1976).

In the absence of a reliable and specific technique for the cytochemical localization of choline acetyltransferase, perhaps the easiest way to study the ultrastructure of cholinergic neuromuscular junctions in molluscs is to look for the endings of identified cholinergic motoneurons such as LD_G or LD_{HI} of the *Aplysia* abdominal ganglion. That the result might not be the expected accumula-

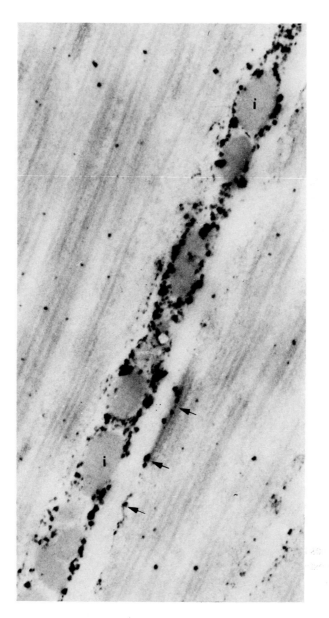

Fig. 18.   Longitudinal section in the ABRM of *Mytilus*. The tissue was stained only by the Karnovsky (1964) technique for ChE. Most of the ChE activity is localized in glio-interstitial tissue (i). Some enzyme activity is present over the muscle plasma membrane (arrows). All this activity is eserine $10^{-4}M$ sensitive. (15,500X.) (From Hemming et al., 1983, by permission of Springer Verlag.)

tion of numerous clear synaptic vesicles is suggested by the studies of Gillette and Pomeranz (1975) and Thompson et al. (1976).

B. Monoamines

Biogenic amine-containing nerve profiles have been localized by formaldehyde-induced fluorescence in various molluscan muscles (see references in Cardot, 1979; Ducros, 1979). Used under proper experimental conditions, particularly in combination with spectrofluorimetry, this method yields reliable results. It allows the investigator to decide if the fluorophore is due, for example, to noradrenalin (NA), dopamine (DA), or 5-hydroxytryptamine (5-HT). These three amines have been localized in peripheral nerves of molluscs (Cardot, 1979). The authors generally assume that the localized amine is used as a neurotransmitter, because this localization is most often associated with physiological and pharmacological evidence.

Ultrastructural methods permit a closer look at molluscan neuromuscular junctions. By means of electron microscopic autoradiography using [^{3}H]5-HT, Taxi and Gautron (1969) were able to identify probable tryptaminergic nerve endings on cardiac muscle cells of *Aplysia*. The neurites contained clear synaptic vesicles, associated with varying amounts of 90-nm, dense-core vesicles. The same method, but with uptake of [^{3}H]DA, was used by Brisson and Collin (1980) on the genital tract of *Bulinus*. They could not see differentiated neuromuscular junctions, but the labeled nerve varicosities contained clear or rather finely granulated vesicles (like the small, dense-core vesicles of Fig. 16). Surprisingly, in the *Octopus* salivary gland, [^{3}H]NA uptake was not detected in the presumed noradrenergic nerve endings, but was confined essentially to muscle cells (Martin and Barlow, 1975).

The chromaffin reaction adapted to electron microscopy was used with glutaraldehyde fixation to demonstrate monoamines, or 5-HT alone after the catecholamine had been inactivated by formalin prefixation (Wood, 1967). Figure 19 represents an application of the Wood method to nerve terminals in *Mytilus* ABRM. With a similar technique, Bogusch (1972a) demonstrated the presence of large (100 nm) and smaller monoamine-containing, dense-core vesicles in the finer nerves of the penis retractor muscle of *Helix;* the presence of both 5-HT and DA was established by other methods.

Finally, false neurotransmitters were also used to identify monoaminergic neuromuscular junctions in molluscs. Sathananthan induced the degeneration of aminergic nerves in the ABRM of *Mytilus* with 5,6-DHT (1976) and, in the gut of *Mytilus,* with 5,7-DHT (1977). The author claimed that he could morphologically separate the presumed serotoninergic neurites containing large granular vesicles (56–168 nm) from the presumed dopaminergic neurites characterized by small granular vesicles (20–64 nm), but both types were affected by the false neurotransmitter (see discussion in Hemming et al., 1983).

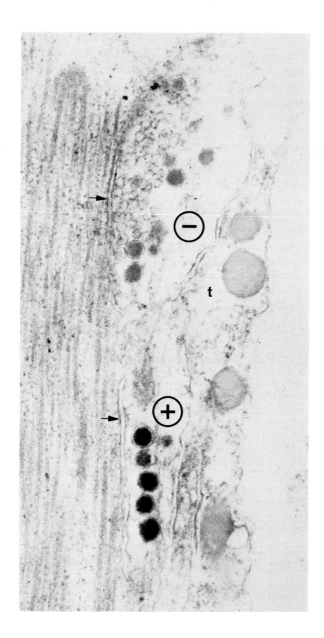

Fig. 19. Longitudinal section in the ABRM of *Mytilus*. The tissue was treated by the Wood (1967) technique for monoamine localization and very lightly stained by lead citrate. One nerve ending contains large, monoamine-positive, dense-core vesicles (+); another contains only monoamine-negative, dense-core vesicles and clear synaptic vesicles (−). It is not possible to tell if these two endings contain different neurotransmitters, because neurites with a mixture of positive and negative dense-core vesicles are also found. It is likely, however, that the positive ending represented here contains 5-HT. The arrows point to differentiated synaptic zones. t, teloglia. (60,000X.) (From Hemming et al., 1983, by permission of Springer Verlag.)

C. Conclusion

These few data can hardly be generalized, but it is likely that in molluscs, at least, the morphology of the vesicular content is very similar in neuromuscular junctions and nerve endings of the ganglionic neuropils. For example, tryptaminergic endings often contain a predominant population of large, dense-core vesicles (approximately 100 nm in diameter; see Jourdan and Nicaise, 1970; Pentreath, 1976), but catecholaminergic and nonaminergic endings also contain similar inclusions. In addition, those vesicular inclusions are not a static entity: Their opacity and diameter evolve (Shkolnik and Schwartz, 1980), and in the terminal they are generally associated with small, clear "synaptic vesicles," the function of which has not been clarified. It is possible that catecholaminergic endings are likewise partially characterized by small granular vesicles, as in vertebrates (Burnstock and Iwayama, 1971). There is very inconclusive information on the ultrastructure of cholinergic neuromuscular junctions in molluscs. Finally, it is likely that there are other types of neuromuscular transmitters, possibly characterized by nerve endings with a different type of vesicle population (Price and McAdoo, 1979; Sathananthan and Burnstock, 1976).

References

Amsellem, J., and Nicaise, G. (1976). Distribution of the glio-interstitial system in molluscs. II. Electron microscopy of tonic and phasic muscles in the digestive tract of *Aplysia* and other opistobranchs. *Cell Tiss. Res.* **165**, 171–184.

Amsellem, J., and Nicaise, G. (1980). Ultrastructural study of muscle cells and their connections in the digestive tract of *Sepia officinalis. J. Submicrosc. Cytol.* **12**, 219–231.

Amsellem, J., Nicaise, G., and Baux, G. (1973). Distribution du système gliointerstitiel chez les mollusques: Les parties tonique et phasique du muscle adducteur postérieur chez *Anodonta. Experientia* **29**, 1274–1276.

Atsumi, S. (1978). Sarcoplasmic reticulum and intracellular calcium localization at rest and during contraction in *Mytilus* smooth muscle. *Arch. Histol. Jpn.* **41**, 239–258.

Atsumi, S., and Sugi, H. (1976). Localization of calcium-accumulating structures in the anterior byssal retractor muscle of *Mytilus edulis* and their role in the regulation of active and catch contractions. *J. Physiol.* **257**, 549–560.

Barrantes, F. J. (1970). The neuromuscular junctions of a pulmonate mollusc. I. Ultrastructural study. *Z. Zellforsch.* **104**, 205–212.

Bennett, M. V. L., and Goodenough, D. A. (1978). Gap junctions, electrotonic coupling and intercellular communication. *Neurosci. Res. Program Bull.* **16**, 373–486.

Bogusch, G. (1972a). Zur Innervation des glatten Penisretraktormusckels von *Helix pomatia:* Allgemeine Histologie und Histochemie des monoaminergen Nervensystems. *Z. Zellforsch.* **126**, 383–401.

Bogusch, G. (1972b). Zur Histochemie der Cholinesterase im Penisretraktor von *Helix pomatia. Z. Mikrosk. Anat. Forsch.* **85**, 35–49.

Bone, Q., and Ryan, K. P. (1973). The structure and innervation of locomotor muscles of salps (Tunicata: Thalicea). *J. Mar. Biol. Ass. U.K.* **53**, 873–883.

Bowden, J. (1958). The structure and innervation of Lamellibranch muscle. *Int. Rev. Cytol.* **7**, 295–335.

Brink, P. R., Kensler, R. W., and Dewey, M. M. (1979). The effect of lanthanum on the nexus of the anterior byssus retractor muscle of *Mytilus edulis* L. *Am. J. Anat.* **154**, 11–26.

Brisson, P., and Collin, J. P. (1980). Systèmes aminergiques des mollusques gastéropodes pulmonés. IV. Paraneurones et innervation catécholaminergiques de la région du carrefour des voies génitales; études autoradiographiques. *Biol. Cell.* **38**(3), 211–220.

Burnstock, G., and Iwayama, T. (1971). Fine structural identification of autonomic nerves and their relation to smooth muscle. *Prog. Brain Res.* **34**, 389–404.

Cambridge, G. W., Holgate, J. A., and Sharpe, J. A. (1959). A pharmacological analysis of the contractile mechanism of *Mytilus* muscle *J. Physiol.* **148**, 451–464.

Carasso N., and Favard, P. (1966). Mise en évidence du calcium dans les myonèmes pédonculaires de ciliés péritriches. *J. Microsc.* **5**, 759–770.

Cardot, J. (1979). Les monoamines chez les mollusques. I. Les catécholamines: Biosynthèse, mise en place et inactivation. *J. Physiol. (Paris)* **75**, 689–713.

Chalazonitis, N. (1969). Differentiation of membranes in axonal endings in the neuropile of *Helix*. *Symp. Int. Soc. Cell Biol.* **8**, 229–243.

Chiang, P. K., Bourgeois, J. G., and Bueding, E. (1972). Histochemical distribution of acetylcholinesterase in the nervous system of the snail *Biomphalaria glabrata*. *Int. J. Neurosc.* **3**, 47–60.

Cloney, R. A., and Florey, E. (1968). Ultrastructure of cephalopod chromatophore organs. *Z. Zellforsch.* **89**, 250–280.

Deane, H. W., and Twarog, B. M. (1957). Histology of an invertebrate smooth muscle. *Anat. Rec.* **128**, 538–539.

Dewey, M. M., Walcott, B., Colflesh, D. E., Terry, H., and Levine, R. J. C. (1977). Changes in thick filament length in *Limulus* striated muscle. *J. Cell Biol.* **75**, 366–380.

Dorsett, D. A., and Roberts, J. B. (1980). A transverse tubular system and neuromuscular junctions in a molluscan unstriated muscle. *Cell Tiss. Res.* **206**, 251–260.

Ducros, C. (1979). Synapses of cephalopods. *Int. Rev. Cytol.* **56**, 1–22.

Elekes, K. (1978) Ultrastructure of synapses in the central nervous system of Lamellibranch molluscs. *Acta Biol. Acad. Sci. Hung.* **29**, 139–154.

Florey, E. (1969). Ultrastructure and function of cephalopod chromatophores. *Am. Zool.* **9**, 429–442.

Florey, E., and Kriebel, M. E. (1969). Electrical and mechanical responses of chromatophore muscle fibers of the squid, *Loligo opalescens*, to nerve stimulation and drugs. *Z. Vgl. Physiol.* **65**, 98–130.

Foh, E. (1969). Die Auswirkung passiver Dehnungen auf die Struktur des glatten Musculus Retractor Penis von *Helix pomatia* L. *Z. Zellforsch.* **93**, 414–433.

Foh, E., and Bogusch, G. (1969). Die Nervennetze im Penisretraktor von *Helix pomatia*. *Z. Zellforsch.* **93**, 439–446.

Gabella, G. (1972). Fine structure of the myenteric plexus in the guinea-pig ileum. *J. Anat.* **111**, 69–97.

Germino, N. I., Castellano, M. A., and Gerard, G. (1968). Histochemical detection of several enzymes in *Cryptomphallus* and *Blaptica* and its relation to neuromuscular transmission. *Comp. Biochem. Physiol.* **24**, 711–716.

Gillette, R., and Pomeranz, B. (1975). Ultrastructural correlates of interneuronal function in the abdominal ganglion of *Aplysia californica*. *J. Neurobiol.* **6**, 463–474.

Gilloteaux, J. (1975). Innervation of the anterior byssal retractor muscle (ABRM) in *Mytilus edulis* L. II. Ultrastructure of the glio-interstitial cells. *Cell Tiss. Res.* **161**, 511–519.

Gilloteaux, J. (1976). Les connections intercellulaires d'un muscle lisse: ultrastructure du muscle rétracteur antérieur du byssus (ABRM) de *Mytilus edulis* L. (Mollusca Pelecypoda). *Cytobiologie* **12**, 457–472.

Gilloteaux, J. (1978). Innervation du muscle rétracteur antérieur du byssus (ABRM) de *Mytilus edulis* L. et de *Mytilus galloprovincialis* Lmk. V. Localisation cytochimique d'activités cholinestérasiques. *Histochemistry* **55**, 209–224.

Graziadei, P. (1966) . The ultrastructure of the motor nerve endings in the muscle of cephalopods. *J. Ultrastruct. Res.* **15**, 1–13.

Hanson, J., and Lowy, J. (1960). Structure and function of the contractile apparatus in the muscle of invertebrate animals. *In* "The Structure and Function of Muscle" (G. H. Bourne, ed.), pp. 265–335. Academic Press, London.

Hanson, J., and Lowy, J. (1961). The structure of the muscle fibres in the translucent part of the adductor of the oyster *Crassostrea angulata*. *Proc. R. Soc., Ser. B* **154**, 173–196.

Hawkins, W. E., Howse, H. D., and Sarphie, I. G. (1980). Ultrastructure of the heart of the oyster *Crassostrea virginica* Gmelin. *J. Submicrosc. Cytol.* **12**, 3, 359–374.

Hayes, R. L., and Kelly, R. E. (1969). Dense bodies of the contractile system of cardiac muscle in *Venus mercenaria*. *J. Morphol.* **127**, 151–161.

Hemming, F. (1981). Etude du tissu glio-interstitiel du muscle antérieur rétracteur du byssus de *Mytilus*. Distribution, cytochimie ultrastructurale et microanalyse. Thèse Doct. 3° cycle, N° 1104. Université Claude Bernard, Villeurbanne, France.

Hemming, F., and Nicaise, G. (1982). Environment-dependent development of glial tissue. *Brain Res.* **245**, 127–130.

Hemming, F. J., Aramant, R., and Nicaise, G. (1983). Simultaneous ultrastructural localization of serotonin and cholinesterases in *Mytilus* byssal retractor muscle (A.B.R.M.). *Histochemistry* **77**, 495–510.

Heumann, H. G. (1969). Calciumakkumulierende Strukturen in einem glatten Wirbellosenmuskel. *Protoplasma* **67**, 111–115.

Heyer, C. B., Kater, S. B., and Karlsson, U. (1973). Neuromuscular systems in molluscs. *Am Zool.* **13**, 247–270.

Hill, R. B., Greenberg, M. J., Irisawa, H., and Nomura, H. (1970). Electromechanical coupling in a molluscan muscle, the radula protractor of *Busycon canaliculatum*. *J. Exp. Zool.* **174**, 331–348.

Hill, R. B., and Sanger, J. W. (1974). Anatomy of the innervation and neuromuscular junctions of the radular protractor muscle of the whelk, *Busycon canaliculatum*. *Biol. Bull.* **147**, 369–385.

Hoyle, G. (1964). Muscle and neuromuscular physiology. *In* "Physiology of Mollusca" (K. M. Wilbur and C. M. Yonge, eds.), Vol. 1. Academic Press, New York.

Hoyle, G., and McNeill, P. A. (1968). Correlated physiological and ultrastructural studies on specialized muscles. Ib. Ultrastructure of white and pink fibers of the levator of the eyestalk of *Podophtalmus vigil* (Weber). *J. Exp. Zool.* **167**, 487–522.

Hoyle, G., McAlear, J. H., and Selverston, A. (1965). Mechanism of supercontraction in a striated muscle. *J. Cell Biol.* **26**, 621–640.

Hoyle, G., McNeill, P. A., and Selverston, A. I. (1973). Ultrastructure of barnacle giant muscle fibers. *J. Cell Biol.* **56**, 74–91.

Huddart, H., Hunt S., and Oates, K. (1977). Calcium movements during contraction in molluscan smooth muscle, and the loci of calcium binding and release. *J. Exp. Biol.* **68**, 45–56.

Hunt, S. (1972). The fine structure of the smooth muscle in the hypobranchial gland of the gastropod *Buccinum undatum* L. *Tissue & Cell.* **4**, 479–492.

Hunt, S. (1981). Molluscan visceral muscle fine structure. General structure and sarcolemmal organization in the smooth muscle of the intestinal wall of *Buccinum undatum* L. *Tissue & Cell* **13**, 283–297.

Irisawa, H., Irisawa, A., and Shigeto, N. (1973). Physiological and morphological correlation of the functional syncytium in the bivalve myocardium. *Comp. Biochem. Physiol. A*, **44**, 207–220.

Jensen, H., and Tjønneland, A. (1977). Ultrastructure of the heart muscle cells of the cuttlefish *Rossia macrosoma* (delle Chiage) (Mollusca Cephalopoda). *Cell Tiss. Res.* **185,** 147–158.

Josephson, R. K. (1975). Extensive and intensive factors determining the performance of striated muscle. *J. Exp. Zool.* **194,** 135–154.

Jourdan, F., and Nicaise, G. (1970). Cytochimie ultrastructurale de la sérotonine dans le système nerveux central de l'Aplysie. *7th Int. Cong. Elect. Microsc.* **3,** 677–678 Grenoble, France.

Karnovsky, M. J. (1964). The localization of cholinesterase activity in rat cardiac muscle by electron microscopy. *J. Cell Biol.* **23,** 217–232.

Kelly, R. E., and Hayes, R. L. (1969). The ultrastructure of smooth cardiac muscle in the clam, *Venus mercenaria. J. Morphol.* **127,** 163–176.

Lanzavecchia, G. (1977). Morphological modulations in helical muscles (Aschelminthes and Annelida). *Int. Rev. Cytol.* **51,** 133–186.

Lowy, J., Millman, B. M., and Hanson, J. (1964). Structure and function in smooth tonic muscles of lamellibranch molluscs. *Proc. R. Soc., Ser. B* **160,** 525–536.

Margulis, B. A., Bobrova, I. F., Mashanski, V. F., and Pinaev, G. P. (1979). Major myofibrillar protein content and the structure of mollusc contractile apparatus. *Comp. Biochem. Physiol. A* **64,** 291–298.

Martin R., and Barlow, J. J. (1975). Muscle and gland cell degeneration in the *Octopus* posterior salivary gland after 6-hydroxydopamine administration. *J. Ultrastruct. Res.* **52,** 167–168.

McCaman, R. E., and McCaman, M. W. (1976). Biology of individual cholinergic neurons in the invertebrate central nervous system. *In* "Biology of cholinergic function" (A. M. Goldberg, I. Hanin, eds.), pp. 485–514. Raven Press, New York.

McKenna, O. C., and Rosenbluth, J. (1973). Myoneural and intermuscular junctions in a molluscan smooth muscle. *J. Ultrastruct. Res.* **42,** 434–450.

Millman, B. M. (1967). Mechanism of contraction in molluscan muscle. *Am. Zool.* **7,** 583–591.

Millman, B. M., and Bennett, P. M. (1976). Structure of the cross-striated adductor muscle of the scallop. *J. Mol. Biol.* **103,** 439–468.

Moon, T. W., and Hulbert, W. C. (1975). The ultrastructure of the mantle musculature of the squid *Symplectoteuthis oualaniesis. Comp. Biochem. Physiol. B,* **52,** 145–149.

Morrison, C. M., and Odense, P. H. (1974). Ultrastructure of some pelecypod adductor muscles. *J. Ultrastruct. Res.* **49,** 228.

Nicaise, G. (1969). Détection histochimique de cholinestérases dans les cellules gliales et interstitielles des Doridiens. *C. R. Soc. Biol.* **163,** 2600–2604.

Nicaise, G. (1973). The gliointerstitial system of molluscs. *Int. Rev. Cytol.* **34,** 251–332.

Nicaise, G., Hernandez-Nicaise, M. L., and Malaval, L. (1982). Electron microscopy and X-ray microanalysis of calcium-binding sites on the plasma membrane of *Beroe* giant smooth muscle fibre. *J. Cell Sci.* **55,** 353–364.

Nisbet, R. H., and Plummer, J. M. (1966). Further studies on the fine structure of the heart of Achatinidae. *Proc. Malac. Soc. London* **37,** 199–208.

Nisbet, R. H., and Plummer, J. M. (1968). The fine structure of cardiac and other molluscan muscle. *Symp. Zool. Soc. London* **22,** 193–211.

Nunzi, M. G., Franzini-Armstrong, Cl. (1981). The structure of smooth and striated portions of the adductor muscle of the valves in a scallop. *J. Ultrastruct. Res.* **76,** 134–148.

Pentreath, V. W. (1976). Ultrastructure of the terminals of an identified 5-hydroxytryptamine-containing neurone marked by intracellular injection of radioactive 5-hydroxytryptamine. *J. Neurocytol.* **5,** 43–61.

Pentreath, V. W., Berry, M. S., and Cobb, J. L. S. (1975). Nerve ending specializations in the central ganglia of *Planorbis corneus. Cell Tiss. Res.* **163,** 99–110.

Peretz, B., and Estes, J. (1974). Histology and histochemistry of the peripheral neural plexus in the *Aplysia* gill. *J. Neurobiol.* **5**, 3–20.

Pfenninger, K. H. (1971). The cytochemistry of synaptic densities. II. Proteinaceous components and mechanism of synaptic connectivity. *J. Ultrastruct. Res.* **35**, 451–474.

Plesch, B. (1977a). An ultrastructural study of the musculature of the pond snail *Lymnaea stagnalis* L. *Cell Tiss. Res.* **180**(3), 317–340.

Plesch, B. (1977b). An ultrastructural study of the innervation of the musculature of the pond snail *Lymnaea stagnalis* L. with reference to peripheral neurosecretion. *Cell Tiss. Res.* **183**, 353–370.

Prescott, L., and Brightman, M. W. (1976). The sarcolemma of *Aplysia* smooth muscle in freeze fracture preparations. *Tissue and Cell.* **8**, 241–258.

Price, C. H., and McAdoo, D. J. (1979). Anatomy and ultrastructure of the axons and terminals of neurons R3–R14 in *Aplysia. J. Comp. Neurol.* **188**, 647–678.

Richardot, M., and Wautier, J. (1971). Une structure intermédiaire entre muscle lisse et muscle strié. La fibre musculaire du bulbe buccal de *Ferrissia wautieri* (Moll. Basomm. Ancylidae). *Z. Zellforsch.* **115**, 100–109.

Rogers, D. C. (1969). Fine structure of smooth muscle and neuromuscular junctions in the foot of *Helix aspersa. Z. Zellforsch.* **99**, 315–335.

Rosenbluth, J. (1963). Fine structure of epineurial muscle cells in *Aplysia californica. J. Cell Biol.* **17**(2), 455–460.

Rosenbluth, J. (1972). Obliquely striated muscle. *In* "The Structure and Function of Muscle" (G. H. Bourne, ed.), Vol. I, pp. 389–420. Academic Press, New York.

Ruegg, J. C. (1971). Smooth muscle tone. *Physiol. Rev.* **51**, 201–248.

Sanger, J. W., and Hill R. B. (1972). Ultrastructure of the radula protractor of *Busycon canaliculatum.* Sarcolemmic tubules and sarcoplasmic reticulum. *Z. Zellforsch.* **127**, 314–322.

Sanger, J. W., and Hill, R. B. (1973). The contractile apparatus of the radular protractor muscle of *Busycon canaliculatum. Proc. Malac. Soc., Lond.* **40**, 335–341.

Sathananthan, A. H. (1976). Degeneration of monoamine nerves in anterior byssus retractor muscles of Mytilus induced by 5,6 dihydroxytryptamine. *Cell Tiss. Res.* **172**, 425–429.

Sathananthan, A. H. (1977). Degeneration of possible serotonergic nerves in the myenteric plexus of *Mytilus* induced by 5,7 dihydroxytryptamine (5,7-DHT). *Cell Tiss. Res.* **179**, 393–399.

Sathananthan, A. H., and Burnstock, G. (1976). Evidence for a non cholinergic, non adrenergic innervation of *Venus* clam heart. *Comp. Biochem. Physiol.* **C55**, 111–118.

Shkolnik, L. J., and Schwartz J. H. (1980). Genesis and maturation of serotonergic vesicles in identified giant cerebral neuron of *Aplysia. J. Neurophysiol.* **43**, 929–944.

Sobieszek, A. (1973). The fine structure of the contractile apparatus of the anterior byssus retractor muscle of *Mytilus edulis. J. Ultrastruct. Res.* **43**, 313–343.

Sugi, H., and Suzuki, S. (1978). Ultrastructural and physiological studies on the longitudinal body wall muscle of *Dolabella auricularia.* I. Mechanical response and ultrastructure. *J. Cell Biol.* **79**, 454–466.

Suzuki, S., and Sugi, H. (1978). Ultrastructural and physiological studies on the longitudinal body wall muscle of *Dolabella auricularia.* II. Localization of intracellular calcium and its translocation during mechanical activity. *J. Cell Biol.* **79**, 467–478.

Szent-Györgyi, A. G., Cohen, C., and Kendrick-Jones, J. (1971). Paramyosin and the filaments of molluscan "catch" muscle. II. Native filaments: isolation and characterization. *J. Mol. Biol.* **56**, 239–258.

Taxi, J., and Gautron, J. (1969). Données cytochimiques en faveur de l'existence de fibres nerveuses sérotoninergiques dans le coeur de l'Aplysie, *Aplysia californica. J. Microsc.* **8**, 627–636.

Thompson, E. B., Schwartz, J. H., and Kandel, E. R. (1976). A radioautographic analysis in the light and electron microscope of identified *Aplysia* neurons and their processes after intrasomatic injection of L(^{3}H) fucose. *Brain Res.* **112**, 251–281.

Treherne, J. E., Carlson, A. D., and Gupta, B. L. (1969). Extraneuronal sodium store in central nervous system of *Anodonta cygnea. Nature* **223**, 377–379.

Twarog, B. M. (1967). The regulation of catch in molluscan muscle. *J. Gen. Physiol.* **50**, 157–169.

Twarog, B. M. (1976). Aspects of smooth muscle function in molluscan catch muscle. *Physiol. Rev.* **56**, 829–838.

Twarog, B. M. (1977). Dissociation of calcium dependent reactions at different sites: Lanthanum block of contraction and relaxation in a molluscan smooth muscle. *In* "Excitation-Contraction Coupling in Smooth Muscle" (R. Casteels, T. Godfraind, and J. C. Ruegg, eds.), pp. 261–271. Elsevier-North Holland. Amsterdam.

Twarog, B. M., Dewey, M. M., and Hidaka, T. (1973). The structure of *Mytilus* smooth muscle and the electrical constants of the resting muscle. *J. Gen. Physiol.* **61**, 207–221.

Wabnitz, R. W. (1975). Functional states and fine structure of the contractile apparatus of the penis retractor muscle (PRM) of *Helix pomatia* L. *Cell Tiss. Res.* **156**, 253–265.

Wabnitz, R. W. (1976). Excitation-contraction coupling and intracellular calcium pool in the penis retractor muscle of *Helix pomatia. Comp. Biochem. Physiol.* **54**, 75–80.

Watts, J. A., and Pierce, S. K. (1978). Acetylcholinesterase: A useful marker for the isolation of sarcolemma from the bivalve (*Modiolus demissus demissus*) myocardium. *J. Cell Sci.* **34**, 193–208.

Wood, J. (1967). Cytochemical localization of 5 hydroxytryptamine (5HT) in the central nervous system. *Anat. Rec.* **157**, 343.

Yamamoto, M. (1977). Electron microscopic studies on the innervation of the smooth muscle and the interstitial cell of Cajal in the small intestine of the mouse and bat. *Arch. Histol. Jpn.* **40**, 171–202.

Zs-Nagy, I. (1964). A histochemical study of cholinesterases on the adductor muscle of the fresh water mussel (*Anodonta cygnea* L.) *Ann. Inst. Biol. Tihany Hung. Acad. Sci.* **31**, 153–157.

Zs-Nagy, I., and Salanki, J. (1970). The fine structure of neuromuscular and intermuscular connections in the adductors of *Anodonta cygnea* L. (Mollusca, Pelecypoda). *Ann. Inst. Biol. Tihany Hung. Acad. Sci.* **37**, 131–143.

Zs-Nagy, I., Salanki, J., and Garamvolgyi, N. (1971). The contractile apparatus of the adductor muscles in *Anodonta cygnea* L. (Mollusca, Pelecypoda). *J. Ultrastruct. Res.* **37**, 1–16.

2

Neuromuscular Transmission and Excitation–Contraction Coupling in Molluscan Muscle

YOJIRO MUNEOKA

Faculty of Integrated Arts and Sciences
Hiroshima University
Higashi-senda-machi
Hiroshima, Japan

BETTY M. TWAROG

Department of Biology
Bryn Mawr College
Bryn Mawr, Pennsylvania

I. Introduction and Perspective

This chapter reviews the current state of knowledge of neuromuscular transmission and excitation–contraction coupling in molluscs. Physiological properties common to many phyla are described, as well as some which may be unique to molluscs. Significant physiological questions that remain unanswered are set forth in detail.

Systematic investigations of the neural control of muscle function in molluscs were few until recently. Elegant studies of excitatory innervation, as well as a few on inhibitory innervation and on modulation, have now been published.

In molluscs, as in other phyla, excitatory nerves activate brief phasic responses or prolonged tonic contractions, depending on specific properties of the muscle fibers and the innervating neurons. These contractions, whether phasic or

THE MOLLUSCA, VOL. 4
Physiology, Part 1

tonic, relax when excitation ceases. Some mollusc muscles display a prolonged contraction called *catch* which is not tonic in the usual sense because it persists long after excitation has ended.

The contractile mechanism is activated by calcium ions (Ca^{2+}), but the sources of activating Ca^{2+} are not so well documented as in vertebrate skeletal muscles. There is abundant morphological and physiological evidence that intracellular stores provide Ca^{2+} for activation of contraction. There is also good evidence that translocation of Ca^{2+} from outside the muscle fiber or from superficial sites on the membrane is involved directly or indirectly in excitation–contraction coupling. The importance of these components in the excitation–contraction coupling process apparently varies from muscle to muscle within species as well as between different species.

Individual muscle fibers may be innervated by a single excitatory neuron, or by two or more excitatory neurons, each of which releases a different transmitter. Dual innervation of individual muscle fibers by inhibitory and excitatory neurons as well as by relaxing and excitatory neurons, also occurs. Because muscle fibers in molluscs are often electrically coupled, the possibility that responses recorded in a single muscle fiber reflect postjunctional responses elicited in coupled fibers must be ruled out in each case before concluding that innervation is polyneuronal. Putative excitatory neurotransmitters include acetylcholine (ACH), glutamate, dopamine, serotonin, and the neuropeptide FMRFamide. The ionic basis of excitatory junction potentials (EJPs) and muscle action potentials is poorly understood.

Except for nervous inhibition of rhythmic activity in the heart, peripheral neuromuscular inhibition has not been systematically investigated in molluscs. Probable inhibitory neurotransmitters include serotonin, octopamine, dopamine, ACh, and ATP.

Modulation is a process whereby the level of excitability in a cell is changed, altering the response to neurotransmitters, without necessarily having any detectable action on the resting membrane. Neuromodulator substances (*local hormones*) are thought to alter metabolism via a second messenger. Serotonin, dopamine, and some peptides are implicated as modulators in molluscs and are under active investigation in a number of laboratories.

Catch is a phenomenon first observed and most highly developed in certain specialized mollusc muscles. During catch contraction, muscles are resistant to stretch, and remain so after excitation has ceased and intracellular free Ca^{2+} has returned to resting levels. Catch apparently depends upon attached actin–myosin cross-bridges that do not cycle. It seems probable that cross-bridge attachment and breaking are controlled, in molluscs, by phosphorylation and dephosphorylation of the paramyosin core of the thick filaments on which myosin molecules are arrayed. Understanding how paramyosin phosphorylation controls myosin bridge attachment and cycling may well contribute to the elucidation of myosin bridge cycling.

Catch is activated by excitatory nerves which release ACh (or the neuropeptide Phe-Met-Arg-Phe-NH$_2$ FRMFamide) and terminated by relaxing nerves which release serotonin. Although catch has been observed in phyla other than molluscs, control of catch by relaxing neurons is a feature unique to molluscs. It is of note that serotonin not only relaxes catch but modulates excitability. It is of biological interest that the increase in excitability induced by serotonin enhances the transition from the passive catch state to a state of phasic activity and that the unusual role of serotonin in relaxing catch may have evolved in connection with its more typical role as a modulator.

The findings outlined in this chapter suggest that further studies on control of contractility in molluscs will provide a wealth of novel data and concepts.

II. Neural Control of Muscle Function

A. Phasic Contraction

1. Excitation

a. Innervation. Each muscle fiber may be innervated by a single excitatory neuron, for example, the columellar and the posterior jugalis muscle of *Helisoma* (Kater et al., 1971). Often, because muscle fibers may be electrically coupled, the possibility cannot be ruled out that the response recorded in any single muscle fiber reflects postjunctional responses elicited in other muscle fibers by one or more neurons. Electrical coupling between muscle fibers has been shown or suggested in many molluscan muscles (Cohen et al., 1978; Florey and Kriebel, 1969; Gilloteaux, 1976; McKenna and Rosenblueth, 1973; Orkand and Orkand, 1975; Twarog et al., 1973).

The circular muscle of the squid *Loligo* is innervated by two kinds of excitatory nerve fibers, small and giant fibers. The former produce slow, small contractions, whereas the latter produce fast, large ones (Prosser and Young, 1937; Wilson, 1960; Young, 1938). Innervation by two kinds of excitatory nerves is also suggested in other muscles, e.g., the octopus mantle (Wilson, 1960), the retractor of the mantle and the anterior adductor muscle of *Mya* (Pumphrey, 1938), and the pharyngeal retractor muscle of snail (Ramsey, 1940; Sato et al., 1960). In these muscles, however, it is less certain whether individual muscle fibers are innervated by two kinds of excitatory nerves.

Polyneural excitatory innervation is frequently seen in molluscan muscles; individual muscle fibers are innervated by two or more excitatory neurons, each of which releases a different neurotransmitter (Carew et al., 1974).

Carew et al. (1974) have shown that a single muscle fiber in the gill muscle of *Aplysia* generates EJPs in response to stimulation of two identified motoneurons, the cholinergic LDG_1 neuron and the noncholinergic L_7 neuron (Fig. 1A). Cohen et al. (1978) have obtained similar results from the accessory radula closer (ARC) muscle of *Aplysia* by stimulating each of two identified cholinergic motoneurons, B_{15} and B_{16} (Fig. 1B). These results indicate polyneural excitatory innervation of individual muscle fibers. Polyneural innervation is probable in many molluscan muscles: the anterior byssus retractor muscle (ABRM) of *Mytilus* (Twarog, 1967a), the fast adductor of the scallop (Mellon, 1968), squid chromatophore muscle fibers (Florey and Kriebel, 1969), the radula protractor muscle of *Rapana* (Kobayashi, 1972a), the lower extrinsic protractor muscle of *Aplysia* (Orkand and Orkand, 1975), the penis retractor muscle of *Aplysia* (Rock et al., 1977), and the pharynx levator muscle of snail (Peters, 1979a). In these muscles, the amplitude of the EJPs increases with increased intensity of stimulation of the nerve.

b. Muscle Responses.   In the buccal mass of *Helisoma,* the muscle potential, motoneuron spike, and muscle contraction have a one-to-one relationship (Heyer et al., 1973; Kater et al., 1971). In the circular muscle of the mantle of *Loligo,* a single stimulus to the giant axon produces an all-or-none contraction (Prosser and Young, 1937), suggesting that the electrical response of the muscle fibers is an all-or-none action potential. Using the sucrose-gap method, Kobayashi (1974a) showed that, in low magnesium ion (Mg^{2+}) solution, the radula protractor muscle of *Rapana* generates an action potential in response to a single neural stimulus.

Summation and facilitation of EJPs are commonly observed in molluscan muscles (e.g., Carew et al., 1974; Kobayashi, 1974a; Orkand and Orkand, 1975) (see also Fig. 1). However, there may be some exceptions. Mellon (1968) reported that the fast adductor muscle of the scallop *Aequipecten* did not show neuromuscular facilitation, although it showed summation. In most cases, action potentials are generated as a result of summation and facilitation of EJPs, although in the ARC muscle of *Aplysia,* muscle action potentials are not elicited by repetitive neural stimulation even though EJPs facilitate and summate (Cohen *et al.,* 1978). In some muscles, the action potentials do not overshoot, e.g., the ABRM of *Mytilus* (Twarog, 1967b) and the pharynx levator muscle of snail (Peters, 1979a), but in other muscles, the action potentials do overshoot, e.g., the fast adductor muscle of scallop (Mellon, 1968), the fast portion of the posterior adductor muscle (Mellon and Prior, 1970), and the incurrent siphonal valve muscle (Prior, 1975) of *Spisula.*

The action potential of the ABRM is blocked by manganese ion (Mn^{2+}) or lanthanum ion (La^{3+}) but not by tetrodotoxin (Hidaka and Goto, 1973; Mizonishi, 1977; Twarog, 1967b; Twarog and Hidaka, 1971). The action potential can

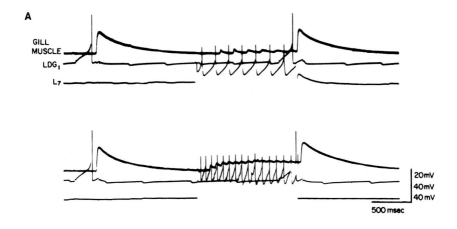

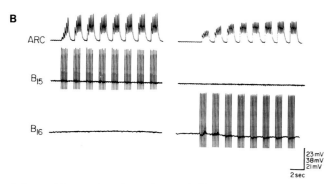

Fig. 1. Polyneural innervation of a single muscle fiber. (**A**) The innervation of a gill muscle fiber in the efferent vein of *Aplysia californica*. The top trace of each pair shows EJPs recorded intracellularly from the fiber; the bottom traces show action potentials recorded intracellularly from LDG$_1$ and L$_7$, respectively. (**B**) The innervation of an ARC muscle fiber of *A. californica*. The top trace of each pair shows EJPs recorded intracellularly from the muscle fiber; the bottom traces show action potentials recorded intracellularly from B$_{15}$ and B$_{16}$, respectively. (**A**: Carew et al., 1974; **B**: Cohen et al., 1978.)

be evoked by electrical stimulation in sodium ion (Na$^+$)-deficient artificial seawater (ASW), and its amplitude increases with an increase in external Ca^{2+} concentration (Hidaka, 1972; Twarog and Hidaka, 1971). Using microelectrodes, Mizonishi (1977) observed that action potentials arising from depolarizing ACh potentials were not elicited in Ca^{2+}-deficient ASW, although the amplitude of the ACh potential was increased in that solution. Thus, the action potential of the ABRM is apparently generated by an increased permeability to Ca^{2+} (Hidaka, 1972; Twarog and Hidaka, 1971; Twarog and Muneoka, 1973). In contrast, Sugi and Yamaguchi (1976) observed, using an oil-gap method, that in ASW containing less than 10 mM Mg^{2+}, removing external Ca^{2+} caused rhythmic, spike-like electrical potentials superimposed on a gradual decline of membrane potential. Although this was taken as evidence that the ABRM may generate Na spikes, the observed potentials may have been generated in fibers within the muscle bundle by intercellular Ca^{2+} that remained after washing out the extramuscular Ca^{2+}. The muscle fibers of the ABRM are electrically interconnected by gap junctions (Twarog et al., 1973). Thus, depolarization of the membrane of outer muscle fibers produced by the removal of external Ca^{2+} would stimulate inner muscle fibers to generate Ca spikes if Ca^{2+} remains in the spaces between the inner fibers.

The amplitude of action potentials in the radula protractor muscle of *Rapana* in response to electrical stimulation is dependent on the external Ca^{2+} concentration (Kobayashi, 1972a). The response is blocked by 10 mM Mn^{2+} but not abolished in tetrodotoxin and low Na^+ solution, suggesting that the generation of action potential depends on inward Ca^{2+} current. However, in low Ca^{2+} ASW containing normal Na^+, the muscle responds to a brief electrical stimulus with a large, prolonged depolarization. The membrane depolarizes for more than 20 sec and then abruptly repolarizes. The depolarization is not blocked by 30 mM Mn^{2+} but is completely inhibited in zero Na^+ solution, suggesting that the membrane of the muscle is capable of responding by a change in Na^+ permeability when Ca^{2+} is removed from the bathing solution (Kobayashi, 1972a).

c. Excitatory Neurotransmitters.   It is generally believed that ACh is an excitatory neurotransmitter in many molluscan muscles. Cholinergic excitatory mechanisms in molluscan muscles have been reviewed by Gerschenfeld (1973) and Leake and Walker (1980). However, recent investigations suggest that other substances, such as glutamate, dopamine, serotonin, and FMRFamide may also be excitatory transmitters at the neuromuscular junctions of molluscs.

The ABRM of *Mytilus* responds to applied ACh with membrane depolarization and contraction, both of which can be inhibited by cholinergic blocking agents and enhanced by the anticholinesterase eserine. Banthine and propantheline (probanthine) are the most effective cholinergic blocking agents, whereas curare is more effective than atropine (Cambridge et al., 1959; Twarog, 1954, 1959, 1960a). Carbachol, a cholinergic agent not hydrolyzed by cholinesterase, produces a contraction which is not potentiated by eserine (Muneoka et al., 1979b). EJPs are also blocked by banthine (Twarog, 1967b) and enhanced by eserine. The reversal potential of the ACh-induced depolarization is very close to the calculated value for that of the EJP, being -13 and -12 mV, respectively (Hidaka and Twarog, 1977). An ACh-like substance can be detected in the muscle by bioassay, and a specific cholinesterase is present (Twarog, 1954). Electron microscopic studies have shown nerve terminals with clear vesicles in the muscle (Gilloteaux, 1978; McKenna and Rosenbluth, 1973). They are 400 Å in diameter and resemble those found in cholinergic terminals in vertebrates. Cytochemical localization of cholinesterase activity has been reported (Gilloteaux, 1978).

ACh applied to the ABRM generates a slow depolarization upon which action potentials are superimposed (Mizonishi, 1977; Twarog, 1960a). The amplitude of the slow ACh potential decreases if external Na^+ is decreased and the potential is blocked or almost blocked in zero Na^+ solution (Hidaka and Goto, 1973; Hidaka and Twarog, 1977; Mizonishi, 1977). The relationship between the resting membrane potential and the ACh potential is also dependent upon the external K^+ concentration. Reducing external chloride ion (Cl^-) causes no appreciable change in the ACh potential. Thus, it seems probable that Na^+ and potassium ion (K^+), but not Cl^-, contribute to generation of the ACh potential in the ABRM (Hidaka and Twarog, 1977).

The radula protractor muscles of prosobranch gastropods *Busycon* (Hill, 1958, 1970; Hill et al., 1970) and *Rapana* (Kobayashi, 1972b; Kobayashi and Muneoka, 1980; Kobayashi and Shigenaka, 1978) are probably innervated by cholinergic excitatory nerves. In both muscles, depolarization and contraction by ACh are markedly depressed in zero Na^+ ASW, suggesting that the depolarizing response to ACh depends mainly on Na^+ inflow (Hill and McDonald-Ordzie, 1979; Kobayashi and Shigenaka, 1978). Kobayashi and Hashimoto (1980) have suggested that, in the *Rapana* muscle, an increase in membrane permeability to Cl^- plays a role in generation of the ACh potential.

The ARC muscles of *Aplysia* are innervated by excitatory nerves originating in the buccal ganglion. Two of these neurons are identified as B_{15} and B_{16}. They are probably cholinergic (Cohen et al., 1978): The muscle responds to ACh with contraction; the cholinergic blocking agent hexamethonium blocks the EJPs evoked by the stimulation of B_{15} and B_{16}; both cells synthesize ACh from injected choline; histochemical reactions for cholinesterase activity reveal a network of nerve fibers that arises at the cartilage and covers the entire muscle.

The regulation of gill movement in *Aplysia* is controlled by the abdominal ganglion. The ganglion has at least six motoneurons to the gill (Kupfermann et al., 1974). Two of these, LDG_1 and LDG_2, are thought to be cholinergic (Carew et al., 1974).

The pharynx levator muscle of *Helix* responds to ACh with depolarization. *d*-Tubocurarine blocks the response to ACh and lowers the amplitude of EJPs in response to neural stimulation (Peters, 1979b). There are many other muscles which are capable of responding to ACh with contraction: the penis retractor muscle of *Helix* (Wabnitz, 1980; Wabnitz and Wachtendonk, 1976), the pharyngeal retractor muscle of *Helix* (Kerkut and Leake, 1966), the mantle of the pelecypod *Spisula* (Wilson, 1969), the intestine of *Mercenaria* (Greenberg and Jegla, 1963) and *Spisula* (Nystrom, 1967), the opaque part of the adductor muscle of the oyster *Crassostrea* (Millman, 1964), the anterior extrinsic protractor of *Aplysia* (Taraskevich et al., 1977), and the abdominal aorta of *Aplysia* (Liebeswar et al., 1975).

The radial muscle fibers of the chromatophores of squid respond to ACh with tonic contractions or pulsations. However, the action of ACh seems to be a presynaptic one, and the excitatory neurotransmitter is not yet established (Florey, 1966; Florey and Kriebel, 1969).

There have been several reports showing that molluscan muscles can respond to L-glutamate. The pharyngeal retractor muscle of *Helix* contracts in response to glutamate (Kerkut and Leake, 1966; Kerkut et al., 1965). Glutamate is released into the perfusate when the nerve to the muscle is stimulated (Kerkut et al., 1965). If the brain has been incubated in [^{14}C]glucose, [^{14}C]alanine, or [^{14}C]glutamate, then labeled glutamate appears at the muscle some 20 min after the brain is stimulated (Kerkut et al., 1967). The pharynx levator muscle of *Helix* responds not only to ACh (Peters, 1979b) but also to glutamate (Peters, 1978) with depolarization. Consideration of glutamate as the transmitter is reinforced by the observation that eserine does not affect the ACh response or EJPs (Peters, 1979b). The radula retractor muscle of *Rapana* also responds to both ACh and glutamate with contraction. At high concentrations such as 10^{-4} and 10^{-3} M, glutamate can produce larger contractions than ACh. Here, also, eserine enhances neither ACh contraction nor twitch contraction due to nerve stimulation (Muneoka and Kobayashi, 1980), and glutamate may well be the excitatory neurotransmitter. The gill muscle of *Aplysia* is controlled by a noncholinergic excitatory neuron, L_7, in addition to cholinergic neurons (Fig. 2). The neuron L_7 may be glutaminergic; glutamate concentrations as low as 10^{-10} M produce a contraction (Carew et al., 1974). The anterior extrinsic protractor of *Aplysia* can also respond to glutamate with contraction (Taraskevich et al., 1977). L-Glutamate has excitatory effects on the fast and slow adductor muscles of *Pecten maximus (L.)*, as well as on mantle and fin muscle from the cephalopods *Sepia officinalis*, *Alloteuthis subulata* (Lamarck), and *Loligo forbesi* (Steenstrop), and these effects are reversibly blocked by glutamate antagonists (Bone and Howarth, 1980). Thus, L-glutamate may be an excitatory transmitter in certain gastropod, lamellibranch, and cephalopod muscles.

The gill muscle of *Aplysia* responds to dopamine with contraction (Ruben et al., 1979; Swann et al., 1978a, 1978). Dopamine is known to be present in the gill on the basis of both biochemical (Carpenter et al., 1971) and fluorescence histochemical studies (Peretz and Estes, 1974; Swann et al., 1978a). The motor neurons L_9, which innervate the gill from the abdominal ganglion, are thought to be dopaminergic (Swann et al., 1978a). These facts suggest that dopamine may be the excitatory transmitter released from L_9 neurons (Swann et al., 1978a).

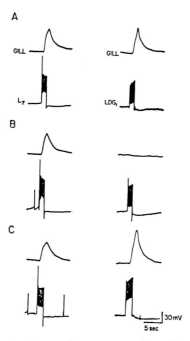

Fig. 2. Effect of hexamethonium on the contraction of the gill muscle in response to L_7 and LDG_1 stimulation in *Aplysia californica*. The top trace of each pair shows a photocell record of gill contraction; the bottom trace shows action potentials recorded intracellularly from L_7 and LDG_1. (**A**) In normal seawater; (**B**) in 10^{-3} M hexamethonium; (**C**) after washing the hexamethonium. Note that the contraction in response to L_7 stimulation is not affected by hexamethonium. (Carew et al., 1974.)

However, at low concentrations such as 5×10^{-7} M, dopamine markedly enhances the contraction of the gill muscle in response to the stimulation of LDG_1 or L_7 (Swann et al., 1978b). Therefore, dopamine might be a modulatory transmitter in the gill muscle rather than an excitatory one. This point will be discussed later. The anterior extrinsic protractor muscle of *Aplysia* is capable of responding not only to ACh and L-glutamate but also to dopamine. Dopamine and ACh are about equally potent in producing contraction (Taraskevich et al., 1977). Newby (1973) has observed excitatory action of dopamine on the siphon muscle of *Aplysia*. Wabnitz (1980) has reported that at low concentrations (10^{-7} M or less), dopamine produces tonic contraction in the rhythmically contracting penis retractor muscle of *Helix*, but has suggested that its effects are presynaptic.

FMRFamide is a cardioexcitatory neuropeptide first found in ganglia of the clam *Macrocallista*. It produces contraction in the radula protractor of *Busycon* at concentrations of 2×10^{-9} M or greater (Price and Greenberg, 1977a, 1977b). Benzoquinonium blocks ACh contraction of the muscle but has no effect on FMRFamide contraction, indicating that the peptide does not act via a cholinergic mechanism (Price and Greenberg, 1980). FMRFamide also causes contraction in the radula muscles of a gastropod *Fusinus* but not in the radula muscles of *Rapana* (Y. Muneoka, unpublished observations). FMRFamide induces a catch contraction in the ABRM of *Mytilus* (Painter and Greenberg, 1981). The physiological role of FMRFamide remains unknown, but it should be considered as an excitatory neurotransmitter or a hormone in molluscan neuromuscular systems in future studies.

2. Inhibition

Except for nervous inhibition of rhythmic activity in the heart, peripheral neuromuscular inhibition has not been systematically investigated in molluscs. Banks (1975) recorded inhibitory junctional potentials (IJPs) in the lower extrinsic protractor muscle of *Aplysia* by stimulating a neuron in the buccal ganglion. Later, he identified the inhibitory neuron using anatomical and electrophysiological techniques (Banks, 1978). Carew et al. (1974) observed spontaneous IJPs in the gill muscle of *Aplysia*. Blankenship et al. (1977) observed spontaneous IJPs in the penis retractor muscle of *Aplysia*. The neurotransmitter which produces these IJPs is not known, although Ruben et al. (1979) have shown that serotonin inhibits spontaneous and stimulus-evoked contractions of the gill muscle of *Aplysia*.

In the radula retractor muscle of *Rapana,* serotonin and octopamine hyperpolarize the muscle fiber membrane and inhibit twitch contraction and EJPs. The presence of biogenic amines in the muscle has been demonstrated by bioassay (Muneoka and Kobayashi, 1980). Thus, serotonin and octopamine may be inhibitory transmitters in the muscle. However, inhibitory innervation has not yet been shown. Histamine also inhibits twitch contraction of the radula retractor, but the site of its action appears to be presynaptic (Kobayashi et al., 1981).

Peters (1979b) observed that serotonin hyperpolarizes the membrane of the pharynx levator muscle fibers of *Helix*. In the ABRM of *Mytilus,* dopamine hyperpolarizes the muscle fiber membrane (Hidaka et al., 1977). Again, however, an inhibitory innervation has not been demonstrated. The mechanism of hyperpolarization by dopamine in the ABRM is not clear. Hidaka et al. (1977) suggested that it may activate an electrogenic Na^+ pump mechanism in the ABRM (Yamaguchi and Twarog, 1980). In the pharyngeal retractor muscle of *Helix,* serotonin hyperpolarizes the fiber membrane (Lloyd, 1980c); it seems not to be an inhibitory transmitter but rather to modulate contraction and relaxation (Lloyd, 1980b, 1980c).

In some molluscs, the gut is innervated by excitatory and inhibitory nerves (Burnstock et al., 1967; Prosser et al., 1965). Pharmacological evidence suggests that some of the inhibitory nerves release ACh (Brunstock *et al.,* 1967;

Greenberg and Jegla, 1963; Phillis, 1966b). In *Tapes* intestine, dopamine may be an inhibitory transmitter (Dougan and McLean, 1970).

Taraskevich et al. (1977) have shown that ATP inhibits spontaneous contractions of the anterior extrinsic protractor of *Aplysia*. An inhibitory neurotransmitter role for ATP should be considered.

3. Modulation

Certain substances, for instance the peptides, have no detectable action on the resting membrane of neurons but rather change the level of excitability so that the response of these neurons to neurotransmitters is altered. The term *neuromodulation* has been used to describe this type of action. The substances which produce this type of effect are called *neuromodulators,* or *local hormones,* and it has been speculated that the changes in excitability involve changes in metabolism due to interaction of the modulator with the adenylcyclase system.

It was noted very early that serotonin, which relaxes catch in *Mytilus* ABRM (see below), increases excitability without altering the resting membrane potential (Twarog, 1954). Serotonin lowers the threshold for spike discharge, increases spike amplitude, and enhances contraction in the ABRM (Hidaka, 1972; Hidaka et al., 1967; Twarog and Muneoka, 1973). That is, serotonin increases the excitability of the muscle (Twarog, 1954, 1966). In the presence of serotonin, ACh (Twarog, 1954, 1960a) or long pulses of cathodal direct current (DC) (Hoyle and Lowy, 1956) often produce rhythmic contraction. It has been thought that the physiological role of serotonin in the ABRM is not only to relax catch but also to enhance active contraction (York and Twarog, 1973). In relation to this, it is interesting that serotonin increases cyclic adenosine monophosphate (cyclic AMP) levels in the ABRM (Achazi et al., 1974; Dölling et al., 1972; Köhler and Lindl, 1980).

Elegant definitive experiments on modulatory neural control of molluscan muscles were reported by Weiss et al. (1978), who showed enhancement of contraction in the ARC muscle of *Aplysia* by a serotonergic neuron. The ARC muscle is innervated by two cholinergic motoneurons located in the buccal ganglion (Cohen et al., 1978) and by a metacerebral cell (Kupfermann and Weiss, 1974) which is known to be a serotonergic neuron (Eisenstadt et al., 1973; Gerschenfeld et al., 1976; Paupardin-Tritsch and Gerschenfeld, 1973; Weinreich et al., 1973). Firing of the metacerebral cell alone produces neither a mechanical nor an electrical response in the muscle fibers. However, brief firing of the metacerebral cell enhances the muscle contraction produced by firing of the cholinergic motoneurons (Fig. 3). Serotonin applied to the muscle also produces a marked increase of the contraction elicited by motoneuron stimulation. The contraction of the muscle produced either by direct electrical stimulation or by application of ACh is enhanced when the metacerebral cell fires, indicating

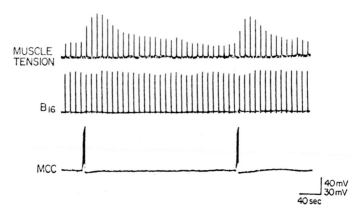

Fig. 3. Contraction of the accessory radula closer muscle of *Aplysia californica* produced by a brief burst of spikes in neuron B_{16}, and enhancement of the contraction by a brief burst of spikes in the metacerebral cell. (MCC). (Weiss et al., 1978.)

that serotonin acts directly on the muscle fibers to enhance the contraction. Firing of the metacerebral cell produces a small increase in the size of EJPs from the motoneurons. In one identified motoneuron, the time course of enhancement of the EJP and enhancement of contraction is parallel, but in another identified motoneuron, it is not. Thus, the metacerebral cell increases the force of contraction of the ARC muscle by an effect that appears to involve a nonelectrogenic action on the excitation–contraction coupling of the muscle (Kupfermann et al., 1979; Weiss et al., 1978). Weiss et al., (1979) have offered evidence that this nonelectrogenic action is mediated by cyclic AMP.

The gill motoneuron L_9 in the abdominal ganglion of *Aplysia* may be dopaminergic (Swann et al., 1978a). Applied dopamine has at least three effects on the gill muscle: It produces muscle contraction, enhances contraction induced by firing of other identified motoneurons (Swann et al., 1978b), and stimulates cyclic AMP production (Kebabian et al., 1977). The concentration of dopamine required for producing contraction is higher than that required for bringing about enhancement (Swann et al., 1978b). Lukowiak (1979) found that contraction of the gill muscle induced by firing the identified excitatory motoneuron L_7 is markedly enhanced by stimulating L_9. He proposed that L_9 is a modulatory neuron. However, dopamine could also be considered the excitatory transmitter used by L_9 (Ruben et al., 1979; Swann et al., 1978a). There is a nerve net in the gill of *Aplysia* (Peretz and Estes, 1974), and this makes the analysis of neural control of the gill muscle complex (Ruben et al., 1979).

Enhancement of muscle contraction by biogenic amines has been reported in other muscles. Contraction and EJPs of the radula protractor muscle of *Rapana* are markedly enhanced by serotonin and octopamine (Kobayashi and Muneoka,

1980). The action of octopamine is blocked by phentolamine. The contraction of the radula protractor muscle of *Busycon* is also enhanced by serotonin and tryptamine (Hill, 1970). It is interesting that at high concentrations serotonin produces rhythmic contraction in these muscles (Hill, 1970; Kobayashi, 1974b), just as it does in quiescent molluscan hearts. Dopamine acts similarly on the gill muscle of *Aplysia*.

In *Busycon,* as in *Mytilus*, serotonin applied in the presence of ACh to a contracted muscle produces relaxation as well as rhythmic contraction (Hill, 1970; Hill et al., 1970; Twarog, 1954). In the pharyngeal retractor muscle of *Helix*, the neurotransmitter which accelerates relaxation is thought to be serotonin: At low concentrations, serotonin enhances contraction in this muscle, an effect again very similar to that seen in ABRM. Relaxing transmitters (discussed below) very frequently enhance excitation. The transition from the catch state to phasic activity is thus facilitated.

Peptides should also be considered as modulators in molluscs. For example, a cardioactive peptide of molecular weight about 7000 has been isolated from *Helix*, which at high concentrations produces contraction in the pharyngeal retractor muscle and at low concentrations enhances contraction. This peptide is not present in the muscle (Lloyd, 1980a, 1980b, 1980c), and thus cannot be considered a neurotransmitter, but may serve as a modulatory hormone.

B. Catch Contraction

1. Excitation and Relaxation

Molluscan smooth muscles may perform a contraction, known as *catch*, that continues long after activation has ended. This has been reviewed by Twarog (1976, 1979). When stimulated by long, cathodal DC pulses (Winton, 1937) which act directly on the muscle membrane (Cambridge et al., 1959) or by ACh (Twarog, 1954) or FRMFamide (Painter and Greenberg, 1981) or neurally, in the presence of an agent which antagonizes serotonin, the isolated ABRM contracts and tension is sustained long after stimulation has ceased. The state of sustained tension is called *catch*. During catch, the active state is absent (Jewell, 1959; Johnson and Twarog, 1960; Twarog, 1960b), the membrane potential is at the resting level, and there are no action potentials (Hidaka and Goto, 1973; Twarog, 1954, 1960b). When the muscle in catch is stimulated by alternating current (AC) or by brief, repetitive pulses, which activate intramuscular nerve fibers, or via the nerve trunk in the absence of inhibitors, an additional contraction is elicited which relaxes rapidly after the stimulation has ceased (Twarog, 1979; Winton, 1937). When serotonin is applied to the muscle in the catch state, or when the muscle is stimulated by brief, repetitive pulses or via the nerve trunk in the presence of an agent which antagonizes ACh, pure, rapid relaxation is elicited (Twarog, 1954, 1979). Such relaxation can also be evoked by various other monoamines, but serotonin is the most effective (Muneoka et al., 1978c; Twarog, 1959; Twarog and Cole, 1973). Several lines of evidence suggest that the nervous control of the ABRM proceeds as follows: The muscle is innervated by at least two kinds of nerves, cholinergic excitatory and serotonergic relaxing nerves; the firing of the excitatory nerves evokes catch contraction, and the firing of the relaxing nerves relaxes catch but does not inhibit active contraction; thus, simultaneous firing of both nerves elicits a contraction which relaxes rapidly (Twarog, 1967a, 1976). When sustained contraction is evoked by DC stimulation, the initial brief firing of intramuscular relaxing nerves is not sufficient to induce relaxation. Because AC and brief pulses stimulate repetitively intramuscular excitatory and relaxing nerves, they produce a phasic contraction (cf. Bullard, 1967; Cambridge et al., 1959; Johnson, 1962; Kinosita et al., 1974; Nagahama et al., 1974; Takahashi, 1960; York and Twarog, 1973). Although neither excitatory nor relaxing neurons have been identified as yet, appropriate neural stimulation, together with the use of ACh and serotonin antagonists, can selectively produce phasic contraction, sustained contraction, or pure catch-relaxation (Bullard, 1967; Fujimoto, 1980; Takahashi, 1960; Twarog, 1979; van Nieuwenhoven, 1947).

2. Relaxing Neurotransmitters

There is an impressive body of evidence in favor of serotonin as relaxing neurotransmitter in the ABRM. Serotonin is present in the muscle (Twarog, 1954; York and Twarog, 1973). Fluorescence histochemistry shows abundant yellow varicose fibers in the muscle (Gilloteaux, 1977; McLean and Robinson, 1978; Twarog, 1976). Dense-cored vesicles, typical of amine-containing nerve endings have been demonstrated (Gilloteaux, 1976; McKenna and Rosenbluth, 1973; Twarog, 1968).

Serotonin is released from the muscle in response to nerve stimulation (Satchell and Twarog, 1978). Serotonin mimics the effects of nerve-mediated relaxation of catch (Twarog, 1954, 1979). Mersalyl, which selectively blocks serotonin relaxation, blocks the relaxation due to nerve stimulation (Twarog et al., 1977). A monoamine oxidase is present in the muscle (Blaschko and Hope, 1957). Prolonged treatment with p-chlorophenylalanine, an inhibitor of serotonin synthesis, decreases the rate of relaxation due to nerve stimulation (Muneoka et al., 1979a).

It has long been known that the adductor muscles of bivalves display catch; in fact, the term was coined with reference to adductor muscles (Pavlov, 1885; von Uexküll, 1912; see also Hoyle, 1957, 1964). Pavlov (1885) first described the function of relaxing nerves. In the adductor muscle of Anodonta, serotonin is implicated as a catch-relaxing transmitter (Hiripi, 1968; Salánki and Hiripi, 1970; Salánki and Lábos, 1969).

Serotonin relaxation of catch in the ABRM is a nonelectrogenic phenomenon; the amine does not change the resting membrane potential (Hidaka et al., 1967). Although serotonin decreases the membrane input resistance as measured with an intracellular electrode (Hidaka et al., 1967), it is unlikely that the amine causes a selective increase in conductance to Na^+, K^+, or Cl^- (Muneoka et al., 1977). It should be noted that serotonin has no or little effect on the membrane resistance measured by the double-sucrose gap method (Hidaka, 1972), and it may be that the decrease in membrane input resistance observed with microelectrodes pertains specifically to the gap junctions between cells rather than the muscle membrane as a whole.

There is a line of evidence indicating that the ABRM is innervated by dopaminergic nerves (McLean and Robinson, 1978; Satchell and Twarog, 1979; Twarog, 1976). Because dopamine, as well as serotonin, relaxes catch (Hidaka et al., 1977), it may also be a relaxing transmitter. However, dopamine is not blocked by mersalyl, which blocks relaxation due to neural stimulation and serotonin. Further, methysergide and 2-bromo-lysergic acid diethylamide are more effective in blocking dopamine than is serotonin but are not effective in blocking relaxation due to neural stimulation (Twarog et al., 1977). Thus, it cannot be ruled out that both serotonin and dopamine are utilized as relaxing transmitters, but it is probable that serotonin is the more important. In contrast to serotonin, dopamine evokes muscle fiber hyperpolarization in addition to catch relaxation and has been considered to be a possible inhibitory neurotransmitter.

Catch tension in the ABRM can be relaxed not only by serotonin and dopamine but by many other monoamines. Muneoka et al. (1978a, 1978b, 1978c) have suggested that some of them, such as phenylethanolamine and hexylamine, bring about relaxation by acting on the relaxing nerve terminals to release transmitter. It has also been suggested that levodopa (L-dopa) and L-5-hydroxytryptophan relax catch by acting at the relaxing nerve terminals. Muneoka et al. (1979a) give evidence that these amine precursors are taken up into the nerve terminals, decarboxylated to dopamine and serotonin, respectively, and released with endogenous relaxing transmitter to relax catch.

Increase in the relaxation rate by nerve stimulation has also been observed in the pharyngeal retractor muscle of *Helix* (Lloyd, 1980b). The neurotransmitter which accelerates relaxation is thought to be serotonin (Lloyd, 1980a, 1980b). In this case, serotonin hyperpolarizes the muscle fiber membrane (Lloyd, 1980b, 1980c). Of the two cardioactive peptides obtained from *Helix* (Lloyd, 1978), one is a small molecular weight (about 1000) peptide, and its action on the pharyngeal retractor muscle is similar to that of serotonin (Lloyd, 1980a, 1980b, 1980c).

The tonic contraction of the penis retractor muscle of *Helix* in response to ACh is relaxed by serotonin (Wabnitz, 1976) and dopamine (Wabnitz, 1980) even in the presence of ACh. The presence of serotonin in the muscle has been shown

(Wabnitz, 1976). Here again, a role for serotonin as a relaxing neurotransmitter is strongly indicated.

C. Autorhythmic Contraction

1. Pacemaker Activity

As mentioned above, in many molluscan muscles the fibers are electrically connected by gap junctions through which action potentials can propagate. If myogenic, rhythmic action potentials are elicited at one locus in a muscle, myogenic, rhythmic contraction of the muscle is transmitted throughout. Such contractions characterize most mollusc cardiac muscles. Myogenic regular beating can be generated in most bivalve hearts by stretching and in gastropod hearts by increasing perfusion pressure (Hill and Irisawa, 1967).

In the mammalian heart, the pacemaker cells where the cardiac action potential is first generated are located at the sinoatrial node. In the sinoatrial node cells, slow depolarization precedes each action potential. In contrast, the mollusc heart has a diffuse myogenic nature (Irisawa, 1978; Krijgsman and Divaris, 1955). In the oyster, a slow depolarizing potential can be recorded from almost all ventricular muscle fibers, and even a small piece of the ventricular wall musculature manifests rhythmic activity (Ebara et al., 1976). However, to produce a coordinated contraction of the ventricle, there should be a locus which becomes active before other sites in the ventricle. Ebara and Kuwasawa (1975) sought such a locus by inserting microelectrodes at two separate sites in the oyster ventricle; they found that the locus wandered from one site to another. They proposed that the overall rhythm of the ventricle is not determined by the activity of a pacemaker locus alone but is influenced by the activity of the follower fibers (Ebara and Kuwasawa, 1975).

2. Innervation

The myogenic activity is neurally modulated. In bivalves and gastropods, the heart is innervated by nerve fibers from the visceral ganglia (Carlson, 1905; Divaris and Krijgsman, 1954; Mayeri et al., 1974; Ripplinger, 1957). Electrical stimulation of visceral nerves or ganglia brings about acceleration and/or inhibition of the heartbeat (Hill and Welsh, 1966). This is because the heart is innervated by two kinds of nerves, one which accelerates the heartbeat and another which inhibits it. The former is called an *excitatory nerve,* but its physiological role is not exactly comparable to that of an excitatory nerve which innervates somatic muscle. The role of cardioexcitatory nerves is to modify myogenic activity rather than to produce contraction. In this sense, the cardioexcitatory nerve may be comparable to modulatory nerves which enhance muscle contraction.

Stimulation of cardioexcitatory nerves produces EJPs in cardiac muscle fibers (Kuwasawa, 1967; Kuwasawa and Hill, 1973a, 1973b; Kuwasawa and Matsui, 1970; Mayeri et al., 1974). In *Dolabella* and *Aplysia,* the EJPs facilitate and summate (Kuwasawa, 1967), and in *Busycon* they summate but do not facilitate (Kuwasawa and Hill, 1973b). Kuwasawa and Matsui (1970) observed that individual low-frequency stimuli to excitatory nerves elicited corresponding action potentials in the *Dolabella* heart. Kuwasawa and Hill (1973b) observed that such stimuli produced corresponding contractions in the quiescent heart of *Busycon.* Thus, in these cases, the molluscan cardioexcitatory nerve is comparable to the ordinary excitatory nerve. However, the cardiac muscle of *Busycon* cannot be tetanized by nerve stimulation, although it can be tetanized by direct stimulation (Kuwasawa and Hill, 1973b). In *Dolabella,* summation of EJPs increases spontaneous action potentials by overall membrane depolarization of the cardiac muscle (Kuwasawa and Matsui, 1970).

3. Neurotransmitters and Modulators

Serotonin is thought to be a cardioexcitatory neurotransmitter in molluscs (Welsh, 1971). An excitatory action of serotonin on heart has been shown in many molluscs (Bacq et al., 1952; Erspamer and Ghiretti, 1951; Gaddum and Paasonen, 1955; Greenberg, 1960a, 1960b; Hill, 1958, 1974; Lloyd, 1980a; Welsh, 1953, 1956), although there are several species in which serotonin inhibits the heart activity (Greenberg, 1965, 1969). Serotonin has been identified in cardiac tissue (Hiripi and Salánki, 1973; Kerkut and Cottrell, 1963; Welsh and Moorehead, 1959). In gastropods, liberation and reaccumulation of serotonin in cardiac nerves have been shown (S.-Rózsa and Perényi, 1966; Taxi and Gautron, 1969). Methysergide (UML), an antagonist of serotonin in the *Venus* heart (Wright et al., 1962), depresses the excitatory effect of neural stimulation (Loveland, 1963). Mayeri et al. (1974) identified two heart-excitor motoneurons, LD_{HE} and RB_{HE}, in the abdominal ganglion in *Aplysia.* Liebeswar et al. (1975) made observations which indicate that RB_{HE} is serotonergic: Serotonin applied directly to the heart mimics the motor effect produced by firing RB_{HE}; cinanserin, a serotonin-blocking agent, blocks the effects of serotonin and RB_{HE} firing on the heart (Fig. 4); RB_{HE} can synthesize significant quantities of serotonin from tryptophan. The presence of serotonin in the *Aplysia* heart has been shown (Carpenter et al., 1971; Chase et al., 1968; Taxi and Gautron, 1969).

The effects of serotonin on the membrane potential in cardiac muscle fibers are different in different species. In *Dolabella,* serotonin at 10^{-8}–$10^{-5} M$ activates the quiescent heart with a slow, small depolarization and at 10^{-7}– $10^{-3} M$ prolongs the duration of the action potential, with a consequent increase in force (Hill, 1974). In the *Mytilus* heart, serotonin induces or augments rhythmic activity, with only a small depolarization and no detectable change in relative membrane resistance (Irisawa et al., 1973). In the *Helix* heart, serotonin in-

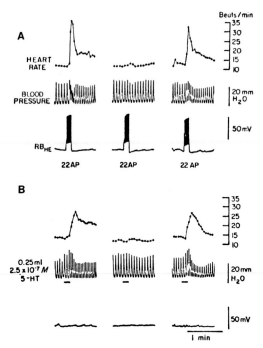

Fig. 4. Effects of (**A**) firing the heart-excitor neuron RB$_{HE}$ and (**B**) injecting serotonin on the heartbeat rate and blood pressure in *Aplysia californica,* and cinanserin blockade of the effects. Records in the left column were taken before, those in the middle column during, and those in the right column after perfusion with cinanserin (4 × 10^{-5} M). Small bars below the blood pressure records in (**B**) indicate a pulse of serotonin (2.5 ml of a 2.5 × 10^{-7} M serotonin infusion). The top trace in each pair shows the heartbeat rate, and the middle and bottom traces show blood pressure and RB$_{HE}$ firing, respectively. (Liebeswar et al., 1975.)

creases the amplitude and duration of the action potential but decreases the frequency of potential generation (Kiss and S.-Rózsa, 1972). Also in the *Helix* heart, serotonin at low concentrations (10^{-9}–10^{-7} M) depolarizes the membrane, whereas at high concentrations (10^{-5}–10^{-4} M) it hyperpolarizes (Kiss and S.-Rózsa, 1975). The depolarization by serotonin in the *Helix* heart is blocked by nicotine and morphine, whereas the hyperpolarization is blocked most effectively by atropine and *d*-tubocurarine, suggesting that there are two types of serotonin receptors on the *Helix* cardiac cell membrane (Kiss and S.-Rózsa, 1978).

The excitatory action of serotonin on the hearts may be mediated by cyclic AMP. It has been shown that serotonin activates adenylate cyclase and increases the intracellular cyclic AMP levels in the hearts of the bivalves, *Mercenaria* and *Macrocallista*. These changes occur at the same rate as the development of the positive inotropic effect (Higgins, 1974). In the hearts of some other molluscs, *Aplysia* (Mendelbaum et al., 1979), *Anodonta,* and *Helix* (Wollemann and S.-Rózsa, 1975), it has also been shown that serotonin increases cyclic AMP levels.

Some phenylethylamine analogs, such as dopamine and octopamine, may also play a role in the neuromuscular mechanism in the molluscan heart. Dopamine and octopamine are important phenylethylamines in the invertebrate nervous system (Gerschenfeld, 1973; Hicks, 1977; Kerkut, 1973; Leake and Walker, 1980). Dopamine has both excitatory and inhibitory actions on molluscan hearts (Chong and Phillis, 1965; de Rome et al., 1980; Dougan et al., 1975; Greenberg, 1960a; Kiss and S.-Rózsa, 1972; Liebeswar et al., 1975; Phillis, 1966a). Sathananthan and Burnstock (1976), on the basis of pharmacological, fluorescence histochemical, and electron microscopic evidence, have suggested the presence of dopaminergic nerve terminals in the heart of an unidentified species of venus clam. Octopamine has an excitatory action on the *Tapes* heart (de Rome et al., 1980; Dougan and Wade, 1978a, 1978b). In the *Tapes* heart, phentolamine antagonizes the excitatory action of octopamine but not dopamine and serotonin, and methysergide antagonizes the action of serotonin but not octopamine and dopamine (de Rome et al., 1980). These facts indicate that the heart has specific receptors for these phenylethylamines. It has been shown that dopamine activates cardiac adenylate cyclase in *Anodonta* and *Helix* (Wolleman and S.-Rózsa, 1975).

Cardioexcitor substances other than serotonin are present in the nervous system of molluscs (Hill and Welsh, 1966). Jaeger (1964) first found such a substance, a relatively large, stable molecule. Frontali et al. (1967) fractionated active ganglion extracts on Sephadex G-15 columns and obtained four peaks of cardioexcitor activity, designated A, B, C, and D. Agarwal et al. (1972) showed that peak C substance has a very potent cardioexcitor effect and is present in the ganglia of representative species from the four major classes of molluscs. Later, Price and Greenberg (1977a, 1977b,) purified peak C substance from the ganglia of *Macrocallista* and determined its structure. It is a tetrapeptide amide, Phe-Met-Arg-Phe-NH_2, and is generally referred to as *FMRFamide*. The effect of FMRFamide on molluscan heart is very close to that of serotonin, but Price and Greenberg (1980) have shown that these substances act at separate receptors. They have shown that, in the *Mercenaria* heart, methysergide blocks serotonin but not FMRFamide (Fig. 5). FMRFamide, as well as serotonin, increase the intracellular cyclic AMP levels in *Mercenaria* heart and the adenylate cyclase activity of a myocardial membrane fraction (Higgins et al., 1978). The physiological role of FMRFamide is not known. It may be a cardioexcitatory neurotransmitter or hormone. Price and Greenberg (1977a) suggested that FMRFamide might act in long-term regulation of muscular rhythmicity and tone.

It has been shown that adenosine compounds, particularly ATP, stimulate the heart in some molluscs (S.-Rózsa, 1969; Sathananthan and Burnstock, 1976).

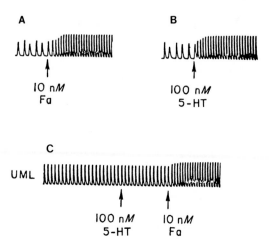

Fig. 5. Effect of methysergide (UML) on the mechanical responses of the *Mercenaria mercenaria* heart to FMRFamide (Fa) and serotonin (5-HT). (**A, B**) Control responses to FMRFamide and serotonin, respectively; (**C**) responses in the presence of methysergide (10^{-5} M). (Price and Greenberg, 1980.)

The presence of purinergic nerve terminals has been suggested by Sathananthan and Burnstock (1976).

It is generally believed that ACh is an inhibitory neurotransmitter in molluscan heart. However, ACh can cause not only inhibition but also excitation or both inhibition and excitation, depending on the species and, in some cases, on the concentration of ACh (Greenberg, 1965; Pilgrim, 1954). In the hearts of some mytilids, *Mytilus edulis* and *Modiolus demissus granosissimus*, ACh shows an excitatory action, whereas in the heart of venerids, it is usually inhibitory. Kuwasawa (1979) showed that in *Dolabella*, ACh usually causes depolarization in the ventricle and hyperpolarization in the auriculoventricular valve. In bivalve hearts, the excitatory action of ACh is antagonized by *d*-tubocurarine and the inhibitory action by benzoquinonium (Greenberg, 1970). The ionic basis for ACh excitation in the hearts of *M. edulis* (Elliott, 1980; Shigeto, 1970) and *M. demissus granosissimus* (Wilkens and Greenberg, 1973) is probably a depolarization resulting from an increase in Na^+ conductance. However, Elliott (1980) has shown that ACh produces a biphasic response consisting of rapid depolarization and slow hyperpolarization in the oyster, *Crassostrea virginica*.

The cardioinhibitory effect of ACh and the presence of ACh in heart muscle have been shown in many molluscs (Hill and Welsh, 1966; Welsh, 1971). Cholinesterase activity in the heart has been investigated (Greenberg et al., 1980; Roop and Greenberg, 1967, 1976; Watts and Pierce, 1978). Blockage by the ACh antagonist benzoquinonium of the cardioinhibitory effect of nerve stimulation has been shown in several molluscs (Loveland, 1963; MacKay and Gelperin, 1972; Phillis, 1966a; S-Rózsa and Zs-Nagy, 1967; Welsh, 1957). Mayeri et al. (1974) identified two heart inhibitor motoneurons, LD_{HI_1} and LD_{HI_2}, in the abdominal ganglion of *Aplysia*. Results obtained by Liebeswar et al. (1975) indicate that the transmitter released from the identified inhibitory neurons is ACh: ACh is the only substance tested which mimics the effect of firing of LD_{HI} on heart; benzoquinonium blocks effects of both ACh and firing of LD_{HI} on heart; LD_{HI} can synthesize ACh from injected choline.

Shigeto (1970) reported that in the heart of the oyster *Crassostrea gigas*, ACh hyperpolarized the muscle fiber membrane and inhibited spontaneous spiking, and the hyperpolarization was abolished in Cl^--free ASW. A similar action of ACh and its dependence on Cl^- were observed in the heart of the mussel *Modiolus demissus demissus* by Irisawa et al. (1973). However, with the hearts of three bivalve species (*Mercenaria mercenaria, Mytilus edulis,* and *C. virginica*), Elliott (1980) observed three different types of membrane potential changes (a slow hyperpolarization in *M. mercenaria,* a rapid depolarization sometimes followed by a slower hyperpolarization in *M. edulis,* and a biphasic response consisting of a rapid depolarization and slower hyperpolarization in *C. virginica*) and showed that the hyperpolarizing response of the cardiac muscle fibers in all three species was not altered by Cl^--free solution but was affected by changes in external K^+. Kuwasawa (1979) also suggested that the hyperpolarization by ACh of the cell membrane of auriculoventricular valve of *Dolabella* is K^+ dependent.

IJPs evoked by the stimulation of cardioinhibitory nerves have been recorded in some molluscs (Kuwasawa, 1967; Kuwasawa and Hill, 1972, 1973a, 1973b). Kuwasawa (1979) investigated the effect of IJPs on spontaneous action potentials in the auriculoventricular valve of *Dolabella*. He showed that stimulation of inhibitory nerves induced discrete IJPs as well as overall hyperpolarization; with an increase in IJP frequency, the action potential decreased in amplitude and then disappeared. He also reported that the IJPs disappeared when the membrane was maximally hyperpolarized by 10^{-5} *M* ACh, suggesting that equilibrium potential for IJPs was very close to E_{ACh}, the equilibrium potential for ACh.

III. Excitation–Contraction Coupling

A. Sources of Ca^{2+} in Activation

It is well established that contraction of vertebrate skeletal muscles is induced by a rise of the myoplasmic free Ca^{2+} level and that the source of the Ca^{2+} is internal stores (Ebashi and Endo, 1968; Sandow, 1965; Weber, 1966). In mollusc muscles, the source of activator Ca^{2+} is not as well documented as in vertebrate skeletal muscles, although it is clear that the contractile mechanism is activated by Ca^{2+}. In molluscan muscles, the most important source of activator Ca^{2+} seems to be internal stores. However, external Ca^{2+} plays a role in excitation–contraction coupling in mollusc muscles, as it does in some vertebrate smooth and cardiac muscles (Kuriyama et al., 1977): Ca^{2+} often carries the charge which generates the action potential. Translocation of Ca^{2+} from outside and/or from superficial sites on the membrane into inner sites during the action potential may cause release of internal stores of Ca^{2+} in response to membrane depolarization. Alternatively, translocated Ca^{2+} may act directly on the Ca^{2+} storage sites, causing Ca^{2+} release as suggested in the vertebrate skeletal muscles (Frank, 1980). The translocated Ca^{2+} may itself serve as activator Ca^{2+} in producing contraction, and it may also be required to replenish internal Ca^{2+} stores.

Sugi and his collaborators have presented strong evidence that the source of activator Ca^{2+} for contraction of the longitudinal body wall muscle of *Dolabella* is internal stores (Sugi and Suzuki, 1978a; Suzuki and Sugi, 1978). Potassium contracture of the muscle is not affected by Mn^{2+} and low pH but is reduced by procaine, which is known to inhibit Ca^{2+} release from the internal sites in vertebrate skeletal muscle. The K^+ contracture is not readily eliminated in calcium ion-deficient (0 Ca^{2+}) solution. Rapid cooling, which produces a contraction of vertebrate skeletal muscle by affecting the Ca^{2+}-storing capacity of the sarcoplasmic reticulum (Lüttgau and Oetliker, 1968; Sakai, 1965), produces contraction even in high K^+ or 0 Ca^{2+} solution. In this muscle, the plasma membrane is invaginated into the muscle fiber, forming tubular structures. Sarcoplasmic reticulum is closely apposed to the plasma membrane and the tubules, and bridgelike structures are observed between the sarcoplasmic reticulum and the invaginated tubule membrane (Sugi and Suzuki, 1978a). Electron microscopic observation of pyroantimonate precipitate in the muscle fibers and electron-probe X-ray microanalysis of calcium in the precipitate suggest that, in fibers fixed during rest, calcium is localized at the periphery (i.e., along the inner surface of the plasma membrane, at the membrane of the invaginated tubule, and at the sarcoplasmic reticulum) and that, in the fibers fixed during contraction, calcium is diffusely distributed in the myoplasm (Suzuki and Sugi, 1978).

In contrast to the K^+ contracture of the longitudinal body wall muscle of *Dolabella,* the K^+ contracture of the ABRM of *Mytilus* is eliminated in 0 Ca^{2+} ASW (Leenders, 1967; Sugi and Yamaguchi, 1976) and is markedly depressed by Mn^{2+} or low pH, but is not affected by procaine (Sugi and Yamaguchi, 1976). The contracture is totally blocked by La^{3+} (Muneoka and Twarog, 1977). Influx of $^{45}Ca^{2+}$ increases during K contracture (Hagiwara and Nagai, 1970; Tameyasu and Sugi, 1976a). The depolarized ABRM immersed in isotonic KCl contracts when 0.1 mM or more Ca^{2+} is introduced into the external KCl solution; the depolarized muscle fiber membrane is permeable to Ca^{2+} (Muneoka, 1974). It has been concluded that the K^+ contracture in ABRM is dependent upon entry of external Ca^{2+} (Hagiwara and Nagai, 1970; Muneoka and Twarog, 1977; Sugi and Suzuki, 1978b; Sugi and Yamaguchi, 1976; Tameyasu and Sugi, 1976a; Twarog and Muneoka, 1973). The source of activator Ca^{2+} for the K^+ contracture is the external Ca^{2+}. The possibility that stored Ca^{2+} is released during K^+ depolarization cannot be ruled out, although the relative contribution of released Ca^{2+} must be small. If this is the case, the release of Ca^{2+} could be triggered by Ca^{2+} entering during K^+ depolarization.

There are vesicles closely apposed to the ABRM fiber membrane. These vesicles are thought to correspond to the sarcoplasmic reticulum (Twarog, 1967a). Many investigators have provided evidence that structures in the ARBM take up and store Ca^{2+}. Atsumi and Sugi (1976) have shown that Ca^{2+} is stored in the surface vesicles. Further, they have shown that Ca^{2+} is also stored at the inner surface of the plasma membrane and in the mitochondria. Heumann (1969) has shown that calcium oxalate can be precipitated in the lumen of the vesicles. Stössel and Zebe (1968) have shown that a fraction isolated from the muscle can accumulate Ca^{2+} in the presence of ATP, and Gogjian and Bloomquist (1977) reported that $^{45}Ca^{2+}$ uptake by the vesicular fraction obtained from the muscle is accelerated in the presence of oxalate and is inhibited in the presence of caffeine.

The ABRM contracts in response to caffeine (Rüegg et al., 1963), which is known to release stored Ca^{2+} in vertebrate skeletal muscle. The caffeine contracture of the ABRM is not associated with membrane depolarization (Muneoka and Mizonishi, 1969), and it is markedly depressed by procaine (Muneoka, 1969). In 0 Ca^{2+} ASW, the initial test response to caffeine is as great as the control response in normal ASW; this is so even when the muscle is perfused in 0 Ca^{2+} for hours prior to testing. Further applications of caffeine, however, produce little or no response until normal ASW is reintroduced (Muneoka and Mizonishi, 1969; Twarog and Muneoka, 1973). The time required for complete restoration of caffeine contracture in normal ASW is quite long (Muneoka and Mizonishi, 1969), unless the muscle is exposed briefly to high K^+ ASW, which increases Ca^{2+} influx and accelerates the restoration (Twarog and Muneoka, 1973). La^{3+} prevents the restoration, presumably by blocking Ca^{2+} entry (Muneoka and Twarog, 1977). These facts suggest that caffeine releases internal stores of Ca^{2+} which are not subject to depletion by prolonged soaking in 0 Ca^{2+} ASW and that, after the stored Ca^{2+} is released by caffeine, entry of external Ca^{2+} is required in order to replenish the stores.

As mentioned before, ACh produces slow depolarization of the ABRM fiber membrane mainly due to increase in sodium ion conductance ($G_{Na}+$) and calcium ion conductance ($G_{Ca}2+$), and the depolarization usually triggers a further increase in $G_{Ca}2+$ which generates the action potential. In sodium ion-deficient (0 Na^+) ASW, the Na^+-dependent depolarization in response to ACh is not elicited and, hence, action potentials are not triggered. Even in 0 Na^+ ASW, however, ACh can produce contraction if the external Ca^{2+} concentration is more than 10 mM. The ACh contraction in 0 Na^+ ASW is Ca^{2+} dependent; contraction tension increases with increase in external Ca^{2+} concentration. The tension increases to a maximum on raising the Ca^{2+} concentration and returns to the control level within 5 min when normal Ca^{2+} levels are restored (Muneoka, 1973). ACh contraction of the ABRM is markedly depressed in normal ASW containing 5 mM La^{3+} (Muneoka and Twarog, 1977; Twarog and Muneoka, 1973), in 0 Ca^{2+} ASW (Muneoka, 1966; Sugi and Yamaguchi, 1976), and in ASW containing a Ca^{2+} antagonist, D-600 or verapamil (Y. Muneoka, unpublished). From these facts, Twarog and Muneoka (1973) (see also Twarog, 1979) proposed that ACh activates as follows: ACh increases muscle fiber membrane conductances to Na^+, K^+, and Ca^{2+}; increased conductance to Na^+ and Ca^{2+} brings about depolarization and a further increase in Ca^{2+} conductance, thus augmenting the influx of activating Ca^{2+}. The depolarization triggers Ca^{2+} spikes, and additional Ca^{2+} enters from outside; the spikes not only lead to Ca^{2+} influx but also liberate stored Ca^{2+}. In other words, ACh contraction is hypothesized to depend on Ca^{2+} entering from outside as well as being released from stored sites.

Sugi and his collaborators have proposed a hypothesis of ACh activation in the ABRM which differs from that proposed by Twarog and Muneoka, namely, that ACh contraction is exclusively due to release of intracellularly stored Ca^{2+} (Sugi and Suzuki, 1978b; Sugi and Yamaguchi, 1976; Tameyasu and Sugi, 1976a). They point out that in contrast to the K^+ contracture, ACh contraction is not affected by Mn^{2+} and low pH but is reduced by procaine, and that for a given amount of depolarization, ACh produces contraction tensions larger than those produced by high K^+. Tameyasu and Sugi (1976a) observed that the rate of $^{45}Ca^{2+}$ uptake (measured by the La^{3+} method) is not significantly changed in ASW containing ACh, a result which strongly supports their hypothesis of ACh activation. However, using a specific Ca^{2+} electrode, Bloomquist and Curtis (1975b) measured Ca^{2+} in the ASW bathing of ABRM and reported that activation in response to ACh is associated with uptake of Ca^{2+} by the muscle. Atsumi and Sugi (1976) have shown that, in the muscle fixed at rest, electron-opaque pyroantimonate precipitate is localized at the peripheral membrane structures and that, in the muscle fixed at the peak of mechanical response to ACh, the precipitate is found to be diffusely distributed in the myoplasm. These observations are consistent with their view that ACh contraction is mainly due to the release of stored Ca^{2+}.

In the hypothesis proposed by Sugi and his collaborators, ACh liberates stored Ca^{2+} by acting directly on internal structures, and this ACh action is not related to membrane depolarization (Sugi and Yamaguchi, 1976). Indeed, in ABRM, as in vertebrate smooth muscles, ACh can produce contraction without membrane potential changes under some conditions: ACh produces contractions in high K^+ ASW and 0 Na^+ ASW. This might suggest a direct action of ACh on internal stores. But these contractions require external Ca^{2+} (Muneoka, 1973, 1974). It is more probable that ACh increases the membrane permeability to Ca^{2+} and that activation requires Ca^{2+} influx from outside. Further supporting this view is the observation that the contractions are depressed by transition divalent metal cations including cobalt ion (CO^{2+}) and nickel ion (Ni^{2+}). In short, if ACh releases stored Ca^{2+} by a mechanism other than membrane potential change, translocation of external and/or superficial Ca^{2+} into inner sties by ACh is probably involved in the Ca^{2+} release. This is consistent with the report by Sugi and Yamaguchi that the Ca^{2+}-releasing action of ACh requires some amount of bound Ca^{2+} at or near the fiber membrane. All authors agree that binding of ACh to its receptors does not require external Ca^{2+}, because ACh can produce both a slow and a large depolarization in 0 Ca^{2+} ASW (Muneoka, 1966; Sugi and Yamaguchi, 1976).

In summary, ACh contraction seems to be induced by Ca^{2+} released from stored sites as well as by Ca^{2+} entering from outside, although the relative contribution to activation of contraction by released and entering Ca^{2+} remains in question. In relation to the Ca^{2+}-releasing action of ACh in the ABRM, it is interesting that ACh acts on the excitation–contraction coupling mechanism in vertebrate skeletal muscle (Frank, 1963, 1980).

DC stimulation of ABRM produces rapid membrane depolarization and, at the onset, action potentials. It is considered that the potential changes cause not only an entry of Ca^{2+} through the membrane but also a release of stored Ca^{2+} (Muneoka and Twarog, 1977; Twarog and Muneoka, 1973). Calcium ions translocated from outside or from superficial sites to inner sites during DC stimulation might also release stored Ca^{2+}. However, even if this is the case, it seems probable that the DC depolarization alone can release stored Ca^{2+}, because DC contraction is never blocked by long-period soaking of the muscle in 0 Ca^{2+} ASW or in ASW containing La^{3+}.

In the ARC muscle of *Aplysia*, nerve stimulation does not elicit muscle action potentials (Cohen et al., 1978). Thus, contraction of the muscle is probably produced in response to EJPs and/or the direct action of the excitatory transmitter ACh on the Ca^{2+} translocation mechanism. In contrast, in the ABRM of *Mytilus*, contraction in response to nerve stimulation depends on firing an action potential, and EJPs are not associated with contraction (Twarog, 1967b).

Wabnitz (1976) has reported the effects of Ca^{2+} antagonism and Ca^{2+} deprivation on contractions of the penis retractor muscle of *Helix*. Contractions of the muscle in response to ACh and KCl are eliminated in 0 Ca^{2+} solution and in solutions containing D-600. Contraction in response to caffeine consists of two phases: an initial fast phase and a second slower phase. In the solution containing D-600 or in low Ca^{2+} solution, the initial fast caffeine contracture is unaffected but the second slow contracture is suppressed. Wabnitz suggested that the mechanical responses to KCl and ACh depend on external Ca^{2+} for direct or indirect activation of the contractile apparatus and that the initial caffeine contracture is triggered by Ca^{2+} released from intracellular storage sites, whereas the second contracture is caused by influx of external Ca^{2+}. Evidence that Ca^{2+} influx in response to ACh or KCl indirectly activates the contractile elements by releasing stored Ca^{2+} is that responses are depressed by treating the muscle with caffeine prior to ACh or KCl and that complete recovery of responsiveness to ACh or KCl is obtained 40–60 min after removal of the caffeine (Wabnitz, 1976).

The involvement of Ca^{2+} entering from outside as well as Ca^{2+} released from stored sites in the excitation–contraction coupling process is also seen in the radula protractor muscle of *Busycon*. Contractions of the muscle in response to KCl and ACh are eliminated in 0 Ca^{2+} ASW, but contractions in response to caffeine are not (Hill and McDonald-Ordzie, 1979; Hill et al., 1970). In this muscle, tryptamine potentiates potassium contracture of the muscle without effecting a membrane potential change. Hill et al. (1970) propose that tryptamine potentiates the contracture by increasing Ca^{2+} influx through the fiber membrane. An effect of indoleamines on Ca^{2+} mobility or excitation–contraction coupling is also suggested in the ABRM of *Mytilus* (Twarog and Muneoka, 1973) and in the ARC muscle of *Aplysia* (Weiss et al., 1978).

There is additional morphological and physiological evidence that intracellular stored Ca^{2+} is involved in the excitation–contraction coupling of molluscan muscles (e.g., Elekes et al., 1973; Huddart et al., 1977; Sanger and Hill, 1972), and, in addition, there are many suggestions that Ca^{2+} entering from outside is involved directly and/or indirectly in the coupling (e.g., Kidokora et al., 1974; Kiss, 1977; Kiss and S.-Rózsa, 1973; Nomura, 1965; Ozeki, 1964). Unquestionably, the importance of these components in the excitation–contraction coupling process will be found to vary from muscle to muscle within species, as well as between different species.

B. Catch and Its Regulation

It was mentioned above that mollusc slow adductor muscles and byssus retractor muscle show a special phenomenon called *catch* (Hoyle, 1964, and Chapter 3, this volume). In the catch muscle, excitation not only induces a rise of intracellular free Ca^{2+} levels, which activates contraction, but also results in a catch state after excitation has ceased. During catch, the muscles are resistant to stretch and relax very slowly; thus, in catch muscles *in vivo*, a very low frequency of motor excitation can maintain a contracted state with little energy expenditure (Baguet and Gillis, 1968) and, hence, without fatigue (Lowy, 1953). The catch state is abolished by release of a neurotransmitter from relaxing nerve terminals. Since Winton (1937) first introduced it, ABRM of *M. edulis* has been used for the studies on catch. The muscle is long and thin; hence, it is convenient for physiological and pharmacological studies. In addition, the animals can be collected at the seashore in all seasons and can easily be kept in the laboratory.

1. Activation

With one exception, all activating stimuli applied to the ABRM result in a catch contraction after stimulation ceases unless the relaxing nerves are simultaneously stimulated, or unless the relaxing mediator serotonin is present at the time of the stimulation. DC pulses or brief application of ACh or caffeine all induce catch. Stimulation of the muscle nerve in the presence of a serotonin antagonist induces catch. In apparent contrast, the contraction in response to high K^+ ASW is characteristically phasic. However, catch is elicited if a serotonin antagonist is present, suggesting that high K^+ releases serotonin from nerve terminals (Muneoka et al., 1978c; Twarog et al., 1977). In 0 Na^+ ASW, as in high K^+ ASW, contractions are phasic: ACh contraction in 0 Na^+ ASW relaxes when ACh is washed out; catch tension produced in normal ASW is relaxed by replacing the external solution with 0 Na^+ ASW (Muneoka, 1973; Twarog and Muneoka, 1973). The relaxation of catch by 0 Na^+ ASW is blocked in the presence of a serotonin antagonist (Muneoka et al., 1978c). Thus, it is likely that relaxation of catch in 0 Na^+ ASW is brought about, at least in part, by its action on the relaxing nerve terminals (Muneoka et al., 1978c).

Responses in 0 Ca^{2+}, 0 Mg^{2+} ASW constitute an exception: Although catch tension is not relaxed by removing Ca^{2+} from the external solution when Mg^{2+} is present (Twarog, 1967c), it cannot be sustained or elicited in 0 Ca^{2+}, 0 Mg^{2+} ASW, a solution in which the membrane depolarizes (Muneoka and Mizonishi, 1969; Sugi, 1971; Twarog and Muneoka, 1973). The muscle soaked in 0 Ca^{2+}, 0 Mg^{2+} ASW responds to caffeine with a phasic contracture (Muneoka and Mizonishi, 1969; Sugi and Yamaguchi, 1976). Here, the evidence does not support the hypothesis that relaxation in 0 Ca^{2+}, 0 Mg^{2+} ASW is a result of releasing relaxing mediator. Rather, Twarog and Muneoka (1973) suggested that the membrane depolarized in 0 Ca^{2+}, 0 Mg^{2+} ASW is permeable to Ca^{2+} and that in this permeable muscle, efflux of intracellular free Ca^{2+} is induced, thus relaxing catch. This implies a hypothesis that not only active contraction but also catch may be controlled by intracellular free Ca^{2+}.

2. Relaxation

a. Do Intracellular Ca^{2+} Levels Control Catch? The idea that catch may be relaxed by a reduction in the intracellular free Ca^{2+} levels was first proposed to explain why excitability of ABRM increases with decrease in its stretch resistance in the presence of serotonin (Twarog, 1966). It had been known that the reduction in intracellular Ca^{2+} in the barnacle muscle fiber results in an increase in its excitability (Hagiwara and Nakajima, 1966). Later, Twarog and Muneoka (1973) proposed that serotonin may decrease the intracellular free Ca^{2+} through the action of a second messenger, cyclic AMP. Many experimental results consistent with the above ideas have been reported: (1) Serotonin increases $^{45}Ca^{2+}$ efflux (Bloomquist and Curtis, 1972, 1975a; Nauss and Davies, 1966). (2) The resistance of the fiber membrane measured using an intracellular microelectrode is decreased by serotonin (Hidaka et al., 1967), whereas the resistance measured using the double-sucrose gap method is not changed (Hidaka, 1972). If serotonin decreases the resistance at gap junctions by reducing the intracellular free Ca^{2+} (see Loewenstein et al., 1967) but does not change the overall resistance of the fiber membrane, the input resistance measured by microelectrodes would be reduced by current flow out of the injected cell into neighboring cells through the gap junctions. (3) Certain types of contraction are depressed rather than enhanced by serotonin: caffeine contracture (Muneoka, 1969), ACh contractions in 0 Na^+ ASW (Muneoka, 1973) and K^+ ASW (Muneoka 1974), and Ca^{2+} contracture of K^+-depolarized muscle (Muneoka, 1974). In all of these, no membrane potential change occurs; the effect of serotonin on excitability plays no role. This depression of active tension development could be attributed to reduction of intracellular Ca^{2+}.

If serotonin relaxes catch by reducing the intracellular free Ca^{2+}, then the Ca^{2+} levels during the catch state must be higher than during rest. The active state is absent during catch, so logic suggests that intracellular free Ca^{2+} concentrations are below the level required for activation. Direct evidence against the maintenance of high levels of Ca^{2+} during catch is the observation that prolonged exposure to 0 Ca^{2+} ASW does not relax (Twarog, 1967c). Even stronger evidence was obtained by Atsumi and Sugi (1976) that the Ca^{2+} levels during catch are as low as during the resting state. They showed that in the ABRM fixed at the peak of the mechanical response to ACh, electron-opaque pyroantimonate precipitate is found to be diffusely distributed in the myoplasm and that, in the muscle fixed during the catch state, the precipitate is found at the peripheral structures, as in the resting muscle. Further evidence against maintenance of high Ca^{2+} levels during catch has been found in chemically skinned fibers. Baguet and Marchand-Dumont (1975) observed that the sustained contraction established in the high Ca^{2+} solution (10^{-6} M) persists under some conditions even when the muscle is returned to low Ca^{2+} solution (10^{-6} M); they take this as evidence that catch tension can be maintained at low intracellular Ca^{2+} levels. Marchand-Dumont and Baguet (1975) have further observed that both serotonin and cyclic AMP relax the sustained contraction of the chemically skinned fibers only in low Ca^{2+} solution. They concluded that the relaxing effect of serotonin is not produced by decreasing intracellular Ca^{2+} levels but by a cyclic AMP-mediated action on the contractile system which causes release of bound Ca^{2+}.

Bone, Q., and Howarth, J. V. (1980). The role of L-Glutamate in neuromuscular transmission in some molluscs. *J. Mar. Biol. Assoc. U. K.* **60,** 619–626.

Bullard, B. (1967). The nervous control of the anterior byssus retractor muscle of *Mytilus edulis*. *Comp. Biochem. Physiol.* **23,** 749–759.

Burnstock, G., Greenberg, M. J., Kirby, S., and Willis, A. G. (1967). An electrophysiological and pharmacological study of visceral smooth muscle and its innervation in an invertebrate *Poneroplax albida. Comp. Biochem. Physiol.* **23,** 407–429.

Cambridge, G. W., Holgate, J. A., and Sharp, J. A. (1959). A pharmacological analysis of the contractile mechanism of *Mytilus* muscle. *J. Physiol. London* **148,** 451–464.

Carew, T. J., Pinsker, H., Robinson, K., and Kandel, E. R. (1974). Physiological and biochemical properties of neuromuscular transmission between identified motoneurons and gill muscle in *Aplysia. J. Neurophysiol.* **37,** 1020–1040.

Carlson, A. J. (1905). Comparative physiology of the invertebrate heart. I. The innervation of the heart. *Biol. Bull.* **8,** 123–170.

Carpenter, D. O., Breese, G. R., Schanberg, S. M., and Kopin, I. J. (1971), Serotonin and dopamine: Distribution and accumulation in *Aplysia* nervous and non-nervous tissues. *Int. J. Neurosci.* **2,** 49–56.

Chase, T. N., Breese, G. R., Carpenter, D. O., Schanberg, S. M., and Kopin, I. J. (1968). Stimulation induced release of serotonin. *Adv. Pharmacol.* **6,** 351–364.

Chong, G. C., and Phillis, J. W. (1965). Pharmacological studies on the heart of *Tapes watlingi:* A mollusc of the family veneridae. *Br. J. Pharmacol.* **25,** 481–496.

Cohen, C., Szent-Györgyi, A. G., and Kendrick-Jones, J. (1971). Paramyosin and the filaments of molluscan "catch" muscles. I. Paramyosin: Structure and assembly. *J. Mol. Biol.* **56,** 223–237.

Cohen, J. L., Weiss, K. R., and Kupfermann, I. (1978). Motor control of buccal muscles in *Aplysia. J. Neurophysiol.* **41,** 157–180.

Cole, R. A., and Twarog, B. M. (1972). Relaxation of catch in a molluscan smooth muscle. Effects of drugs which act on the adenyl cyclase system, Vol. 1. *Comp. Biochem. Physiol. A* **43,** 321–330.

De Rome, P. J., Jamieson, D. D., Taylor, K. M., and Davies, L. P. (1980). Ligand-binding and pharmacological studies on dopamine and octopamine receptors in the heart of the bivalve mollusc, *Tapes watlingi. Comp. Biochem. Physiol. C* **67,** 9–16.

Divaris, G. A., and Krijgsman, B. J. (1954). Investigations into the heart function of *Cochlitoma (=Achatina) zebra. Arch. Int. Physiol.* **62,** 211–233.

Dölling, B., Achazi, R. K., Zebe, E., and Ahlert, U. (1972). 3′,5′-Adenosinmonophosphat im Byssusretraktormuskel der Miesmuschel *Mytilus edulis. Naturwissenschaften* **59,** 313.

Dougan, D. F. H., and McLean, J. R. (1970). Evidence for the presence of dopaminergic nerves and receptors in the intestine of a mollusc, *Tapes watlingi. Comp. Gen. Pharmac.* **1,** 33–46.

Dougan, D. F. H., and Wade, D. N. (1978a). Action of octopamine agonists and stereoisomers at a specific octopamine receptor. *Clin. Exp. Pharmacol.* **5,** 333–339.

Dougan, D. F. H., and Wade, D. N. (1978b). Different blockade of octopamine and dopamine receptors by analogues of clozapine and metoclopramide. *Clin. Exp. Pharmacol.* **5,** 341–349.

Dougan, D. F. H., Wade, D. W., and Mearrick, P. T. (1975). Excitatory and inhibitory receptors for dopamine in the molluscan heart. *Proc. Aust. Physiol. Pharmacol. Soc.* **6,** 49–50.

Ebara, A., and Kuwasawa, K. (1975). The initiating site for spontaneous electrical activity in the oyster ventricle. *Annot. Zool. Jpn.* **48,** 219–226.

Ebara, A., Kuwasawa, K., and Kuramoto, T. (1976). Slow depolarizing potential recorded from oyster ventricle during spontaneous electrical activity. *Sci. Rep. Tokyo Kyoiku Daigaku Sect. B* **16,** 149–157.

Ebashi, S., and Endo, M. (1968). Ca ion and muscle contraction. *Prog. Biophys. Mol. Biol.* **18**, 123–183.

Eisenstadt, M., Goldman, J., Kandel, E. R., Koike, H., Koester, J., and Schwartz, J. H. (1973). Intrasomatic injection of radioactive percursors for studying transmitter synthesis in identified neurons of *Aplysia californica*. *Proc. Natl. Acad. Sci. U. S. A.* **70**, 3371–3375.

Elekes, K., Kiss, T., and S.-Rózsa, K. (1973). Effect of Ca-free medium on the ultrastructure and excitability of the myocardial cells of the snail, *Helix pomatia* L. *J. Mol. Cell. Cardiol.* **5**, 133–138.

Elliott, E. J. (1980). Three types of acetylcholine response in bivalve heart muscle cells. *J. Physiol. London* **300**, 283–302.

Erspamer, V., and Ghiretti, F. (1951). The action of enteramine on the heart of molluscs. *J. Physiol. London* **115**, 470–481.

Florey, E. (1966). Nervous control and spontaneous activity of the chromatophores of a cephalopod, *Loligo Opalescens*. *Comp. Biochem. Physiol.* **18**, 305–324.

Florey, E., and Kriebel, M. E. (1969). Electrical and mechanical responses of chromatophore muscle fibers of the squid, *Loligo opalescens*, to nerve stimulation and drugs. *Z. Vgl. Physiol.* **65**, 98–130.

Frank, G. B. (1963). Utilization of bound calcium in the acetycholine contracture of frog skeletal muscle. *J. Pharmacol. Exp. Ther.* **139**, 261–268.

Frank, G. B. (1980). The current view of the source of trigger calcium in excitation-contraction coupling in vertebrate skeletal muscle. *Biochem. Pharmacol.* **29**, 2399–2406.

Frontali, N., Williams, L., and Welsh, J. H. (1967). Heart excitatory and inhibitory substances in molluscan ganglia. *Comp. Biochem. Physiol.* **22**, 833–841.

Fujimoto, M. (1980). Relaxing effect of nervous stimulation on contraction of anterior byssus retractor muscle of *Mytilus edulis*. *J. Sci. Hiroshima Univ. Ser. B Div. 1* **28**, 1–10.

Gaddum, J. H., and Paasonen, M. K. (1955). The use of some molluscan hearts for estimation of 5-hydroxytryptamine. *Br. J. Pharmacol.* **10**, 474–483.

Gerschenfeld, H. M. (1973). Chemical transmission in invertebrate central nervous systems and neuromuscular junctions. *Physiol. Rev.* **53**, 1–119.

Gerschenfeld, H. M., Hamon, H., and Paupardin-Tritsch, D. (1976). Release and uptake of 5-hydroxytryptamine (5-HT) by a single 5-HT containing neurone. *J. Physiol. London* **260**, 29–30.

Gilloteaux, J. (1976). Intercellular connections of a smooth muscle: ultrastructure of the anterior byssal retractor muscle (ABRM) of *Mytilus edulis* L. (mollusca pelecypoda). *Cytobiologie* **12**, 457–472.

Gilloteaux, J. (1977). Innervation du muscle rétracteur antérieur du byssus (ABRM) de *Mytilus edulis* L. III. Localisation histochemique des terminaisons nerveuses à 5-hydroxtrytamine. *Histochemistry* **51**, 343–351.

Gilloteaux, J. (1978). Innervation du muscle rétracteur antérieur du byssus (ABRM) de *Mytilus edulis* L. et de *Mytilus galloprovincialis* Lmk. V. Localisation cytochimique d'activitiés cholinestérasiques. *Histochemistry* **55**, 209–224.

Gogjian, M. A., and Bloomquist, E. (1977). Calcium uptake by a subcellular membrane fraction of anterior byssus retractor muscle. *Comp. Biochem. Physiol. C* **58**, 97–102.

Greenberg, M. J. (1960a). The responses of the *Venus* heart to catecholamines and high concentrations of 5-hydroxytryptamine. *Brt. J. Pharmacol.* **15**, 365–374.

Greenberg, M. J. (1960b). Structure-activity relationship of tryptamine analogues on the heart of *Venus mercenaria*. *Brt. J. Pharmacol.* **15**, 375–388.

Greenberg, M. J. (1965). A compendium of responses of bivalve hearts to acetylcholine. *Comp. Biochem. Physiol.* **14**, 514–539.

Greenberg, M. J. (1969). The role of isoreceptors in the neurohormonal regulation of bivalve hearts.

In "Comparative Physiology of the Heart: Current Trends" (F. McCann, ed), pp. 232–248. Experientia Supp. 15, Brukhauser, Stuttgart, Germany.

Greenberg, M. J. (1970). A comparison of acetylcholine structure-activity relations on the hearts of bivalve molluscs. *Comp. Biochem. Physiol.* **33,** 259–294.

Greenberg, M. J., and Jegla, T. C. (1963). The action of 5-hydroxytryptamine and acetylcholine on the rectum of the Venus clam *Mercenaria mercenaria. Comp. Biochem. Physiol.* **9,** 275–290.

Greenberg, M. J., Roop, T., and Painter, S. D. (1980). The relative contributions of the receptors and cholinesterases to the effects of acetylcholine on the hearts of bivalve molluscs. *Gen. Pharmacol.* **11,** 65–74.

Hagiwara, E., and Nagai, T. (1970). ^{45}Ca movement at rest and during potassium contracture in *Mytilus* ABRM. *Jpn. J. Physiol.* **20,** 72–83.

Hagiwara, S., and Nakajima, S. (1966). Effects of intracellular Ca ion concentration upon the excitability of the muscle fiber membrane of a barnacle. *J. Gen. Physiol.* **49,** 807–818.

Heumann, H. G. (1969). Calciumakkumulierende strukturen in einem glatten wirbellosenmuskel. *Protoplasma* **67,** 111–115.

Heyer, C. B., Kater, S. B., and Larlsson, U. L. (1973). Neuromuscular systems in molluscs. *Am. Zool.* **13,** 247–270.

Hicks, T. P. (1977). The possible role of octopamine as a synaptic transmitter: a review. *Can. J. Physiol. Pharmacol.* **55,** 137–152.

Hidaka, T. (1972) Electrical and mechanical responses of *Mytilus* smooth muscle recorded by the double sucrose gap method. *Zool. Mag. Tokyo* **81,** 72–74.

Hidaka, T., and Goto, M. (1973). On the relationship between membrane potential and tension in *Mytilus* smooth muscle. *J. Comp. Physiol.* **82,** 357–364.

Hidaka, T., and Twarog, B. M. (1977). Neurotransmitter action on the membrane of *Mytilus* smooth muscle. I. Acetylcholine. *Gen. Pharmacol.* **8,** 83–86.

Hidaka, T., Osa, T., and Twarog, B. M. (1967). The action of 5-hydroxytryptamine on *Mytilus* smooth muscle. *J. Physiol. London* **192,** 869–877.

Hidaka, T., Yamaguchi, H., Twarog, B. M., and Muneoka, Y. (1977). Neurotransmitter action on the membrane of *Mytilus* smooth muscle. II. Dopamine. *Gen. Pharmac.* **8,** 87–91.

Higgins, W. J. (1974). Intracellular actions of 5-hydroxytryptamine on the bivalve myocardium. I. Adenylate and guanylate cyclases. *J. Exp. Zool.* **190,** 99–110.

Higgins, W. J., and Greenberg, M. J. (1974). Intracellular actions of 5-hydroxytryptamine on the bivalve myocardium. II. Cyclic nucleotide-dependent protein kinases and microsomal calcium uptake. *J. Exp. Zool.* **190,** 305–316.

Higgins, W. J., Price, D. A., and Greenberg, M. J. (1978). FMRFamide increases the adenylate cyclase activity and cyclic AMP level of molluscan heart. *Eur. J. Pharmacol.* **48,** 425–430.

Hill, R. B. (1958). The effects of certain neurohumors and of other drugs on the ventricle and radula protractor of *Busycon canaliculatum* and on the ventricle of *Strombus gigas. Biol. Bull.* **115,** 471–482.

Hill, R. B. (1970). Effects of postulated neurohumoral transmitters on the isolated radula protractor of *Busycon canaliculatum. Comp. Biochem. Physiol.* **33,** 249–258.

Hill, R. B. (1974). Effects of 5-hydroxytryptamine on action potentials and on contractile force in the ventricle of *Dolabella auricularia. J. Exp. Biol.* **61,** 529–539.

Hill, R. B., and Irisawa, H. (1967). The immediate effect of changed perfusion pressure and the subsequent adaptation in the isolated ventricle of the marine gastropod *Rapana thomasiana* (prosobranchia). *Life Sci.* **6,** 1691–1696.

Hill, R. B., and McDonald-Ordzie, P. E. (1979). Ionic dependence of the response to acetylcholine of a molluscan buccal muscle: the radula protractor of *Busycon canaliculatum. Comp. Biochem. Physiol. C* **62,** 19–30.

Hill, R. B., and Welsh, J. H. (1966). Heart, circulation, and blood cells. *In* "Physiology of

Mollusca" (K. M. Wilbur and C. M. Yonge, eds.), Vol. 2, pp. 125–173. Academic Press, New York.

Hill, R. B., Greenberg, M. J., Irisawa, H., and Nomura, H. (1970). Electromechanical coupling in a molluscan muscle, the radula protractor of *Busycon canaliculatum*. *J. Exp. Zool.* **174**, 331–348.

Hiripi, L. (1968). Paper chromatographic and fluorometric examination of the serotonin content in the nervous system and other tissues of three freshwater molluscs. *Annal. Biol. Tihany* **35**, 3–11.

Hiripi, L., and Salànki, J. (1973). Seasonal and activity-dependent changes of the serotonin level in the C.N.S. and heart of the snail. *(Helix pomatia* L.). *Comp. Gen. Pharmacol.* **4**, 285–292.

Hoyle, G. (1957). "Comparative Physiology of the Nervous Control of Muscular Contraction." Cambridge Univ. Press, Cambridge, England.

Hoyle, G. (1964). Muscle and neuromuscular physiology. *In* "Physiology of Mollusca" (K. M. Wilbur and C. M. Yonge, eds.), Vol. 1, pp. 313–351. Academic Press, New York.

Hoyle, G., and Lowy, J. (1956). The paradox of *Mytilus* muscle. A new interpretation. *J. Exp. Biol.* **33**, 295–310.

Huddart, H., Hunt, S., and Oates, K. (1977). Calcium movements during contraction in molluscan smooth muscle, and the loci of calcium binding and release. *J. Exp. Biol.* **68**, 45–65.

Irisawa, H. (1978). Comparative physiology of the cardiac pacemaker mechanism. *Physiol. Rev.* **58**, 461–498.

Irisawa, H., Wilkens, L. A., and Greenberg, M. J. (1973). Increase in membrane conductance by 5-hydroxytryptamine and acetylcholine on the hearts of *Modiolus demissus demissus* and *Mytilus edulis* (Mytilidae, bivalvia). *Comp. Biochem. Physiol. A* **45**, 653–666.

Jaeger, C. P. (1964). Regulacâo neurohormonal de atividade cardiaca no *Strophocheilus oblongus musculus*. Thesis, University of Rio Grande do Sul, Porto Alegre, Brazil.

Jewell, B. R. (1959). The nature of the phasic and the tonic responses of the anterior byssal retractor muscle of *Mytilus*. *J. Physiol. London* **149**, 154–177.

Johnson, W. H. (1962). Tonic mechanisms in smooth muscles. *Physiol. Rev.* **42**, 113–159.

Johnson, W. H., and Twarog, B. M. (1960). The basis for prolonged contraction in molluscan muscles. *J. Gen Physiol.* **43**, 941–960.

Kater, S. B., Heyer, C., and Hegmann, J. P. (1971). Neuromuscular transmission in the gastropod mollusc *Helisoma trivolvis:* Identification of motoneurons. *Z. Vgl. Physiol.* **74**, 127–139.

Kebabian, P. R., Kebabian, J. W., Swann, J. W., and Carpenter D. O. (1977). Cyclic AMP in *Aplysia* gill: Increases by putative neurotransmitters. *Neurosci. Abstr.* **3**, 557.

Kerkut, G. A. (1973). Catecholamines in invertebrates. *Br. Med. Bull.* **29**, 100–103.

Kerkut, G. A., and Cottrell, G. A. (1963). Acetylcholine and 5-hydroxytryptamine in the snail brain. *Comp. Biochem. Physiol.* **8**, 53–63.

Kerkut, G. A., and Leake, L. D. (1966). The effect of drugs on the snail pharyngeal retractor muscle. *Comp. Biochem. Physiol.* **17**, 623–633.

Kerkut, G. A., Leake, L. D., Shapira, A., Cowan, S., and Walker, R. J. (1965). The presence of glutamate in nerve-muscle perfusates of *Helix, Carcinus,* and *Periplaneta. Comp. Biochem. Physiol.* **15**, 485–502.

Kerkut, G. A., Shapira, A., and Walker, R. J. (1967). The transport of ^{14}C-labelled material from CNS⇌muscle along a nerve trunk. *Comp. Biochem. Physiol.* **23**, 729–748.

Kidokoro, Y., Hagiwara, S., and Henkart, M. P. (1974). Electrical properties of obliquely striated muscle fiber membrane of *Anodonta* glochidium. *J. Comp. Physiol.* **90**, 321–338.

Kinosita, H., Nagahama, H., and Ueda, K. (1974). Mechanical responses of the anterior byssus retractor muscle of *Mytilus edulis* to repetitive square pulses of direct current. *J. Sci. Hiroshima Univ. Ser. B, Div. 1* **25**, 291–308.

Kiss, T. (1977). Electrical properties of the cardiac muscle cell membrane and its role in the excitation-contraction coupling. *Acta Biochim. Biophys. Acad. Sci. Hung.* **12**, 291–302.

Kiss, T., and S.-Rózsa, K. (1972). Effect of biologically active substances on the spontaneous electrical activity of the heart muscle cells of *Helix pomatia* L. *Annal. Biol. Tihany* **39**, 29–38.

Kiss, T., and S.-Rózsa, K. (1973). The role of mono- and divalent cations in the spike generation of myocardial cells in the snail, *Helix pomatia* L. *Comp. Biochem. Physiol. A* **44**, 173–181.

Kiss, T., and S.-Rózsa, K. (1975). Site of action of 5-hydroxytryptamine on the membrane of heart muscle cells in *Helix pomatia* L. *Annal. Biol. Tihany* **42**, 61–72.

Kiss, T., and S.-Rózsa, K. (1978). Pharmacological properties of 5-HT-receptors of the *Helix pomatia* L. (gastropoda) heart muscle cells. *Comp. Biochem. Physiol. C* **61**, 41–46.

Kobayashi, M. (1972a). Electrical and mechanical activities in the radula protractor of a mollusc, *Rapana thomasiana*. *J. Comp. Physiol.* **78**, 1–10.

Kobayashi, M. (1972b). Prolonged depolarization of a molluscan muscle in Ca-free solution. *J. Comp. Physiol.* **78**, 11–19.

Kobayashi, M. (1974a). Antagonistic action of Ca and Mg on the neuromuscular transmission and excitation in a molluscan muscle (radula protractor). *J. Comp. Physiol.* **94**, 17–24.

Kobayashi, M. (1974b). Facilitation and post-tetanic potentiation of junctional potential in the molluscan nerve-muscle preparation. *Annot. Zool. Jpn.* **47**, 199–205.

Kobayashi, M., and Hashimoto, T. (1980). Effects of low-chloride solutions on the membrane activities of molluscan muscle cells. *Comp. Biochem. Physiol. C* **66**, 87–91.

Kobayashi, M., and Muneoka, Y. (1980). Modulatory actions of octopamine and serotonin on the contraction of buccal muscles in *Rapana thomasiana*. I. Enhancement of contraction in radula protractor. *Comp. Biochem. Physiol. C* **65**, 73–79.

Kobayashi, M., and Shigenaka, Y. (1978). The mode of action of acetylcholine and 5-hydroxytryptamine at the neuromuscular junctions in a molluscan muscle (radula protractor). *Comp. Biochem. Physiol. C* **60**, 115–122.

Kobayashi, M., Muneoka, Y., and Fujiwara, M. (1981). The modulatory actions of the possible neurotransmitters in the molluscan muscles. *In* "Neurotransmitters in Invertebrates" (K. S.-Rózsa, ed.). Pergamon Press, New York.

Köhler, G., and Lindl, T. (1980). Effects of 5-hydroxytryptamine, dopamine, and acetylcholine on accumulation of cyclic AMP and cyclic GMP in the anterior byssus retractor muscle of *Mytilus edulis* L. (mollusca). *Pflüegers Arch.* **383**, 257–262.

Köhler, G., Heilmann, C., Nickel, E. and Florey, E. (1976). Properties of sarcoplasmic reticulum of a mollusc smooth muscle. *Pflüegers Arch.* **365**, supplement R29.

Krijgsman, B. J., and Divaris, G. A. (1955). Contractile and pacemaker mechanisms of the heart of molluscs. *Biol. Rev.* **30**, 1–39.

Kupfermann, I., and Weiss, K. R. (1974). Functional studies on the metacerebral cells of *Aplysia*. *Abstr. Soc. Neurosci.* **4**, 297.

Kupfermann, I., Carew, T. J., and Kandel, E. R. (1974). Local, reflex, and central commands controlling gill and siphon movements in *Aplysia*. *J. Neurophysiol.* **37**, 996–1019.

Kupfermann, I., Cohen, J. L., Mandelbaum, D. E., Schonberg, M. Susswein, A. J., and Weiss, K. R. (1979). Functional role of serotonergic neuromodulation in *Aplysia*. *Fed. Proc.* **38**, 2095–2102.

Kuriyama, H., Ito, Y., and Suzuki, H. (1977). Effects of membrane potential on activation of contraction in various smooth muscles. *In* "Excitation-Contraction Coupling in Smooth Muscle" (R. Casteels, T. Godfraind, and J. C. Rüegg, eds.), pp. 25–35. Elsevier/North-Holland, Amsterdam.

Kuwasawa, K. (1967). Transmission of impulses from the cardiac nerve to the heart in some molluscs (*Aplysia* and *Dolabella*). *Sci. Rep. Tokyo Kyoiku Daigaku, Ser. B* **13**, 111–128.

Kuwasawa, K. (1979). Effects of ACh and IJPs on the AV valve and the ventricle of *Dolabella auricularia. Am. Zool.* **19**, 129–143.

Kuwasawa, K., and Hill, R. B. (1972). Interaction of inhibitory and excitatory junctional potentials in the control of a myogenic myocardium: the ventricle of *Busycon canaliculatum. Experientia* **28**, 800–801.

Kuwasawa, K., and Hill, R. B. (1973a). Junctional potential in molluscan cardiac muscle. *Life Sci.* **12**, 365–372.

Kuwasawa, K., and Hill, R. B. (1973b). Regulation of ventricular rhythmicity in the hearts of prosobranch gastropods. *Neurobiol. Invertebr. Tihany 1971*, 143–165.

Kuwasawa, K., and Matsui, K. (1970). Postjunctional potentials and cardiac acceleration in a mollusc (*Dolabella auricula*). *Experientia* **26**, 1100–1101.

Leake, L. D., and Walker, R. J. (1980). "Invertebrate neuropharmacology." Backie, Glasgow and London.

Leenders, H. J. (1967). Ca-coupling in the anterior byssal retractor muscle of *Mytilus edulis* L. *J. Physiol. London* **192**, 681–693.

Liebeswar, G., Goldman, J. E., Koester, J., and Mayeri, E. (1975). Neural control of circulation in *Aplysia*. III. Neurotransmitters. *J. Neurophysiol.* **38**, 767–779.

Lloyd, P. E. (1978). Distribution and molecular characteristics of cardioactive peptides in the snail, *Helix aspersa. J. Comp. Physiol.* **128**, 269–276.

Lloyd, P. E. (1980a). Biochemical and pharmacological analyses of endogenous cardioactive peptides in the snail, *Helix aspersa. J. Comp. Physiol.* **138**, 265–270.

Lloyd, P. E. (1980b). Modulation of neuromuscular activity by 5-hydroxytryptamine and endogenous peptides in the snail, *Helix aspersa. J. Comp. Physiol.* **139**, 333–339.

Lloyd, P. E. (1980c). Mechanisms of action of 5-hydroxytryptamine and endogenous peptides on a neuromuscular preparation in the snail, *Helix aspersa. J. Comp. Physiol.* **139**, 341–347.

Loewenstein, W. R., Nakas, M., and Scolar, S. J. (1967). Junctional membrane uncoupling: permeability transformations at a cell membrane junction. *J. Gen. Physiol.* **50**, 1865–1892.

Loveland, R. E. (1963). 5-hydroxytryptamine, the probable mediator of excitation in the heart of *Mercenaria (Venus) mercenaria. Comp. Biochem. Physiol.* **9**, 95–104.

Lowy, J. (1953). Contraction and relaxation in the smooth muscle of *Mytilus. J. Physiol. London* **120**, 129–140.

Lowy, J., and Millman, B. M. (1963). The contractile mechanism of the anterior byssus retractor muscle of *Mytilus edulis. Phil. Trans. R. Soc. London B* **246**, 105–148.

Lowy, J., Millman, B. M., and Hanson, J. (1964). Structure and function in smooth tonic muscles of lamellibranch molluscs. *Proc. R. Soc. B* **160**, 525–536.

Lukowiak, K. (1979). L_9 modulation of L_7's elicited gill withdrawal response in *Aplysia. Brain Res.* **163**, 207–222.

Lüttgau, H. C., and Oetliker, H. (1968). The action of caffeine on the activation of the contractile mechanisms in striated muscle fibers. *J. Physiol. London* **194**, 51–74.

MacKay, A. R., and Gelperin, A. (1972). Pharmacology and reflex responsiveness of the heart in the giant garden slug *Limax maximus. Comp. Biochem. Physiol. A* **43**, 877–896.

McKenna, O., and Rosenbluth, J. (1973). Myoneural and intermuscular junctions in a molluscan smooth muscle. *J. Ultrastruct. Res.* **42**, 434–450.

McLean, J. R., and Robinson, J. E. (1978). Histochemical identification of monoaminergic nerve cell bodies in the pedal ganglion which innervates the anterior byssus retractor muscle of the lamellibranch mollusc *Mytilus edulis. J. Anat.* **126**, 640.

Mandelbaum, D. E., Koester, J., Schonberg, M., Weiss, K. R. (1979). Cyclic AMP mediation of the excitatory effects of serotonin on the heart of *Aplysia. Brain Res.* **177**, 388–394.

Marchand-Dumont, G., and Baguet, F. (1975). The control mechanism of relaxation in molluscan catch muscle (ABRM). *Pflüegers Arch.* **354**, 87–100.

Mayeri, E., Koester, J., Kupfermann, I., Liebeswar, G., and Kandel, E. R. (1974). Neural control of circulation in *Aplysia*. 1. Motoneurons. *J. Neurophysiol.* **37**, 458–475.

Mellon, DeF. Jr. (1968). Junctional physiology and motor nerve distribution in the fast adductor muscle of the scallop. *Science* **160**, 1018–1020.

Mellon, DeF. Jr. and Prior, D. J. (1970). Components of a response programme involving inhibitory and excitatory reflexes in the surf clam. *J. Exp. Biol.* **53**, 711–725.

Millman, B. M. (1964). Contraction in the opaque part of the adductor muscle of the oyster (*Crassostrea angulata*) *J. Physiol. London* **173**, 238–262.

Mizonishi, T. (1977). Effects of external Na and Ca ion concentration on response of anterior byssal retractor muscle of *Mytilus edulis* to acetylcholine application. *Hiroshima J. Med. Sci.* **26**, 1–8.

Muneoka, Y. (1966). Effects of $CaCl_2$ on the response of anterior byssal retractor muscle of *Mytilus edulis* to acetylcholine. *Biol. Bull. Hiroshima Univ.* **33**, 25–33.

Muneoka, Y. (1969). Effects of some divalent cations, serotonin, and procaine on the caffeine contracture of anterior byssal retractor muscle of *Mytilus edulis*. *Zool. Mag. Tokyo* **78**, 127–133.

Muneoka, Y. (1973). Calcium-dependent acetylcholine contracture of a molluscan catch muscle in sodium-free solution. *Comp. Gen. Pharmacol.* **4**, 277–284.

Muneoka, Y. (1974). Mechanical responses in potassium-depolarized smooth muscle of *Mytilus edulis*. *Comp. Biochem. Physiol. A* **47**, 61–70.

Muneoka, Y., and Kobayashi, M. (1980). Modulatory actions of octopamine and serotonin on the contraction of buccal muscles in *Rapana thomasiana*. II. Inhibition of contraction in radula retractor. *Comp. Biochem. Physiol. C* **65**, 81–86.

Muneoka, Y., and Mizonishi, T. (1969). Effects of changes in external calcium concentration on the caffeine contracture of anterior byssal retractor muscle of *Mytilus edulis*. *Zool. Mag. Tokyo* **78**, 101–107.

Muneoka, Y., and Twarog, B. M. (1977). Lanthanum block of contraction and of relaxation in response to serotonin and dopamine in molluscan catch muscle. *J. Pharmacol. Exp. Ther.* **202**, 601–609.

Muneoka, Y., Cottrell, G. A., and Twarog, B. M. (1977). Neurotransmitter action on the membrane of *Mytilus* smooth muscle. III. Serotonin. *Gen. Pharmacol.* **8**, 93–96.

Muneoka, Y., Shiba, Y., and Kanno, Y. (1978a). Effect of propranolol on the relaxation of molluscan smooth muscle: Possible inhibition of serotonin release. *Hiroshima J. Med. Sci.* **27**, 155–161.

Muneoka, Y., Shiba, Y., and Kanno, Y. (1978b). Effects of neuroleptic drugs on the relaxing action of various monoamines in molluscan smooth muscle. *Hiroshima J. Med. Sci.* **27**, 163–171.

Muneoka, Y. Shiba, Y., Maetani, T., and Kanno, Y. (1978c). Further study on the effect of mersalyl, an organic mercurial, on relaxing response of a molluscan smooth muscle to monoamines. *J. Toxicol. Sci.* **3**, 117–126.

Muneoka, Y., Shiba, Y., and Kanno, Y. (1979a). Relaxation of *Mytilus* smooth muscle by 5-hydroxytryptophan and dopa. *Hiroshima J. Med. Sci.* **28**, 123–132.

Muneoka, Y., Twarog, B. M., and Kanno, Y. (1979b). The effects of zinc ion on the mechanical responses of *Mytilus* smooth muscle. *Comp. Biochem. Physiol. C* **62**, 35–40.

Muneoka, Y., Twarog, B. M., and Mikawa, T. (1978c). Relaxation of *Mytilus* smooth muscle in low sodium. *Comp. Biochem. Physiol. C* **61**, 267–273.

Nagahama, H., Tanaka, Y., and Tazumi, M. (1974). Mechanical responses of the anterior byssus retractor muscle of *Mytilus edulis* to direct-current stimulation. *J. Sci. Hiroshima Univ. Ser. B Div. 1* **25**, 309–325.

Nauss, K., and Davies, R. E. (1966). Changes in inorganic phosphate and arginine during the development, maintenance and loss of tension in the anterior byssal retractor muscle of *Mytilus edulis*. *Biochem. Z.* **345**, 173–187.

Newby, N. A. (1973). Habituation to light and spontaneous activity in the isolated siphon of *Aplysia:* Pharmacological observation. *Comp. Gen. Pharmacol.* **4**, 91–100.

Nomura, H. (1965). Potassium contracture and its modification by cations in the heart muscle of the marine mollusc, *Dolabella auricula. Jpn. J. Physiol.* **15**, 253–269.

Nystrom, T. A. (1967). Mechanism of contraction in molluscan muscle. *Am. Zool.* **7**, 583–591.

Orkand, P. M., and Orkand, P. K. (1975). Neuromuscular junctions in the buccal mass of *Aplysia:* Fine structure and electrophysiology of excitatory transmission. *J. Neurobiol.* **6**, 531–548.

Ozeki, M. (1964). Effect of calcium and sodium ions on potassium induced contracture and twitch tension development of the retractor pharynx muscle of a snail. *Jpn. J. Physiol.* **14**, 155–164.

Painter, S. D., and Greenberg, M. J. (1981). Separating the catch from the contracture. *J. Gen. Physiol.* **78**, 24a.

Paupardin-Tritsch, D., and Gerschenfeld, H. M. (1973). Transmitter role of serotonin in identified synapses in *Aplysia* nervous system. *Brain Res.* **58**, 529–534.

Pavlov, J. (1885). Wie die Muschel ihre Schaale öffnet. *Pflüegers Arch.* **37**, 6–31.

Peretz, B., and Estes, J. (1974). Histology and histochemistry of the peripheral neural plexus in the *Aplysia* gill. *J. Neurobiol.* **5**, 3–19.

Peters, M. (1978). Effect of glutamate and some other amino acids on the membrane potential of muscle cells in the pharynx-levator muscle of *Helix pomatia. Comp. Biochem. Physiol. C*#**61**, 223–227.

Peters, M. (1979a). Motor innervation of the pharynx levator muscle of the snail, *Helix pomatia:* Physiological and histological properties. *J.Neurobiol.* **10**, 137–152.

Peters, M. (1979b). Responses of snail muscle fibers to acetylcholine and serotonin. *Comp. Biochem. Physiol. C* **62**, 181–185.

Phillis, J. W. (1966a). Innervation and control of a molluscan (*Tapes*) heart. *Comp. Biochem. Physiol.* **17**, 719–739.

Phillis, J. W. (1966b). Regulation of rectal movement in *Tapes watlingi. Comp. Biochem. Physiol.* **17**, 909–928.

Pilgrim, R. L. C. (1954). The action of acetylcholine on the heart of lamellibranch molluscs. *J. Physiol. London* **125**, 208–214.

Price, D. A., and Greenberg, M. J. (1977a). Structure of a molluscan cardioexcitatory neuropeptide. *Science* **197**, 670–671.

Price, D. A., and Greenberg, M. J. (1977b). Purification and characterization of a cardioexcitatory neuropeptide from the central ganglia of a bivalve mollusc. *Prep. Biochem.* **7**, 261–281.

Price, D. A., and Greenberg, M. J. (1980). Pharmacology of the molluscan cardioexcitatory neuropeptide FMRFamide. *Gen. Pharmacol.* **11**, 237–241.

Prior, D. J. (1975). A study of the electrophysiological properties of the incurrent siphonal valve muscle of the surf clam, *Spisula solidissima. Comp. Biochem. Physiol. A* **52**, 607–610.

Prosser, C. L., and Young, J. Z. (1937). Responses of muscles of the squid to repetitive stimulation of the giant nerve fibers. *Biol. Bull.* **73**, 237–241.

Prosser, C. L., Nystrom, R. A., and Nagai, T. (1965). Electrical and mechanical activity in intestinal muscle of several invertebrate animals. *Comp. Biochem. Physiol.* **14**, 53–70.

Pumphrey, R. J. (1938). The double-innervation of muscles in the clam (*Mya arenaria*). *J. Exp. Biol.* **15**, 500–505.

Ramsay, J. A. (1940). A nerve-muscle preparation from the snail. *J. Exp. Biol.* **15**, 96–115.

Ripplinger, J. (1957). Contribution à l'étude de la physiologie du coeur et de son innervation extrinséque chez l'Escargot (*Helix pomatia*). *Ann. Sci. Univ. Besançon Zool. Physiol.* **8**, 3–179.

Rock, M. K., Blankenship, J. E., and Lebeda, F. J. (1977). Penis-retractor muscle of *Aplysia:* excitatory motor neurons. *J. Neurobiol.* **8,** 569–579.

Roop, T., and Greenberg, M. J. (1967). Acetylcholinesterase activity in *Crassostrea virginica* and *Mercenaria mercenaria. Am. Zool.* **7,** 737–738.

Roop, T., and Greenberg, M. J. (1976). A comparison of the cholinesterase of an oyster (*Crassostrea virginica*) and a clam (*Macrocallista nimbosa*). *J. Exp. Zool.* **198,** 121–134.

Ruben, P. C., Swann, J. W., and Carpenter, D. O. (1979). Neurotransmitter receptors on gill muscle fibers and the gill peripheral nerve plexus in *Aplysia. Can. J. Physiol. Pharmacol.* **57,** 1088–1097.

Rüegg, J. C. (1971). Smooth muscle tone. *Physiol. Rev.* **51,** 201–248.

Rüegg, J. C., Straub, R. W., and Twarog, B. M. (1963). Inhibition of contraction in a molluscan smooth muscle by thiourea, an inhibitor of the actomyosin contractile mechanism. *Proc. R. Soc. B* **158,** 156–176.

Sakai, T. (1965). The effects of temperature and caffeine on activation of the contractile mechanism in the striated muscle fibers. *Jikeikai Med. J.* **12,** 88–102.

Salánki, J., and Hiripi, L. (1970). Increase of serotonin in the adductors of *Anodonta cygnea* L. (pelecypoda) relaxed by nerve stimulation and in relation to the periodic activity. *Comp. Biochem. Physiol.* **32,** 629–636.

Salánki, J., and Làbos, E. (1969). On the role of cholinergic, adrenergic and tryptaminergic mechanisms in the regulation of a "catch" muscle (*Anodonta cygnea* L.). *Annal. Biol. Tihany* **36,** 77–93.

Sandow, A. (1965). Excitation-contraction coupling in skeletal muscle *Pharmacol. Rev.* **17,** 265–320.

Sanger, J. W., and Hill, R. B. (1972). Ultrastructure of the radula protractor of *Busycon canaliculutum.* Sarcolemic tubules and sarcoplasmic reticulum. *Z. Zellforsch. Mikrosk. Anat.* **127,** 314–322.

Satchell, D. G., and Twarog, B. M. (1978). Identification of 5-hydroxytryptamine (serotonin) released from the anterior byssus retractor muscle of *Mytilus californianus* in response to nerve stimulation. *Comp. Biochem. Physiol. C* **59,** 81–85.

Satchell, D. G., and Twarog, B. M. (1979). Identification of dopamine released from the anterior byssus retractor muscle of *Mytilus californianus* in response to nerve stimulation. *Comp. Biochem. Physiol. C* **64,** 231–235.

Sathananthan, A. N., and Burnstock, G. (1976). Evidence for a non-cholinergic, non-aminergic innervation of the Venus clam heart. *Comp. Biochem. Physiol C* **55,** 111–118.

Sato, M., Tamasige, M., and Ozeki, M. (1960). Electrical activity of the retractor pharynx muscle of the snail. *Jpn. J. Physiol.* **19,** 85–98.

Shigeto, N. (1970). Excitatory and inhibitory actions of acetylcholine on hearts of oyster and mussel. *Am. J. Physiol.* **218,** 1773–1779.

Siegman, M. J., Butler, T. M., Mooers, S. U., and Davies, R. E. (1976). Crossbridge attachment, resistance to stretch, and visco-elasticity in resting mammalian smooth muscle. *Science* **191,** 383–385.

S.-Rózsa, K. (1969). Theory of step-wise excitation in gastropod hearts. *In* "Comparative Physiology of the Heart: Current Trends" (F. V. McCann, ed.), pp. 69–77. Burkhauser, Basel, Switzerland.

S.-Rózsa, K., and Perényi, L. (1966). Chemical identification of the excitatory substance released in *Helix* heart during stimulation of the extracardial nerve. *Comp. Biochem. Physiol.* **19,** 105–113.

S.-Rózsa, K., and Zs.-Nagy, I. (1967). Physiological and histochemical evidence for neuroendocrine regulation of heart activity in the snail *Lymmaea stagnalis* L. *Comp. Biochem. Physiol.* **23,** 373–382.

Stössel, W., and Zebe, E. (1968). Zur intracellulären regulation der kontraktionsaktivität. *Pfluegers Arch.* **302**, 38–56.

Sugi, H. (1971). Contracture of molluscan smooth muscle during calcium deprivation. *Proc. Jpn. Acad.* **47**, 683–688.

Sugi, H., and Suzuki, S. (1978a). Ultrastructural and physiological studies on the longitudinal body wall muscle of *Dolabella auricularia*. I. Mechanical response and ultrastructure. *J. Cell. Biology* **79**, 454–466.

Sugi, H., and Suzuki, S. (1978b). The nature of potassium- and acetylcholine-induced contractures in the anterior byssal retractor muscle of *Mytilus edulis*. *Comp. Biochem. Physiol. C* **61**, 275–279.

Sugi, H., and Yamaguchi, T. (1976). Activation of the contractile mechanism in the anterior byssal retractor muscle of *Mytilus edulis*. *J. Physiol. London* **257**, 531–547.

Suzuki, S., and Sugi, H. (1978). Ultrastructural and physiological studies on the longitudinal body wall muscle of *Dolabella auricularia*. II. Localization of intracellular calcium and its translocation during mechanical activity. *J. Cell. Biol.* **79**, 467–478.

Swann, J. W., Sinback, C. N., and Carpenter, D. O. (1978a). Evidence for identified dopamine motor neurons to the gill of *Aplysia*. *Neurosci. Lett.* **10**, 275–280.

Swann, J. W., Sinback, C. N., and Carpenter, D. O. (1978b). Dopamine-induced muscle contractions and modulation of neuro-muscular transmission in *Aplysia*. *Brain Res.* **157**, 167–172.

Szent-Györgyi, A. G., Cohen, C. and Kendrick-Jones, J. (1971). Paramyosin and the filaments of molluscan "catch" muscles. II. Native filament: Isolation and characterization. *J. Mol. Biol.* **56**, 239–258.

Takahashi, K. (1960). Nervous control of contraction and relaxation in the anterior byssal retractor muscle of *Mytilus edulis*. *Annot. Zool. Jpn.* **33**, 67–84.

Tameyasu, T. (1978). The effect of hypertonic solutions on the rate of relaxation of contracture tension in *Mytilus* smooth muscle. *J. Exp. Biol.* **74**, 197–210.

Tameyasu, T., and Sugi, H. (1976a). Effect of acetylcholine and high external potassium ions on ^{45}Ca movements in molluscan smooth muscle. *Comp. Biochem. Physiol. C* **53**, 101–103.

Tameyasu, T., and Sugi, H. (1976b). The series elastic component and the force-velocity relation in the anterior byssal retractor muscle of *Mytilus edulis* during active and catch contractions. *J. Exp. Biol.* **64**, 497–510.

Taraskevich, P. S., Gibbs, D., Schmued, L., and Orkand, P. K. (1977). Excitatory effects of cholinergic, adrenergic and glutaminergic agonists on a buccal muscle of *Aplysia*. *J. Neurobiol.* **8**, 325–335.

Taxi, J., and Gautron, J. (1969). Données cytochimiques en faveur de l'existence de fibre nerveuses sérotoninergiques dans le coeur l'Aplysie, *Aplysia californica*. *J. Microsc.* **8**, 627–636.

Twarog, B. M. (1954). Responses of a molluscan smooth muscle to acetylcholine and 5-hydroxytryptamine. *J. Cell. Comp. Physiol.* **44**, 141–163.

Twarog, B. M. (1959). The pharmacology of a molluscan smooth muscle *Br. J. Pharmacol.* **14**, 404–407.

Twarog, B. M. (1960a). Effects of acetylcholine and 5-HT on the contraction of a molluscan smooth muscle. *J. Physiol. London* **152**, 236–242.

Twarog, B. M. (1960b). Innervation and activity of a molluscan smooth muscle. *J. Physiol. London* **152**, 220–235.

Twarog, B. M. (1966). Catch and the mechanism of action of 5-hydroxytryptamine on molluscan muscle: A speculation. *Life Sci.* **5**, 1201–1213.

Twarog, B. M. (1967a). The regulation of catch in molluscan muscle. *J. Gen. Physiol.* **50**, 157–169.

Twarog, B. M. (1967b). Excitation of *Mytilus* smooth muscle. *J. Physiol. London* **192**, 857–868.

Twarog, B. M. (1967c). Factors influencing contraction and catch in *Mytilus* smooth muscle. *J. Physiol. London* **192**, 847–856.

Twarog, B. M. (1968). Possible mechanism of action of serotonin on molluscan muscle. *Adv. Pharmacol.* **6B**, 5–16.

Twarog, B. M. (1976). Aspects of smooth muscle function in molluscan catch muscle. *Physiol. Rev.* **56**, 829–838.

Twarog, B. M. (1977). Dissociation of calcium dependent reactions at different sites: lanthanum block of contraction and relaxation in a molluscan smooth muscle. *In* "Exictation-Contraction Coupling in Smooth Muscle" (R. Casteels, T. Godfraind, and J. C. Rüegg, eds.), pp. 261–271. Elsevier/North-Holland, Amsterdam.

Twarog, B. M. (1979). The nature of catch and its control. *In* "Motility in Cell Function" (F. A. Pepe, J. W. Sanger, and V. T. Nachmias, eds.), pp. 231–241. Academic Press, New York.

Twarog, B. M., and Cole, R. A. (1973). Relaxation of catch in a molluscan smooth muscle. II. Effects of serotonin, dopamine and related compounds. *Comp. Biochem. Physiol. A* **46**, 831–835.

Twarog, B. M., and Hidaka, T. (1971). The calcium spike in *Mytilus* muscle and the action of serotonin. *J. Gen. Physiol.* **57**, 252.

Twarog, B. M. and Muneoka, Y. (1973). Calcium and the control of contraction and relaxation in a molluscan catch muscle. *Cold Spring Harbor Symp. Quant. Biol.* **37**, 489–503.

Twarog, B. M., Dewey, M. M., and Hidaka, T. (1973). The structure of *Mytilus* smooth muscle and the electrical constants of the resting muscle. *J. Gen. Physiol.* **61**, 207–221.

Twarog, B. M., Muneoka, Y., and Ledgere, M. (1977). Serotonin and dopamine as neurotransmitters in *Mytilus:* Block of serotonin receptors by an organic mercurial. *J. Pharmacol. Exp. Ther.* **201**, 350–356.

Van Nieuwenhoven, L. M. (1947). An investigation into the structure and function of the anterior byssal retractor muscle of *Mytilus edulis* L. Ph.D. Thesis, University of Utrecht, The Netherlands.

Von Uexküll, J. (1912). Studien über den Tonus. VI. Die Pilgermuschel. *Biol. Z.* **58**, 305–332.

Wabnitz, R. W. (1976). Excitation-contraction coupling and intracellular calcium pool in the penis retractor muscle of *Helix pomatia*. *Comp. Biochem. Physiol. C* **54**, 75–80

Wabnitz, R. W. (1980). Spontaneously occurring neurogenic muscle activity in a molluscan visceral smooth muscle (penis retractor muscle of *Helix pomatia*). II. Tonic contraction-evidence for a modulatory control of the excitatory motor innervation by central nervous structures. *Comp. Biochem. Physiol. C* **66**, 125–136.

Wabnitz, R. W., and von Wachtendonk, D. (1976). Evidence for serotonin (5-hydroxytryptamine) as transmitter in the penis retractor muscle of *Helix pomatia* L. *Experientia* **32**, 707–709.

Watts, J. A., and Pierce, S. K. (1978). Acetylcholinesterase: A useful marker for the isolation of sarcolemma from the bivalve (*Modiolus demissus demissus*) myocardium. *J. Cell. Sci.* **34**, 193–208.

Weber, A. (1966). Energized calcium transport and relaxing factors. *In* "Current Topics in Bioenergetics" (D. S. Sanadi, ed.), pp. 203–254. Academic Press, New York.

Weinrich, D., McCaman, M. W., McCaman, R. E., and Vaughn, J. E. (1973). Chemical, enzymatic and ultrastructural characterization of 5-hydroxytryptamine-containing neurons from the ganglia of *Aplysia californica* and *Tritonia diomedia*. *J. Neurochem.* **20**, 969–976.

Weiss, K. R., Cohen, J. L., and Kupfermann, I. (1978). Modulatory control of buccal musculature by a serotonergic neuron (metacerebral cell) in *Aplysia*. *J. Neurophysiol.* **41**, 181–203.

Weiss, K. R., Mandelbaum, D. E., Schoenberg, M., and Kupfermann, I. (1979). Modulation of buccal muscle contractility by serotonergic metacerebral cells in *Aplysia:* Evidence for a role of cyclic adenosine monophosphate. *J. Neurophysiol.* **42**, 791–803.

Welsh, J. H. (1953). Excitation of the heart of *Venus mercenaria*. *Naunyn-Schmiedebergs Arch. Exp. Pathol. Pharmakol.* **219**, 23–29.

Welsh, J. H. (1956). Neurohormones of invertebrates. I. Cardioregulators of *Cyprina* and *Buccinum*. *J. Mar. Biol. Assoc. U. K.* **35**, 193–201.

Welsh, J. H. (1957). Serotonin as a possible neurohumoral agent: evidence obtained in lower animals. *Ann. N. Y. Acad. Sci.* **66**, 618–630.

Welsh, J. H. (1971). Neurohumoral regulation and the pharmacology of a molluscan heart. *Comp. Gen. Pharamol.* **2**, 423–432.

Welsh, J. H., and Moorhead, M. (1959). Identification and assay of 5-hydroxytryptamine in molluscan tissues by fluorescence method. *Science* **129**, 1491–1492.

Wilkens, L. A., and Greenberg, M. J. (1973). Effects of acetylcholine and 5-hydroxytryptamine and their ionic mechanisms of action on the electrical and mechanical activity of molluscan heart smooth muscle. *Comp. Biochem. Physiol. A* **45**, 637–651.

Wilson, D. F. (1969). The basis for double contractions and slow relaxations of non-striated muscle in a pelecypode mantle. *Comp. Biochem. Physiol.* **29**, 703–715.

Wilson, D. M. (1960. Nervous control of movement in cephalopods. *J. Exp. Biol.* **37**, 57–72.

Winton, F. R. (1937). The changes in viscosity of an unstriated muscle (*Mytilus edulis*) during and after stimulation with alternating, interrupted and uninterrupted direct currents. *J. Physiol. London* **88**, 492–511.

Wolleman, M., and S.-Rózsa, K. (1975). Effects of serotonin and catecholamines on the adenylate cyclase of molluscan heart. *Comp. Biochem. Physiol. C* **51**, 63–66.

Wright, A. M., Moorhead, M., and Welsh, J. H. (1962). Actions of derivatives of lysergic acid on the heart of *Venus mercenaria*. *Br. J. Pharmacol.* **18**, 440–450.

Yamaguchi, H., and Twarog, B. M. (1980). Electrogenic sodium pumping in *Mytilus* smooth muscle. *Comp. Biochem. Physiol. A* **66**, 265–269.

York, B., and Twarog, B. M. (1973). Evidence for the release of serotonin by relaxing nerves in molluscan muscle. *Comp. Biochem. Physiol. A* **44**, 423–430.

Young, J. Z. (1938). The functioning of the giant nerve fibers of the squid. *J. Exp. Biol.* **15**, 170–185.

3

Biochemical and Structural Aspects of Molluscan Muscle

P. D. CHANTLER*

Department of Biology
Brandeis University
Waltham, Massachusetts

*Present address: Department of Anatomy, Eastern Pennsylvania Psychiatric Research Institute Division, Medical College of Pennsylvania, Philadelphia, Pennsylvania 19129.

THE MOLLUSCA, VOL. 4
Physiology, Part 1

I. Introduction

Molluscan muscles have been the subject of experimentation for well over a century, and their study has aided considerably our understanding of muscle as a whole. One of the early attractions of molluscan muscle was the capacity of certain bivalve adductor muscles to maintain a large contractile force (in life, keeping the hinged shell closed) for long periods of time with little expenditure of energy. Such muscles, [named *Sperrmuskel* by Von Uexkuell (1912) and translated as *catch muscle* by Bayliss (1924)], attracted the attention of many great nineteenth-century physiologists, including Pavlov, who showed that stimulation of the efferent nerves of *Anodonta* catch muscle caused the tonically contracting muscle to relax (Pavlov, 1885). Ironically, despite the manifold advances in our knowledge of muscle mechanism and control during the past 30 years, the biochemical origins of catch remain obscure.

In discussing the biochemistry of muscular contraction, it is difficult to avoid making generalizations that rely heavily on evidence obtained from rabbit skeletal muscle. Furthermore, *molluscan* muscle biochemistry is dominated by the large amount of information obtained from the study of scallop adductor muscles. It is possible to extrapolate biochemical results from vertebrate muscles to those of molluscan muscles primarily for two reasons: First, the mechanism of actomyosin-based force generation, exemplified by the sliding-filament mechanism, is ubiquitous; the biochemistry of regulation, an area in which molluscan muscle differs most radically from vertebrate muscle, has been studied in detail in molluscs. Second, structural studies on molluscan muscles comprise a particularly strong area, and the combined techniques of electron microscopy and X-ray diffraction serve to unify the field by showing that the underlying filament structures, in vertebrate and invertebrate muscles alike, are very similar.

A. Movement

The variety of actions encompassed by molluscan muscles is great. They range from the prolonged attachment of the limpet to the quick, darting movements of the squid and other cephalopods. The catch muscle of the bivalves are among the strongest muscles known, generating tensions of up to 15 kg/cm², many times greater than those of vertebrate skeletal muscle; indeed, just to counteract the force of the elastic ligament, which acts so as to close the shell, a pressure of 1 kg/cm² must be overcome in some bivalves.

Movement is a common attribute of gastropods and cephalopods. Freshwater, sea and land snails, and slugs all possess a well-developed muscular foot. Cephalopods swim by jet propulsion using powerful syphon retractor muscles.

Although movement is not a particularly well-developed function in the Bivalvia, there are some exceptions. In *Nuculanidae* and *Septibranchia,* water and detritus may be expelled suddenly from the mantle cavity by contraction of the adductor muscle (Purchon, 1968). This action is well developed in some species, such as *Chlamys opercularis* and *Aequipecten irradians,* in which vigorous contractions by the phasic adductor muscle cause rapid ejection of water from the mantle cavity, thus enabling the animal to "swim" by jet propulsion. Such actions are used during escape from predators, or when the animal searches for a new habitat or, in some cases, for nest building and rock boring (Purchon, 1968; Rees, 1957; Yonge, 1936). Alternatively, many species of bivalve molluscs (e.g., *Mytilus edulis*) remain attached to the substratum by a byssus gland or "foot" throughout their lives. The free-living molluscs have apparently evolved from ancestors that exhibited byssal attachment (Purchon, 1968).

In addition, all these animals possess internal muscles not necessarily related to movement of the whole animal or its shell. These include the muscles operating between the body wall and complex stomachs in all molluscs, the radula retractor muscles and the penis retractor muscles of gastropods.

B. Essential Preliminaries

The sliding filament theory is accepted by most researchers today as the conceptual framework for muscle contraction. It is most easily visualized in striated muscles, but it should be emphasized that the principle holds true at the molecular level in all muscle cells and is the key to actomyosin-based force generation in nonmuscle cells as well.

The sliding filament theory describes how contraction occurs by cross-bridge cycling between two sets of interdigitating filaments (Huxley, 1966; Huxley and Hanson, 1954, 1960; Huxley and Niedergerke, 1954), each set of filaments remaining at a constant length throughout the process. This is illustrated in Fig. 1. The cross-striations seen in the light microscope are due to the underlying filament pattern: Bipolar myosin-containing thick filaments give rise to the anisotropic A band; actin-containing thin filaments comprise the isotropic I band. Z lines, containing the proteins actin, α-actinin, desmin, and vimentin (Lazarides, 1980; Masaki et al., 1967), delineate the sarcomere, which is the fundamental contractile unit of muscle. Actin filaments change their orientation at the Z lines in order to have the correct polarity for interaction with myosin. In smooth muscles, dense bodies act as attachment points for oriented actin filaments instead of Z lines, which are absent. Contraction proceeds as cross-bridges (myosin heads) detach from actin and then reattach in a cyclical process, causing the ratio of the A-band length to the sarcomere length to increase.

The energy for contraction comes from hydrolysis of ATP. MgATP is hydrolyzed on the myosin molecule, there being one active site on each of the two myosin heads. In the absence of actin, the rate of hydrolysis is slow, being limited by isomerization of the myosin-product complex and, at lower tempera-

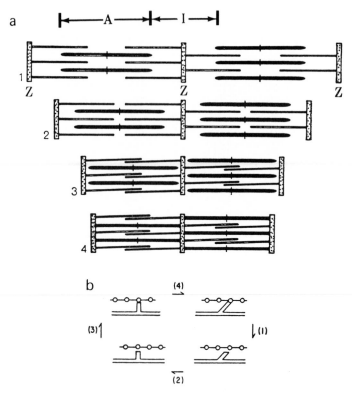

Fig. 1. (a) Schematic illustrating the sliding filament mechanism. Each sarcomere, deline-ated by Z lines, contains two sets of interdigitating filaments. The thick, myosin-containing filaments lie at the centre of the sarcomere. Thin, actin-containing filaments are attached to Z discs at one end and overlap partially with thick filaments when the sarcomere is at rest length. (**a1**). The A band, I band, and Z line are marked. The orientation of myosin molecules in each thick filament changes at the center of the A band. The orientation of actin filaments is opposed on either side of the Z line. This is because the actin and myosin in each half-sarcomere are oriented so that their mutual interaction causes force generation, with directionality from the Z line toward the center of each sarcomere. As contraction proceeds (1–4), the proportion of each sarcomere occupied by the A band increases, at the expense of the I band, until (if contraction proceeds far enough) the A band occupies the entire sarcomere and the I band disappears. Note that myosin heads are not shown. (**b**) Schematic illustrating cross-bridge movement during a single cycle. A single myosin head is shown extending from a thick filament backbone to an actin filament (chain of circles). The head can assume one of two angular conformations. In this simple model, there are two attached states and two detached states. The power stroke is step 4. The cycle could just as easily be shown going in the reverse direction. (**a** reproduced, with permission of the publisher, from Huxley, 1972.)

tures, by the release of ADP (Bagshaw and Trentham, 1974). The rate of hydrolysis at low ionic strength is enhanced 10- to 200-fold by the presence of F actin, this actin-activated ATPase being the *in vitro* analog of contraction. In principle, it should be possible to relate this ATPase mechanism to cross-bridge cycling involving attachment and detachment steps, as well as a power stroke (Fig. 1b).

Actomyosin interaction is controlled by calcium, the ubiquitous secondary messenger of muscle contraction. Calcium is released, under nervous control, from membranous folds within the sarcoplasmic reticulum that make intimate contact with each sarcomere. In general, calcium may be sensitized by one of two different control systems, or both, depending on the source of the muscle: thin filament (actin-linked) control and thick filament (myosin-linked) control (Lehman and Szent-Gyorgyi, 1975). These alternatives are illustrated in Fig. 17. Thin filament-controlled muscles require tropomyosin and troponin for calcium-dependent regulation; calcium-specific sites are present on troponin, which binds exclusively to the thin filament. Thick filament-controlled muscles do not require troponin: Calcium-specific binding sites are present on each myosin molecule. Most muscles possess both forms of regulation. However, vertebrate skeletal muscles lack thick filament control; thin filament control is substantially, if not completely, lacking in many molluscan muscles. More will be said of this in Section VI.

II. Muscle Organization and Fiber Types

Most types of muscle fiber found in the animal kingdom are also represented in the Mollusca. These include cross-striated (large adductor of the scallop), oblique-striated (translucent adductor of the oyster), helical smooth (cephalopod mantle and syphon retractor), paramyosin smooth (lamellibranch catch adductor), and classic smooth (gastropod penis and pharynx retractor) muscles.

The smooth muscles described here are defined morphologically in that no banding pattern is seen when the muscle is observed in the light microscope or under low power in the electron microscope. Striated muscles are also defined morphologically as those exhibiting a banding pattern, seen in both the light and electron microscopes, that arises from the underlying filament structure. One of the major problems in molluscan muscle biochemistry during the late 1950s and early 1960s was whether or not the muscle proteins were present in discrete, discontinuous filaments, so as to enable operation by a sliding filament mechanism. Discontinuous, actin-containing filaments have now been shown to be separate entities from discontinuous, myosin-containing filaments in all molluscan muscle types, and the sliding filament theory is as firmly established in molluscs as it is in vertebrates, thanks mainly to the work of Jean Hanson and Jack Lowy (Hanson and Lowy, 1957, 1959, 1961, 1964).

A considerable effort has been devoted to the biochemistry of bivalve adductor muscles; therefore, it will not be amiss here to clarify certain gross aspects of adductor muscles from various sources. Most lamellibranch molluscs possess two adductor muscles. The large one, often the dominant feature in the animal body, is relatively fast (with contraction times comparable to those of many vertebrate skeletal muscles) and capable of phasic contractions. Depending on the species, this muscle is either translucent or colored; furthermore, it may be cross-striated (e.g., scallops), obliquely striated (e.g., oysters) or of the paramyosin smooth type (e.g., mussels). The second, smaller, white adductor is the tonic catch muscle (although it can also produce phasic responses upon appropriate stimulation; see Chapter 2, this volume). In those species that possess only one adductor muscle, the adductor is invariably a catch muscle. Catch muscles are always of the paramyosin type. The catch adductor muscles of some species (e.g., *Pecten maximus*) show a gradual transition of fiber type, from translucent smooth to paramyosin catch, across the body of the adductor muscle (Ruegg, 1961).

Certain molluscan fiber types have been studied in considerable detail, and these are discussed under the following headings. The following generalizations can be made. Most, if not all, molluscan muscles contain paramyosin. The degree of order present within molluscan muscle fibers is poor when compared with vertebrate skeletal muscle; however, a limited degree of order is seen in cross-striated fibers and, in certain places in smooth muscles, where small groups of actin filaments form aligned semicrystalline arrangements (Lowy and Vibert, 1967; Millman and Bennett, 1976; Sobieszek, 1973). Molluscan thick filaments display a wide range of diameters, depending upon the particular muscle; in addition, molluscan thick filaments are often much longer than their 1.5-μm vertebrate counterparts. All molluscan muscle cells are uninucleate (Hoyle, 1957).

A. Cross-Striated

In the past, cross-striated fibers were considered somewhat unusual in molluscan muscles, but they are now being found with increasing frequency and are typified by the large adductor muscles of various scallops, e.g., *A. irradians*, *Hinnites*, *P. maximus*, and *Placopecten magellanicus*. The overall appearance in the electron microscope is reminiscent of that of vertebrate skeletal muscle (Fig. 2) (Millman and Bennett, 1976). Each muscle cell of the scallop striated adductor muscle is relatively small and ribbon-like, compared to vertebrate skeletal muscle cells, and contains only a single "myofibril." Although the sarcoplasmic reticulum is complex, there are no invaginations of the sarcolemma and therefore

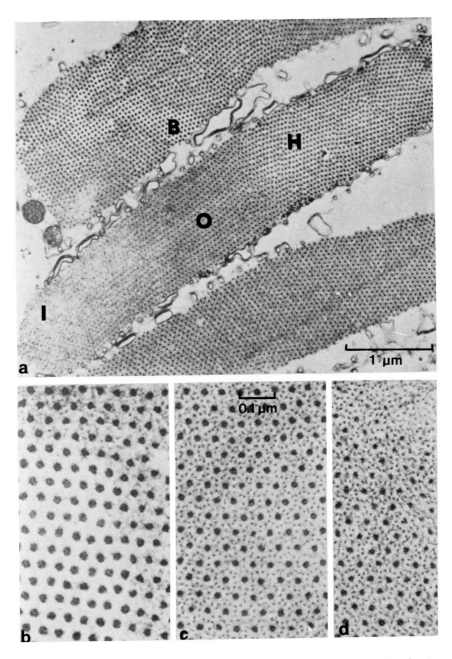

Fig. 2. Scallop striated adductor muscle. (a) Low-magnification transverse section showing the elongated profile of the muscle fibers and the appearance of the filaments at different

no transverse tubular system in this muscle (Sanger, 1971); the small fiber diameter (1–3 μm) renders such a system unnecessary. In contrast to vertebrate skeletal muscle, there is no evidence of M bridges (which transversely connect the centers of adjacent thick filaments in many cross-striated muscles), and the Z lines separating adjacent sarcomeres do not show the same regular order that is present in vertebrate fast twitch muscles (Millman and Bennett, 1976). In addition, the lateral alignment of Z lines in adjacent myofibrils is very easily disturbed. This is partly due to the much larger size of the "myofibrillar unit" that is bounded by membrane in *Pecten* muscle (Philpott et al., 1960) but could also be due to decreased amounts of intermediate (100 Å) filament proteins that hold adjacent Z discs together in the same plane in vertebrate striated muscle (Lazarides, 1980). Each myosin-containing thick filament (about 20 nm backbone diameter and 1.7 μm long in *P. maximus*) is packed in a hexagonal array, 60 nm from its neighbor, and is surrounded by approximately 12 actin-containing

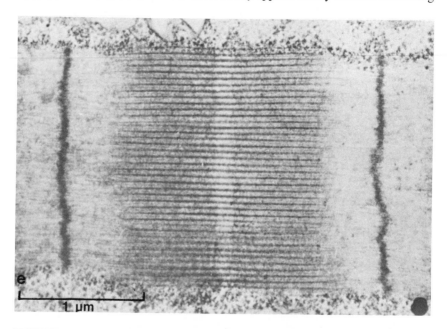

regions in the sarcomere. B, bare zone; H, part of the H zone where the thick filaments show projections; I, I band; O, region where the thin and thick filaments overlap. (25,000X.) **(b)** High-magnification transverse section of the bare zone region of the sarcomere. (110,000X. Scale as in **c.**) **(c)** High-magnification transverse section of the overlap region of the sarcomere. (Magnification as in **b.** Bar represents 0.1 μm.) **(d)** High-magnification transverse section of the beginning of the I region of the sarcomere where the thick filaments have a decreasing diameter. (Magnification and scale as in **c.**) **(e)** Longitudinal section. (36,000X.) (Photographs reproduced, with permission of the publisher, from Millman and Bennett, 1976.)

thin filaments, each about 7 nm in diameter (Fig. 2c, d). Each halo of the 12 thin filaments is shared by six thick filaments. In common with all molluscan adductor muscles, each thick filament contains a central core of paramyosin (Hardwicke and Hanson, 1971; Millman and Bennett, 1976), albeit in much smaller amounts (~7% of the thick filament by weight) than in smooth adductor muscles (Philpott et al., 1960; Szent-Gyorgyi et al., 1973)—such small amounts, in fact, that no paramyosin was detected in early low angle X-ray diffraction patterns of *Pecten* muscle (Bear, 1945; Schmitt et al., 1947).

The above results are known to apply to the fast adductors of the scallops *A. irradians, P. maximus,* and *P. magellanicus* but may also apply to other species of scallop. Not all fast adductor muscles are cross-striated, and it is therefore difficult to correlate cross-striation with speed. Similarly, it is difficult to correlate cross-striations with swimming capability, for the adult *P. maximus,* unlike *Aequipecten irradians,* cannot swim (Purchon, 1968). In addition to the cross-striated adductor, some, but not all lamellibranch and gastropod heart muscles, are cross-striated (Hoyle, 1957; Krijgsman and Divaris, 1955; Marceau, 1905). The radula retractor muscle of some gastropods is also cross-striated (Hanson and Lowy, 1960).

B. Oblique-Striated

Oblique-striated muscle shows the typical banding pattern of a striated muscle in the light microscope. However, the bands are at a small angle (about 10°) to the fiber axis, thus giving rise to oblique as opposed to cross-striation; the angle increases during shortening. As these oblique striations exist throughout the fiber, not just at the periphery, order in the central region breaks down and the thick filaments appear to anastomose (Hanson and Lowy, 1961; Millman, 1967). This gives rise to the appearance of "double-oblique striation" when viewed in the phase-contrast microscope. The dense lines seen in double-oblique striations are A bands composed of filaments which lie parallel to the fiber axis but are grouped into arrays that lie at an acute angle to that axis. These muscles also differ from cross-striated muscles in that they lack Z lines; instead, the actin filaments of the I bands are attached to dense bodies, this being more typical of a vertebrate smooth muscle (Hanson and Lowy, 1961).

Oblique-striated muscles are somewhat unusual in molluscs, being more typical of nematodes (Epstein et al., 1974; Hirumi et al., 1971; Rosenbluth, 1965). There are certain striking molluscan examples however, such as the fibers of the translucent adductor of *Crassostrea angulata* or *Anodonta.* Sections through the translucent adductor of *C. angulata* are seen in Fig. 3. The separate thin and thick filaments are clearly seen, as are dense bodies. The thick filaments are arranged approximately hexagonally, and each one is surrounded by a halo of 12 actin filaments (Hanson and Lowy, 1961). Every halo of thin filaments is shared

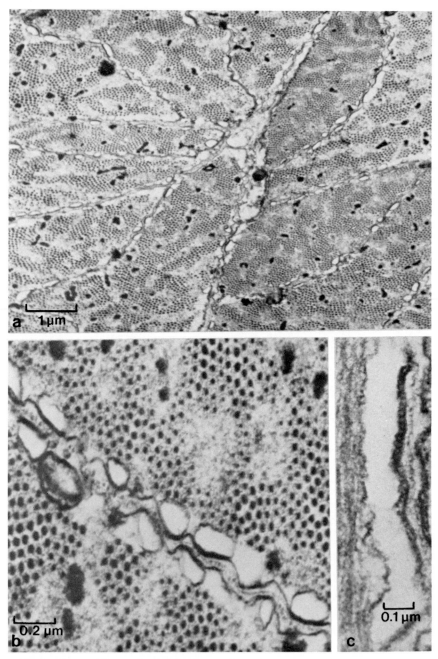

Fig. 3. Oyster oblique striated adductor muscle. (a) Low-magnification transverse section. Note the branched A bands and the conspicuous black dense bodies lying in the I-band regions of the sarcomere. (15,000X.) (b) High-magnification transverse section showing the boundary between two fibers. (60,000X.) (c) High-magnification longitudinal section of the boundary between two muscle fibers showing two large vesicles and the fine structure of the dense layer

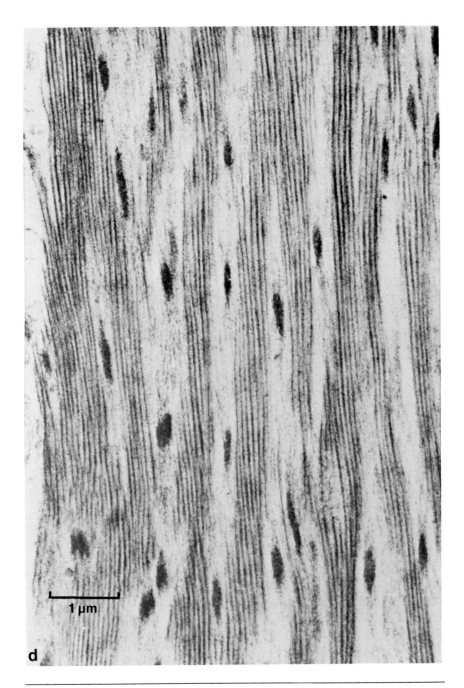

under the cell membrane. (80,000X.) (**d**) Low-magnification longitudinal section through part of one fiber in a stretched glycerol-extracted muscle. Note the branched A bands and the spindle-shaped dense bodies in the I bands. (20,000X.) (Figures reproduced, by permission of the publisher, from Hanson and Lowy, 1961.)

by six thick filaments. Each thick filament is about 40 nm in diameter at the filament center and 5–15 μm long (Millman, 1967). The range of diameters seen varies from 15 to 60 nm depending on where the section is cut. In general, the diameter increases from the end of the A band toward the middle of the sarcomere. The thick filaments of *C. angulata* are paramyosin-rich structures containing 16–20% paramyosin expressed as a percentage of total protein in the muscle (Philpott et al., 1960; Ruegg, 1961; Szent-Gyorgi et al., 1971). This figure is still less than the percentage of paramyosin found in the opaque catch muscle from the same animal (22–39%). Actin filaments, found in the I bands, are present as typical 7-nm-diameter filaments and are attached to dense bodies, probably only at one end. The dense bodies are spindle-shaped, about 0.7 μm long and 0.15 μm in diameter; they lie parallel to the fiber axis (Fig. 3d). In the periphery of the fiber, the dense bodies are fairly regularly spaced at about 2-μm intervals (Hanson and Lowy, 1961). These peripheral dense bodies apparently make contact with the plasma membrane.

C. Helical Smooth

Helical smooth muscles are found in annelids (Hanson, 1957) and cephalopods, where they have been known for a long time (Ballowitz, 1892; Hanson and Lowy, 1957; Marceau, 1905). These are well-ordered muscle cells (5–10 μm in diameter) in which groups of unstriated fibrils form distinct units, separated from each other by sarcoplasm, and these fibrils wind a helical course about the periphery of the cell, there being a central core of mitochondria, as well as a single nucleus (Fig. 4) (Hanson and Lowy, 1957; Kawaguti, 1962; Millman, 1967). Packing within the fibril is such that a cross section in the light microscope gives the appearance of striations; this double-oblique striation is also observed in transverse sections by phase-contrast microscopy (Fig. 4a) and has led some authors to describe these muscles as having regular obliquely striated fibers (Millman, 1967). The angle between the fiber axis and the helical core of the fibril is usually about 5–10 degrees in the mantle and syphon retractor muscles of *Loligo* and in the funnel retractor muscles of *Octopus vulgaris* and *Sepia officinalis* (Hanson and Lowy, 1960); upon contraction, this angle can increase to as much as 60 degrees in glycerol-extracted fibers treated with ATP (Hanson and Lowy, 1957).

The thick filaments are 1–2 μm long and 10–14 nm thick and lie about 35 nm apart in a hexagonal array (Hanson and Lowy, 1957; Millman, 1967), although this distance seems to be short and may be due to specimen shrinkage during dehydration. Each thick filament is surrounded by an irregular arrangement of actin-containing thin filaments; there are frequently more than six thin filaments around each thick filament (Hanson and Lowy, 1957).

The isometric contractions of cephalopod muscles resemble those of frog sartorius muscle rather than those of other types of molluscan smooth muscle (Lowy and Millman, 1962). Whether or not this is due to reduced amounts of paramyosin in these muscles does not seem to have been quantitatively tested; it is known, however, that paramyosin is found in octopus arm muscles (Hanson and Lowy, 1960) and squid pharynx retractor and mantle muscles (Lehman and Szent-Gyorgyi, 1975; Tsuchiya et al., 1980). Very little detailed structural work has been done on these muscles.

D. Paramyosin Smooth

Although many types of molluscan muscle contain paramyosin, it is only those muscles capable of the physiological function known as *catch* that fall within the paramyosin smooth category. Such muscles are characterized by an extremely high proportion of paramyosin to myosin (2–10 : 1 by mass) and thick filaments of great length (25–100 μm) and diameter (20–100 nm) that contain paramyosin as the core protein (Cohen and Szent-Gyorgyi, 1971; Hanson and Lowy, 1964; Millman, 1967; Ruegg, 1961; Szent-Gyorgyi et al., 1971). Two main sources of paramyosin smooth muscle are known: All opaque, white adductor muscles are catch muscles, as are some byssal or foot retractor muscles.

One catch muscle that has been particularly well characterized is the anterior byssus retractor muscle (ABRM) of *M. edulis*. This muscle exhibits typical features of a smooth muscle: the absence of striations and the presence of dense bodies. By using fixation conditions designed to preserve structure maximally, Sobieszek (1973) has produced some beautiful electron micrographs of this muscle which tell us a lot about its ultrastructure (Fig. 5a–d). Thin filaments are present in well-ordered hexagonal arrays (Fig. 5c,d) with 12-nm interfilament spacing, originally predicted from X-ray diffraction studies on this muscle (Lowy and Vibert, 1967). The ratio between the number of thin and thick filaments seen in cross section is about 17 : 1 (Sobieszek, 1973). The center-to-center distance between neighboring thick filaments is 70 ± 10 nm. Thick filaments are 20–30 μm long and 60–70 nm in diameter at their centers (Millman, 1967; Sobieszek, 1973). Because of thick filament tapering at both ends, any cross section reveals a range (10–75 nm) of thick filament diameters (Fig. 5c,d). At maximal width, each thick filament is surrounded by about 20 thin filaments. Dense bodies are distributed uniformly throughout the muscle; they exhibit a range of sizes (mean values: 1.5 μm length; 0.12 μm diameter). From 60 to 80 thin filaments originate, with opposite polarity, on either side of each dense body (Sobieszek, 1973; Szent-Gyorgyi et al., 1971), except for those dense bodies that are attached to the cell membrane at regular intervals: Here, thin filaments are attached only on one side (Sobieszek, 1973). The sarcoplasmic reticulum of the ABRM is well developed but, like the scallop striated adductor, does not exhibit invaginations of the sarcolemma (Heumann and Zebe, 1968).

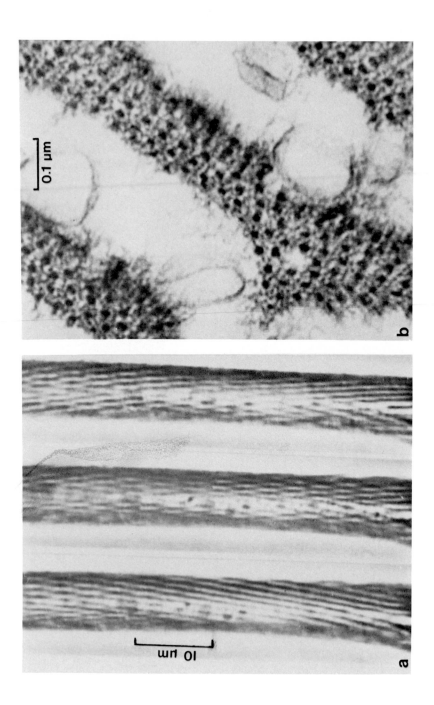

90

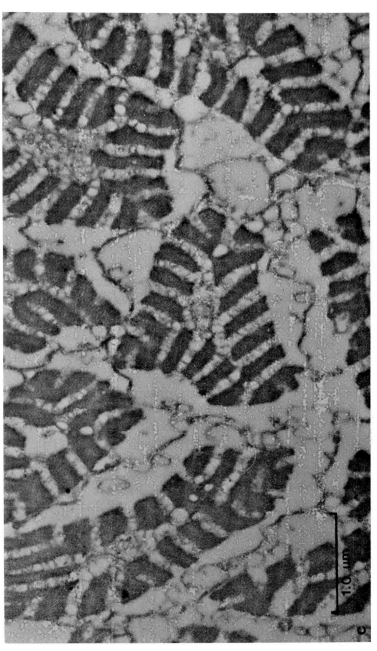

Fig. 4. Squid helical smooth mantle muscle. **(a)** Light micrographs at different focal levels through a squid fiber at rest length, fixed in formaldehyde. Note the helical path of the fibrils and the appearance of double-oblique striation when both sides of the helix are in focus. (2000X.) **(b)** High-magnification transverse section through fibrils. Note the lamellar shape of the fibril. (106,000X.) **(c)** Low-magnification transverse section. Note the triangular shape of the fiber cross section, containing the ribbon-like fibrils and surrounding a central mitochondrion. (27,000X.) (Photographs reproduced, with permission of the publisher, from Hanson and Lowy, 1957.)

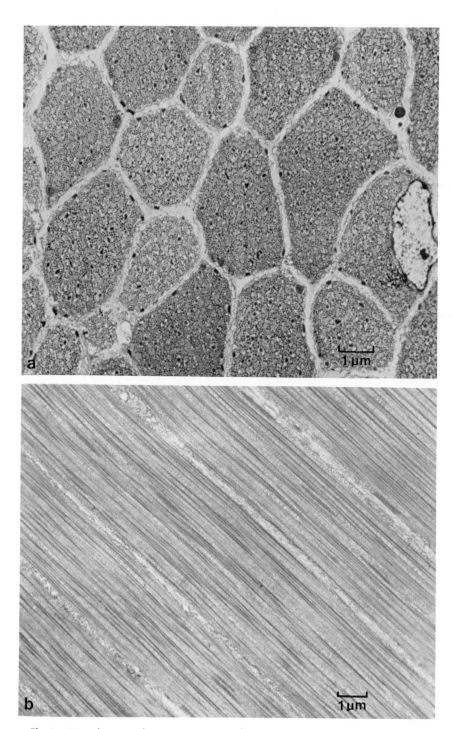

Fig. 5. Mussel anterior byssus retractor muscle—a paramyosin smooth muscle. (a) Low-magnification transverse section. (7600X.) (b) Low-magnification longitudinal section. (8700X.) (c) High-magnification transverse section. Note the prominent thick filaments; the large, more diffuse dense bodies; and the locally ordered thin filaments. (67,000X.) (d) High-magnification

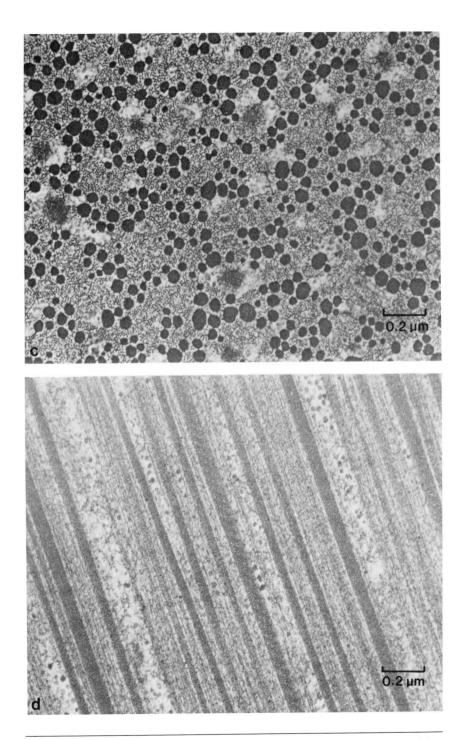

longitudinal section. (68,000X.) (Photographs reproduced, with permission of the publisher, from Sobieszek, 1973.)

E. Classic Smooth

The term *classic smooth muscle* was used by Hanson and Lowy (1960) and refers to the similarity in morphology of these molluscan muscles to such classic vertebrate smooth muscles as taenia coli or uterine muscle (see, e.g., Shoenberg and Needham, 1976). In reality, any molluscan smooth muscle that does not fall within one of the aforementioned categories is usually classified as classic smooth. Such muscles include the pharynx retractor and penis retractor muscles of *Helix pomatia* (Hanson and Lowy, 1960). However, considerable advances have been made in the biochemistry of vertebrate smooth muscle during the past 5 years, especially in the field of regulation (see Section VI); as yet, it is unknown whether any molluscan classic smooth muscles are similar to vertebrate smooth muscles in any detailed way. Indeed, of the molluscan classic muscles examined in any detail, all reveal characteristic differences from vertebrate smooth muscles.

Invaginations of the sarcolemma, forming an extensive tubule system, exist in some classic smooth muscle cells and are particularly well characterized in the phasic radula protractor muscle of *Busycon canaliculatum* (Sanger and Hill, 1972).

III. The Thin Filament

All molluscan thin filaments contain actin and tropomyosin. Some may contain troponin. The properties of individual proteins will be discussed briefly, followed by a description of how they fit together to form an intact thin filament.

A. Actin

Actin is one of the best-known proteins; its distribution is ubiquitous (Pollard and Weihing, 1974), and it is involved in force generation in muscle and non-muscle cells alike. Nevertheless, it should be stated at the outset that very little of the work on actin chemistry has been performed directly on molluscan actin. It is assumed, by analogy, that most of the properties of vertebrate skeletal actin are also representative of molluscan actin. This is a reasonable assumption for actin is a highly conserved protein (Elzinga et al., 1973; Vandekerckhove and Weber, 1978), and extensive structural studies on molluscan actin (Craig et al., 1980; Lowy and Vibert, 1967; Nakamura et al., 1979; Vibert et al., 1972, 1978) have shown the structure of these actins to be remarkably similar. The complete sequences of actin from rabbit skeletal muscle (Elzinga et al., 1973), chicken skeletal actin, and chicken gizzard actin (Vandekerckhove and Weber, 1979), together with nearly complete sequences from three nonmuscle actins (Vandekerckhove and Weber, 1978), are now known. Most of the 25 amino acid differences (out of a total of 374) between skeletal actin and nonmuscle actins are conservative nonpolar substitutions. Furthermore, regions of change are not randomly distributed but are confined to distinct sequences (Vandekerckhove and Weber, 1978); the implication here is that the actin–actin interface and the areas of contact between actin and the many proteins with which it interacts are conserved with impeccable fidelity.

In muscle cells, the physiologically important form of actin is the polymerized protein, F actin (Fig. 6a). F actin can be considered as a double string of beads, 7–8 nm wide, helically organized about a central axis (Fig. 6c). As such, its structure can be defined in two ways: either as two interwound helical strands, each with a pitch of about 75 nm, or in terms of the genetic helix. The genetic helix is the left-handed, short-pitch (5.9 nm) helix formed as one moves about a unidirectional axis from protomer to adjacent protomer (Fig. 6c) (Hanson and Lowy, 1963). Protomers, in the two long-pitch strands are about 180 degrees out of phase with each other.

F actin can be depolymerized *in vitro* into its individual globular subunits (G actin, 42,300 MW) by dialysis against ATP, and trace amounts of Ca^{2+} or Mg^{2+} at low ionic strength. G actin possesses 1 mole of ATP and one divalent cation per mole of protein; both the ATP (Martonosi et al., 1960) and the cation (Barany et al., 1962) are freely exchangeable. Upon polymerization to F actin by raising the ionic strength, the nucleotide is hydrolyzed on the actin to ADP (Straub and Feuer, 1950). Polymerization occurs by self-assembly. The exchangeability of nucleotide (Bender et al., 1974; Martonosi et al., 1960) and cation (Barany et al., 1962) in F actin is reduced considerably when compared with G actin. Actin is sometimes found constrained in monomeric form in nonmuscle cells (Lindberg et al., 1979); this is brought about by binding to low molecular weight proteins such as profilin. If a role for monomeric actin exists in muscle cells, it is probably during thin filament turnover, and one might predict the presence of profilin-like proteins to protect the newly biosynthesized actin monomer from premature polymerization en route to the polymerization site. A small amount of free G actin is always present, however, as a consequence of the G–F equilibrium.

Although all actins run identically on SDS-acrylamide gels as 45-kdalton proteins, isoelectric focusing gels in 9.5 M urea are capable of resolving actins of differing charge (Garrels and Gibson, 1976; Gordon et al., 1977; Rubenstein and Spudich, 1977; Whalen et al., 1976). Thus, skeletal and cardiac muscle actins run as a single band (α-actin), whereas smooth muscle and nonmuscle actins are less acidic and run as two bands (β- and γ-actins) in varying proportions, depending on the source. Figure 7 shows that actin from both the striated and smooth adductors of the scallop, *A. irradians*, runs as a single band of decreased acidity, compared with rabbit skeletal muscle actin (P. D. Chantler and G.

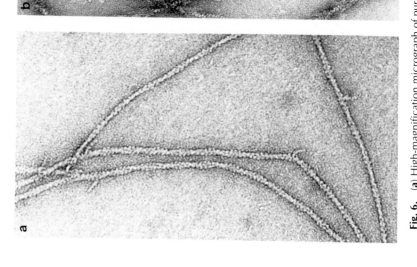

Fig. 6. (a) High-magnification micrograph of purified scallop actin, partially decorated with Ca·Mg·S-1 (only three S-1 molecules are visible in this micrograph). (~290,000X.) (b) High-magnification micrograph of purified scallop actin fully decorated with scallop Ca·Mg·S-1. (~290,000X.) (c) Diagram illustrating the structure of an actin helix with 28 subunits in 13 turns of the genetic helix. (Photographs courtesy of R. Craig and P. Vibert.)

Waller, unpublished observations). This band corresponds to that of β-actin, similar to actin from *S. officinalis* and other invertebrate actins (deCouet et al., 1980).

Identity of mobility on an isoelectric focusing gel does not imply identity of sequence. Thus, the main component of chicken gizzard actin and one of the nonmuscle cytoplasmic actins run identically on isoelectric focusing gels as γ-actin, yet there are more than 20 amino acid replacements between these two actins (Vandekerckhove and Weber, 1978). The position of actin on isoelectric focusing gels appears to be determined entirely by the N-terminal four amino acid residues (Vandekerckhove and Weber, 1979). In the absence of sequence data, one must be cautious in concluding that there is a closer relationship of molluscan actin to vertebrate nonmuscle or smooth muscle actins as opposed to skeletal muscle actin.

The positions of the five cysteine residues found in rabbit skeletal actin (Elzinga et al., 1973) appear to be preserved in molluscan actins. The reagent 2-nitro-5-thiocyanobenzoic acid cleaves peptides primarily at cysteine residues. Using this procedure, Nakamura and colleagues (1979) showed that actins from the striated adductor of the scallop *Patinopecten yessoensis,* the fast muscle of the abalone *Haliotis discus,* and the mantle of the squid *Todarodes pacificus* yielded peptides similar to those of actins from vertebrate, protochordate, and nematode muscles. The amino acid composition of these actins was also remarkably similar (Nakamura et al., 1979).

B. Tropomyosin

Tropomyosin has been found in all muscles without exception. There is exactly one tropomyosin molecule for every seven actin protomers in vertebrate skeletal muscle (O'Brien et al., 1975; Potter, 1974). In scallop myofibrils, actin/tropomyosin ratios obtained vary from 5 (Szent-Gyorgyi, 1976) to 7 (Chantler and Szent-Gyorgyi, 1980). Tropomyosin is an essential component of thin filament regulation; as the regulation of molluscan muscle is primarily determined by thick filament regulation, tropomyosin being a dispensable component (see Section VI), the role of tropomyosin in molluscan muscle is uncertain.

Muscle tropomyosin is an extremely asymmetric fibrous protein, about 41 nm long and 2 nm in diameter, with a molecular weight of 66–70 kdaltons (Phillips et al., 1979; Weber and Murray, 1973). The molecule is a two-chain α-helical coiled coil that forms end-to-end filaments as it winds around the two grooves of the actin long-pitch helix; the two chains are probably in register (Caspar et al., 1969; McLachlan et al., 1975; Phillips et al., 1979). Many muscles contain polymorphic forms of tropomyosin that can be separated by SDS-gel electrophoresis or by hydroxyapatite chromatography: Thus, rabbit skeletal muscles contain α- and β-tropomyosin in the ratio 4 : 1 (Cummins and Perry, 1973). Such polymorphism has not been detected in molluscan muscles. The amino acid

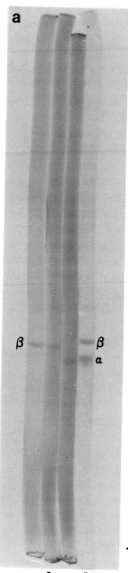

Fig. 7. (a) Isoelectric focusing gels of actin from rabbit skeletal muscle, scallop striated adductor muscle, and scallop smooth adductor (catch) muscle. (a) Scallop striated adductor muscle actin. ~0.5 μg loading. (b) Scallop smooth adductor muscle actin. ~0.5 μg loading. (c) Rabbit striated muscle actin. ~1.0 μg loading. (d) Scallop striated adductor muscle actin (~0.5 μg) plus scallop smooth adductor muscle actin (~0.5 μg) plus rabbit skeletal muscle actin (~1 μg). (b) Two-dimensional gels. Isoelectric focusing (pH 4–10 gradient) in the horizontal dimension; SDS-gel electrophoresis in the vertical dimension. Scallop striated adductor actin (~0.5 μg) plus scallop smooth adductor actin (~0.5 μg). (c) Scallop striated adductor actin (~0.5 μg) plus scallop smooth adductor actin (~0.5 μg) plus rabbit skeletal muscle actin (~1.0 μg). Note that both scallop actins migrate as a single spot upon two-dimensional electrophoresis as β-actin. Gel standards are phosphorylase b (97 kdaltons), bovine serum albumin (68 kdaltons), aldolase (40 kdaltons), and lysozyme (14 kdaltons). (Data from G. Waller and P. D. Chantler, unpublished observations.)

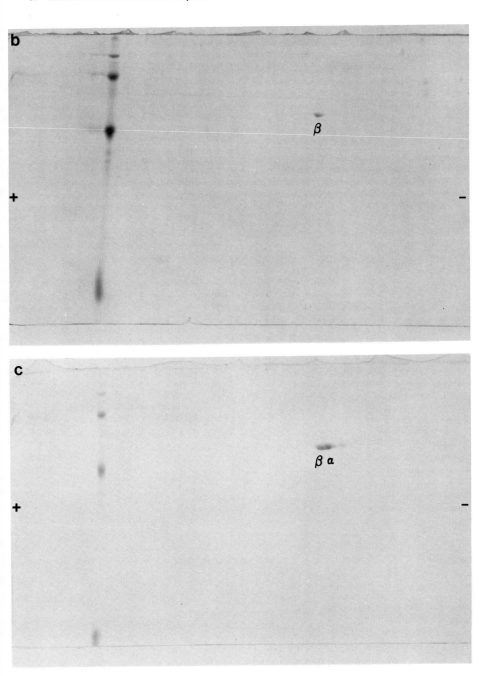

sequence of rabbit skeletal α-tropomyosin is known (Stone and Smillie, 1978), and it was found that nonpolar residues exist, on average, every 3.5 residues throughout long stretches of sequence, in accordance with Crick's prediction for a coiled coil (Crick, 1953). The steric blocking model (Section VI) of thin filament regulation proposes that tropomyosin rolls farther into the actin groove when calcium switches the muscle on, thus removing a physical block to myosin head attachement; consistent with this notion, sequence analysis has revealed 14 (as opposed to 7) quasi-equivalent zones on tropomyosin (McLachlan and Stewart, 1976).

Tropomyosin was the first fibrous protein to be crystallized (Bailey, 1948). Tropomyosin crystals have an open net structure and are 95% water (Caspar et al., 1969; Phillips et al., 1979); the molecules in the crystal are bonded in a head-to-tail fashion, with a nine-residue overlap between molecules (Phillips et al., 1979). Tactoids, or paracrystals, can be formed from tropomyosin by elevation of the divalent cation concentration (Cohen and Longley, 1966). Such paracrystals give information about the underlying structure of the molecules they comprise; tropomyosin paracrystals from various sources are very similar, including paracrystals from the oyster, *Crassostrea commercialis;* from the abalone, *Notohaliotis ruber* (Millward and Woods, 1970; Woods and Pont, 1971); and from the adductor and foot muscles of *Anodonta pacifica* (Tien-chin et al., 1965). All show a 40-nm period similar to that of skeletal tropomyosin.

Molluscan tropomyosins are similar to vertebrate tropomyosins in many physical properties. Subunit molecular weights are in the range 34 ± 3 kdaltons (Woods and Pont, 1971). Although soluble at high ionic strength, tropomyosin polymerizes to give a highly viscous solution as the ionic strength falls below 0.1 (Tsao et al., 1951). This property is a critical function of the carboxyl terminus of tropomyosin, for if the nine-residue overlap region is digested away by carboxypeptidase A treatment, the 64-kdalton product cannot polymerize at low ionic strength or bind to F actin even in the presence of excess magnesium (Mak and Smillie, 1981). An inability to polymerize at low ionic strength is also a feature of nonmuscle tropomyosins (Cote et al., 1978). Nonmuscle tropomyosins, incidentally, are of lower molecular weight (~60,000) than muscle tropomyosins, although this does not affect their actin binding ability (Cohen and Cohen, 1972; Fine and Blitz, 1975).

Molluscan muscle tropomyosins differ from vertebrate tropomyosin in several respects. The amino acid compositions of tropomyosin from *C. commercialis, N. ruber,* and *Pecten alba* were found to be similar to those of other invertebrate tropomyosins, although they differed substantially from vertebrate muscle tropomyosin, most notably in the contents of lysine, asparagine, and threonine (Woods, 1976; Woods and Pont, 1971). There is some evidence that a disulfide bridge exists in native molluscan tropomyosins (Woods and Pont, 1971), but one should be cautious in making such an appraisal in view of the known ease of air oxidation in neighboring cysteine residues on adjacent strands of vertebrate muscle tropomyosin (Lehrer, 1975).

Although the thermal denaturation profiles (monitored as molar ellipticity at 222 nm, a measure of the α-helix content) of some molluscan tropomyosins were similar to those of their nonmolluscan counterparts, the profile for scallop (*P. alba*) tropomyosin showed a lower thermal stability than the others (Woods, 1976). Scallop tropomyosin unfolded as a single transition (the temperature required for 50% denaturation, $T_{\frac{1}{2}} = 30.8°C$); it is interesting to note that this temperature corresponds to the lower of the two transitions observed with rabbit tropomyosin ($T_{\frac{1}{2}(I)} = 30.8°C$; $T_{\frac{1}{2}(II)} = 54.1°C$), suggesting a region of substantial homology.

Scallop (*A. irradians, P. magellanicus*) tropomyosin dissociates completely from actin at temperatures $\geq 25°C$ or in 0.6 M NaCl (2 mM MgATP) (Chantler and Szent-Gyorgyi, 1980; Newman and Carlson, 1980), possibly reflecting a difference in the critical overlap region between adjacent tropomyosin molecules on the thin filament compared with skeletal tropomyosin.

C. Troponin

Troponin, originally isolated by Ebashi (Ebashi et al., 1968), is the other key regulatory protein in vertebrate skeletal muscle and other thin filament-regulated muscles. It binds calcium, released from the sarcoplasmic reticulum (under nervous control), and thereby initiates contraction. As the role, and even the presence, of troponin in molluscan muscle have been brought into question, I will not discuss the individual subunits in any great detail; the reader is referred to two comprehensive reviews on muscle: Weber and Murray (1973) and Harrington (1979).

Troponin (~80 kdaltons) is composed of three components: troponin C (18 kdaltons) which possesses the calcium-specific binding sites; troponin I (21–32 kdaltons) which is instrumental in inhibiting the actomyosin interaction; and troponin T (31–52 kdaltons), which affixes the complex to tropomyosin. The stoichiometry of troponin : tropomyosin:actin is $1 : 1 : 7$ (Ebashi et al., 1969; Potter, 1974). The stoichiometry of subunits within the troponin complex was shown to be $T : I : C = 1 : 1 : 1$ (Potter, 1974), and this value is generally accepted; however, this ratio has been questioned recently, both in vertebrate (Sperling et al., 1979) and invertebrate (Lehman et al., 1976) muscles, and a $T : I : C$ ratio $= 1 : 2 : 1$ suggested.

Although it was originally thought that troponin did not exist in molluscan muscles (Kendrick-Jones et al., 1970), small amounts of material of appropriate molecular weight to be putative troponin I and C components were consistently found in association with molluscan thin filament preparations (Szent-Gyorgyi, 1976). These components have now been seen in muscles from various members of Pelecypoda, Gastropoda, and Cephalopoda (Lehman, 1981; Lehman and Ferrell, 1980; Lehman and Szent-Gyorgyi, 1975). However, only trace quantities are seen in *P. magellanicus* and *Chlamys icelandica* (Lehman, 1981; Szent-Gyorgyi, 1976). Troponin T has not been identified unambiguously in any mollusc.

If the presence of troponins I and C are diagnostic of a full troponin complex in molluscan muscle, two main questions arise: How much troponin is there? Is it functional? Densitometry of fast-green–stained polyacrylamide gels has revealed tropomyosin : troponin C : troponin I ratios of 1.0 : 0.4 : 0.2 in the case of washed *Aequipecten* thin filaments, according to Szent-Gyorgyi (1976); a ratio of 1.0 : 1.0 : 0.6 was obtained by Lehman et al. (1980) using similar preparations. Both groups agree that some molluscan muscles possess only trace amounts of troponin. One confusing circumstance is the presence of calmodulin in many molluscan muscles, including scallop and octopus (Lehman et al., 1980; Molla et al., 1980; Yazawa et al., 1980). Calmodulin is a ubiquitous, multifunctional, calcium-binding protein that has 50% sequence homology with troponin C (Cheung, 1980; Klee et al., 1980; Means and Dedman, 1980). The role of calmodulin is that of a calcium-binding regulator protein; it binds to other proteins, such as phosphodiesterase or adenyl cyclase, and confers calcium sensitivity on their activity. Calmodulin behaves similarly to troponin C upon electrophoresis in SDS or urea-containing polyacrylamide gels, and the two proteins can be easily confused unless a full characterization is performed. *Aequipecten* calmodulin apparently can be distinguished from troponin C on both amino acid composition and electrophoretic grounds (W. Lehman, personal communication).

A functional whole troponin has not yet been isolated from molluscan muscle, evidence for troponin T is nonexistent, and, from the above, evidence for troponin C can be ambiguous. The evidence for troponin in molluscs, therefore, comes mainly from troponin I. The amino acid composition of scallop troponin I has been determined (Lehman et al., 1980). Similar to that of vertebrate troponin I, it is a relatively basic protein, although there are pronounced differences in the amounts of some amino acids (threonine, methionine, phenylalanine, and arginine) between these proteins (i.e., a factor of more than 2) (Lehman et al., 1980; Wilkinson, 1974). This protein appears to function in a similar manner to vertebrate troponin I and can hybridize with vertebrate troponin C to form a complex that confers calcium sensitivity on the vertebrate actomyosin interaction in the presence of vertebrate tropomyosin (Goldberg and Lehman, 1978). Furthermore, antibodies to troponin I specifically stain thin filaments when applied to *Aequipecten* myofibrils (Lehman et al., 1980).

Using troponin I as an indicator of the presence of molluscan troponin, stoichiometries to tropomyosin range from tropomyosin : troponin = 1 : 0.08–0.15 (*Placopecten*) to 1 : 0.2–0.6 (*Aequipecten*), depending on the experimental source and on the chosen stoichiometry of the troponin subunits. In other words, molluscan troponin may be present in about one-half the quantity required for full thin filament control in the most favorable case (*Aequipecten*) but in only 10% of that required in the least favorable case. Despite the functionality of scallop troponin I, there is no evidence from *in vitro* solution or fiber studies for a functional troponin complex in molluscan muscles (see Section VI). As the proportions of the subunits remain constant during thin filament preparation (Lehman et al., 1980), it is unlikely that troponin components are washed off during manipulations. It is plausible that in these thick filament-regulated muscles, troponin components represent vestigial peptides, no longer of functional use *in situ,* yet individually still capable of forming functional molecules when hybridized with vertebrate components.

D. Thin Filament Structure

The combined techniques of X-ray diffraction, electron microscopy, and optical diffraction have revealed that the symmetry of the actin filaments, and the nature of tropomyosin attachment in molluscan thin filaments, are similar to those found in other muscles. In individual scallop F-actin filaments, a wide distribution of symmetries exists, illustrating the flexibility of the actin helix (Egelman et al., 1982); at the center of this distribution, 28 actin protomers are present in 13 turns of the genetic helix (2.1538 protomers per turn). This represents one complete turn of the actin long-pitch helix in 76.8 nm, with an apparent repeat every 38.4 nm. The precise number of protomers per turn of genetic helix depends on the exact supramolecular form of actin (e.g., filaments, bundles, paracrystals) and the nature of the preparation (intact fiber, glycerinated or fixed preparations); these variables should be taken into account when making comparisons. The length of native molluscan thin filaments appears, from light-scattering studies, to be close to 1.0 μm (Newman and Carlson, 1980).

X-ray diffraction of relaxed molluscan muscles (Lowy and Vibert, 1967) indicates crossovers on the actin long-pitch helix every 37.0 nm (living ABRM from *M. edulis,* and dried specimens from the adductor muscle of *Pecten chlamys* and the pharynx retractor muscle from *H. pomatia*). Millman and Bennett (1976) obtained a crossover of 38.0 nm, by X-ray diffraction techniques, for thin filaments from *P. maximus* and *P. magellanicus.* Such measurements are made in two ways: calculated either from the separation of the 5.9- and 5.1-nm reflections, as described by Huxley and Brown (1967), or from direct measurement of the position of the first actin layer line. Within the precision limits of these techniques, all these molluscan thin filaments appear indistinguishable. Adjacent protomers in each long-pitch strand are separated by 5.46 nm, and the two strands are half-staggered by 2.73 nm—values subject to only minor variations in different muscles.

Tropomyosin lies in the two long-pitch grooves of the actin filament (O'Brien et al., 1971). Low-angle X-ray diffraction patterns of living ABRM muscle clearly show an intensity change on one of the actin layer lines upon activation of the muscle (Vibert et al., 1972). This is consistent with the movement of tropomyosin deeper into the actin groove upon activation of the muscle, a feature typical of thin filament-regulated muscles, but somewhat intriguing here considering the reduced amounts of troponin in the ABRM muscle.

Evidence for the absence of troponin in scallop adductor muscles is also seen in a comparison of electron micrographs from scallop and vertebrate muscles. A banding pattern of 38.0 nm periodicity within the I band, which typifies the presence of troponin (Ohtsuki et al., 1967), is absent in micrographs from scallop fibers (Millman and Bennett, 1976). However, a series of sharp meridional reflections, with approximately a 38–nm period, are visible in the X-ray diffraction patterns of scallop muscle fibers (Millman and Bennett, 1976). These are probably the result of the bulge produced by the nine-residue overlap between adjacent tropomyosin molecules in the long-pitch groove. Such a distortion would index on the same lattice as the troponin complex, for the specificity of troponin binding is determined by the tropomyosin binding component, troponin T.

IV. The Thick Filament

Most molluscan thick filaments are composed of two proteins: paramyosin and myosin. Paramyosin is found at the core of molluscan thick filaments. A discussion of the biochemistry of the individual proteins will precede structural details about their location within the thick filament.

A. Myosin

All muscle myosin molecules have a molecular weight of 450,000–500,000 and have similar structural and physical properties. Extensive physical, chemical, and biochemical studies by many investigators have established that myosin is a highly asymmetric hexameric molecule composed of one pair of high-molecular-weight heavy chains (200 kdaltons each) and two pairs of low-molecular-weight light chains (15–30 kdaltons each). A schematic diagram of the myosin molecule is seen in Fig. 8 together with a rotary shadowed picture of a scallop myosin molecule; this bipartite structure contains two globular heads (12–20 nm, long dimension) and a 140-nm-long tail where the two heavy chains come together as coiled-coil α-helices. Each globular head contains the actin binding site (Lowey et al., 1969) and the ATPase site (Young, 1967) of myosin, as well as the principal binding determinants for one member from each of the

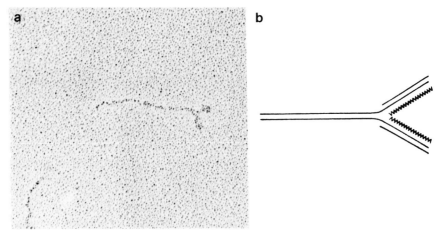

Fig. 8. (a) Rotary-shadowed micrograph of a scallop myosin molecule. (b) Schematic diagram of a scallop myosin molecule, illustrating the two types of light chain located on each globular head. (**a**. Photograph courtesy of Paula Flicker.)

two pairs of light chains (Weeds and Lowey, 1971). The long coiled-coil tail of myosin is responsible for the ability of myosin to aggregate into thick filaments at low ionic strength (I < 0.2) (Harrison et al., 1971).

All myosin molecules possess two kinds of light-chain, termed *regulatory* and *essential*. The origins of these epithets will become clear in a moment. Although of low molecular weight, light-chains are elongated structures (Alexis and Gratzer, 1978; Hartt and Mendelson, 1979; Stafford and Szent-Gyorgyi, 1978); a long axis of at least 10 nm in solution opens up the possibility that the light-chains could have an *in situ* topography that extends from the neck region of myosin to the tip of the globular head. The nearly complete insensitivity of the circular dichroic spectra of scallop light-chains to pH, ionic strength, and temperature (4–70°C) suggests that these subunits are composed of very stable regions of structure which probably exist when bound to myosin (Stafford and Szent-Gyorgyi, 1978).

Myosins from molluscan muscles are regulatory myosins: The ability to interact with actin is controlled by a switch on the myosin molecule itself. This is not true of vertebrate skeletal myosin. In molluscan myosins, this switch is a calcium-specific site structurally determined by both the myosin heavy-chain and the regulatory light-chain (Chantler and Szent-Gyorgyi, 1980), there being one site per myosin head. This site is capable of specific calcium binding when calcium is present at micromolar concentrations, even in the presence of millimolar magnesium (Chantler and Szent-Gyorgyi, 1980; Kendrick-Jones et al., 1970). This calcium-specific site should not be confused with the nonspecific divalent cation site found on all myosin molecules (Bagshaw and Kendrick-

Jones, 1979) and localized on the regulatory light chains (Alexis and Gratzer, 1978; Bagshaw and Kendrick-Jones, 1979; Chantler and Szent-Gyorgyi, 1978; Kuwayama and Yagi, 1977; Werber et al., 1972a). The calcium-specific switch, typical of molluscan myosins and found in addition to the nonspecific binding sites, is not localized on the isolated light chain (Asakawa et al., 1981; Bagshaw and Kendrick-Jones, 1979; Chantler and Szent-Gyorgyi, 1978; Kendrick-Jones, 1976). The sequence of a scallop regulatory light chain (*P. maximus*) is now known (Kendrick-Jones and Jakes, 1977). The nonspecific divalent cation binding site is localized close to the N-terminus; this has been shown through sequence homology with the regulatory light chain from rabbit skeletal myosin (Jakes et al., 1976), secondary structure prediction (P. D. Chantler, unpublished observations, using the methods of Chou and Fasman, 1977) and by an elegant spectroscopic technique involving interaction of the metal binding site with a spin label bound to an adjacent cysteine residue (Bagshaw and Kendrick-Jones, 1980).

The scallop myosin molecule is virtually unique in that both of its regulatory light chains can be removed completely, leaving the essential light chains in place; this is brought about by a short 10 mM EDTA treatment at 25–35°C and is a fully reversible process (Fig. 9) (Chantler and Szent-Gyorgyi, 1980). A similar treatment at lower temperature (0–10°C) removes only one regulatory light chain (Szent-Gyorgyi et al., 1973). The ability of myosin to regulate the actin-activated MgATPase in response to calcium is lost irrespective of whether one or two regulatory light-chains have been removed from the myosin molecule (Chantler and Szent-Gyorgyi, 1980; Kendrick-Jones et al., 1976). This loss of regulation is known as *desensitization;* more will be said of this in Section VI.

All myosins contain one type of light chain that can rebind and restore a regulatory function to 0°C-desensitized scallop myofibrils (from which only one light chain has been removed)—hence the name *regulatory* light chain. The other type of light chain, the *essential* light chain, gets its name from early experiments in which any attempt to dissociate this subunit from myosin resulted in complete inactivation of the myosin (Dreizen and Gershman, 1970); this was probably due to rapid denaturation of the heavy chain under the conditions used (Leger and Marotte, 1975). It is now possible to remove essential light chains without drastic effects on ATPase activity (Sivaramakrishnan and Burke, 1982; Wagner and Giniger, 1981); the role of the essential light chain is still ill-defined.

SDS-gel electrophoretic analysis of some muscle myosins (e.g., rabbit skeletal myosin) indicates three or more kinds of light chain in these muscles. In all cases in which this occurs, the molar ratio of regulatory light chains to essential light chains is 1.0. Such results demonstrate the presence of myosin isozymes in these muscles; not only are the isozymes present in single muscle fibers (Weeds et al., 1975), but they are manifest in individual thick filaments, apparently at random throughout the structure (Silberstein and Lowey, 1981). Isozymic forms of myo-

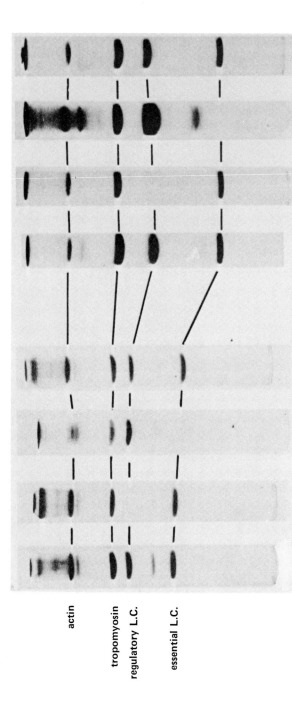

Aequipecten   **Placopecten**

actin

tropomyosin
regulatory L.C.

essential L.C.

1   2   3   4   5   6   7   8

Fig. 9. Removal and readdition of regulatory light chains by a 10-min treatment with 10 mM EDTA at 30°C (*Placopecten magellanicus*) or 35°C (*Aequipecten irradians*). Urea-containing gel electrophoresis. Gels 1–4, *Aequipecten*; 5–8, *Placopecten*; 1 and 5, intact myofibrils; 2 and 6, myofibrils remaining after EDTA treatment at the specified temperature; 3 and 7, first supernatant after the EDTA treatment (dialyzed to remove EDTA and then lyophylized and resuspended in water); gels 4 and 8, myofibrils obtained after incubation overnight with a one molar excess of *Aequipecten* regulatory light chains, followed by washing to remove any excess light chain. (Photograph reproduced, with permission of the publisher, from Chantler and Szent-Gyorgyi, 1980.)

sin have not been found so far within molluscan muscles of uniform fiber type. Two kinds of myosin have been demonstrated in *Patinopecten yessoensis* catch adductor muscle (Kondo and Morita, 1981), but these are probably present in separate muscle cells, for their distribution corresponds to that of fiber types in the smooth adductor (Ruegg, 1961).

The ATPase activity of myosin alone may be measured under a variety of conditions and varies accordingly. ATP alone is not a substrate for myosin; it must be liganded to a cation. ATPase measurements are often named after the ligand associated with the nucleotide: Ca^{2+}, Mg^{2+}, or K^+-ATPase. The exact rates obtained for these different ATPase measurements are also functions of pH, ionic strength, and temperature; this means that it is often difficult to compare measurements made in different laboratories because one or more parameters may not match exactly. The precise values of the Ca^{2+}, Mg^{2+}, or K^+-ATPase for a particular myosin under a given set of conditions are typical of that myosin. In general, in $0.6\ M$ KCl, 2.0 mM ATP, pH 7.5, 25°C, molluscan myosins show a CaATPase (10 mM Ca^{2+}) of more than 2/sec, a K^+-EDTA ATPase (2 mM EDTA) of less than 3/sec, and an MgATPase (5 mM Mg^{2+}) of about 0.7/sec. Particularly detailed measurements are given for the scallops *A. irradians* and *P. magellanicus* (Chantler and Szent-Gyorgyi, 1980), the clam *Mercenaria mercenaria* (Szent-Gyorgyi et al., 1971), the abalone *Haliotis discus* (Azuma et al., 1975) and the squid *Ommastrephes sloani pacificus* (Tsuchiya et al., 1978). The MgATPase of clam (*Meretrix Lusoria*) foot myosin alone has been shown to have a calcium-sensitive turnover rate (Ashiba et al., 1980). This sensitivity should not be confused with the much greater rates (plus Ca^{2+}) and sensitivities attained in the presence of actin.

It has been known for a long time that sulfhydryl-specific reagents affect the K^+-EDTA and Ca^{2+}-ATPases of myosin (Sekine and Kielly, 1964). Modification of vertebrate skeletal myosin by a two-molar excess of *N*-ethyl maleimide (NEM) or *p*-chloromercuribenzoate (PCMB), in the presence of nucleotide or PP_i at 0°C, leads to full inhibition of the K^+-EDTA ATPase and to an activation of the Ca^{2+}-ATPase, provided the latter is assayed at temperature of 25°C or higher (Reisler et al., 1974; Schaub et al., 1975; Sekine and Kielly, 1964). Although similar experiments on scallop myosin knock out the K^+-EDTA ATPase, no enhancement of the Ca^{2+}-ATPase is seen; instead, it is always inactivated by sulfhydryl reagents (P. D. Chantler, unpublished observations) (Fig. 10). A similar situation is seen in the reaction of Abalone (*H. discus*) myosin with PCMB (Azuma et al., 1975). There are two alternative explanations to account for these findings: Either one or more of the characteristic sulfhydryl groups of vertebrate myosin are lacking in molluscan myosins, or, if present, they are situated in a different microenvironment. Whether or not these sensitive sulfhydryl groups are located in a cleft or "jaws" on the molluscan myosin

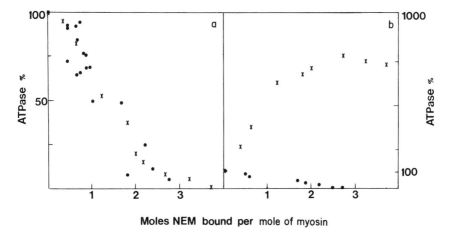

Fig. 10. The effect of the binding of N-ethylmaleimide (NEM) on the K^+-EDTA and Ca^{2+} ATPase activity of rabbit skeletal and scallop striated adductor myosin. Modification performed in 0.6 M KCl, 50 mM Tris, 5 mM Mg^{2+}, 2.5 mM pyrophosphate, pH 7.6, 0°C [myosin] = 1.0–2.0 mg/ml. (**a**) K^+-EDTA ATPase: 10 mM EDTA, 1.0 M KCl, pH 7.5; (**b**) Ca^{2+} ATPase: 10 mM Ca^{2+}, 0.5 M KCl, pH 7.5. (Crosses: data taken from Schaub et al., 1975, for rabbit skeletal myosin, with permission of the publisher. Circles: results using scallop myosin, obtained under conditions identical to those of the rabbit myosin results, from P. D. Chantler, unpublished observations.)

surface, as has been suggested in the case of vertebrates (Wells and Yount, 1979), is not yet known.

Hydrolysis of MgATP at the active site in several vertebrate myosins is accompanied by an increase in tryptophanyl fluorescence (Werber et al., 1972b); the enhancement is partly due to the binding and partly to the hydrolysis of the nucleotide (Chock et al., 1979). Although uv-difference spectroscopy has been used to monitor a similar, albeit smaller, change in *P. yessoensis* myosin (Kondo et al., 1979), zero change has been observed with another scallop myosin, *P. magellanicus,* when hydrolysis was monitored by circular dichroism, uv-difference spectroscopy, electron paramagnetic spin resonance techniques, and intrinsic or extrinsic fluorescence (Chantler and Szent-Gyorgyi, 1978; Chantler et al., 1981a). Apparently the fluorescence change seen in vertebrate myosin, which has proved so useful in elucidating elementary steps of the myosin and actomyosin ATPase, is fortuitous, and unnecessary for the workings of the mechanism.

Proteolytic digestion of the myosin molecule has been a key tool in understanding the structure and function of myosin. Soluble myosins heads (subfragment-1, or S-1) can be prepared using either papain (Lowey et al., 1969) or chymotrypsin (Weeds and Taylor, 1975). A larger, soluble subfragment, heavy meromyosin (HMM) containing both heads and a short portion of the tail, can be obtained by tryptic digestion (Mihalyi and Szent-Gyorgyi, 1953). Whereas molluscan HMM has a calcium-sensitive, actin-activated ATPase, molluscan S-1s are calcium insensitive. Papain digestion of HMM yields S-1 together with the short section of rod that connects the myosin head to the tail. This structure, subfragment-2 (S-2)(60 kdaltons), is soluble despite its α-helical structure (Lowey et al., 1969) and, in the thick filament, can probably move away from the filament backbone in response to the motion of the myosin heads during cross-bridge cycling. Two forms of S-1 have been obtained by papain digestion of *A. irradians* myosin. Digestion in the presence of EDTA yields a subfragment that lacks a regulatory light chain and possesses only an 8 kdalton fragment of the essential light chain (Stafford et al., 1979). Digestion in the presence of divalent cations yields a subfragment that possesses both kinds of light chain, although the regulatory light chain is slightly nicked (Stafford et al., 1979); divalent cations protect the light chains from further digestion (Bagshaw, 1977; Szent-Gyorgyi et al., 1973). These fragments will prove particularly useful in the elucidation of the molluscan myosin and actomyosin ATPase mechanisms.

Finally, it may be noted that in the past, many workers referred to the unstable nature of molluscan myosins (see, e.g., Barany and Barany, 1966; Kondo et al., 1979; Lehman and Szent-Gyorgyi, 1975). It is the opinion of this author, however, that such instability is due to denaturation during preparation. When properly prepared, molluscan myosins retain their activity and their ability to interact with actin in a calcium-sensitive manner for at least 1 week. Furthermore, the myosin can be stored indefinitely in 40% saturated ammonium sulfate (Wallimann and Szent-Gyorgyi, 1981a). The preferred method of preparation has been described (Chantler and Szent-Gyorgyi, 1978); it is a rapid method that produces pure myosin and is based on the ammonium sulfate fractionation procedures of Focant and Huriaux (1976).

B. Paramyosin

Paramyosin has been isolated from the muscles of at least seven invertebrate phyla (Winkelman, 1976), including molluscs and arthropods; it is therefore not unique to molluscan muscles, although it is found in virtually all of them. The distribution of paramyosin is limited to invertebrates, however, for its presence does not appear to have progressed further along the evolutionary tree than members of the protochordata (Yung-shui and Tsu-hsun, 1979).

The paramyosin molecule is composed of two polypeptide chains wound around each other as a coiled coil, to form a rodlike molecule $\cong 130$ nm long and 2.0 nm in diameter (Kendrick-Jones et al., 1969; Lowey et al., 1963; McCubbin and Kay, 1968; Olander, 1971). Paramyosin is therefore somewhat similar in structure, and has properties similar to, those of tropomyosin and the rod portion of myosin (Lowey, 1965). The two chains of paramyosin are probably identical and are oriented in parallel (Weisel and Szent-Gyorgyi, 1975). The molecular weights of paramyosins from various sources have been reported to vary in the range 200–226 kdaltons (Winkelman, 1976). It has been shown, however, in the case of *M. mercenaria* paramyosin, that muscle extracts contain a metal ion-requiring protease which removes a 5-kdalton fragment from the C-terminal (Yeung and Cowgill, 1976) end of each paramyosin polypeptide chain (Stafford and Yphantis, 1972). The product of this cleavage is then susceptible to further attack by serine proteases. When suitable precautions are taken to overcome proteolysis (inclusion of 10 mM EDTA and 0.3 mM PMSF in the extraction buffer), paramyosins from various sources, including *M. mercenaria* opaque adductor, *M. edulis* ABRM, and *A. irradians* striated adductor, appear to have the same molecular weight, 220,000 (Elfvin et al., 1976; Stafford and Yphantis, 1972). Molluscan paramyosins are immunologically similar to each other (Elfvin et al., 1976) and have a similar amino acid composition, typified by a Lys/Arg ratio <1.0 (Tsuchiya et al., 1980; Winkelman, 1976).

Paramyosin is an exceptionally resilient molecule and is often prepared via an ethanol denaturation step. Guanidine hydrochloride denaturation profiles of paramyosin, monitored by optical rotatory dispersion or uv-difference spectroscopy, indicate that melting of the coiled coil occurs in a series of discrete steps rather than by a single cooperative transition (Halsey and Harrington, 1973; Olander, 1971; Riddiford, 1966) suggestive of regions of differing conformational stability along the molecule. Halsey and Harrington (1973) have shown that about one-third of the amino terminal segment of the molecule can be removed by trypsin or pepsin treatment, leaving behind a completely helical rod segment characterized by a single high melting temperature (T_m = 64°C).

The solubility properties of paramyosin are remarkably pH dependent: As the pH drops below 7.0, paramyosin will fall out of solution and form paracrystals over a very small pH range (~0.25 pH units), the transition occurring at lower pH values as a function of increasing ionic strength (Johnson et al., 1959). In the presence of excess divalent cations (50 mM), paramyosin forms paracrystals possessing regular 14.5- and 72.5-nm periodicities (Cohen and Szent-Gyorgyi, 1971; Kendrick-Jones et al., 1969). This latter periodicity, although not always present, is diagnostic of paramyosin and has enabled the length of the paramyosin molecule to be determined with high precision as discussed in Section IV,C.

Paramyosin was shown to inhibit selectively the actin-activated MgATPase, leaving other activities, intrinsic to the active site, alone (Szent-Gyorgyi et al., 1971). No such inhibition was seen by Nonomura (1974), however. This discrepancy is now understood: It depends on how the experiment is performed. If paramyosin and myosin are mixed together at high ionic strength and then diluted rapidly in the presence of ATP and actin so as to precipitate simultaneously the protein and initiate the assay, a hyperbolic inhibition of the actin-activated ATPase, as a function of paramyosin concentration, is obtained (Epstein et al., 1976), a result that is qualitatively similar to that of Szent-Gyorgyi et al. (1971). Alternatively, if paramyosin and myosin are mixed at high ionic strength and then dialyzed against low ionic strength overnight, or, if one prepares myosin-bound paramyosin paracrystals in advance, these *in vitro* complexes are activated by actin to the same extent as that of myosin alone (Epstein et al., 1976; Nonomura, 1974). In addition, under those conditions in which inhibition occurs, the extent of inhibition is dependent upon the actin concentration, there being competition (in the presence of MgATP) between paramyosin and actin for myosin (Epstein et al., 1976). It would appear, therefore, that two forms of interaction exist between paramyosin and myosin, only one of which affects the actin-activated ATPase.

The function of paramyosin is still not entirely clear. Synthetic structures that are formed slowly from myosin and paramyosin most nearly mimic the filaments found *in vivo;* thus, from the above, one would predict no direct effect of paramyosin on the actin-activated ATPase of myosin *in vivo*. However, such an effect may be possible in thick filaments possessing large amounts of paramyosin and capable of catch; evidence for this will be considered later (Section VI). As a core protein, paramyosin is not essential for thick filament formation because molluscan myosins can form synthetic thick filaments without it. The filaments that form in the absence of paramyosin, however, have been reported to be no longer than vertebrate myosin filaments (Ikemoto and Kawaguti, 1967; Szent-Gyorgyi et al., 1971); that is, the absence of paramyosin prevented these filaments from reaching their natural *in vivo* length. A good correlation has been obtained, albeit for only a few species, between *in vivo* thick filament length and paramyosin content (Levine et al., 1976). This correlation has also been noted in nematode body-wall muscles, where it is possible to observe directly the effects of paramyosin absence by a suitable choice of mutants lacking paramyosin (Mackenzie and Epstein, 1980). The involvement of paramyosin in thick filament length determination is also supported by *in vitro* assembly experiments (Ikemoto and Kawaguti, 1967). Although it seems likely that paramyosin is necessary for formation of the exceptionally long (20–30 μm) thick filaments of molluscan paramyosin smooth muscles, the simple beauty of the above observations is marred by a report that claims attainment of long (6–9 μm) thick filaments *in vitro* from nematode myosin in the complete absence of paramyosin (Harris and Epstein, 1977); furthermore, only a poor correlation could be found between thick filament diameter and paramyosin content (Levine et al., 1976; Szent-Gyorgyi et al., 1971). Possibly other components are involved in length determination.

The question of whether increasing paramyosin content enables a muscle to cope with increasing tensions has also been addressed; the answer is not straight-forward. There does appear to be an overall correlation of thick filament length with maximum tension development, but this breaks down in specific cases. For example, the correlation is not true of noncatch invertebrate muscles as a group (Levine et al., 1976). It is possible that maximum tension development can sometimes be independent of filament length, for as Levine and her colleagues point out, it is not clear that the option of increasing filament length is the only way a muscle fiber can increase its tension-developing capabilities for a given muscle length. Nevertheless, as a first approximation, one might expect the tension per unit area to be proportional to the number of cross-bridges per half-sarcomere and therefore to thick filament length, all things being equal—an idea first promulgated by Huxley and Niedergerke (1954).

One clue to the function of paramyosin may be the observation that the molecule can be phosphorylated both *in vitro* and *in vivo* (Achazi, 1979; Cooley et al., 1979). The implications of this phenomenon are discussed further in Section VI.

C. Thick Filament Structure

In common with other muscle myosin thick filaments, molluscan thick fila-ments are bipolar structures (Fig. 14b): They show a central bare zone where myosin molecules are packed in an antiparallel manner. Farther away from the bare zone, on either side of it, myosin molecules are packed in a parallel arrange-ment. Molluscan thick filaments are often much longer (2–40 μm) and thicker (20–80 nm) than their vertebrate counterparts, the largest of the thick filaments coming from muscles capable of prolonged tonic contraction. Furthermore, para-myosin is usually found as a core protein in thick filaments. Hence, a description of molluscan thick filament structure divides naturally into two parts: a brief description of the paramyosin core, and a description of the surrounding myosin surface lattice.

1. The Paramyosin Core

Paramyosin exists in an organized array at the core of most molluscan thick filaments (Szent-Gyorgyi et al., 1971). Low-angle X-ray diffraction techniques (Bear, 1944; Bear and Selby, 1956) indicated a paracrystalline structure that could be described by a two-dimensional net (Fig. 11). The exact manner of packing, or the number of filaments between nodes, could not be deduced from these results alone, although it was anticipated that paramyosin molecules lie parallel to one set of internodal lines (Fig. 11). The X-ray data were equally consistent with the arrangement of molecules in a crystalline or helicoidal man-ner along these axes—a problem described as the *helix–net ambiguity* (Bear and

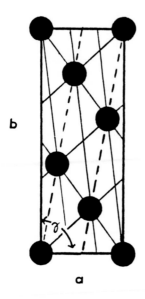

Fig. 11. The Bear–Selby net. Solid circles represent the nodes of the paramyosin net. Various planes connecting these nodes are illustrated. There are five equivalent locations in this nonprimitive unit cell: at the origin and at ($\frac{1}{5}$, $\frac{2}{5}$), ($\frac{2}{5}$, $\frac{4}{5}$), ($\frac{3}{5}$, $\frac{1}{5}$), and ($\frac{4}{5}$, $\frac{3}{5}$).

Selby, 1956). The net pattern and dimensions (b $\cong$ 72 nm, a $\cong$ 19–30 nm; the a distance varies among species and between wet and dry sample states) were in good agreement with negatively stained images taken of fibrils from the white adductor muscles of *M. mercenaria,* first seen by Hall et al. in 1945. A pronounced 14.5-nm banding pattern contained features that showed a true repeat every 72 nm. This checkerboard pattern, obtained with far greater clarity by later workers (Cohen et al., 1971; Elliott, 1979; Hanson and Lowy, 1964; Szent-Gyorgyi et al., 1971), immediately suggests the two-dimensional netlike array and is typical of native filaments from which the surface layer of myosin has been removed. Under favorable conditions, the individual nodes display a triangular aspect that allows the polarity of the filament to be determined; native thick filaments, analyzed in this manner, change their polarity only once, at the center of the filament (Szent-Gyorgyi et al., 1971).

An ostensibly different form of packing for paramyosin molecules was observed initially in *in vitro* aggregates of paramyosin (Hodge, 1952) and later, with greater clarity, in paracrystals produced by divalent cation precipitation of paramyosin in the presence of 1 *M* urea or a small amount of chaotropic agent (Kendrick-Jones et al., 1969). Such paracrystals (Fig. 12a,b) have a distinctive banding pattern and lack discrete nodes. The main features consist of alternate light and dark bands of characteristic length and 72.5-nm periodicity, and these may be arranged in bipolar (Fig. 12a) or polar (Fig. 12b) structures, the polarity being seen by inspection of minor dark striations within the main light band. The importance of these structures lies in the fact that they allow one to obtain the molecular dimensions of paramyosin with great precision, and they give insight into the structural design of native thick filaments. Thus, the simplest interpretation of the polar array (Fig. 12b) is obtained by considering the negative staining pattern in terms of alternating *overlaps* (where stain is largely excluded) and *gaps* (where stain has penetrated). This arrangement is seen in Fig. 12f,g, and it can be seen that the general solution to the molecular length from such a pattern must be of the form $l = nP + L$, where l is the molecular length, P is the period, L is the length of the overlap region (light band), and n is an integer. The value of n, in this case, is constrained by the results from hydrodynamic and light-scattering data (Lowey et al., 1963) and must be unity, giving the length of the paramyosin molecule as 127.5 nm ($P = 72.5$ nm, $L = 55$ nm) (Kendrick-Jones et al., 1969). The fine structure of the pattern varies, depending on the source of the paramyosin, and is characteristic of that source. All other banding patterns can be explained using the concepts of gaps and overlaps.

It soon became apparent that both the checkerboard and the banding pattern types of negatively stained images could be obtained both in "synthetic" paracrystals (Cohen et al., 1971) (see Fig. 12) and in native paramyosin cores (Elliott, 1979; Hanson and Lowy, 1964) (see Fig. 12), suggesting that both patterns were manifestations of the underlying "Bear–Selby" net. *In vitro* studies on paramyosin paracrystals revealed transitional forms (Fig. 12) (Cohen et al., 1971); direct rotation of a native paramyosin filament about its axis, whereas in place on the electron microscope stage, finally revealed the identical origins of the two types of pattern on a single filament (Elliott, 1979). This latter observation suggested a solution to the helix–net ambiguity: The underlying arrangement was probably crystalline. This conclusion was verified by three-dimensional reconstruction of a paramyosin filament (Dover and Elliott, 1979) and by tilting accurately cut transverse sections of molluscan paramyosin cores in the electron microscope (Bennett and Elliott, 1981), where, at favorable angles of tilt, striations are revealed (Fig. 13) corresponding to particular internodal planes (Fig. 11).

The core of molluscan thick filaments is therefore composed of paramyosin molecules packed in a crystalline array—at least in the two dimensions corresponding to the Bear–Selby net. There are gaps between head-to-tail arrangements of molecules, which can fill with stain, giving rise to characteristic negatively stained images (Cohen and Szent-Gyorgyi, 1971). Order along the axis perpendicular to the two-dimensional net is not total. This confers an imperfect crystalline (paracrystalline) property on the core which is manifested, for example, in the curved form of the striations seen in tilted transverse sections during electron microscopy (Fig. 13).

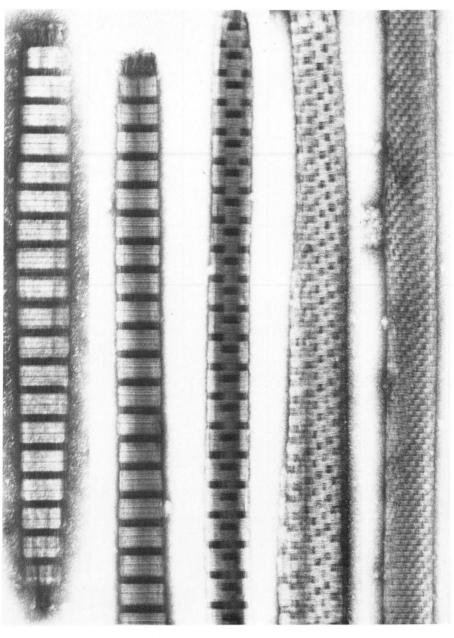

A   B   C   D   E

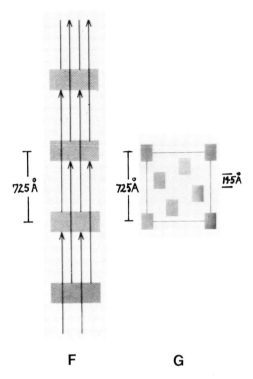

F   G

Fig. 12. (a) Bipolar paramyosin filament. Paramyosin from the white adductor muscle of *Mercenaria mercenaria* precipitated from 50 mM Tris, pH 8.0, 1.0 M urea with 50 mM barium chloride. The fine striations within the light bands show that the polarity of the arrays is opposite in the two halves of the filament. The band pattern is nonpolar (dihedral) at the center. (b) Paramyosin filament showing a simple polar array. Paramyosin from the translucent adductor of *M. mercenaria* precipitated from 50 mM Tris, pH 8.0, with 50 mM $MgCl_2$. (c) Paramyosin filament showing a transitional form. Paramyosin from the smooth adductor muscle of *Aequipecten irradians* precipitated with 50 mM barium chloride from 50 mM Tris, pH 8.0, 50 mM KSCN. The subfilament in the center is staggered by $\frac{2}{3}$ of a period. This translation generates the net pattern. (d) Paramyosin filament showing the synthetic net pattern. Paramyosin from the smooth adductor of *Placopecten magellanicus* precipitated with 50 mM barium chloride from 50 mM Tris, pH 8.0, 0.1 M KSCN. (e) Molluscan thick filament from which myosin has been removed. Paramyosin core from the white adductor of *M. mercenaria*. Myosin removed with 0.4 M KCl, 5 mM ATP, pH 6.0. The nodes have a roughly triangular shape and are oriented oppositely in the two halves of the filament. The helical sense of the nodes is the same in both parts of the filament. (f) Two-dimensional representation of molecular packing of paramyosin molecules in the basic polar array. (g) Unit cell of the paramyosin net pattern. Compare this with the Bear–Selby net (Fig. 11). (Figures reproduced, with permission of the publisher, from Cohen and Szent-Gyorgyi, 1971.)

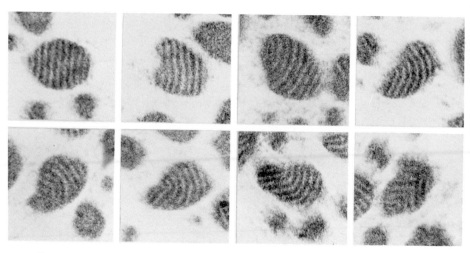

Fig. 13. Transverse sections of embedded smooth adductor of *Ostrea edulis* (130,000X). A selection of micrographs taken on a tilted grid illustrating transverse striations in the paramyosin core with varying degrees of regularity. Viewing angles are along the 301 or 201 planes of Fig. 11. (Photograph reproduced, with permission of the publisher, from Bennett and Elliott, 1981.)

2. The Arrangement of Myosin

In the large thick filaments of molluscan muscles, myosin is found as a surface layer surrounding the paramyosin core. This conclusion arises from the simple demonstration that it is easily possible to extract myosin from thick filaments while leaving the paramyosin core intact (Hardwicke and Hanson, 1971; Kahn and Johnson, 1960; Szent-Gyorgyi et al., 1971), whereas it is not possible to disrupt the paramyosin core without releasing myosin and destroying its natural arrangement. In favorable cases, projections can be seen on the surface of native thick filaments (Elliott, 1974; Hanson and Lowy, 1964; Nonomura, 1974), and these have been attributed to myosin. The surface layer of myosin masks observation of the checkerboard pattern of paramyosin: A 14.5-nm banding pattern is predominant. This repeat (which is typical of myosin from many species, including those that lack paramyosin; Squire, 1973) is identical to the apparent repeat distance in the paramyosin core ($72.5 \times \frac{1}{5}$; see Fig. 12) and thus poses the question: Does the underlying structure of the paramyosin core determine the surface arrangement of myosin in molluscan thick filaments, or is the 14.5-nm repeat an intrinsic property of myosin alone? The observation that pure molluscan myosins form synthetic thick filaments of similar length (1.0–2.0 μm) to those formed from vertebrate skeletal myosin, in the absence of paramyosin, suggests that the former hypothesis may be true in paramyosin-containing muscles; this would not be inconsistent with general theories of filament formation that have been proposed (Squire, 1973). It is claimed, however, that nematode myosin can form long thick filaments ($\sim$10 μm) in the absence of paramyosin *in vitro* (Harris and Epstein, 1977), although the presence of paramyosin is necessary for such formation *in vivo* (Mackenzie and Epstein, 1980). Other arguments in favor of a role for paramyosin in thick filament length determination have been presented above. Whatever the answer, two points are clear: First, the surface area available per myosin molecule on a native molluscan thick filament is of the order required to allow myosin to cover the surface completely (Szent-Gyorgyi et al., 1971; Cohen, 1982); second, if paramyosin–myosin interactions are required for thick filament structure in the native filament, such an interaction is not always necessary for, and does not preclude, myosin–myosin interactions (see also Section VI).

The axial translation between successive layers of myosin cross-bridges is within 2% of 14.5 nm in all thick filaments. However, the exact rotational symmetry of these cross-bridges about the thick filament varies from species to species and is still the subject of debate in vertebrate skeletal muscle. If one assumes that, in all thick filaments, each myosin molecule has a local environment and a geometric relationship similar to those of adjacent molecules within the filament, than a relationship can be inferred between the rotational symmetry and the diameter of the filament (Squire, 1973). The different helices arising from differing rotational symmetries give rise to very similar diffraction patterns; information regarding the diameter can be extracted only from the detailed profiles of the reflections. A comparison of low-angle X-ray diffraction patterns from three invertebrate muscles with each other, and with calculated diffraction patterns from thick filament models with differing symmetries and cross-bridge configurations, was undertaken by Wray et al., (1975). The diffraction patterns from these muscles offered an advantage over those from vertebrate and insect muscles in that diffraction from the myosin showed sampling from the hexagonal lattice only on the equator; this allowed the rotational symmetry to be determined directly from the reflection profile in the case of *Limulus* and *Placopecten* muscles. A six- or sevenfold rotational symmetry was predicted for *Placopecten* thick filaments with some evidence of axial splaying of the cross-bridges, which appeared to be tilted away from the normal to the filament axis in the relaxed muscle (Wray et al., 1975). Despite the differing diameters of *Limulus,* *Homarus,* and *Placopecten* thick filaments, the cross-bridges appeared to lie at similar radii from the filament axis in all three muscles. It appears that a consequence of a universal packing theory for myosin molecules is that individual cross-bridges in relaxed muscle take up precise yet varied orientations with respect to the filament axis, depending upon the particular rotational symmetry (Wray, 1981). This may be the consequence of an underlying subfilament struc-

a

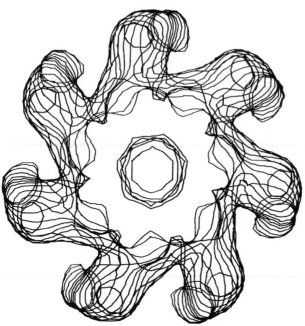

Fig. 14. (**a**) A thick filament from striated scallop muscle (*Placopecten magellanicus*) in three-dimensional reconstruction. This view down the axis shows the seven cross-bridges at one level, each probably representing an unresolved pair of myosin heads. The elongated bridges lie almost parallel to the filament axis and are slewed only slightly around the surface; they are seen here in almost an end-on view. Detail in the central backbone region of the filament is not revealed at this resolution. (Figure courtesy of P. Vibert.) (**b**) Electron micrograph of a negatively stained image of scallop (*P. magellanicus*) thick filament, typical of those used for the three-dimensional reconstruction. Note the central bare zone and the prominent striations on either side. (83,500X. Bar represents 0.1 μm.) (**a**: Figure courtesy of P. Vibert; **b**: photograph courtesy of R. Craig and P. Vibert.)

ture within the backbone of thick filaments; evidence for this comes mainly from intermediate-angle X-ray diffraction studies on crustacean muscle, but it is likely to be a more general phenomenon (Wray, 1979).

Following the procedure of Kensler and Levine, who obtained high-resolution electron micrographs and optical diffraction patterns of *Limulus* thick filaments (Kensler and Levine, 1982; Stewart et al., 1981), Craig and Vibert obtained detailed structural information on *Placopecten* thick filaments. An electron micrograph of a *Placopecten* thick filament (Fig. 14b) shows considerable detail (micrograph and reconstruction kindly given to me, prior to publication, by Drs. Craig and Vibert); the bare zone is clearly visible, and it is even possible to see small axial perturbations of the bridges at alternate 14.5-nm levels along the filament, a feature predicted from earlier X-ray diffraction results (Wray et al., 1975). Optical diffraction of this micrograph and three-dimensional reconstruction yields a detailed cross section (Fig. 14a) which shows the sevenfold rotational symmetry of the cross-bridges. These high-resolution electron micrographs of *Limulus* and *Placopecten* thick filaments substantiate and extend the earlier predictions derived from X-ray diffraction of these muscles (Millman and Bennett, 1976; Wray et al., 1975).

V. The Interaction between Actin and Myosin

A. In the Absence of Nucleotide

The interaction of actin with myosin in the absence of nucleotide produces so-called rigor complexes. A detailed study of this interaction has been made possible by the use of soluble myosin subfragments. The binding of S-1 or HMM to actin is endothermic (Chantler and Gratzer, 1976; Highsmith, 1978) and fairly strong. Most authors agree that the binding constant of S-1 for F actin, in the absence of nucleotide (25°C, pH 7.0, I $\cong$ 0.15), is of the order $3–10 \times 10^6/M$ (Greene, 1981; Greene and Eisenberg, 1980; Margossian and Lowey, 1978; Marston and Weber, 1975). The binding constant of HMM for actin has been the subject of considerable argument; whereas some authors have obtained constants 10-fold higher than the acto–S-1 values (Eisenberg et al., 1972; Highsmith, 1978; Margossian and Lowey, 1978), others assert that $K_B \cong 5 \times 10^9/M$ under similar conditions (Greene, 1981; Greene and Eisenberg, 1980). The discrepancy arises not so much from observation but from manipulation of the data so as to account for the statistical factors inherent in a double-headed molecule binding to two unoccupied adjacent sites. It is clear, however, that the second head of HMM does make some contribution to the free energy of binding, although it is considerably less than a value computed from classical thermodynamics using values obtained for the acto–S-1 binding constant (Chantler and Gratzer, 1976).

Theoretical considerations suggest that the binding constant between HMM and actin should be about 10^3 higher than that between S-1 and actin (Taylor, 1979) thereby supporting the stronger constant of $5 \times 10^9/M$. It is of interest to note that the principal binding domain is entirely preserved in the actin monomer; the binding constant for the monomeric actin–S-1 interaction is $10^6/M$ at 25°C (I = 0.1) (Chantler and Gratzer, 1976).

Decoration of F actin with myosin subfragments gives rise to the typical arrowhead pattern (Fig. 6b) seen in the electron microscope. This configuration is often referred to as the *45-degree state*, this angle being particularly favored from analysis of rigor patterns in insect flight muscle (Reedy, 1967). The exact angle may in fact be species and/or muscle dependent; elegant spectroscopic studies, using electron paramagnetic resonance spectra to probe the orientation of spin-labeled myosin heads in glycerinated rabbit psoas muscle fibers, favor an angle of 68 degrees between the head axis and the fiber axis (Thomas and Cooke, 1980). Whatever the exact angle, it is clear that rigor patterns exhibit a narrow distribution range of orientations.

Similar rigor decoration patterns occur with scallop proteins (Fig. 6b). The attached state persists, and the angle of attachment remains unchanged, irrespective of the presence or absence of calcium (Craig et al., 1980). Whether or not the rigor binding constant of a regulatory myosin and F actin is perturbed by calcium is not yet known; it is evident, however, that the binding constant must be greater than $10^6/M$, whatever the level of free calcium.

B. In the Presence of Nucleotide

The interaction between myosin and actin in the presence of MgATP lies at the heart of muscle contraction. Unfortunately, apart from measurements of the steady-state rates of turnover in a variety of actomyosin preparations, very little is known of the actin-activated ATPase mechanism in molluscan muscle. Actin-activated ATPase rates (in the presence of calcium) of molluscan actomyosin preparations vary from the relatively slow rates seen in gastropod preparations ($\leqq 0.15$ μmol/min/mg myosin) to the high rates typical of *Placopecten* ($\geqq 1.00$ μmol/ min/mg myosin)(Lehman and Szent-Gyorgyi, 1975). Actin activation of various subfragments of scallop myosin have yielded V_{max} values of 3.5 μmol/min/mg S-1 (Ca·Mg·S-1) and 4.4 μmol/min/ mg S-1 (for EDTA S-1) (Stafford et al., 1979), and 1.0 μmol/min/mg (for HMM)(Chantler et al., 1981b). These V_{max} values are 20–100 times higher than the rate of hydrolysis of myosin alone and are comparable with rates of hydrolysis in the muscle itself. K_m values of these subfragments for actin range from 0.2 to 0.5 mg/ml.

The enhancement of the myosin ATPase by F actin is not a property of the isolated actin monomer: G actin activates the myosin ATPase by a factor of two at most (Chantler and Gratzer, 1976; Estes and Gershman, 1978; Offer et al.,

1972). Although possibly a conformational effect, this low activation may result from the need for each myosin head to interact with two actin protomers in F actin in order to achieve maximal rate enhancement (Chantler and Gratzer, 1976), with the second site making an insignificant contribution to the binding. Support for this concept has come from recent cross-linking experiments in which a single S-1 could be directly shown to cross-link two actin protomers in F actin (Mornet et al., 1981); this covalent complex was capable of hydrolyzing ATP at rates close to those of V_{max}.

Our picture of the actomyosin ATPase mechanism, both the solution kinetics and their relationship to physiological data on tension development and contraction, has advanced considerably during the past decade. Although the kinetic mechanism and its relationship to muscle physiology have been the subject of several detailed yet lucid reviews (Adelstein and Eisenberg, 1980; Eisenberg and Greene, 1980; Taylor, 1979), it may not be amiss to adumbrate the main points here.

The steady-state, actin-activated ATPase rate shows a hyperbolic dependence on actin concentration (Eisenberg and Moos, 1968; Szentkiralyi and Oplatka, 1969). Data are often recorded in the form of double reciprocal plots, with the straight line through the points yielding the maximal actin-activated ATPase rate (V_{max}) as it intersects the ordinate and the actin concentration required to achieve $V_{max/2}$ (K_{app}) from extrapolation to the abscissa. K_{app} is found to be much weaker than the binding constant of S-1 or HMM to actin, K_b being measured in the absence of ATP (Eisenberg and Moos, 1968, 1970). This simple relationship, typical of classical Michaelis–Menten kinetics, masks a far more complex mechanism; it follows, however, that a stringent test of any proposed mechanism must be that it manifests itself in this straightforward manner in the steady state.

The overall ATPase mechanism is divided into a number of elementary steps. Rates and equilibrium constants for many of these steps have been determined by several groups, by a variety of steady- and pre-steady-state techniques. The rate of nucleotide cleavage (Fig. 15, steps 5 and 6) is much faster than the rate-limiting step. This cleavage results in rapid formation of enzyme-bound phosphate seen easily during the pre-steady state, a phenomenon known as the *early phosphate burst* (Hayashi and Tonomura, 1970; Kanazawa and Tonomura, 1965; Lymn and Taylor, 1970). The equilibrium constant for this step of the order of 1–10, so that there is a minimal decrease in free energy (Bagshaw and Trentham, 1973). In the presence of low concentrations of actin, actomyosin dissociation precedes hydrolysis (Lymn and Taylor, 1971). This observation gave rise to the now classical Lymn–Taylor scheme (Fig. 15a), in which the hydrolysis step was throught to occur on myosin alone and the rate-limiting step was that of product release after reattachment of actin to $M \cdot ADP \cdot P_i$. Such a scheme seemed to mesh well with physiological requirements, mandating, as it did, the dissociation of actin from myosin per round of ATP hydrolysis. It was anticipated that hydro-

a

b

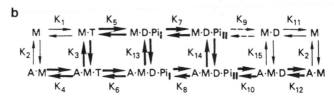

Fig. 15. Kinetic schemes for the hydrolysis of ATP by myosin and actomyosin. (**a**) Lymn–Taylor model, (**b**) attached state model. The optimal pathway for the actomyosin ATPase cycle in each scheme is shown by the heavy solid lines. The dashed arrows indicate the rate-limiting steps in the myosin ATPase cycles in the absence of actin. M, a single myosin cross-bridge head (or S-1, *in vitro*); A, an actin protomer in the F actin filament; T, ATP; D, ADP. It is likely that all the steps involving the binding or release of ATP, ADP, and P_i in these schemes are, in fact, at least two-step processes consisting of the formation of a collision intermediate followed by a conformational change. For simplicity, all these two-step processes are shown as single steps. (Figure adapted, with permission of the publisher, from Adelstein and Eisenberg, 1980.)

lysis would change the conformation of myosin, producing a new angle of attachment; release of products after actin reattachment was envisaged as the power stroke, with the myosin returning to its initial configuration.

More recently, Eisenberg and collaborators have shown that at high actin concentration, as in muscle, hydrolysis need not require actomyosin dissociation (Stein et al., 1979). The mechanism favored by these authors is seen in Fig. 15b. As can be seen, hydrolysis can occur on the dissociated myosin (as in the Lymn–Taylor scheme) or in the associated state. It is currently thought that more than 50% of the hydrolysis occurs via the associated pathway.

Each intermediate in the myosin ATPase cycle (Fig. 15b, top line) is thought to be capable, in the presence of actin, of equilibrium with an associated intermediate in the actomyosin pathway (Fig. 15b, bottom line). The rates of association for each pair of intermediates are roughly the same ($\sim$5 $\times$ 10^6/M/sec) and therefore the variation in binding constants ($K_b \cong 3 \times 10^5$/M in the presence of ADP; $K_b \cong 10^4$/M in the presence of ATP, or ADP plus P_i) probably reflects differences in the rates of dissociation (Adelstein and Eisenberg, 1980). Nucleotide association and dissociation occur, on myosin or actomyosin, by a two-step process—formation of a collision intermediate followed by a conformational change during association, the reverse for dissociation.

The rate-limiting step of the actomyosin ATPase mechanism is still the subject of some controversy, although there is general agreement that the slow step is an attached step (Adelstein and Eisenberg, 1980; Taylor, 1979). Stein et al. (1979) suggest that the slow step precedes phosphate release and postulate that it is step 8 (Fig. 15b)—which is necessary to account for the fact that the concentration of actin required to attain half-maximal binding in the presence of nucleotide is four times greater than K_{app}, the actin concentration required for half-maximal ATPase activity. On the other hand, Lymn and Taylor (1971) originally suggested that phosphate release (step 10, Fig. 15b) was rate limiting; this rate has still not been measured directly, although it is generally assumed that phosphate release will be accompanied by a large drop in free energy (Adelstein and Eisenberg, 1980).

How, then, is it possible to relate a kinetic scheme, such as that shown in Fig. 15b, to the concept of cross-bridge cycling during muscle contraction? Our ideas about cross-bridge action have, until recently, been dominated by two seminal manuscripts by A. F. Huxley (1957) and H. E. Huxley (1969). In its simplest form, the cross-bridge (myosin head) may be considered to attach at a preferential angle (Θ_1) and to detach at a different angle (Θ_2). Simple rotation of the head about a fulcrum near the neck region of myosin (where S-1 joins S-2) will move the head from Θ_1 to Θ_2 and will displace the actin-bound apex of the head 10–15 nm in an axial direction toward the center of the sarcomere—the expected distance of movement per single crank of the cycle. This simple picture can be related to the attached states of Fig. 15b by assuming that each associated intermediate of the cycle has a *preferential* angle of attachment, Θ_n, yet can, in principle, attach at any angle, with the probability of attachment or detachment being determined by a free energy profile. This approximates to a two-state model if different intermediates can be grouped, according to angular preference, into two sets, e.g., $AMT \cong AMDP_{iI} \equiv \Theta_1$ and $AM \cong AMD \equiv \Theta_2$. The important point is that the same free energy minimum for some intermediates (Θ_1) represents conformations at a different angle to the free energy minimum for other intermediates (Θ_2), and consequently to the axial position (x). Thus, Θ can be directly related to the distance of displacement. The positions of the free energy curves relative to an x axis (Fig. 16), and hence the optimal pathway through the cycle, will be determined by the chemical nature of the transitions between intermediates. Thus, adopting the scheme of Eisenberg and Greene (1980), the large free energy drop due to ATP binding to AM, and the large free energy drop due to P_i release from $AM \cdot D \cdot P_{iII}$, will be utilized most efficiently by following the free energy contours and orientations shown in Fig. 16. The nondissociating pathway also has important implications for the mechanism of action of the highly efficient insect oscillatory flight muscles (Cox and Kawai,

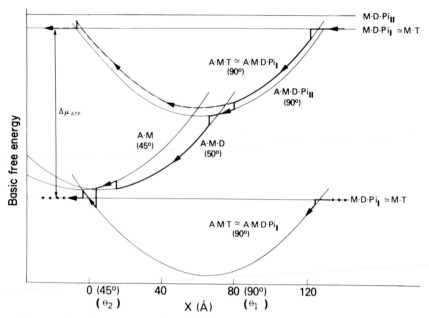

Fig. 16. Basic free energy profile for the attached-state cross-bridge model. The relative basic free energies of the cross-bridge states are plotted as a function of their axial position, x. The angle of the attached cross-bridge to actin at $x = 0$ and $x = 80$Å is indicated in parentheses on the abscissa. The preferred angles of the different cross-bridge states are shown beneath the given states. The heavy solid line represents the path of a nearly optimal cross-bridge cycle during isotonic contraction. The new cycle is seen to start again on the lower left-hand side of the figure. The heavy dashed line shows the path of the cross-bridge if it does not make the transition $A \cdot M \cdot D \cdot Pi_{II} - A \cdot M \cdot D \cdot Pi_{III}$. (Figure adapted, with permission of the publisher, from Eisenberg and Greene, 1980.)

1981) and possibly for catch. For more information, the interested reader is referred to the following comprehensive treatises: Hill, (1974), Eisenberg and Hill, (1978), Taylor, (1979), Tregear and Marston, (1979), and Eisenberg and Greene (1980).

VI. Regulation

A. Types of Regulation: Comparative Aspects

Contraction is controlled by a number of different regulatory mechanisms that vary according to the source of the muscle or nonmuscle force-generating complex (Fig. 17). In broad terms, the type of regulation can be subdivided in terms

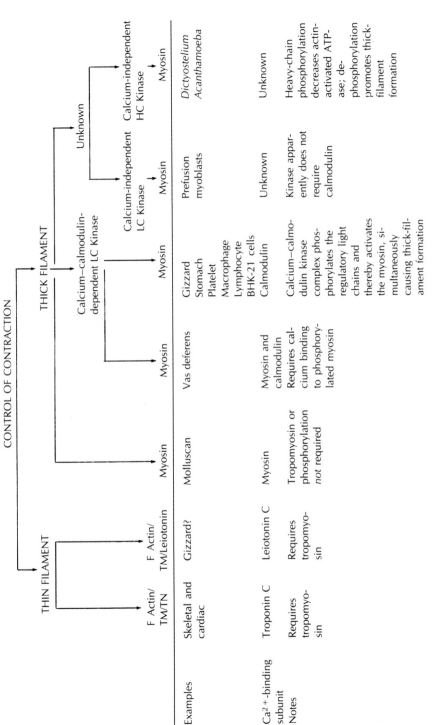

CONTROL OF CONTRACTION

	THIN FILAMENT			THICK FILAMENT			
	F Actin/ TM/TN	F Actin/ TM/Leiotonin	Myosin	Myosin	Calcium–calmodulin-dependent LC Kinase Myosin	Calcium-independent LC Kinase → Myosin	Calcium-independent HC Kinase → Myosin
Examples	Skeletal and cardiac	Gizzard?	Molluscan	Vas deferens	Gizzard Stomach Platelet Macrophage Lymphocyte BHK-21 cells	Prefusion myoblasts	*Dictyostelium Acanthamoeba*
Ca^{2+}-binding subunit	Troponin C	Leiotonin C	Myosin	Myosin and calmodulin	Calmodulin	Unknown	Unknown
Notes	Requires tropomyosin	Requires tropomyosin	Tropomyosin or phosphorylation *not required*	Requires calcium binding to phosphorylated myosin	Calcium–calmodulin kinase complex phosphorylates the regulatory light chains and thereby activates the myosin, simultaneously causing thick-filament formation	Kinase apparently does not require calmodulin	Heavy-chain phosphorylation decreases actin-activated ATPase; dephosphorylation promotes thick-filament formation

Fig. 17. A contemporary view of the different types of regulatory mechanisms operating in various cells in order to control the interaction of actin with myosin in the presence of ATP.

of whether the control proteins are present on the thin or thick filaments. Most muscles possess both thick and thin filament control (Lehman and Szent-Gyorgyi, 1975).

Thick filament control was first discovered in molluscan muscle (Kendrick-Jones et al., 1970). These authors found that thin filaments, prepared from *Mercenaria* or *Aequipecten* muscles, were incapable of bestowing calcium sensitivity on preparations of rabbit myosin, and that calcium-specific binding sites were present on molluscan but not rabbit myosins. A simple test was developed to distinguish between thin and thick filament control in washed myofibril preparations (Lehman and Szent-Gyorgyi, 1975). Addition of pure actin to washed myofibrils would probe for myosin-linked regulation; if the system was myosin controlled, the actin would still be unable to activate the myosin, whereas if actin-linked regulation existed, this would be circumvented by pure actin and activation would ensue. This test was also supplemented by a competitive myosin activation assay which probed for actin-linked control by addition of pure rabbit myosin to washed myofibril preparations. Using these tests, Lehman and Szent-Gyorgyi were able to classify broadly numerous phyla according to thin and/or thick filament regulation and, in particular, established that the type of regulation of contraction present in molluscan muscles was solely myosin linked.

These competition tests did not distinguish between phosphorylatory and non-phosphorylatory control mechanisms. Thick filament control by a phosphorylatory mechanism was first discovered in platelets (Adelstein and Conti, 1975) and has now been found in a variety of nonmuscle and vertebrate smooth muscle cells. Myosin becomes activated by phosphorylation of a single serine residue on each regulatory light-chain (Jakes et al., 1976). This process is ultimately controlled by calcium, for calcium binds to calmodulin and calcium–calmodulin then binds to, and activates, the myosin light-chain kinase which specifically phosphorylates myosin (Adelstein and Eisenberg, 1980). Light-chain phosphorylation also confers thick filament assembly on smooth muscle and thymocyte myosins (Scholey et al., 1980; Suzuki et al., 1978). Dephosphorylation and inactivation of actin activation is brought about by a light-chain phosphatase. At physiological ionic strength, the dephosphorylated smooth muscle myosin monomer folds over on itself to form an unusual conformation with a rodlike tail region in close juxtaposition to the globular heads (Trybus et al., 1982). Although actin activation in these systems is usually switched on directly by light-chain phosphorylation (Fig. 17), it has been reported that in vas deferens smooth muscle, calcium is still needed to activate the phosphorylated myosin (Chacko et al., 1977).

In retrospect, it is remarkable that the competition tests worked as well as they did, for a positive result in favor of myosin-linked regulation in a system that operates by a phosphorylatory control mechanism would require retention of calmodulin and light-chain kinase in the washed myofibril preparations, not to mention phosphatase if reversibility was to be demonstrated. Although thick filament regulation by a phosphorylatory mechanism has been found in the body wall muscles of the sea cucumber, *Parastichopus californicus* (Kerrick and Bolles, 1982) and in doubly regulated *Limulus* telson muscles (Sellers, 1981), it is not known if this is generally true for invertebrate muscles. Thick filament regulation in molluscan adductor muscles definitely does not operate by a phosphorylatory pathway, for determined efforts in this and other laboratories have failed to reveal substantial phosphorylation of molluscan myosins using a variety of light-chain kinase preparations (Adelstein et al., 1979; Frearson et al., 1976; Jakes et al., 1976; J. R. Sellers, unpublished observations). Consistent with this conclusion, phenothiazines, a group of calcium-calmodulin specific drugs, failed to reduce tension in functionally skinned scallop fiber preparations under conditions in which tension was abolished in vertebrate smooth muscle fibers by a similar treatment (Kerrick et al., 1981). Furthermore, molluscan thick filaments are stable even at millimolar MgATP concentrations (Scholey et al., 1980), consistent with the idea that molluscan thick filament assembly does not depend on light-chain phosphorylation. In the absence of evidence to the contrary, however, it is plausible that phosphorylation-dependent modes of regulation could exist in some molluscan smooth muscles, such as those of stomach and heart.

Although many investigators assert that light-chain phosphorylation is a major control mechanism in vertebrate smooth muscle, although not necessarily the only one, this is not the consensus. Ebashi and co-workers claim that the phosphorylation of smooth muscle light-chains does not correlate with superprecipitation (Mikawa et al., 1977). Instead, they present evidence that another protein, leiotonin, is the calcium-binding protein switch, located on the thin filament in these muscles (Mikawa et al., 1978). The existence of leiotonin at this time may be considered controversial. Its possible presence in molluscan smooth muscles has not been investigated. Another means of controlling contraction, which has so far been found only in slime molds and amoebae, is phosphorylation of the myosin heavy chain by a heavy-chain kinase (Fig. 17) (Kuczmarski and Spudich, 1980; Maruta and Korn, 1977; Rahmsdorf et al., 1978). Heavy-chain phosphorylation produces disaggregation of thick filaments and a decrease in actin-activated ATPase activity, in contrast to the light-chain phosphorylatory control mechanism described above. In both cases, filament assembly correlates with ATPase activity.

The current debate on the presence or absence of troponin in molluscan muscles has been detailed in Section III. As stated above, however, there is no evidence for functional thin filament control in molluscan muscles. This is true irrespective of whether one tests washed myofibrils by competition assay procedures (Lehman and Szent-Gyorgyi, 1975) or examines the effect of light-chain removal on tension development in skinned scallop fiber bundles (Simmons and

Szent-Gyorgyi, 1978, 1980). In the former case, no evidence for thin filament control was detected; in the latter case, a state of calcium-insensitive maximal tension development was achieved solely by removal of regulatory light chains: Full calcium-dependent tension generation was restored simply by readdition of regulatory light chains.

In thin filament regulation, a body of evidence suggests that troponin and tropomyosin prevent myosin heads from interacting with actin, in the absence of calcium, by the now classical steric blocking mechanism. Tropomyosin sits in one of two positions within the actin groove. One position, at the edge of the groove, is inhibitory, whereas the other, closer to the center of the groove, allows actomyosin interaction; the exact position depends on the presence or absence of bound calcium on the associated troponin molecule (Huxley, 1973). Although this model has undergone considerable refinement since its conception (Taylor and Amos, 1981), the basic idea of a physical steric blocking of attachment in the relaxed state has been retained by structuralists. It follows, therefore, that if it were possible somehow to affix tropomyosin in a nonblocking position, myosin binding sites on actin would be exposed. This situation was approached experimentally some time ago when the actin-activated ATPase of rabbit myosin subfragments was examined at low ATP concentrations (<20 μM) in the absence of calcium and in the presence of troponin and tropomyosin (Bremel and Weber, 1972). It was found that myosin heads that did not possess bound nucleotide would bind to the thin filament as rigor complexes, presumably pushing tropomyosin away from its inhibitory position and thereby allowing those heads containing bound nucleotide to become activated. Thus, relaxation was reversed at low ATP concentrations in the continued absence of calcium. Such behavior would not be anticipated for muscles controlled solely by myosin, for tropomyosin is not necessary for such regulation. It was therefore somewhat surprising to read reports of just such a reversal of relaxation in actomyosin preparations from scallop and squid (Konno et al., 1981; Toyo-oka, 1979). The ATPase data of these authors are at odds with many other observations on scallop actomyosin, including similar experiments made under similar conditions by others (Knox et al., 1983, and references therein). Apparent observed increases in superprecipitation at low ATP concentrations (Konno et al., 1981; Toyo-oka, 1979) turn out to be due to large-scale swelling and shrinking of the scallop myofibrils, unaccompanied by ATPase activity (Knox et al., 1983). Such observations underline the difficulties of using turbimetric measurements with organized fibrillar material.

Which step in the actomyosin ATPase cycle is subject to regulation? If regulation were an all-or-none phenomenon, then one might expect the relaxed state to be dominated by the longest-lived intermediate of the myosin ATPase pathway. For rabbit skeletal myosin at ambient temperature, this intermediate is $M \cdot ADP \cdot P_i$ (Fig. 15) (Bagshaw and Trentham, 1974); consistent with this, ADP and P_i are found bound to myosin in relaxed vertebrate skeletal muscles (Marston, 1973). The steady-state intermediate in relaxed scallop adductor myofibrils was also shown to contain bound ADP (Marston and Lehman, 1974). It is not clear, however, that the relaxed state is necessarily an all-or-none phenomenon. If the association reactions (Fig. 15b) are rapidly reversible, it still may be possible to account for the ease of passive stretch seen in relaxed muscle, while at the same time having a partially associated state. If this were the case, the long-lived intermediates mentioned above might also involve $A \cdot M \cdot ADP \cdot P_i$. In this regard, Chalovich and collaborators (Chalovich et al., 1981; Chalovich and Eisenberg, 1982) have performed *in vitro* experiments in a thin filament regulated system that calls into question the steric blocking model. They found that the association constant for S-1 binding to the actin–troponin–tropomyosin complex, in the presence of ATP, was independent of the presence or absence of calcium ($K_B \cong 2.0 \times 10^4/M$ at 25°C). This means that tropomyosin was not interfering directly with the attachment of S-1 onto actin. Before discarding all evidence in favor of the steric blocking model, it may be stated that there is a way out of this enigma that should satisfy kineticists, structuralists, and theorists alike. Models of cross-bridge action (e.g., Fig. 16) often depict the myosin head undergoing an angular change while bound to actin (e.g., Huxley and Simmons, 1971). It is possible, therefore, that the inhibitory position of tropomyosin does not block the initial *binding* of S-1 onto actin but that it physically blocks a subsequent *rotation* of the bound head on the actin, this rotation being a necessary step in the force-generating pathway. Only time and subsequent experiments will tell whether this compromise situation is correct.

The binding constant of calcium-sensitive scallop HMM for pure F actin in the presence of MgATP is changed little by the presence or absence of calcium (J. M. Chalovich, P. D. Chantler, E. Eisenberg, and A. G. Szent-Gyorgyi, unpublished observations). Thus one could envisage a mechanism for myosin-linked regulation that has analogies with the steric blocking mechanism; for instance, the light chain might physically block a rotation of the myosin head when the head was bound to actin. Such an impression is probably incorrect. It is possible to label various regulatory light chains with a variety of photosensitive cross-linkers; these modified light chains are then added back to regulatory light-chain–denuded scallop myofibrils prior to photolysis under defined conditions. When such experiments are performed under "rest" conditions (0.1 mM EGTA; 1.0 mM ATP), no cross-linking between actin and regulatory light chains can be detected using antiactin antibodies and antibodies against the various light chains (A. G. Szent-Gyorgyi, personal communication). Such cross-linking would be expected if the light chain physically blocked a rotation of the myosin head on the actin surface.

It is possible that the essential light chains are also involved in the machinery of regulation in molluscan muscles. Not only have cross-linking studies revealed that the regulatory and essential light chains are in close apposition to each other, at least at one location (Walliman et al., 1982), but also that antibodies (both monovalent and divalent) to the scallop essential light chain specifically increase the actin-activated ATPase ($-Ca^{2+}$) activity of scallop myofibrils while leaving the actin-activated ATPase ($+Ca^{2+}$) rate unaffected (Wallimann and Szent-Gyorgyi, 1981b). As the actin-activated rate in the absence of calcium is considered to be a true measure of regulation (Chantler and Szent-Gyorgyi, 1980), this effect appears to implicate the essential light chain in regulation. In addition, the spatial relationship of the essential light chain to the regulatory light chain is altered between the rest and rigor states (A. G. Szent-Gyorgyi, personal communication). Photochemical cross-linking studies indicate that the regulatory and essential light chains are in close proximity during rigor, yet this juxtaposition is abolished in the presence of MgATP (and in the absence of calcium). Such a movement is probably small, for no conformational change has been detected in spectroscopic studies performed under these conditions (Chantler and Szent-Gyorgyi, 1978; Chantler et al., 1981a).

Finally, at the risk of overemphasis, let me reiterate that tropomyosin is not required for, and may not influence, myosin-linked regulation in molluscs (Chantler and Szent-Gyorgyi, 1980; Kendrick-Jones et al., 1970). In some species, tropomyosin appears to be capable of potentiating actin-activated activities. Such a situation is observed with *Limulus* myosin (Lehman and Szent-Gyorgyi, 1972; Sellers, 1981). Activation of *Limulus* myosin is controlled by light-chain phosphorylation (Sellers, 1981). Such potentiation has not been observed in molluscan preparations. One continuing puzzle is the increase in the second layer line intensity of X-ray diffraction patterns from tonically contracting and rigor specimens of *Mytilus* ABRM muscle compared with the relaxed pattern (Vibert et al., 1972). This has been interpreted as a movement of tropomyosin in the actin groove; it is qualitatively similar to the movement of tropomyosin depicted in the steric blocking model discussed above, except that in the relaxed state, the ABRM tropomyosin is not thought to be as far out of the groove as in the vertebrate structure. If this interpretation is correct, then the movement of tropomyosin in response to actin attachment by intermediates of the ATPase pathway must be passive and inconsequential; otherwise, the importance of tropomyosin should be manifested as a modulation of the regulatory process in molluscan muscles—which is not seen. Tropomyosin may well have functions other than the steric blocking action ascribed to it in vertebrate muscles, modulation of the flexibility of actin being one possibility (Oosawa et al., 1973).

B. Cooperative Aspects

Researchers have pondered the significance of the two-headedness of myosin for some time. However, firm evidence for cooperativity between each of the two heads has been presented only for regulatory myosins during operation of their regulatory function; most of this evidence comes from the study of scallop myosin.

An early pointer to cooperativity in scallop myosin–the sharing of a single regulatory light chain between two myosin heads (Szent-Gyorgyi et al., 1973)—proved erroneous; the scallop myosin molecule, like all myosins, possesses two regulatory light chains (Kendrick-Jones et al., 1976). The concept of cooperativity remained, however, primarily due to two key observations described in the earlier paper. First, scallop S-1, although capable of being regulated by the rabbit troponin–tropomyosin complex, would not by itself bestow calcium sensitivity on the actin-activated ATPase in the presence of actin alone (Szent-Gyorgyi et al., 1973). This suggested something fundamentally different between the double-headed myosin molecule as opposed to the single-headed subfragment. Second, removal of one regulatory light chain per molecule of scallop myosin resulted in the inability of the myosin molecule to maintain the off ($-Ca^{2+}$) state (Kendrick-Jones et al., 1976); it was "desensitized."

Further evidence in support of cooperativity was soon forthcoming. Readdition of regulatory light chains from a variety of vertebrate and invertebrate muscles to 0°C-desensitized scallop myosin completely restored calcium sensitivity (Kendrick-Jones et al., 1976). However, not all regulatory light chains restored calcium-specific sites (Section IV); it therefore appeared that complete calcium sensitivity could be restored even if there were only one calcium-specific site per myosin molecule. Many aspects of cooperativity became testable after it became possible to reversibly remove *both* regulatory light chains from each scallop myosin molecule (Chantler and Szent-Gyorgyi, 1980) (Figure 9). Readdition of regulatory light chains to 30°C-desensitized scallop myofibrils restored calcium-specific sites (Fig. 18) and the turnover rate of the actin-activated MgATPase ($+Ca^{2+}$)(Fig. 19) in a linear manner (Chantler and Szent-Gyorgyi, 1980). However, the actin activated MgATPase ($-Ca^{2+}$) rate was restored in a biphasic manner (Fig. 19) (Chantler and Szent-Gyorgyi, 1980), an observation consistent with the earlier finding that removal of 1 mol of regulatory light chain per mole myosin resulted in complete desensitization. This biphasic restoration of relaxation, as monitored by the actin-activated ATPase ($-Ca^{2+}$), is consistent with a model in which two regulatory light chains are required to maintain the off state in scallop myosin and in which readdition of light chains to regulatory light-chain–denuded myosin occurs in a negatively cooperative manner (Chantler and Szent-Gyorgyi, 1980). This latter point assumes that there is no intrinsic difference in affinity for light chain between the two heads of scallop myosin. This assumption was shown to be correct using labeled regulatory light chains; the two light-chain binding sites on myosin are equivalent, and differences in affinity appear to be the result of an interaction between the bipartite halves of the myosin molecule (Sellers et al., 1980). S-1 populations, on the

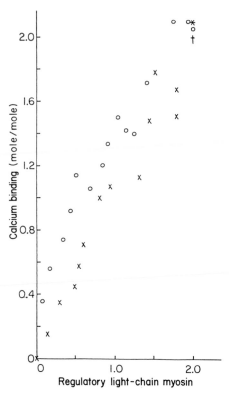

Fig. 18. Restoration of calcium binding upon readdition of regulatory light-chains to 30°C-desensitized *Placopecten* (x) or 35°C-desensitized *Aequipecten* (o) myofibrils. Calcium binding was measured by the double isotope labeling method at a free calcium concentration of 2.1 × 10⁻⁶M in the presence of 1 mM $MgCl_2$ (pH 7.0; 0°C). *,†, values for intact *Placopecten* and *Aequipecten* myofibrils, respectively. (Figure reproduced, with permission of the publisher, from Chantler and Szent-Gyorgyi, 1980.)

other hand, bind regulatory light chains with a single high affinity (Stafford et al., 1979).

The finding that single-headed myosin molecules retain their calcium sensitivity (Stafford et al., 1979) pointed to the neck region (S-2) as the critical region for cooperativity. Rotary-shadowed images of myosin molecules and S-1 subfragments (Flicker et al., 1981), as well as three-dimensional reconstructions of decorated scallop thin filaments (Vibert and Craig, 1982), confirm that light chains extend down to the S-2 region. If both regulatory and essential light chains are proximal in the S-2 region, and if the location of the calcium-specific site (a function of the regulatory light chain and the heavy chain of myosin, possibly also involving the essential light chain) is also here, then the beginnings of a structural basis for myosin-linked regulation become apparent: Six polypeptide chains, three from each half of the myosin, will be in close juxtaposition at the apex of the fork between the two myosin heads, the region where, from the above scenario, calcium exercises its profound effect. Preliminary evidence in favor of this picture comes from studies concerning the proteolytic digestion of scallop myosin subfragments (Szentkiralyi, 1982). These studies suggest the formation of a complex containing both types of light-chain together with a ~14-kdalton fragment from the heavy chain, the exact origin of which is not known. Regulatory light chains dissociate from this complex in the absence of divalent cations, indicating that the nonspecific metal ion binding site is located on this fragment; whether or not the calcium-specific site is located here as well is not known at present.

Although all the above evidence is consistent with cooperativity between the two heads of myosin, necessary to maintain the off state, there is an alternative explanation. Removal of a regulatory light chain could leave a "sticky patch" on the residual heavy chain, which binds to and fouls up the regulatory functioning

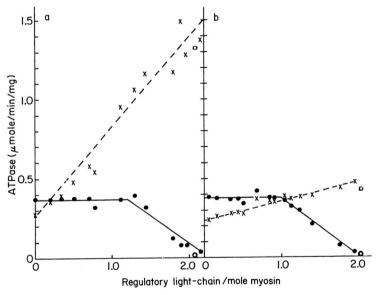

Fig. 19. Actin-activated Mg^{2+}-ATPase ($\pm Ca^{2+}$) as a function of regulatory light-chain content for (**a**) *Placopecten* and (**b**) *Aequipecten* myofibrils. Light-chain/myosin ratios were obtained by densitometry and planimetry of fast green–stained, urea-containing gels. ● in the presence of 0.1 mM EGTA; x, in the presence of 0.1 mM EGTA, 0.2 mM Ca^{2+}. ○ □, values for intact myofibrils. (Figure reproduced, with permission of the publisher, from Chantler and Szent-Gyorgyi, 1980.)

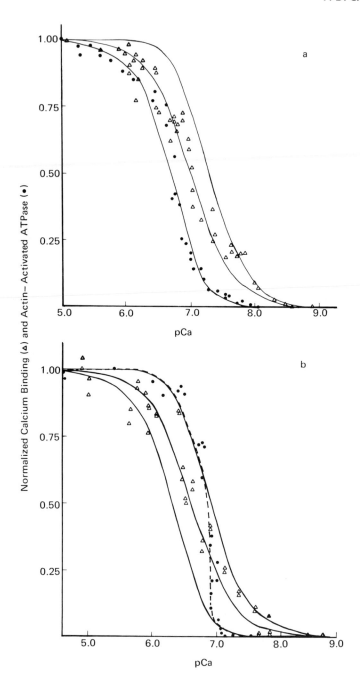

of the second head. This explanation was first put forward by Kendrick-Jones and Jakes (1977) and has since been elaborated further by Bagshaw (1980). Although this *clumping* model can, in principle, explain the above observations as manifestations of aberrant myosins (desensitized or proteolyzed preparations), it predicts that such cooperativity will not occur on intact myosin. On the other hand, the model of Chantler and Szent-Gyorgyi (1980) suggests that two regulatory light chains are necessary for regulation in intact myosin and therefore predicts that events at the two calcium-specific sites must somehow reflect this cooperativity. In particular, there must be a direct relationship in a regulatory myosin between calcium binding and ensuing actin-activated ATPase activity.

This prediction was tested directly by performing calcium-binding and actin-activated ATPase measurements under nearly identical conditions as a function of the free calcium ion concentration (Chantler et al., 1981b). Calcium binding is not cooperative and could be fitted by two identical stoichiometric constants of about $10^7/M$ (pH 7.5). One can envisage, in the extreme cases, three separate models relating calcium binding to ATPase activity, two depending on whether one or two calcium ions are required to bind per molecule in order to switch that molecule on and one in which the heads act independently, calcium binding by one head switching only that particular head on. These predictive models for ATPase activity are illustrated by the solid lines in Fig. 20, with their exact position being determined by the nature of the calcium-binding curve. In the case of HMM, the actin-activated ATPase closely follows the predicted path for the model that requires two calcium ions to bind per molecule of HMM in order to switch that molecule on (Fig. 20a). Because HMM retains both of its regulatory light chains, and because the two calcium-specific binding sites are located on separate heads, these results directly support a model for cooperativity between the two halves of the bipartite myosin molecule (Chantler and Szent-Gyorgyi, 1980; Chantler et al., 1981b) and do not support a clumping hypothesis (Bagshaw, 1980; Kendrick-Jones and Jakes, 1977).

Fig. 20. Calcium dependence of actin-activated ATPase activity (●) and calcium binding (△) for (a) *Aequipecten* HMM and (b) myosin. The solid lines represent model-dependent predictions for the actin-activated ATPase activity derived from the binding data. Upper curve: model in which one calcium ion bound per myosin molecule is sufficient for activation. Middle curve: model in which one calcium ion bound per molecule activates only the head to which it is bound. This line also represents the best fit to the calcium-binding data. Lower curve: model in which two calcium ions must bind per myosin molecule in order to switch on that molecule. Note the transition between models in the case of the scallop myosin ATPase data. Calcium-binding and actin-activated ATPase measured under similar conditions. For HMM: 3.0 mM Mg^{2+}, 2.0 mM ATP, 0.1 mM EGTA, 30 mM NaCl, pH 7.5, 25°C. Best fit to the calcium-binding data obtained using stoichiometric constants $K_1 = K_2 = 10^7/M$. For myosin: 3.0 mM Mg^{2+}, 2.0 mM ATP, 0.1 mM EGTA, 20 mM NaCl, pH 7.5, 25°C. Best fit to calcium-binding data obtained using $K_1 = K_2 = 4 \times 10^6/M$. (Figures adapted, with permission of the publisher, from Chantler et al., 1981b.)

When such an experiment is performed with actomyosin or myofibrils, an interesting phenomenon occurs. Although the ATPase activity initially (i.e., at low free calcium concentrations) follows the two-site model, the rate takes off explosively at higher free calcium concentrations and, crossing the binding curve, finally follows the model in which only one calcium ion is required to switch that molecule on (Fig. 20b). Tension generation by fiber bundles follows the same curve (Chantler et al., 1981b). The fundamental difference in results between Figs. 20a and 20b is that HMM is soluble under the conditions employed (low ionic strength), whereas myosin is not and exists as thick filaments. It was proposed, therefore (Chantler et al., 1981b), that the results in Fig. 20b demonstrate intermolecular cooperativity within the thick filament as well as intramolecular cooperativity between the myosin heads. Initially, two calcium ions must bind to a myosin molecule within the filament in order to switch that molecule on (as in the case of HMM); however, at $\leq 50\%$ site occupancy of the thick filament, some sort of phase transition is effected throughout each filament which enables remaining myosin molecules in the filament to be switched on by binding only one calcium ion. The result is a mode of regulation that exhibits a hypersharp calcium switch over a very small range of free calcium ion concentrations ($\cong 0.5$ pCa units).

The intra- and intermolecular cooperativity manifested by scallop actomyosin, myofibrils, and fiber bundles is entirely a result of the intrinsic properties of the scallop myosin molecule and probably occurs in all myosin-regulated molluscan muscles. Neither tropomyosin nor paramyosin is required for this cooperativity. It is possible to distinguish these cooperative phenomena from those exhibited by vertebrate skeletal muscle proteins in which tropomyosin is an absolute requirement and troponin is the calcium sensor. In fully regulated vertebrate preparations, soluble subfragments still show hypersharp activity curves as a function of free calcium concentration (Murray and Weber, 1981). In any case, vertebrate myosins are not regulatory myosins and do not possess calcium-specific binding sites (Bagshaw and Kendrick-Jones, 1979; Sellers et al., 1980). It is possible, therefore, that cooperative aspects of the calcium response are restricted to thick filaments in myosin-regulated muscles and to thin filaments during thin filament regulation.

Finally, in this section on cooperativity, it should be emphasized that the cooperative phenomena under discussion pertain to the *regulation* of the actomyosin ATPase mechanism. Although much energy has been expended, especially with vertebrate preparations, in trying to demonstrate head–head cooperativity during ATP hydrolysis, such efforts have usually yielded negative results (Reisler, 1980; Taylor, 1979). Preliminary studies with scallop myosin suggest that nucleotide binds to the two heads in a noncooperative manner (Chantler et al., 1981a). Hence, head–head cooperativity may be limited to the control functions of regulatory myosins; in the fully active molecule, the heads could be essentially independent.

C. Catch

Molluscan paramyosin smooth muscles are capable of sustaining large tensions for long periods of time with little expenditure of energy. Such muscles, during tonic contraction, are said to exhibit catch, which persists long after the active state has ceased. Catch muscles can be manipulated by pharmacological agents: Serotonin causes relaxation of catch and acetylcholine initiates catch contraction; when serotonin and acetylcholine are present simultaneously, phasic contraction results (Twarog, 1979). The physiological and pharmacological details of catch are discussed in detail in Chapter 2, this volume. The intention here is simply to discuss the possible biochemical mechanisms of catch.

Most proposed mechanisms of catch have tried to take into account its paramyosin-rich character. Some early theories viewed catch as being entirely a manifestation of changes within the paramyosin core that are completely dissociated, in a functional sense, from actomyosin. Such theories (Johnson et al., 1959; Ruegg, 1961, 1971) viewed catch as resulting from paramyosin crystallization or from fusion between adjacent paramyosin filaments or subfilaments (between different thick filaments or within the same filament, respectively); these theories now appear to be structurally and biochemically unsound. On the other hand, it is certain that catch cannot be explained solely in terms of catch muscle actomyosin ATPase kinetics, a proposition discussed earlier by others (Lowy and Millman, 1963; Lowy et al., 1964). One reason (among many) is that actin-activated rates of scallop catch adductor actomyosin are only a factor of two slower than equivalent rates in striated adductor actomyosin. Instead, it seems likely that the catch state involves formation of actomyosin links which are somehow preserved, in the presence of nucleotide, because of some structural change in the paramyosin core. This mechanism, involving at least two forms of paramyosin–myosin interaction, was first proposed by Szent-Gyorgyi and Cohen (Cohen and Szent-Gyorgyi, 1971; Szent-Gyorgyi et al., 1971) and has been refined recently by the suggestion of Cohen (1982) that the distinction between catch and noncatch thick filaments has a simple structural basis. Cohen has proposed that all noncatch muscles have thick filaments within which myosin molecules are arranged as subfilaments, whereas catch muscles possess a monomolecular layer of myosin. Implicit in this suggestion is the concept that thick filaments are made up of subfilaments in most muscles; there is some experimental evidence for this (Maw and Rowe, 1980; Tien-chin et al., 1966; Wray, 1979, 1981). Thus, myosin–myosin interactions are predominant in noncatch muscles, whereas only myosin–paramyosin interactions are possible for myosin molecules in catch muscles. The details of this hypothesis still await direct proof.

The fact that paramyosin aggregates can undergo a phase transition over a markedly small pH range (Johnson et al., 1959), together with observations that paramyosin is capable of *in vivo* and *in vitro* phosphorylation (Achazi, 1979; Cooley et al., 1979), suggest a means for the control of catch. When paramyosin is isolated from untreated or serotonin-treated *M. edulis* ABRM muscles and assayed for its ability to act as a substrate for cytoplasmic or catch muscle–associated protein kinases, it is found that paramyosin extracted from serotonin-treated muscles incorporates two to four times as much covalently bound phosphate as paramyosin from untreated (or acetylcholine-treated) muscle (Achazi, 1979). This observation is consistent with the concept of a greater degree of paramyosin phosphorylation within the muscle during catch, for new phosphorylation can occur only at a vacant site. Although these measurements do not provide evidence regarding the maximal amount of phosphate incorporated per paramyosin molecule *in vivo,* the amount of phosphorylation, as gauged by the above experimental techniques, changed from 0.1 (untreated or acetylcholine-treated) to 0.4 (serotonin-treated) moles of phosphate incorporated per mole of paramyosin in some experiments. The hypothesis of Cohen (1982) that catch muscles possess a monomolecular surface layer of myosin molecules suggests that only phosphorylation of paramyosin molecules on the surface of the core need be important for determining catch. Taking a range of diameters and paramyosin/myosin molar ratios of catch muscles from the literature, it turns out that one would expect a change of only 0.1–0.3 moles of phosphate per mole of paramyosin between relaxation and maximal onset of catch, provided that only the surface layer of paramyosin core is phosphorylated during catch. Even less would be required if the mechanism for the core transition were cooperative. The data of Achazi (1979) are therefore consistent with the hypothesis that phosphorylation of paramyosin can initiate catch. The preference shown by myosin and paramyosin to aggregate with a 14.3-nm repeat, the paracrystalline nature of the paramyosin core and its extreme solubility properties, the potential for electrostatic interactions inherent in phosphate groups and the known importance of the neck region of myosin in modulating the activity of the heads all suggest the machinery necessary to connect cause and action. Unfortunately, confirmation of this hypothesis by direct determination of paramyosin-bound phosphate from relaxed and catch-state muscles has so far proved unsuccessful (Achazi, 1979). The derth of this vital information forces one to conclude that the above scheme is unproven; one must continue to consider alternatives to Achazi's model. The biochemistry of catch is certainly not restrictive at the present time.

A number of observations suggest that relaxation of the catch state by serotonin involves a second messenger, cAMP (Achazi et al., 1974; Cole and Twarog, 1972). One possibility, consistent with the above model, might be that the mechanism of catch control has analogies with the control of smooth muscle. In smooth muscle, myosin becomes activated by light-chain phosphorylation; however, the light-chain kinase is also capable of phosphorylation by a cAMP-dependent protein kinase; phosphorylation of the light-chain kinase decreases its affinity for calmodulin, resulting in smooth muscle relaxation (Adelstein and Eisenberg, 1980). Serotonin-induced relaxation of catch muscles may therefore operate by activation of a cAMP-dependent kinase that phosphorylates para-myosin kinase, decreasing its activity. The situation may be more complicated, however: Not only were the results of Achazi's experiments affected only mini-mally by addition of cAMP, but a recent characterization of serotonin-activated accumulation of cAMP in the ABRM muscle of *M. edulis* (Kohler and Lindl, 1980) indicated a 30-fold discrepancy in the dose levels required for half-maxi-mal stimulation of relaxation by serotonin as opposed to cAMP. Clearly, further work is required.

VII. Conclusions and Perspectives

Our understanding of the structure and biochemistry of molluscan muscle has increased dramatically during the past two decades. Despite the variety of muscle types and patterns of organization in molluscs, which at times seem to defy well-used epithets such as *smooth* or *striated,* it is clear that the sliding filament mechanism underlies and constrains all molluscan muscle structures. As more and more muscles are subjected to detailed morphological analysis, it is possible that distinctions between molluscan smooth and striated muscles will become blurred even further.

Considerable progress has been made in elucidating the structures of the two main components of the sliding filament mechanism, the thick and thin fila-ments. One is struck by the similarities between these structures to their counter-parts in other muscles rather than by the detailed differences that distinguish, say, a scallop adductor thick filament from a thick filament from chicken pec-toralis muscle. In the absence of a functional test, only refined structural analysis can distinguish vertebrate thin filaments from their molluscan counterparts. Thick filament structures vary to a greater extent, but even here the similarities transcend the differences.

These structural similarities arise from the use of a limited number of highly conserved, self-assembling, protein-building blocks. Rabbit skeletal muscle ac-tin shows fewer than 30 amino acid substitutions from calf thymus or bovine brain actins in its long sequence of 374 amino acids (Vandekerckhove and Weber, 1978), a remarkable degree of conservation. Although comparative se-quence data of myosins and paramyosins are at a much less advanced stage than those of actin, it is clear that self-assembly of these proteins produces remarkably similar structures, whatever the specific source of the protein.

Against this backdrop of familiar proteins and structures, biochemical dif-ferences peculiar to molluscan muscles stand out clearly. Regulation of contrac-tion by direct calcium binding to myosin is a mode of regulation which, if not unique to molluscs, is certainly restrictive in its distribution. The direct demon-stration of cooperativity between the two heads of scallop myosin during regulation (Chantler et al., 1981b) opens the door to a study of the molecular details of myosin action not possible with myosins from many other species. It has a predictive value as well: Probably all regulatory myosins operate through such a cooperative mechanism, irrespective of whether light-chain phosphorylation or calcium binding switches on activity. Preliminary data suggest that smooth muscle myosin, whose activity is controlled by light-chain phosphorylation, is also controlled in a cooperative manner (Ikebe et al., 1982; Persechini and Hartshorne, 1981; Sellers and Adelstein, 1982); both heads of myosin need to be phosphorylated before actin activation ensues. Whether or not molluscan muscles contain functional troponin *in vivo* is, from the foregoing, a somewhat controversial issue. If the proponents of thin filament regulation in molluscs are to win any new supporters, they must demonstrate the existence of a *functional whole* troponin complex.

One area of molluscan muscle biochemistry likely to yield interesting results in the near future is the *in vitro* kinetic mechanism of the actin-activated ATPase. Because *regulation* of activity, as well as the active sites themselves, is a property of the myosin molecule, the subject should, in principle, be simpler to study than its vertebrate counterpart—in which tropomyosin movements and thin filament cooperativity cloud the issue. In any case, building upon the vast edifice of knowledge gained from studies on the vertebrate actomyosin ATPase mechanism should make the task easier. A future problem concerns the *in vivo* mechanism of tension generation and contraction. Ironically, there is every reason to expect that this problem may be solved first in molluscan muscle. The ability reversibly to remove regulatory light chains from scallop myosin constrained in fiber bundles (Simmons and Szent-Gyorgyi, 1978), together with the ability of fluorescently labeled regulatory light-chains to restore fully *in vitro* calcium sensitivity to desensitized preparations (Chantler and Szent-Gyorgyi, 1978), make the scallop fibers ideal subjects for future mechanochemical coupling experiments. The inability of current probes on regulatory light chains to monitor changes during ATPase activity will, hopefully, be a temporary technical problem.

Similar approaches using modified regulatory light chains, or alternative methods such as the recent successful use of caged ATP (McCray et al., 1980; Goldman et al., 1982), may well aid future studies on the catch mechanism. Although likely, the concept that catch is due to maintenance of an associated actomyosin state, long after the active state has passed, remains to be tested directly. Fluorescent derivatives of caged ATP as probes of intermediates in the ATPase cycle may well prove useful. Direct phosphate analysis is also required if one is to substantiate the attractive hypothesis that the onset of catch is controlled by paramyosin phosphorylation modulated in some way by cAMP (Achazi, 1979).

In Section I, I apologized for the fact that certain aspects of molluscan muscle biochemistry had not been studied directly and that great reliance had to be placed on analogies with studies in vertebrates. It should be apparent that this is a two-way street, however. In certain areas—the biochemistry of regulation, for example—our knowledge is most detailed in molluscan muscle. I think the potential is present for scallop adductor muscles to be the most comprehensively understood muscles, at the molecular level, before this decade is through.

Acknowledgments

I wish to thank Drs. Gaku Ashiba, Carolyn Cohen, Andrew Szent-Gyorgyi, Betty Twarog, and Peter Vibert, and Ms. Meg Titus for critically reading an earlier draft of this manuscript. I thank Ms. Noreen Francis and Ms. Guillermina Waller for help with some supplemental experiments. I am indebted to Drs. Pauline Bennett, Carolyn Cohen, Roger Craig, Evan Eisenberg, Arthur Elliott, Paula Flicker, Hugh Huxley, Jack Lowy, Barry Millman, Marcus Schaub, Apollinary Sobieszek, and Peter Vibert for supplying me with photographs and allowing me to use their data.

References

Achazi, R. K. (1979). Phosphorylation of molluscan paramyosin. *Pfluegers Arch.* **379,** 197–201.

Achazi, R. K., Dolling, B., and Haakshorst, R. (1974). 5-HT-induzierte Erschlaffung und cyclisches AMP bei einem glatten Molluskenmuskel. *Pfluegers Arch.* **349,** 19–27.

Adelstein, R. S., and Conti, M. A. (1975). Phosphorylation of platelet myosin increases actin-activated myosin ATPase activity. *Nature* **256,** 597–598.

Adelstein, R. S., and Eisenberg, E. (1980). Regulation and kinetics of the actin - myosin - ATP interaction. *Ann. Rev. Biochem.* **49,** 921–956.

Alexis, M. N., and Gratzer, W. B. (1978). Interaction of skeletal myosin light-chains with calcium ions. *Biochemistry* **17,** 2319–2325.

Asakawa, T., Yazawa, Y., and Azuma, N. (1981). Light-chains of abalone myosin. UV - absorption difference spectrum and resensitization of desensitized scallop myosin. *J. Biochem.* **89,** 1805–1814.

Ashiba, G., Asada, T., and Watanbe, S. (1980). Calcium regulation in clam foot muscle. *J. Biochem.* **80,** 837–844.

Azuma, N., Asakura, A., and Yagi, K. (1975). Myosin form molluscan abalone, *Haliotis discus.* Isolation and enzymatic properties. *J. Biochem.* **77,** 973–981.

Bagshaw, C. R. (1977). On the location of the divalent metal binding sites and the light-chain subunits of vertebrate myosin. *Biochemistry* **16,** 59–67.

Bagshaw, C. R. (1980). Divalent metal ion binding and subunit interactions in myosins: A critical review. *J. Muscle Res. Cell Motil.* **1,** 255–277.

Bagshaw, C. R., and Kendrick-Jones, J. (1979). Characterization of homologous divalent metal ion binding sites of Vertebrate and Molluscan myosins using electron paramagnetic resonance spectroscopy. *J. Mol. Biol.* **130,** 317–336.

Bagshaw, C. R., and Kendrick-Jones, J. (1980). Identification of the divalent metal ion binding domain of myosin regulatory light-chains using spin-labelling techniques. *J. Mol. Biol.* **140,** 411–433.

Bagshaw, C. R., and Trentham, D. R. (1973). The reversibility of adenosine triphosphate cleavage by myosin. *Biochem. J.* **133,** 323–328.

Bagshaw, C. R., and Trentham, D. R. (1974). The characterization of myosin-product complexes and of product release steps during the magnesium-ion dependent adenosine phosphatase reaction. *Biochem. J.* **141**, 331–348.

Bailey, K. (1948). Tropomyosin, a new asymmetric protein of the muscle fibril. *Biochem. J.* **43**, 271–278.

Ballowitz, E. (1892). Ueber den feineren Bau der Muskelsubstanzen 1). Muskelfaser der Cephalopoden. *Arch. Mikr. Anat.* **29**, 291–324.

Barany, M., and Barany, K. (1966). Myosin from the striated adductor of scallop (*Pecten arradians*). *Biochemische Z.* **345**, 37–56.

Barany, M., Finkelman, F., and Theratill-Antony, T. (1962). Studies on the bound calcium of actin. *Arch. Biochem. Biophys.* **93**, 28–45.

Bayliss, W. M. (1924), "Principles of general physiology", 4th edition. Longmans, Green and Co., London.

Bear, R. S. (1944). X-ray diffraction on protein fibres. Feather rachis, porcupine quill tip and clam muscle, Vol. 2. *J. Amer. Chem. Soc.* **66**, 2043–2050.

Bear, R. S. (1945). Small angle x-ray diffraction studies on muscle. *J. Am. Chem. Soc.* **67**, 1625–1626.

Bear, R. S., and Selby, C. C. (1956). The structure of paramyosin fibrils according to x-ray diffraction. *J. Biophys. Biochem. Cytol.* **2**, 55–69.

Bender, N., Fasold, H., and Rack, M. (1974). Interaction of rabbit muscle actin and chemically modified actin with ATP, ADP and protein. *FEBS. Lett.* **44**, 209–212.

Bennett, P. M., and Elliott, A. (1981). The structure of the paramyosin core in molluscan thick filaments. *J. Muscle Res. Cell Motil.* **2**, 65–81.

Bremel, R. D., and Weber, A. (1972). Cooperation within actin filament in vertebrate skeletal muscle. *Nature (London) New Biol.* **238**, 97–101.

Caspar, D. L. D., Cohen, C., and Longley, W. (1969). Tropomyosin: Crystal structure, poymorphism and molecular interactions. *J. Mol. Biol.* **41**, 87–107.

Chacko, S., Conti, M. A., and Adelstein, R. S. (1977). Effect of phosphorylation of smooth muscle myosin on actin activation and calcium regulation. *Proc. Nat. Acad. Sci.* **74**, 129–133.

Chalovich, J. M., and Eisenberg, E. (1982). Inhibition of actomyosin ATPase activity by troponin–tropomyosin without blocking the binding of myosin to actin. *J. Biol. Chem.* **257**, 2432–2437.

Chalovich, J. M., Chock, P. B., and Eisenberg, E. (1981). Mechanism of action of troponin. tropomyosin. *J. Biol. Chem.* **256**, 575–578.

Chantler, P. D., and Gratzer, W. B. (1976). The interaction of actin monomers with myosin heads and other muscle proteins. *Biochemistry* **15**, 2219–2225.

Chantler, P. D., and Szent-Gyorgyi, A. G. (1978). Spectroscopic studies on invertebrate myosins and light-chains. *Biochemistry* **17**, 5440–5448.

Chantler, P. D., and Szent-Gyorgyi, A. G. (1980). Regulatory light-chains and scallop myosin: Full dissociation, reversibility and cooperative effects. *J. Mol. Biol.* **138**, 473–492.

Chantler, P. D., Marsh, D. J., and Martin, S. R. (1981a). The myosin ATPase mechanism does not require a conformationally sensitive aromatic residue. *J. Muscle Res. Cell Motil.* **2**, 453–466.

Chantler, P. D., Sellers, J. R., and Szent-Gyorgyi, A. G. (1981b). Cooperativity in scallop myosin. *Biochemistry* **20**, 210–216.

Cheung, W. Y. (1980). Calmodulin plays a pivotal role in cellular regulation. *Science* **207**, 19–27.

Chock, S. P., Chock, P. B., and Eisenberg, E. (1979). The mechanism of the skeletal muscle myosin ATPase. *J. Biol. Chem.* **254**, 3236–3243.

Chou, P. Y., and Fasman, G. D. (1977). Prediction of the secondary structure of proteins from their amino acid sequence. *Adv. Enzymol.* **47**, 45–148.

Cohen, C. (1982). Matching molecules in the catch mechanism. *Proc. Natl. Acad. Sci. U.S.A.* **79,** 3176–3178.

Cohen, I., and Cohen, C. (1972). A tropomyosin-like protein from human platelets. *J. Mol. Biol.* **68,** 383–387.

Cohen, C., and Longley, W. (1966). Tropomyosin paracrystals formed by divalent cations. *Science* **152,** 794–796.

Cohen, C., and Szent-Gyorgyi, A. G. (1971). Assembly of myosin filaments and the structure of molluscan catch muscles. *In* "Contractility of Muscle Cells and Related Processes" (R. J. Podolsky, ed.), pp. 23–26. Prentice-Hall, Princeton, New Jersey.

Cohen, C, Szent-Gyorgyi, A. G., and Kendrick-Jones, J. (1971). Paramyosin and the filaments of molluscan 'Catch' muscle. *J. Mol. Biol.* **56,** 223–237.

Cole, R. A., and Twarog, B. M. (1972). Relaxation of catch in a molluscan smooth muscle. Effect of drugs which act on the adenyl cyclase system, Vol. 1. *Comp. Biochem. Physiol.* **A43,** 321–330.

Cooley, L. B., Johnson, W. H., and Krause, S. (1979). Phosphorylation of paramyosin and its possible role in the catch mechanism. *J. Biol. Chem.* **254,** 2195–2198.

Cote, G., Lewis, W. G., and Smillie, L. B. (1978). Non-polymerizability of platelet tropomyosin and its NH_2 and CO_2H terminal sequences. *FEBS. Lett.* **91,** 237–241.

Cox, R. N., and Kawai, M. (1981). Alternate energy transduction routes in chemically skinned rabbit psoas muscle fibres: A further study of the effect of MgATP over a wide concentration range. *J. Muscle Res. Cell Motil.* **2,** 203–214.

Craig, R., Szent-Gyorgyi, A. G., Beese, L., Flicker, P., Vibert, P., and Cohen, C. (1980). Electron microscopy of thin filaments decorated with a calcium-regulated myosin. *J. Mol. Biol.* **140,** 35–55.

Crick, F. H. C. (1953). The packing of α-Helices: Simple coiled-coils. *Acta Crystallogr.* **6,** 689–697.

Cummins, P., and Perry, S. V. (1973). The subunits and biological activity of polymorphic forms of tropomyosin. *Biochem. J.* **133,** 765–777.

deCouet, H. C., Mazander, K. D., and Groschel-Stewart, U. (1980). A study of invertebrate actins by isoelectric focusing and immunodiffusion. *Experientia* **36,** 404–405.

Dover, S. D., and Elliott, A. (1979). Three-dimensional reconstruction of a paramyosin filament. *J. Mol. Biol.* **132,** 340–341.

Dreizen, P., and Gershman, L. C. (1970). Relationship of structure to function in myosin. Salt denaturation and recombination experiments, Vol. 2. *Biochemistry* **9,** 1689–1693.

Ebashi, S., Kodama, A., and Ebashi, F. (1968). Troponin. *J. Biochem.* **64,** 465–477.

Ebashi, S., Endo, M., and Ohtsuki, I. (1969). Control of muscle contraction. *Q. Rev. Biophys.* **2,** 351–384.

Egelman, E. H., Francis, N., and DeRosier, D. J. (1982). F-actin is a helix with a random variable twist. *Nature (London)* **298,** 131–135.

Eisenberg, E., and Greene, L. E. (1980). The relation of muscle biochemistry to muscle physiology. *Ann. Rev. Physiol.* **42,** 293–309.

Eisenberg, E., and Hill, T. L. (1978). A cross-bridge model of muscle contraction. *Prog. Biophys. Molec. Biol.* **33,** 55–82.

Eisenberg, E., and Moos, C. (1968). The Adenosine triphosphatase activity of acto-heavy meromyosin. A kinetic analysis of actin activation. *Biochemistry* **7,** 1486–1489.

Eisenberg, E., and Moos, C. (1970). Actin activation of heavy meromyosin adenosine triphosphatase. *J. Biol. Chem.* **245,** 2451–2456.

Eisenberg, E., Dobkin, L., and Kielley, W. (1972). Binding of actin to heavy mero-myosin in the absence of ATP. *Biochemistry* **11,** 4657–4660.

Elfvin, M., Levine, R. J. C., and Dewey, M. M. (1976). Paramyosin in invertebrate muscles. Identification and localization, Vol. 1. *J. Cell Biol.* **71**, 261–272.

Elliott, A. (1974). The arrangement of myosin on the surface of paramyosin filaments in the white adductor of *Crassostrea angulata Proc. Soc. Ser. B* **186**, 53–66.

Elliott, A. (1979). Structure of molluscan thick filaments: A common origin for diverse appearance. *J. Mol. Biol.* **132**, 323–340.

Elzinga, M., Collins, J. H., Kuehl, W. H., and Adelstein, R. S. (1973). Complete amino acid seqence of actin of rabbit skeletal muscle. *Proc. Nat. Acad. Sci.* **70**, 2687–2691.

Epstein, H. F., Waterston, R. H., and Brenner, S. (1974). A mutant affecting the heavy chain of myosin in *Caenorhabditis elegans. J. Mol. Biol.* **90**, 291–300.

Epstein, H. F., Aronow, B. J., and Harris, H. E. Myosin-Paramyosin cofilaments: Enzymatic interactions with F-actin. *Proc. Nat. Acad. Sci.* **73**, 3015–3019.

Estes, E. J., and Gershman, L. C. (1978). Activation of heavy meromyosin adenosine triphosphatase by various states of Actin. *Biochemistry* **17**, 2495–2499.

Fine, R. E., and Blitz, A. L. (1975). A chemical comparison of tropomyosins from muscle and non-muscle cells. *J. Mol. Biol.* **95**, 447–454.

Flicker, P., Wallimann, T., and Vibert, P. (1981) Location of regulatory light-chains in scallop myosin. *Biophys. J.* **33**, 279a.

Focant, B., and Huriaux, F. (1976). Light chains of carp and pike skeletal muscle myosins. Isolation and characterization of the most anodic light chain on alkaline pH electrophoresis. *FEBS. Lett.* **65**, 16–19.

Frearson, N., Focant, B. W. W., and Perry, S. V. (1976). Phosphorylation of a light chain component of myosin from smooth muscle. *FEBS. Lett.* **63**, 27–32.

Garrels, J. I., and Gibson, W. (1976). Identification and characterization of multiple forms of actin. *Cell* **9**, 793–805.

Goldberg, A., and Lehman, W. (1978). Troponin-like proteins from muscles of the scallop, *Aequipecten irradians. Biochem. J.* **171**, 413–418.

Goldman, Y. E., Hibberd, M. G., McCray, J. A., and Trentham, D. R. (1982). Relaxation of muscle fibres by photolysis of caged ATP. *Nature (London)* **300**, 701–705.

Gordon, D. J., Boyer, J. L., and Korn, E. D. (1977). Comparative biochemistry of nonmuscle actins. *J. Biol. Chem.* **252**, 8300–8309.

Greene, L. E. (1981). Comparison of the binding of heavy meromyosin and myosin subfragment 1 to F-actin. *Biochemistry* **20**, 2120–2126.

Greene, L. E., and Eisenberg, E. (1980). The binding of heavy meromyosin to F-actin. *J. Biol. Chem.* **255**, 549–554.

Hall, C. E., Jakus, M. A., and Schmitt, F. O. (1945). The structure of certain muscle fibrils as revealed by the use of electron stains. *J. Appl. Phys.* **16**, 459–465.

Halsey, J. F., and Harrington, W. F. (1973). Substructure of paramyosin. Correlation of helix stability, trypsin digestion kinetics and amino acid composition. *Biochemistry* **12**, 693–701.

Hanson, J. (1957). The structure of the smooth muscle fibres in the body wall of the earthworm. *J. Biophys. Biochem. Cytol.* **3**, 111–122.

Hanson, J., and Lowy, J. (1957). Structure of Smooth Muscle. *Nature* **180**, 906–909.

Hanson, J., and Lowy, J. (1959). Evidence for a sliding filament mechanism in Tonic smooth muscles of lamellibranch molluscs. *Nature* **184**, 286–287.

Hanson, J., and Lowy, J. (1960). Structure and function of the contractile apparatus in the muscles of invertebrate animals. *In* "Structure and Function of Muscle" (G. H. Bourne, ed.), pp. 265–335. Academic Press, New York.

Hanson, J., and Lowy, J. (1961). The structure of muscle fibres in the translucent part of the oyster, *Crassostrea angulata. Proc. R. Soc. Ser. B* **154**, 173–196.

Hanson, J., and Lowy, J. (1963). The structure of F-actin and actin filaments isolated from muscle. *J. Mol. Biol.* **6**, 46–60.

Hanson, J., and Lowy, J. (1964). The structure of molluscan tonic muscles. *In* "The Biochemistry of Muscle Contraction" (J. Gergely, ed.), pp. 400–411. Little, Brown and Co., Boston, Massachusetts.

Hardwicke, P. M. D., and Hanson, J. (1971). Separation of thick and thin myofilaments. *J. Mol. Biol.* **59**, 509–516.

Harrington, W. F. (1979). Proteins of contractile systems. *In* "The Proteins" (H. Neurath, and R. Hill, eds.), pp. 245–409. Academic Press, New York.

Harris, H. E., and Epstein, H. F. (1977). Myosin and paramyosin of *Caenorhabditis elegans:* Biochemical and structural properties of wild type and mutant proteins. *Cell* **10**, 709–719.

Harrison, R. G., Lowey, S., and Cohen, C. (1971). Assembly of myosin. *J. Mol. Biol.* **59**, 531–535.

Hartt, J. E., and Mendelson, R. A. (1979). X-ray scattering of myosin light-chains in solution. *Biophys. J.* **25**, 71a.

Hayashi, Y. and Tonomura, Y. (1970). On the active site of myosin A - adenosine triphosphatase. *J. Biochem.* **68**, 665–680.

Heumann, H. G., and Zebe, E. (1968). Uber die Funktionsweise glatter Muskelfasern. Elektron-mikroskopische Untersuchungen am Byssus retraktor (ABRM) von *Mytilus edulis. Z. Zellforsch.* **85**, 534–551.

Highsmith, S. (1978). Heavy meromyosin binds actin with negative cooperativity. *Biochemistry* **17**, 22–26.

Hill, T. L. (1974). Theoretical formalism for the sliding filament model of contraction of striated muscle, Part 1. *Prog. Biophys. Mol. Biol.* **28**, 267–340.

Hirumi, H., Ruski, D. J., and Jones, N. O. (1971). Primitive muscle cells of nematodes: Morphological aspects of platymyarian and shallow coelomyarian muscles in two plant parasitic nematodes *Trichodorus christiei* and *Longidorus elongatus. J. Ultrastruct. Res.* **34**, 517–543.

Hodge, A. J. (1952). A new type of periodic structure obtained by reconstitution of paramyosin from acid solutions. *Proc. Nat. Acad. Sci.* **38**, 850–855.

Hoyle, G. (1957), "Comparative physiology of the nervous control of muscular contraction." Cambridge University Press, Cambridge, England.

Huxley, A. F. (1957). Muscle structure and theories of contraction. *Prog. Biophys. Biophys. Chem.* **7**, 255–318.

Huxley, A. F., and Niedergerke, R. (1954). Structural changes in muscle during contraction. *Nature* **173**, 971–973.

Huxley, A. F., and Simmons, R. M. (1971). Proposed mechanism of force generation in striated muscle. *Nature (London)* **233**, 533–538.

Huxley, H. E. (1966). The fine structure of striated muscle and its functional significance. *In* "The Harvey Lectures", Series 60, pp. 85–118. Academic Press, New York.

Huxley, H. E. (1969). The mechanism of muscle contraction. *Science* **164**, 1356–1366.

Huxley, H. E. (1972). Molecular basis of contraction in cross-striated muscle. *In* "The Structure and Function of Muscle" (G. H. Bourne, ed.), pp. 301–387. Academic Press, New York.

Huxley, H. E. (1973). Structural changes in the actin and myosin-containing filaments during contraction. *Cold Spring Harbor Symp. Quant. Biol.* **37**, 361–376.

Huxley, H. E., and Brown, W. (1967). The low angle x-ray diagram of vertebrate striated muscle and its behavior during contraction and rigor. *J. Mol. Biol.* **30**, 383–434.

Huxley, H. E., and Hanson, J. (1954). Changes in the cross-striation of muscle during contraction and stretch and their structural interpretation. *Nature* **173**, 973–976.

Huxley, H. E., and Hanson, J. (1960). The molecular basis of contraction in cross-striated muscles.

In "Structure and Function of Muscle" (G. H. Bourne, ed.), pp. 205–227. Academic Press, New York.

Ikebe, M., Ogihara, S., and Tonomura, Y. (1982). Non-linear dependence of actin-activated Mg^{++}-ATPase activity on the extent of phosphorylation of gizzard myosin and HMM. *J. Biochem.* **91**, 1809–1812.

Ikemoto, N., and Kawaguti, S. (1967). Elongating effect of tropomyosin A on the thick myofilaments in the long-sracomere muscle of the horse-shoe crab. *Proc. Jpn. Acad.* **43**, 974–979.

Jakes, R., Northrop, F., and Kendrick-Jones, J. (1976). Calcium binding regions of myosin regulatory light-chains. *FEBS. Lett.* **70**, 229–234.

Johnson, W. H., Kahn, J. S., and Szent-Gyorgyi, A. G. (1959). Paramyosin and contraction of 'Catch Muscles'. *Science* **130**, 160–161.

Kahn, J. S., and Johnson, W. H. (1960). The localization of myosin and paramyosin in the myofilaments of the byssus retractor muscle of *Mytilus edulis*. *Arch. Biochem. Biophys.* **86**, 138–143.

Kanazawa, T., and Tonomura, Y. (1965). The pre-steady state of the myosin adenosine triphosphate system. *J. Biochem.* **57**, 604–615.

Kawaguti, S. (1962). Arrangement of myofilaments in the oblique striated muscles. *In* "Proc. 5th Intern. Congress for Electron Micros. Mll." Academic Press, New York.

Kendrick-Jones, J. (1976). *In* "Myosin-Linked Calcium Regulation", (Heilmeyer et al., eds.), pp. 122–136. Mosbach Colloquiam, Springer-Verlag. Germany.

Kendrick-Jones, J., and Jakes, R. (1977). Myosin-linked Regulation - a chemical approach. *In* "International Symposium on Myocardial Failure" (Rieker *et al.*, ed.), pp. 28–40. Tergensee, Munich, W. Germany.

Kendrick-Jones, J., Cohen, C., Szent-Gyorgyi, A. G., and Longley, W. (1969). Paramyosin: Molecular length and assembly. *Science* **163**, 1196–1198.

Kendrick-Jones, J., Lehman, W., and Szent-Gyorgyi, A. G. (1970). Regulation in molluscan muscles. *J. Mol. Biol.* **54**, 313–326.

Kendrick-Jones, J., Szentkiralyi, E. M., and Szent-Gyorgyi, A. G. (1976). Regulatory light-chains in myosins. *J. Mol. Biol.* **104**, 747–775.

Kensler, R. W., and Levine, R. J. C. (1982). An electron microscopic and optical diffraction analysis of the structure of *Limulus* telson muscle thick filaments. *J. Cell. Biol.* **92**, 443–451.

Kerrick, W. G., and Bolles, L. L. (1982). Evidence that myosin light-chain phosphorylation regulates contraction in the body wall muscles of the sea cucumber. *J. Cell. Physiol.* **112**, 307–315.

Kerrick, W. G., Hoar, P. E., Cassidy, P. S., Bolles, L., and Malencik, D. A. (1981). Calcium regulatory mechanisms. Functional classification using skinned fibres. *J. Gen. Physiol.* **77**, 177–190.

Klee, C. B., Crouch, T. H., and Richman, P. G. (1980). Calmodulin. *Ann. Rev. Biochem.* **49**, 489–515.

Knox, M. K., Szent-Gyorgyi, A. G., Trueblood, C. E., Weber, A., and Zigmond, S. (1983). The effect of low ATP concentration on relaxation in the myosin-regulated myofibrils from *Aequipecten*. Submitted for publication.

Kohler, G., and Lindl, T. (1980). Effect of 5-hydroxytrptamine, dopamine, and acetylcholine on accumulation of cyclic AMP and cyclic GMP in the anterior byssus retractor muscle of *Mytilus edulis*. *Pfluegers. Arch.* **383**, 257–262.

Kondo, S., and Morita, F. (1981). Smooth muscle of scallop adductor contains at least two kinds of myosin. *J. Biochem.* **90**, 673–681.

Kondo, S., Asakawa, T., and Morita, F. (1979). Difference uv-absorption spectrum of scallop adductor myosin induced by ATP. *J. Biochem.* **86**, 1567–1571.

Konno, K., Arai, K., Yoshida, M., and Watanabe, S. (1981). Calcium regulation in squid mantle and scallop adductor muscles. *J. Biochem.* **89**, 581–589.

Krijgsman, B. J., and Divaris, G. A. (1955). Contractile and pacemaker mechanisms of the heart of molluscs. *Biol. Rev. Cambridge Philos. Soc.* **30**, 1–39.

Kuczmarski, E. R., and Spudich, J. A. (1980). Regulation of myosin self-assembly: Phosphorylation of *Dictyostelium* heavy chain inhibits formation of thick filaments. *Proc. Nat. Acad. Sci.* **77**, 7292–7296.

Kuwayama, H., and Yagi, K. (1977). Separation of low molecular weight components of pig cardiac myosin and myosin subfragment-1, and calcium binding to one of the components (g_2). *J. Biochem.* **82**, 25–33.

Lazarides, E. (1980). Intermediate filaments as mechanical integrators of cellular space. *Nature* **283**, 249–256.

Leger, J. J., and Marotte, F. (1975). The effects of concentrated salt solutions on the structure and enzymatic activity of myosin molecules from skeletal and cardiac muscles. *FEBS. Lett.* **52**, 17–21.

Lehman, W. (1981). Thin-filament-linked regulation in molluscan muscles. *Biochim. Biophys. Acta.* **668**, 349–356.

Lehman, W., and Ferrell, M. (1980). Phylogenetic diversity of troponin-C amino acid composition. *FEBS. Lett.* **121**, 273–274.

Lehman, W., and Szent-Gyorgyi, A. G. (1972). Activation of the adenosine triphosphatase of limulus polyphemus actomyosin by tropomyosin. *J. Gen. Physiol.* **59**, 375–387.

Lehman, W., and Szent-Gyorgyi, A. G. (1975). Regulation of muscular contraction. Distribution of actin control and myosin control in the animal kingdom. *J. Gen. Physiol.* **66**, 1–30.

Lehman, W., Regenstein, J. M., and Ransom, A. L. (1976). The stoichiometry of the components of arthropod thin filaments. *Biochim. Biophys. Acta* **434**, 215–222.

Lehman, W., Head, J. F., and Grant, P. W. (1980). The stoichiometry and location of troponin I and troponin C - like proteins in the myofibril of the Bay Scallop, *Aequipecten irradians. Biochem. J.* **187**, 447–456.

Lehrer, S. S. (1975). Intramolecular crosslinking of tropomyosin via disulphide bond formation: Evidence for chain register. *Proc. Natl. Acad. Sci.* **72**, 3377–3381.

Levine, R. J. C., Elfvin, M., Dewey, M. M., and Walcott, B. (1976). Paramyosin in invertebrate muscles. Content in relation to structure and function, Vol. 2. *J. Cell Biol.* **71**, 273–279.

Lindberg, U., Carlsson, L., Markey, F., and Nystrom, L. E. (1979). The unpolymerized form of actin in non-muscle cells. *Method Achiev. Exp. Pathol.* **8**, 143–170.

Lowey, S. (1965). Comparative study of the α-helical muscle proteins. *J. Mol. Biol.* **7**, 234–244.

Lowey, S., Kucera, J., and Holtzer, A. (1963). On the structure of the paramyosin molecule. *J. Mol. Biol.* **7**, 234–244.

Lowey, S., Slayter, H. S., Weeds, A. G., and Baker, H. (1969). Substructure of the myosin molecule. Subfragments of myosin by enzymic degredation, Vol. 1. *J. Mol. Biol.* **42**, 1–29.

Lowy, J., and Millman, B. M. (1962). Mechanical properties of smooth muscles of cephalopod molluscs. *J. Physiol.* **160**, 353–363.

Lowy, J., and Millman, B. M. (1963). The contractile mechanism of the anterior byssus retractor muscle of mytilus edulis. *Phil. Trans. R. Soc.* **246**, 105–148.

Lowy, J. and Vibert, P. G. (1967). Structure and organization of actin in a molluscan smooth muscle. *Nature* **215**, 1254–1255.

Lowy, J., Millman, B. M. and Hanson, J. (1964). Structure and function in smooth tonic muscles of lamellibranch molluscs. *Proc. R. Soc. Ser. B* **160**, 525–536.

Lymn, R. W., and Taylor, E. W. (1970). Transient state phosphate production in the hydrolysis of nucleoside triphosphate by myosin. *Biochemistry* **9**, 2975–2983.

Lymn, R. W., and Taylor, E. W. (1971). Mechanism of adenosine triphosphate hydrolysis by actomyosin. *Biochemistry* **10**, 4617–4624.

McCray, J. A., Herbette, L., Kihara, T. and Trentham, D. R. (1980). A new approach to time-resolved studies of ATP-requiring biological systems: Laser flash photolysis of caged ATP. *Proc. Nat. Acad. Sci.* **77**, 7237–7241.

McCubbin, W. D., and Kay, C. M. (1968). The subunit structure of fibrous muscle proteins as determined by Osmometry. *Biochim. Biophys. Acta* **154**, 239–241.

Mackenzie, J. M., and Epstein, H. F. (1980). Paramyosin is necessary for determination of Nematode thick filament length *in vivo. Cell* **22**, 747–755.

McLachlan, A. D., and Stewart, M. (1976). The 14-fold periodicity in α-tropomyosin and the interaction with actin. *J. Mol. Biol.* **103**, 271–298.

McLachlan, A. D., Stewart, M., Smillie, L. B. (1975). Tropomyosin coiled-coil interactions: Evidence for an unstaggered structure. *J. Mol. Biol.* **98**, 293–304.

Mak, A. S., and Smillie, L. B. (1981). Non-polymerizable tropomyosin: Preparation, some properties, and F-actin binding. *Biochem. Biophys. Res. Comm.* **101**, 208–214.

Marceau, F. (1905). Recherches sur la structure des Muscles du manteau des Cephalopodes en rapport avec leur mode de contraction. *Trav. Lab. Soc. Sci. pp. 48–65. (Arachon, 8th year).*

Margossian, S. S., and Lowey, S. (1978). Interaction of myosin subfragments with F-actin. *Biochemistry* **17**, 5431–5439.

Marston, S. B. (1973). The nucleotide complexes of myosin in glycerol-extracted muscle fibres. *Biochim. Biophys. Acta* **305**, 397–412.

Marston, S. B., and Lehman, W. (1974). ADP binding to relaxed scallop myofibrils. *Nature* **252**, 38–39.

Marston, S. B., and Weber, A. (1975). The dissociation constant of the actin - heavy-meromyosin subfragment-1 complex. *Biochemistry* **14**, 3868–3873.

Martonosi, A., Gouvea, M. A., and Gergely, J. (1960). The interaction of ^{14}C-labelled nucleotides with actin. *J. Biol. Chem.* **235**, 1700–1706.

Maruta, H., and Korn, E. D. (1977). *Acanthamoeba* cofactor protein is a heavy chain kinase required for actin activation of the MgATPase activity of *Acanthamoeba*. Myosin 1. *J. Biol. Chem.* **252**, 8329–8332.

Masaki, T., Endo, M., and Ebashi, S. (1967). Localization of 6S component of actinin at Z-band. *J. Biochem.* **62**, 330–332.

Maw, M. C., and Rowe, A. J. (1980). Fraying of A-filaments into three subfilaments. *Nature* **286**, 412–414.

Means, A. R., and Dedman, J. R. (1980). Calmodulin - an intracellular calcium receptor. *Nature* **285**, 73–77.

Mihalyi, E., and Szent-Gyorgyi, A. G. (1953). Trypsin digestion of muscle proteins. *J. Biol. Chem.* **201**, 189–219.

Mikawa, T., Nonomura, Y., and Ebashi, S. (1977). Does phosphorylation of myosin light-chain have direct relation to regulation in smooth muscle? *J. Biochem.* **82**, 1789–1791.

Mikawa, T., Nonomura, Y., Hirata, M., Ebashi, S., and Kakiuchi, S. (1978). Involvement of an acidic protein in regulation of smooth muscle contraction by the tropomyosin-leiotonin system. *J. Biochem.* **84**, 1633–1636.

Millman, B. M. (1967). Mechanism of contraction in molluscan muscle. *Am. Zool.* **7**, 583–591.

Millman, B. M., and Bennett, P. M. (1976). Structure of the cross-striated adductor muscle of the scallop. *J. Mol. Biol.* **103**, 439–467.

Millward, G. R., and Woods, E. F. (1970). Crystals of tropomyosin from various sources. *J. Mol. Biol.* **52**, 585–588.

Molla, A., Kilhoffer, M., Ferraz, C., Audemard, E., Walsh, M. P., and Demaille, J. G. (1980). Octopus calmodulin. *J. Biol. Chem.* **256**, 15–18.

Mornet, D., Bertrand, R., Pantel, P., Audemard, E., and Kassab, R. (1981). Structure of the actin-myosin interface. *Nature* **292**, 301–306.

Murray, J. M., and Weber, A. (1981) Cooperativity of the calcium switch of regulated rabbit actomyosin system. *Mol. Cell Biochem.* **35**, 11–15.

Nakamura, T., Yamaguchi, M., and Yanagisawa, T. (1979). Comparative studies on actins from various sources. *J. Biochem.* **85**, 627–631.

Newman, J., and Carlson, F. D. (1980). Dynamic light-scattering evidence for the flexibility of native muscle thin filaments. *Biophys. J.* **29**, 37–48.

Nonomura, Y. (1974). Fine structure of the thick filament in molluscan catch muscle. *J. Mol. Biol.* **88**, 445–455.

O'Brien, E. J., Bennett, P. M., and Hanson, J. (1971). Optical diffraction studies of myofibrillar structure. *Phil. Trans. R. Soc. London.* **261**, 201–208.

O'Brien, E. J., Gillis, J. M., and Crouch, J. (1975). Symmetry and molecular arrangement in paracrystals of reconstituted muscle thin filaments. *J. Mol. Biol.* **99**, 461–475.

Offer, G., Baker, H., and Baker, L. (1972). Interaction of monomeric and polymeric actin with myosin subfragment 1. *J. Mol. Biol.* **66**, 435–444.

Ohtsuki, I., Masuki, T., Nonomura, Y., and Ebashi, S. (1967). Periodic distribution of troponin along the thin filament. *J. Biochem.* **61**, 817–819.

Olander, J. (1971). Substructure of the paramyosin molecule *Biochemistry* **10**, 601–609.

Osawa, F., Fujime, S., Ishiwata, S., and Mihashi, K. (1973). Dynamic property of F-actin and thin filament. *Cold Spring Harbor Symp. Quant. Biol.* **37**, 277–285.

Pavlov, J. (1885). Wie die Muschel ihre Schaale offnet. *Pfluegers Arch. Gesamte Physiol. Menschen, Tiere* **37**, 6–31.

Persechini, A., and Hartshorne, D. J. (1981). Phosphorylation of smooth muscle myosin: Evidence for cooperativity between the myosin heads. *Science* **213**, 1383–1385.

Phillips, G. N., Lattman, E. E., Cummins, P., Lee, K. Y., and Cohen, C. (1979). Crystal structure and molecular interactions of tropomyosin. *Nature* **278** , 413–417.

Philpott, D. E., Kahlbrock, M., and Szent-Gyorgyi, A. G. (1960). Filamentous organization of molluscan muscles. *J. Ultrastruct. Res.* **3**, 254–269.

Pollard, T. D., and Weihing, R. R. (1974). Actin and myosin and cell movement. *Crit. Rev. Biochem.* **2**, 1–65.

Potter, J. D. (1974). The content of troponin, tropomyosin, actin and myosin in rabbit skeletal muscle fibres. *Arch. Biochem. Biophys.* **162**, 436–441.

Purchon, R. D. (1968). ''The biology of the mollusca.'' Pergamon, Oxford, England.

Rahmsdorf, H. J., Malchow, D., and Gerisch, G. (1978). Cyclic AMP-induced phosphorylation in *Dictyostelium* of a polypeptide comigrating with myosin heavy-chains. *FEBS. Lett.* **88**, 322–326.

Reedy, M. K. (1967). Crossbridges and periods in insect flight muscle. *Am. Zool.* **7**, 465–481.

Rees, W. J. (1957). ''The scallop: Studies of a shell and its influence on humankind.'' The Shell Transport and Trading Co., London.

Reisler, E. (1980). On the question of cooperative interaction of myosin heads with F-actin in the presence of ATP. *J. Mol. Biol.* **138**, 93–107.

Reisler, E., Burke, M., Himmelfarb, S., and Harrington, W. F. (1974). Spatial proximity of the two essential sulfhydryl groups of myosin. *Biochemistry* **13**, 3837–3840.

Riddiford, L. M. (1966). Solvent perturbation and ultraviolet optical rotary dispersion studies of paramyosin. *J. Biol. Chem.* **241**, 2792–2802.

Rosenbluth, J. (1965). Ultrastructural organization of obliquely striated muscle fibres in *Ascaris Lumbricoides. J. Cell. Biol.* **25**, 495–515.

Rubenstein, P. A., and Spudich, J. A. (1977). Actin microheterogeneity in chick embryo fibroblasts. *Proc. Nat. Acad. Sci.* **74**, 120–123.

Ruegg, J. C. (1961). On the tropomyosin-paramyosin system in relation to the viscous tone of lamellibranch catch muscle. *Proc. R. Soc. Ser. B* **154**, 224–249.

Ruegg, J. C. (1971). Smooth muscle tone. *Physiol. Rev.* **51**, 201–248.

Sanger, J. W. (1971). Sarcoplasmic reticulum in the cross-striated adductor muscle of the Bay Scallop, *Aequipecten irradians. Z. Zellforsch.* **118**, 156–161.

Sanger, J. W., and Hill, R. B. (1972). Ultrastructure of the radula protractor of *Busycon canaliculatum. Z. Zellforsch.* **127**, 314–322.

Schaub, M. C., Watterson, J. G., and Waser, P. G. (1975). Radioactive labelling of specific thiol groups as influenced by ligand binding. *Hoppe-Seyler's Z. Physiol. Chem.* **356**, 325–339.

Schmitt, F. O., Bear, R. S., Hall, C. E., and Jakus, M. A. (1947). Electron microscope and x-ray diffraction studies of muscle structure. *Ann. N.Y. Acad. Sci.* **47**, 799–809.

Scholey, J. M., Taylor, K. A., and Kendrick-Jones, J. (1980). Regulation of non-muscle myosin assembly by calmodulin-dependent light-chain kinase. *Nature* **287**, 233–235.

Sellers, J. R. (1981). Phosphorylation-dependent regulation of limulus myosin. *J. Biol. Chem.* **256**, 9274–9278.

Sellers, J. R., and Adelstein, R. S. (1982). Cooperativity and the reversible phosphorylation of smooth muscle heavy meromyosin. *Biophys. J.* **37**, 262a.

Sellers, J. R., Chantler, P. D., and Szent-Gyorgyi, A. G. (1980). Hybrid formation between scallop myofibrils and foreign regulatory light-chains. *J. Mol. Biol.* **144**, 223–245.

Sekine, T., and Kielley, W. W. (1964). The enzymic properties of N-Ethylmaleimide modified myosin. *Biochim. Biophys. Acta* **81**, 336–345.

Shoenberg, C. F., and Needham, D. M. (1976). A study of the mechanism of contraction in vertebrate smooth muscle. *Biol. Rev.* **51**, 53–104.

Silberstein, L., and Lowey, S. (1981). Isolation and distribution of myosin isozymes in chicken pectoralis muscle. *J. Mol. Biol.* **148**, 153–189.

Simmons, R. M., and Szent-Gyorgyi, A. G. (1978). Reversible loss of calcium control of tension in scallop striated muscle associated with the removal of regulatory light-chains. *Nature* **273**, 62–64.

Simmons, R. M. and Szent-Gyorgyi, A. G. (1980). Control of tension development in scallop muscle fibres with foreign regulatory light-chains. *Nature* **286**, 626–628.

Sivaramakrishnan, M., and Burke, M. (1982). The free heavy-chain of vertebrate skeletal myosin subfragment, Vol. 1 shows full enzymatic activity. *J. Biol. Chem.* **257**, 1102–1105.

Sobieszek, A. (1973). The fine structure of the contractile apparatus of the anterior byssus retractor muscle of *Mytilus edulis. J. Ultrastruct. Res.* **43**, 313–343.

Sperling, J. E., Feldmann, K., Meyer, H., Jahnke, U., and Heilmeyer, L. M. G. (1979). Isolation, characterization and phosphorylation pattern of the troponin complexes TI_2C and I_2C. *Eur. J. Biochem.* **101**, 581–592.

Squire, J. M. (1973). General model of myosin filament structure. *J. Mol. Biol.* **77**, 291–323.

Stafford, W. F. and Szent-Gyorgyi, A. G. (1978). Physical characterization of myosin light chains. *Biochemistry* **256**, 9274–9278.

Stafford, W. F., and Yphantis, D. A. (1972). Existence and inhibition of hydrolytic enzymes attacking paramyosin in myofibrillar extracts of *Mercenaria mercenaria. Biochem. Biophys. Res. Comm.* **49**, 848–854.

Stafford, W. F., Szentkiralyi, E. M., and Szent-Gyorgyi, A. G. (1979). Regulatory properties of single-headed fragments of scallop myosin. *Biochemistry* **18**, 5273–5280.

Stein, L. A., Schwartz, R. P., Chock, P. B., and Eisenberg, E. (1979). Mechanism of actomyosin adenosine triphosphatase. Evidence that Adenosine 5'-triphosphate Hydrolysis can occur without dissociation of the actomyosin complex. *Biochemistry* **18**, 3895–3909.

Stone, D., and Smillie, L. B. (1978). The amino-acid sequence of rabbit skeletal α-Tropomyosin. *J. Biol. Chem.* **253**, 1137–1148.

Straub, F. B., and Feuer, G. (1950). Adenosinetriphosphate. The functional group of actin. *Biochim. Biophys. Acta* **4**, 455–470.

Stewart, M., Kensler, R. W., and Levine, R. J. C. (1981). Structure of *Limulus* telson muscle thick filaments. *J. Mol. Biol.* **153**, 781–790.

Suzuki, H., Ohnishi, H., Takahashi, K., and Watanabe, S. (1978). Structure and function of chicken gizzard myosin. *J. Biochem.* **84,** 1529–1542.

Szent-Gyorgyi, A. G. (1976). Comparative survey of the regulatory role of calcium in muscle. *In* "Calcium in Biological Systems". *Symp. Soc. Exp. Biol.* **30.** 335–347.

Szent-Gyorgyi, A. G., Cohen, C., and Kendrick-Jones, J. (1971). Paramyosin and the filaments of molluscan 'Catch' muscles. *J. Mol. Biol.* **56,** 239–258.

Szent-Gyorgyi, A. G., Szentkiralyi, E. M., and Kendrick-Jones, J. (1973). The light-chains of scallop myosin as regulatory subunits. *J. Mol. Biol.* **74,** 179–203.

Szentkiralyi, E. M. (1982). Scallop regulatory and essential light-chains complex with the same S-1 heavy-chain peptide fragment. *Biophys. J.* **37,** 39a.

Szentkiralyi, E. M., and Oplatka, A. (1969). On the formation and stability of the enzymically active complexes of heavy meromyosin with actin. *J. Mol. Biol.* **43,** 551–566.

Taylor, E. W. (1979). Mechanism of actomyosin ATPase and the problem of muscle contraction. *Crit. Rev. Biochem.* **6,** 103–165.

Taylor, K. A., and Amos, L. A. (1981). A new model for the geometry of the binding of myosin crossbridges to muscle thin filaments. *J. Mol. Biol.* **147,** 297–324.

Thomas, D. D., and Cooke, R. (1980). Orientation of spin-labelled myosin heads in glycerinated muscle fibers. *Biophys. J.* **32,** 891–906.

Tien-chin, T., Tsu-hsun, K., Chia-mu, P., Yu-shang, C., and Yung-shui, T. (1965). Electron microscopical studies of tropomyosin and paramyosin. *Sci. Sin.* **14,** 91–105.

Tien-chin, T., Yung-Shui, T., Tsu-hsun, K., Chia-hsiu, P., and Zi--xian, L. (1966). The presence of subfilaments in the thick and the paramyosin filaments of muscle. *Kexue Tongbao* **17,** 308–310.

Toyo-oka, T. (1979). Effects of various concentrations of MgATP on the superprecipitation and ATPase activity of scallop striated muscle myosin B. *J. Biochem.* **85,** 871–877.

Tregear, R. T., and Marston, S. B. (1979). The crossbridge theory. *Ann. Rev. Physiol.* **41,** 723–736.

Trinick, J. A. (1981). End-filaments: A new structural element of vertebrate skeletal muscle thick filaments. *J. Mol. Biol.* **151,** 309–314.

Trybus, K. M., Huiatt, T. W., and Lowey, S. (1982). A bent monomeric conformation of myosin from smooth muscle. *Proc. Natl. Acad. Sci. U.S.A.* **79,** 6151–6155.

Tsao, T. C., Bailey, K., and Adair, G. S. (1951). The size, shape and aggregation of tropomyosin particles. *Biochem. J.* **49,** 27–35.

Tsuchiya, T., Yamada, N., Mori, H., and Matsumoto, J. J. (1978). Adenosinetriphosphatase activity of squid myosin. *Nippon Suisan Gakkaishi* **44,** 203–207.

Tsuchiya, T., Fukuhara, S., and Matsumoto, J. J. (1980). Physico-chemical properties of squid paramyosin. *Nippon Suisan Gakkaishi* **46,** 197–200.

Twarog, B. M. (1979). The nature of catch and its control. *In* "First John M. Marshall Symposium in Cell Biology; Motility in Cell Function" (F. A. Pepe, J. W. Sanger, and V. T. Nachmias, eds.), pp. 231–241. Academic Press, New York.

Vandekerckhove, J., and Weber, K. (1978). Actin amino-acid sequences. *Eur. J. Biochem.* **90,** 451–462.

Vandekerckhove, J., and Weber, K. (1979). The amino-acid sequence of actin from chicken skeletal muscle actin and chicken gizzard smooth muscle actin. *FEBS. Lett.* **102,** 219–222.

Vibert, P., and Craig, R. (1982). Three dimensional reconstruction of thin filaments decorated with a calcium-regulated myosin. *J. Mol. Biol.* **157,** 299–319.

Vibert, P. J., Haselgrove, J. C., Lowy, J., and Poulsen, F. R. (1972). Structural changes in actin-containing filaments of muscle. *J. Mol. Biol.* **71,** 757–767.

Vibert, P., Szent-Gyorgyi, A. G., Craig, R., Wray, J., and Cohen, C. (1978). Changes in crossbridge attachment in a myosin-regulated muscle. *Nature* **273,** 64–66.

Von Uexkuell, J. (1912). Studien uber den Tonus. Die Pilgermuschel, Vol. 6. *Z. Biol.* **58,** 305–332.

Wagner, P. D., and Giniger, E. (1981). Hydrolysis of ATP and reversible binding to F-actin by myosin heavy-chains free of all light-chains. *Nature* **292,** 560–562.

Wallimann, T., and Szent-Gyorgyi, A. G. (1981a). An immunological approach to myosin light-chain function in thick filament linked regulation. Characterization, specificity, and cross-reactivity of anti-scallop myosin heavy- and light-chain antibodies by competitive, solid-phase radioimmunoassay. *Biochemistry* **20**, 1176–1187.

Wallimann, T., and Szent-Gyorgyi, A. G. (1981b). An immunological approach to myosin light-chain function in thick filament linked regulation. Effects of anti-scallop myosin light-chain antibodies. Possible regulatory role for the essential light-chain. *Biochemistry* **20**, 1187–1195.

Wallimann, T., Hardwicke, P. M. D., and Szent-Gyorgyi, A. G. (1982). Regulatory and essential light-chain interactions in scallop myosin. *J. Mol. Biol.* **156**, 153–173.

Weber, A., and Murray, J. M. (1973). Molecular control mechanism in muscle contraction. *Physiol. Rev.* **53**, 612–673.

Weeds, A. G., and Lowey, S. (1971). Substructure of the myosin molecule. The light-chains of myosin. *J. Mol. Biol.* **61**, 701–725.

Weeds, A. G., and Taylor, R. S. (1975). Separation of subfragment-1 isozymes from rabbit skeletal muscle myosin. *Nature (London)* **257**, 54–57.

Weeds, A. G., Hall, R., and Spurway, N. C. S. (1975). Characterization of myosin light-chains from histochemically identified fibres of rabbit psoas muscle. *FEBS. Lett.* **49**, 320–324.

Weisel, J. W., and Szent-Gyorgyi, A. G. (1975). The coiled-coil structure: Identity of the two chains of mercenaria paramyosin. *J. Mol. Biol.* **98**, 665–673.

Wells, J. A., and Yount, R. G. (1979). Active site trapping of nucleotides by crosslinking two sulfydryls in myosin subfragment-1. *Proc. Nat. Acad. Sci.* **76**, 4966–4970.

Werber, M. M., Gaffin, S. L., and Oplatka, A. (1972a). Physico-chemical studies on the light-chains of myosin. *J. Mechanochem. Cell Motil.* **1**, 91–96.

Werber, M. M., Szent-Gyorgyi, A. G., and Fasman, G. D. (1972b). Fluorescence studies on heavy-meromyosin substrate interaction. *Biochemistry* **11**, 2872–2883.

Whalen, R. G., Butler-Browne, G. S., and Gros, F. (1976). Protein synthesis and actin heterogeneity in calf muscle cells in culture. *Proc. Nat. Acad. Sci.* **73**, 2018–2022.

Wilkinson, J. M. (1974). The preparation and properties of the components of troponin B. *Biochim. Biophys. Acta* **359**, 379–388.

Winkelman, L. (1976). Comparative studies of paramyosins. *Comp. Biochem. Physiol.* **B55**, 391–397.

Woods, E. F. (1976). The conformational stabilities of tropomyosins. *Aust. J. Biol. Sci.* **29**, 405–418.

Woods, E. F., and Pont, M. J. (1971). Characterization of some invertebrate tropomyosins *Biochemistry* **10**, 270–276.

Wray, J. S. (1979). Structure of the backbone in myosin filaments of muscle. *Nature* **277**, 37–40.

Wray, J. S. (1981). Organization of myosin in invertebrate thick filaments. *J. Gen. Physiol.* **78**, 2a.

Wray, J. S., Vibert, P. J., and Cohen, C. (1975). Diversity of cross-bridge configurations in invertebrate muscles. *Nature* **257**, 561–564.

Yazawa, M., Sakuma, M., and Yagi, K. (1980). Calmodulins from muscles of marine invertebrates, scallop and sea anenome. *J. Biochem.* **87**, 1313–1320.

Yeung, A. T., and Cowgill, R. W. (1976). Structural difference between α-paramyosin and β-paramyosin of *Mercenaria mercenaria*. *Biochemistry* **15**, 4654–4659.

Yonge, C. M. (1936). The evolution of the swimming habit in the lamellibranchia. *Mem. Mus. Hist. Nat. Belg. Ser. 2.* **3**, 78–100.

Young, D. M. (1967). On the interaction of adenosine diphosphate with myosin and its enzymically active fragments. *J. Biol. Chem.* **242**, 2790–2792.

Yung-shui, T., and Tsu-hsun, K. (1979). Paramyosin filaments of amphioxus. *Sci. Sin.* **22**, 1329–1332.

4

Locomotion in Molluscs

E. R. TRUEMAN

Department of Zoology
University of Manchester
Manchester, England

I. Outline

Molluscs move by crawling over or burrowing into substrata and by swimming. The locomotion of different molluscs has been described in a number of classical articles referred to by Morton (1964) in his review of molluscan locomotion, but the more modern experimental and qualitative approach virtually

THE MOLLUSCA, VOL. 4
Physiology, Part 1

commenced with Lissmann's (1945a, 1945b) investigations of snail locomotion. More recently the development of electronic techniques (Hoggarth and Trueman, 1967) has allowed the detection of movement and pressure changes and, together with analysis of ciné film, has led to a much fuller understanding of the dynamics of locomotion (Trueman, 1968a, 1980).

The foot is the principal locomotory organ in those molluscs which crawl or burrow. In chitons and many epifaunal gastropods it has a broad sole, giving the ability to form both a broad disk for firm attachment and a succession of muscular waves for movement. Gastropods, which burrow, have in general retained a large dorsoventrally compressed foot, as in *Bullia* (Trueman and Brown 1976) or *Natica* (Russell-Hunter and Russell-Hunter, 1968; Trueman, 1968b). In the Bivalvia, for example, *Tellina* and *Cardium* (Trueman, 1968a), the foot is characteristically slipper-shaped, which facilitates penetration of substrata, and by dilation becomes anchored so that the body may be drawn down into the sand. The foot of Scaphopoda is similarly shaped but has epipodial skirts which are expanded to obtain anchorage in a manner similar to the dilation of the bivalve foot (Morton, 1964; Trueman, 1968c). In bivalves dwelling on firm surfaces, attachment and progression are attained by the pedal secretion of proteinaceous threads termed the *byssus* (Yonge, 1972). Although pedal lobes, or parapodia, are the most effective means of swimming in Gastropoda, e.g., *Aplysia* and Pteropoda (Morton, 1964), the most powerful swimming is found in the squids and cuttlefish, in which water is expelled as a powerful jet from the mantle cavity (Trueman, 1980). The bivalves also expel water from the mantle cavity by forcibly adducting their valves. This enables them either to swim, as in scallops, or to aid burial by excavating a cavity in the sand into which the shell may be drawn (Trueman, 1975).

In this review, molluscan locomotion will be approached from the aspect of habitat and mechanism used rather than that of their taxonomic affinities. However, crawling is most characteristic of snails and chitons, burrowing is characteristic of bivalves and scaphopods, and swimming has been most effectively developed by coleoid cephalopods.

II. Crawling

A. Introduction

Gastropoda are characterized by a ventral foot, often of relatively large plantar area, which bears the head, shell, and visceral mass. A few snails, for example, *Strombus gigas,* incorporate the use of the shell or operculum into the pattern of locomotion (Miller, 1974b), but the majority creep continuously on the sole.

Creeping allows adhesion to the substratum both to resist external forces, such as gravity or waves, and to provide anchorage during the movement of other regions of the foot. Locomotion is achieved either by ciliary action or by muscular locomotory waves. Mucus plays an important role in locomotion both by cilia and by muscle (Jones, 1975), whereas adhesion or tenacity (Miller, 1974a) of the foot is related both to the area of attachment and to the properties of pedal mucus (Denny, 1980).

B. Ciliary Locomotion

When gastropods are propelled by the cilia on the sole of the foot, movement is a smooth, uniform gliding with no differential motion of regions of the foot. Ciliary locomotion is found in several families of Mesogastropoda and Meogastropoda, many opisthobranchs, and Pulmonata (Miller, 1974b). Almost all basommatophorans and some of the smaller stylommatophorans, such as *Discus rotundatus* and *Zonitoides nitidus* (Elves, 1961), employ cilia for locomotion. The incidence of ciliary locomotion appears to bear little relation to the size of the animal (Clark, 1964), although relatively weak forces must be produced. Foot length commonly exceeds shell length where cilia are employed, and greater pedal size is commonly a feature even within families exhibiting both ciliary and muscular locomotion (Miller, 1974a). Such an increase may be related to the reduction of force required per unit of plantar area with a relatively weak propulsive system. The pedal cilia of *Polinices* (Naticidae) can apparently transport up to 5 g/cm^2, but large individuals resort to muscular locomotory waves when greater speed or power is required (Copeland, 1919, 1922). However, in pulmonates, size is less important then the effective weight of the animal in the medium in which it lives. Basommatophorans are commonly neutrally buoyant because of air contained in the mantle cavity, and the load to be carried by the foot is minimal (Jones, 1975).

There are comparatively few experimental observations on ciliary locomotion, and these are conveniently summarized by Jones (1975). For a given frequency of beat, the longer the cilium the greater will be the velocity at the tip, so that a snail with longer pedal cilia should move faster than one with shorter cilia but the latter should carry a greater load. The pedal cilia of *Lymnea peregra* and *Physa fontinalis* are 10 and 7 μm, respectively, in length (Elves, 1961) and have crawling velocities of 17.5 and 6 cm/min, respectively (Pelseneer, 1935). This is a larger difference than is indicated by the length of the cilia and suggests that other factors must be involved. Pedal cilia are almost certainly under nervous control, for the rates of beating have been observed to slow down as the snail slows. Copeland (1919) has observed that *Nassarius*, with cilia at rest, responds to a food stimulus by the resumption of ciliary beating. The cilia in this genus are unaffected by magnesium sulfate anesthesia and continue beating in apparent isolation from the nervous system, suggesting an inhibitory rather than an excitatory control.

C. Muscular Locomotory Waves

Locomotory waves may be readily observed by viewing the sole of the foot of many species of gastropods through a transparent surface while crawling. Dark transverse bands appear across the foot and move forward in, for example, *Helix*. Two types of waves may be distinguished (Vlès, 1907): direct waves moving in the same direction as the animal and retrograde waves moving in the opposite direction to that of the animal, that is, from front to rear during forward locomotion. The mechanical principles involved are the same as those for the locomotion of worms (Elder, 1980; Gray, 1968; Trueman, 1975). For direct waves, parts of the sole are attached to the substratum at maximum extension, and forward movement occurs when regions of the foot are longitudinally compressed (Fig. 1a). By contrast, when retrograde waves occur, the foot is anchored by regions at their shortest length and forward movement takes place with the sole maximally extended (Fig. 1b). The mechanical basis of crawling using muscular waves or of burrowing by molluscs is that part of the body is firmly anchored to the substratum so as to withstand the backthrust (or static reaction) developed by the extension of another region. The amount of pedal sole anchored to the substratum, and its tenacity during locomotion and at rest are of considerable importance to marine snails on wave-swept shores and to Stylommatophora when climbing. Species of limpets, whose adhesion to rocks when disturbed is

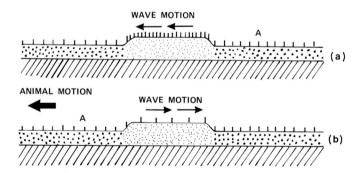

Fig. 1. Characteristics of (**a**) direct and (**b**) retrograde waves in hypothetical sections of the foot of gastropods crawling over slightly flexible or porous substratum. The shortening and extension of the pedal sole occurring at the leading edge of direct and retrograde waves, respectively, are shown diagrammatically by markers. Substratum is crosshatched; mucus is stippled, coarsely where viscous and sole is anchored (A) and finely where more fluid and motion of the pedal wave occur.

well known, are good examples (Warburton, 1976). Branch and Marsh (1978) have observed that the force required to detach various species of *Patella* ranges from 5.18 to 1.95 kg/cm² and that high tenacity may be associated with low mucus secretion. This general locomotory principle is well illustrated by the so-called leaping progression of *S. gigas*. In this species the foot detaches and extends, with the shell resting on the ground as an anchor to resist the backthrust resulting from pedal protraction. Reattachment of the foot is immediately followed by the drawing forward of the shell, with backthrust being developed between the substratum and foot, particularly with respect to the operculum. This type of progression shows anchorage clearly occurring alternatively in distinct regions of the animal's body; but with normal pedal locomotory waves, a similar situation occurs dynamically between different regions of the foot.

The different types of pedal locomotory waves are conveniently summarized by Miller (1974b) (Table I). Both retrograde and direct waves may extend across the width of the foot, being termed *monotaxic*, e.g., in *Chiton* and *Helix* (Fig. 2a,c). Alternatively, they may form two lateral and parallel systems, termed *ditaxic waves*, e.g., in *Patella* (Fig. 2b). An advantage of the differentiation of the foot into two regions is the ability of the limpet to turn around without moving forward, for in this genus retrograde waves may move in opposite

TABLE I

Occurrence of the Principal Types of Pedal Locomotion in Polyplacophora
and Gastropoda[a]

Locomotor type	Taxa
Ciliary	*Hydrobia, Natica, Cassis, Nassarius, Oliva, Conus, Actaeon, Dendronotus,* Lymnaeidae
Muscular, rhythmic	
Retrograde	
Monotaxic	Chitonida, Fissurellidae, Meritidae, *Bullia,* Aplysiidae, Ellobiidae, Otinidae
Ditaxic	Patellacea, Littorinidae, Lacunidae
Direct	
Monotaxic	*Polinices,* Stylommatophora
Ditaxic	Haliotidae, *Trochus, Gibbula, Pomatias*
Composite	Cypraeidae
Arrhythmic	
Distinct terminating	*Cymatium, Murex,* some species of *Conus*
Indistinct, discontinuous	*Terebra, Aporrhais*
Leaping	Strombidae

[a] After Miller (1974b).

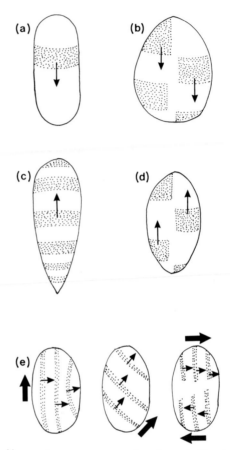

Fig. 2. Diagrams of locomotory waves as seen from beneath the sole of the foot. Animals moving up page except in (e), where motion is indicated by large arrow. (a) Retrograde monotaxic, *Chiton;* (b) retrograde ditaxic, *Patella;* (c) direct monotaxic, *Helix;* (d) direct ditaxic, *Gibbula* (after Trueman and Jones, 1977); (e) composite waves, *Cypraea,* showing lateral monotaxic waves, diagonal monotaxic waves, and turning (after Miller, 1974b). Not drawn to scale. Clear areas of sole represent pedal anchorage; stipple represents locomotory waves moving in the direction of the arrow.

directions on either side of the foot, causing rotation in a similar manner to pulling and pushing simultaneously on the oars of a rowing dinghy. In *Pomatias* the lateral regions of the foot are more clearly differentiated, one side moving forward while the other remains anchored, the whole cycle producing a shuffling gait not unlike that of a terrestrial bipedal vertebrate (Gray, 1968).

Other types of pedal movement described by Miller include the composite pattern, unique to the Cypraeacea, in which the most common pattern is lateral monotaxic waves (Fig. 2e). Other other occasions, the waves may be diagonal or the foot may be divided anteroposteriorly into two or three regions with differently oriented wave patterns, probably allowing for sharp turns. At still other times, different sets of locomotory waves may cross each other without apparent interference in a remarkable display of muscular coordination. The term *arrhythmic pedal locomotion* is used to describe muscular locomotion not organized into distinct rhythmic waves. Miller (1974b) divides this into two categories. (i) distinct arrhythmic terminating waves, e.g., in *Cymatium*, in which irregular patches of the sole move forward in an unpredictable order; and (ii) indistinct arrhythmic movement in which no recognizable means of movement may be observed. The role of the shell in *Aporrhais*, as an anchor for pedal extension, has been likened by Haefelfinger (1968) to that of a crutch.

D. Mechanisms of Direct and Retrograde Waves

1. Introduction

When the foot of a snail, e.g., *Helix*, or slug, e.g., *Agriolimax* (Jones, 1973, 1975), is viewed from beneath, waves appear as darker areas passing forward. Each band represents a region of longitudinal contraction of the sole. Parker (1911) showed that these waves were concavities of the sole by observing the behavior of air bubbles trapped in mucus beneath the sole. Jones and Trueman (1970) allowed snails to crawl over a small hole connected to a pressure transducer to demonstrate both the concavity and a small pressure below ambient during the passage of each wave. More recently, however, Denny (1981) has discussed the role of mucus in gastropod locomotion and suggests that the pedal waves represent not a concavity but a region where the sole is in motion and is not raised off the substratum. He points out that the presence of a hole beneath the foot for the attachment of a transducer or the instantaneous fixation of the foot of slugs, e.g., *Agriolimax* (Jones, 1973) while crawling over a porous surface would serve only to enhance the concavity. In this chapter, pedal waves have been drawn with a somewhat exaggerated cavity for the sake of clarity (Figs. 1, 4, 7).

Experimental evidence showing that where the sole is stationary on the substratum it is applying backthrust so as to pull the body continuously forward in relation to both direct and retrograde waves has been put forward by Lissman (1945a, 1945b) and Jones (1975) (Fig. 3).

2. Direct Waves

Movement of *Helix* over a plate and across a small bridge that is attached to a transducer and recording equipment allowed frictional and reactive forces to be recorded for each pedal wave (Fig. 4). The progress of a snail or slug by direct

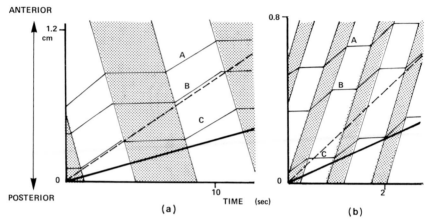

ANTERIOR

POSTERIOR

(a)

TIME (sec)

(b)

Fig. 3. Graphs obtained from ciné film of the forward locomotion of (**a**) *Patella*, and (**b**) *Helix*. Each shows the forward motion (dotted) of three points (A–C) arranged anteroposteriorly on the sole of the foot and the relative compression of the sole (stipple). Compression occurs with a retrograde wave (**a**) when the sole is attached to the substratum and with a direct wave (**b**) when moving forward. The heavy line represents the average speed of the animal and the broken line the average speed of a point in forward motion. (**a**: After Jones and Trueman, 1970; **b**: after Lissmann, 1945a.)

waves depends on the animal applying sufficient force to the ground in a posterior direction (static reaction) to overcome the sliding friction of parts moving forward. The foot moves forward only when slightly raised from the substratum, probably by tension in the anterior oblique muscles (Fig. 5). Where the foot is attached, the posterior oblique muscles must contract to draw the body forward. A continuous motion is achieved, for successive regions of posterior oblique muscles take over this function as the waves move forward along the sole. Where the sole is anchored the thin layer of pedal mucus must act as a glue, whereas in waves of motion the mucus must flow readily. The problem of how an animal with only one foot can walk on glue has largely been solved by Denny (1980, 1981; see also Vol. 1, Chapter 10). By investigations of the physical properties of pedal mucus, he has shown that the mucus beneath the foot will "yield" rapidly from the solid to the liquid phase when subjected to shearing strains of the magnitude experienced beneath the sole of the foot and "heal" again to the gel or solid phase in as little as 0.15 sec. (Fig. 6). The yield–heal characteristics of the pedal mucus are ideally suited to the locomotion of the slug *Ariolimax columbianus*. In this species 12–17 direct waves are present on the foot, and at the leading edge of a wave mucus is stressed to yielding, presumably by contraction of the anterior oblique muscles (Fig. 5); in consequence, the wave moves forward over mucus in its liquid form when little resistance is offered to movement. At the trailing edge of the wave, where the sole becomes attached to the ground, the mucus rapidly heals, so that the foot here rests on mucus in its solid

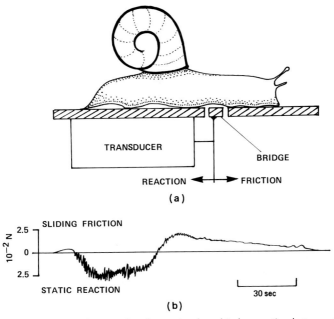

(a)

(b)

Fig. 4. (a) Arrangement for recording forces developed in locomotion between the foot of *Helix pomatia* and the substratum. (b) Recording obtained using wide bridge (about 1 cm); sliding friction occurs when the bridge is pushed to the right (i.e., forward.) Note the oscillation in the trace representing each pedal wave.

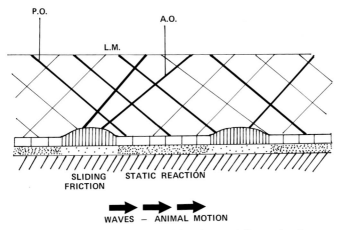

Fig. 5. Diagrammatic section of the foot of the slug, *Agriolimax,* showing passage of two direct waves and accompanying contractions in oblique pedal muscles (A.O. and P.O., anterior and posterior oblique, respectively) lying between a layer of longitudinal muscle (L.M.) and the pedal sole. Contracting muscles are drawn in the heavy line. Mucus (stipple as Fig. 1) beneath the epithelial sole is viscous in the region of static reaction and more fluid beneath moving wave. (After Denny, 1981; Trueman and Jones, 1977.)

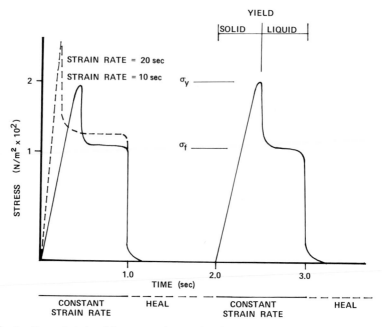

Fig. 6. Properties of pedal mucus under simulated natural conditions. Mechanical properties of *Ariolimax columbianus* pedal mucus at high strains. For alternate periods, the mucus is sheared at a constant strain rate and held stationary. The mucus is thus exposed to conditions analogous to those found under a crawling slug. When it is first sheared, stress is proportional to strain, indicating that the mucus is an elastic solid. At a strain of 5–6 (stress = σy) the mucus yields. With further strain, stress is proportional to strain rate, i.e., the mucus behaves as a liquid. When shearing is stopped, the mucus begins to heal. The healing period of 1 sec represents the time the mucus is unsheared beneath an interwave during locomotion; flow stress, σ_f. (After Denny, 1980.)

form. If the shear strength of the solid mucus (σy, Fig. 6) is sufficient to resist the static reaction produced by wave motion, the slug will crawl forward. The yield–heal cycle of the pedal mucus allows it to act as a material ratchet, facilitating forward and resisting backward motion. This results in effective adhesive locomotion (Denny, 1980) and may be used in the model for locomotion with direct waves (Fig. 5).

3. Retrograde Waves

How the changing fluidity of pedal mucus may be applied to retrograde locomotion is less evident, for the sole of mucus with backward-moving waves has not been investigated experimentally. It is, however, very likely that mucus will occur in the solid and liquid phases in respect to the attached and moving regions, respectively, of the pedal sole. In the species best known in regard to the mechanics of pedal locomotion, *Patella vulgata*, there is very little muscle near

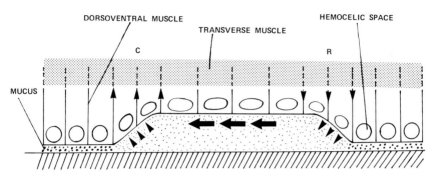

Fig. 7. Diagrammatic section of the foot of *Patella* showing the factors involved in the progression of a retrograde wave (arrows). Hemocoelic spaces are distorted against the transverse muscles as the leading edge of the sole is drawn up by contraction (c) of the dorsoventral muscles, with the lagging edge drawn down (R). Mucus beneath the foot is shown by coarse and fine stippling indicative of viscous and fluid phases, respectively. (After Trueman, 1975.)

the pedal epithelium lying on the longitudinal plane; most muscle connected to the sole runs dorsoventrally (Trueman and Jones, 1977) (Fig. 7). In the model initially put forward in an attempt to explain the mechanism of locomotory waves, a pressure beneath the ambient one was recorded from below each retrograde wave (Jones and Trueman, 1970). However, Denny's (1980) work on pedal mucus reduces the importance of this factor, and it is likely that contraction of the dorsoventral muscles at the leading edge of each wave causes the mucus beneath to yield and become fluid. This fluid mucus must then be healed at the trailing edge of the wave, allowing the sole to move beneath the pedal wave without being greatly withdrawn from the ground. In regard to retrograde waves, the sole must be extended beneath each wave (Fig. 1b), and this cannot be readily accounted for in *Patella* in the absence of longitudinal oblique muscle fibers. It seems likely that in *Patella*, at least, extension is attained by the dorsoventral muscle fibers acting through the numerous blood spaces of the pedal hemocoel. If these are assumed to be incompressible and of constant volume, then contraction of dorsoventral muscles will deform the hemocoelic spaces by compressing them against a thick layer of transverse muscle fibers which lie immediately above. The width of the foot is kept constant by transverse muscle fibers, so that deformation must occur along the antero-posterior axis of the foot, causing extension of the sole during the passage of each wave.

E. Factors Affecting Locomotion

1. Temperature

Environmental temperature has a marked effect on the velocity of locomotion of Pulmonata, e.g., the slug *Limax*, and details of velocity of movement are of little value without data on temperature and humidity (Jones, 1975). The latter may well affect mucus viscosity and thus velocity of motion. Increased temperature increases the frequency of waves but decreases step length in *Limax*. This again may be related to mucus viscosity.

2. Locomotor Type and Size

Miller (1974a) discusses this aspect of locomotion with particular reference to prosobranchs (Table II). The larger-size groups move at the greatest velocities, but the highest speeds relative to size within each locomotory type are found in the smaller individuals. Ciliary locomotion appears to be the fastest means of movement for very small animals in terms of both relative and absolute velocity. Direct and retrograde ditaxic patterns give very similar maxima and, except for leaping motion, swiftest movement. The majority of prosobranchs live on hard surfaces and have good tenacity during movement. Adhesion appears to be maximal for rhythmic pedal waves; indeed, high speed and tenacity are simultaneously attained only by using such wave forms (Miller, 1974a). Long, rhythmic waves, moving much of the foot at once, are probably more economic of energy than numerous small waves resulting in the same velocity, but Miller's experiments show that tenacity is significantly reduced when the wavelength is increased. The optimal wave pattern is a balance between velocity with minimal energy requirements, that is, long steps, and the demand for tenacity. The latter is a particularly important factor on different substrata in wave-washed habitats and in climbing in the terrestrial environment.

The most interesting change in the mode of locomotion occurs in *Helix*, in which more rapid "galloping" locomotion occurs when escaping predators

TABLE II

Maximum Velocities for Different Mechanisms of Crawling in Four Shell Size Classes (cm/sec)[a]

	Shell size (cm)			
Mechanism	<1.5	1.5–4.0	4.1–8.0	>8.0
Ciliary	0.67	0.29	0.53	0.62
Ditaxic				
Retrograde	0.25	0.36	0.58	1.25
Direct	0.18	0.32	1.0	1.3
Composite	—	0.4	0.48	0.25
Arrhythmic				
Terminating	—	0.22	0.14	0.2
Indistinct	0.04	0.05	0.09	0.01
Leaping	—	3.7	1.6	2.6

[a] After Miller (1974a).

(Jones, 1975). Normal movement is by direct waves, but when galloping, the snail lifts the head and part of the foot off the ground by 2–4 mm and pushes them forward. The wave form becomes retrograde, and the snail leaves a dotted slime track instead of the normal continuous one. This form of locomotion may be fairly widespread in terrestrial pulmonates, for a number of reports have been made of dotted slime tracks.

F. Other Crawling Mechanisms

Representatives of both the Bivalvia and the Cephalopoda are adapted for movement over hard substrata. Bivalve spat and epifaunal species, such as members of the Mytilidae (Yonge, 1972) or *Lima lima* (Stanley, 1970), are able to move over surfaces by extension of the narrow, turgid foot, followed by anchorage either by direct adhesion to the substratum, possibly using a mucous film, or by the use of byssus threads (Yonge, 1972). Contraction of the pedal retractor muscles then draws the shell forward over the anchored foot. This process may be repeated rhythmically, as occurs in *Mytilus* (Morton, 1964) or in young *Tridacna* (Yonge, 1936). *Lasaea* and many small Erycinacea wander freely; the narrow foot leaves a mucous attachment trail while a byssus is secreted for fixation intermittently. Members of this genus climb steep surfaces and utilize the secretion of a byssus as a safety line against falling off the substratum (Morton, 1960). Creeping in bivalves follows a rhythm of pedal extension and retraction in association with the alternate anchorage of the shell and pedal sole, respectively. It resembes the discontinuous locomotory pattern described for *Strombus* (Section II,C) rather than that of the passage of muscular waves along the sole of the foot.

A small number of bivalve species are able to leap, e.g., *Neotrigonia margaritacea* (Stanley, 1977) and members of the Cardiidae (Ansell, 1969). The foot of most of these species resembles an L shape and straightens with the contraction of the intrinsic pedal muscles, levering the shell upward.

In a study of the motor performances of some cephalopods (Trueman and Packard, 1968), the relatively poor swimming performance of the octopus was noted. The octopus is essentially bottom-living and crawls by using its powerful arms. In marked contrast to squid, only about 10% of its body weight is represented by the mantle muscles, the most powerful muscles being in the arms. This muscular development may be summarized in the statement that although the jet swimming pulls of small octopuses are equivalent to less than half the body weight, the pulls using five arms are 30–100 times the body weight.

G. Extrusion from the Shell

Extrusion from the shell is a particular problem encountered by gastropods and represents a necessary preliminary to locomotion. The gastropod shell provides protection and support, but when the animal is protracted, it depends largely on the blood as a hydrostatic skeleton for support (Jones, 1975, 1978). Dale (1974) has shown that in *Helix* extrusion is essentially a hydraulic process achieved in a step-line manner, each step being an exaggeration of a normal respiratory cycle. There must also be redistribution of blood to the head during extrusion, although how blood pressure is increased in the cephalic region is not always clear (Brown, 1964; Russell-Hunter and Russell-Hunter, 1968). In the nassarid *Bullia,* the columellar muscle is in part composed of a three-dimensional network of muscle fibers which could well bring about extrusion because recordings of pressure from different body cavities of *Bullia* did not show any fluctuations in any way coincident with extrusion (Trueman and Brown, 1976).

III. Burrowing

A. Introduction

The Bivalvia are the burrowing molluscs *par excellence,* but it should be remembered that this mode of life has also been exploited by members of the Gastropoda, Scaphopoda, and Cephalopoda. Whereas Gastropoda have adapted secondarily to an infaunal habitat and Cephalopoda, e.g., *Sepia,* burrow shallowly be gentle fin movements, the bivalves must be considered to be primitively infaunal, with a bivalved shell and a compressed, bladelike foot with which to thrust ahead into sand or mud (Trueman, 1976).

Many marine molluscs burrow into soft, relatively unstable substrata from sands containing some granules more than 2 mm in diameter through clean, fine sand (diameter about 0.2 mm) to fine estuarine silts. Marine soils may have different physical properties which are important to infaunal animals (Webb, 1969). A dilatant sand becomes hard and more resistant to shear as increased force is applied, whereas thixotropic systems show reduced resistance to increased rates of shear. Thus, anchorage in sand requires a material with dilatant qualities, whereas motion may be facilitated in a thixotropic system (Trueman and Ansell, 1969).

Clark (1964) suggested that all soft-bodied animals burrow in an essentially similar manner based on the formation of two types of anchor, applied alternatively to produce a stepping motion, each step being termed a *digging cycle* (Trueman and Ansell, 1969). The first anchor is produced by dilatation of the body above the distal extremity, termed the *penetration anchor;* this prevents the animal from being pushed out of the sand as it thrusts downward. In bivalves, it is effected by opening the valves by the elasticity of the ligament against the adjacent sand. The second anchor is formed by a terminal dilatation (termed the *terminal anchor*) which allows the body to be drawn into the burrow by contraction of retractor muscles (Fig. 8). To enable such dilatations of the body to take place, animals which burrow have characteristically developed large, fluid-filled cavities, e.g., in *Bullia* (Trueman and Brown, 1976) and *Ensis* (Trueman, 1967). These allow both changes in shape and the transfer of muscular force through a hydraulic system (Trueman, 1975).

Burrowing activity from the surface of the substratum until motion ceases, termed the *digging period,* may be divided into two parts: (i) initial penetration when cycles occur only sporadically; (ii) movement into and through the substratum when cycles follow in regular succession (Fig. 9). In regard to bivalves, burial usually terminates when the siphons extend just to the sand surface. All animals that burrow into a substratum must be able to escape from it, particularly following deposition of additional sand. Bivalves are able to accomplish this by pushing down with the foot to force the shell upward, e.g., *Donax* (Trueman, 1971) and *Cardium* (Ansell, 1967). Snails emerge by burrowing upward, e.g., *Bullia* (Trueman and Brown, 1976), in the same manner as normal progression through the substratum. *Donax denticulatus* and other species of *Donax* utilize their ability to emerge from the sand and to burrow rapidly to effect migrations up or down tropical beaches with the rising and falling tide (Trueman, 1971).

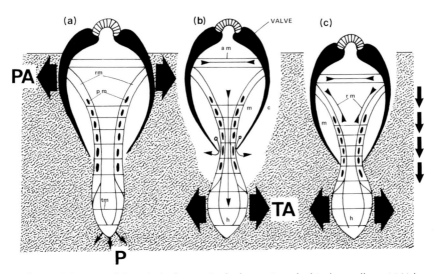

Fig. 8. Diagrams of the principal stages in the burrowing of a bivalve mollusc. (a) Valves press against the sand by an opening thrust of the ligament to provide a penetration anchor (PA) while the foot probes downward (P). (b) Adductor muscles (am) contract, ejecting water from the mantle cavity to form a cavity in the sand (c) and high pressure in the pedal hemocoel (h), causing dilatation forming a terminal anchor (TA). (c) Retractor muscles (rm) contract to pull the shell down toward the foot (arrows). tm, Transverse muscles; pm, protractor muscles; ▶━━━◀, tension in adductor muscle.

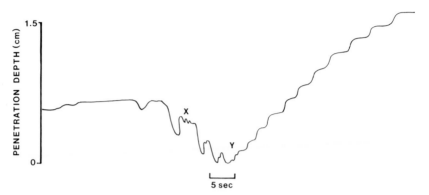

Fig. 9. A complete digging period of the bivalve *Donax vittatus*, recorded by attaching a thread from the posterior of the shell to an isotonic transducer. Two phases are shown, namely, probing of the foot (X) to penetrate the sand with the animal lying on its side, followed at Y, when the shell is drawn into an erect position, by a sequence of digging cycles. Upstrokes represent penetration of the shell into the sand.

B. Bivalvia

1. Initial Penetration

The digging period commences with the clam lying on its side with the foot probing into the sand. The force with which the foot can enter the substratum is limited because the only resistance to backthrust derived from probing is due to the weight of the shell upon the sand. Eventually, after several failures, the foot penetrates far enough to obtain sufficient purchase in the sand for the shell to be drawn erect (Fig. 9, Y). This marks the commencement of the second phase, in which cycles follow in quick succession.

2. Movement into the Substratum

Movement through sand consists of a series of digging cycles which are essentially similar for all burrowing bivalves and are best explained by reference to Fig. 10, in which a typical dimyarian bivalve, e.g., *Tellina*, *Cardium*, or *Anodonta*, is represented. The application of the terminal anchor by adduction of the valves and massive blood flow into the foot is seen clearly in Fig. 8. Representative genera investigated include *Nucula*, *Glycymeris*, *Margaritifera*, *Anodonta*, *Cardium*, *Tellina*, *Donax*, *Mercenaria*, *Mactra*, *Mya*, and *Ensis* (Trueman, 1968a), and in all, the sequence of events is the same during a digging cycle, although the precise duration of each part may vary. Notable differences occur in *Mactra*, in which the rocking motion of the valves brought about by the contraction of first the anterior and then the posterior retractor muscles (Fig. 10) is repeated (Trueman, 1968a); and in bivalves with elongated valves such as

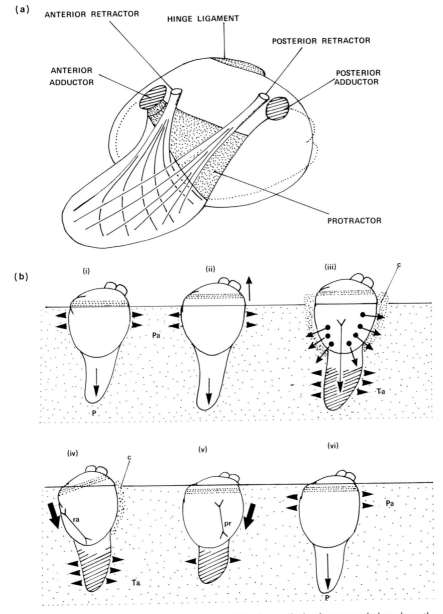

Fig. 10. (a) Sagittal section of a generalized bivalve with the foot extended to show the principal musculature; (b) diagrams of successive stages (**i–vi**) in the burrowing of a typical bivalve, showing penetration (Pa) and terminal (Ta) anchors. The band across the shell shows movement relative to the sand surface. Note the siphonal closure (**ii–iv**), probing of the foot (p), adduction of valves (**iii**) forming a cavity in the sand (c), and blood flow causing pedal dilatation and contraction of the anterior (ra, **iv**) and posterior (pr, **v**) retractor muscles. Heavy arrows indicate the rocking motion of the shell; arrowheads indicate anchorage.

Ensis and *Solenomya* and to a more limited extent in *Donax*, which penetrate the sand along the long axis of the shell, the rocking movement being suppressed, with posterior retraction attaining greater importance (Trueman, 1975).

3. Fluid Dynamics of Burrowing

Digging cycles consist essentially of two successive phases. The first is penetration of the substratum by pedal probing while the valves of the shell form a penetration anchor by the outward thrust of the ligament (Fig. 8a). Probing is carried out by the intrinsic pedal musculature at relatively low pressures, whereas during the second phase much higher pressures are developed.

Anchorage of the foot is achieved by adduction of the valves (Fig. 8b), which (i) brings about removal of the penetration anchor; (ii) generates a pressure pulse (Fig. 11a) which forces blood into the foot, causing pedal dilatation; and (iii) a powerful jet of water from the mantle cavity into the sand. Thus, bivalves have a double fluid-muscle system whereby they not only anchor the foot but facilitate shell movement. Indeed, in some bivalves with broad shells, e.g., *Mercenaria* (Ansell and Trueman, 1967), downward motion is more a passive settlement of the shell into the cavity produced by the water jet than an active pulling of the shell into the sand by means of the pedal retractors. Adduction of the valves generates a pressure pulse in both the hemocoel and the mantle cavity of 5–10 kPa but of longer duration in the hemocoel. It is assumed that blood is prevented from flowing out of the foot by Keber's valve, ensuring pedal dilatation while

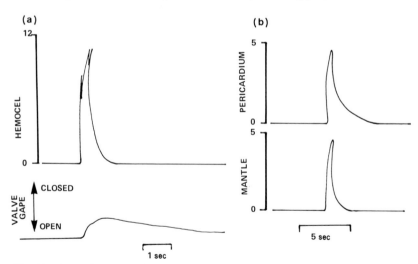

Fig. 11. Internal pressure (kPa) recorded during burrowing in (**a**) *Ensis arcuatus,* from a pedal hemocoel, with valve closure shown beneath, and (**b**), *Margaritifera margaritifera* from mantle and pericardial cavities. Pulses of similar amplitude but different duration occur in (**b**). (From Trueman, 1967, 1968.)

water is ejected from the mantle cavity. When adduction is completed, the mantle cavity pulse terminates, whereas that of the hemocoel persists, principally by pedal retraction, which commonly produces secondary peaks of pressure (Fig. 11a). In *Ensis* the foot increases in volume by 2 ml at adduction, and more than 4 ml of water is expelled from the mantle cavity (Trueman, 1975).

4. Effect of Shell Shape on Burial

Attempts have been made to assess the effect of shell shape on burrowing. These involve the comparison of movement per unit of applied force for dead shells being pushed into the sand with the force developed by the retractor muscles (Nair and Ansell, 1968a), and consideration of morphological adaptations to life in soft substrata (Thayer, 1975). In bivalves with slim shells, e.g., *Tellina,* the forces were relatively similar, whereas in bivalves with more tumid shells, e.g., *Mactra,* the motion during burrowing may be up to 75% greater in life than in with the same force applied to a dead shell. This may be accounted for by the use of both rocking motion and water jets. In an investigation of shell ornamentation and burrowing in the Trigonidae, Stanley (1977) devised a machine to make dead shells burrow with a rocking motion, but it must be appreciated that the experimental burial of dead shells can never be an entirely satisfactory substitute for the digging animal.

Stanley (1970) makes comparisons between species of bivalves by calculating a *burrowing rate index:*

$$\mathrm{BRI} = \sqrt[3]{\frac{\text{mass of animal}}{\text{burrowing time}}} \times 100$$

(Burrowing time is measured from the erect position to the completion of burial.)

Species which burrow rapidly have an index greater than 2, e.g., *Spisula solidissima* 4, *Ensis directus* 5, *Tivela mactroides* 8, *D. denticulatus* 17; and slow burrowers have an index of less than 1, e.g., *Mercenaria mercenaria* 0.8, *Macoma balthica* 0.7, *Anadara ovalis* 0.3, *Lucinia pennsylvanica* 0.03. Stanley (1969, 1970, 1975) has shown that most rapid burrowers have developed shells that are streamlined, e.g., *Tellina, Donax,* and *Ensis,* and that strongly ornamented and thick valves are employed primarily by shallow burrowers for stability near the sediment surface, e.g., *Chione, Cardium,* and *Mercenaria.* Experiments with shells of the Trigonidae indicated that in this group, at least, discordant shell ornamentation tends to aid burial (Stanley, 1977)

A further feature of shells in relation to burrowing is the hinge teeth. It is important that the valves not be rotated about one another, damaging the ligament during motion through the substrata. All burrowing movements are carried out with the valves gaping to allow the foot to be extended; the hinge teeth, situated umbonally, are then the only part of the shell remaining in contact. Stanley (1977) has shown how this occurs in *Neotrigonia,* in which the hinge teeth are relatively enormous and must constitute a physical means of aligning the valves. The possible existence of a proprioceptive sense organ in association with hinge teeth should not be overlooked (Trueman, 1968a).

5. Control of Burrowing

Burial involves the coordination of much of the body musculature, but we know little of the responses which initiate burial and control the digging cycle. Most bivalves live inactively on the sediment surface for long periods when removed from the substratum, but a few species, notably *D. denticulatus* and *Ensis* spp. (Trueman, 1968a), respond to tactile stimulation by digging and therefore are suitable experimental animals. Rotary motion of this species of *Donax* causes the valves to open and the foot to extend, with digging commencing only after a tactile response by the foot to sand on its lateral surfaces (Trueman, 1971). *Macoma balthica* responds to vibrations in the sand by burrowing activity (Mosher, 1972), and Breum (1970) has demonstrated that this and other species may react to wave action by burrowing. During the ensuing digging cycles, the clams do not respond to external stimuli except in regard to the duration of pedal retraction and probing. It seems likely that some parts of the digging cycle may be programmed within the nervous system with a minimum of peripheral feedback, such as siphonal closure and adduction.

In contrast, the duration of probing is probably controlled by pedal stretch receptors which respond to the physical properties of the medium.

C. Scaphopoda

Essentially the same mechanisms and sequence of activities are involved in the burrowing process of *Dentalium* as in Bivalvia (Trueman, 1968c), for in both taxa digging consists of the integration of pedal protraction and retraction with the application of penetration and terminal anchors (Trueman, 1968a). The principal difference is that in Scaphopoda the tubular shell cannot eject mantle cavity water by adduction, and *Dentalium* is unable to use water jets to loosen the sand or such high pressures in the hemocoel for pedal dilatation. It accomplishes the latter by the erection of a fold around the foot, termed the *epipodial lobes,* and achieves an anchorage of comfortable strength to bivalves, taking its size into account. In contrast, the probing force of the foot is weak because the penetration or shell anchor of *Dentalium* is largely limited to its own weight rather than to the outward force of the hinge ligament.

D. Gastropoda

A small number of gastropods are adapted for burrowing. These move either by cilia, e.g., naticids in the initial stages of burrowing (Trueman and Ansell, 1969), arrhythmic, discontinuous locomotion, e.g., *Conus abbreviatus* (Miller,

1974b), or by rhythmic pedal waves, e.g., *Bullia* (Trueman and Brown, 1976) and *Polinices* (Trueman and Ansell, 1969). It is likely that in *Polinices*, as in other naticids, both ciliary and muscular pedal waves are utilized. The muscular, rhythmic waves in *Bullia* are similar to those used in progression over hard substrata. In the basic movement, the whole anterior part of the foot moves forward into the sand, as in a retrograde monotaxic wave, whereas the metapodium, and in some the shell, remain still so as to resemble a penetration anchor; the propodium then dilates to form a terminal anchor, and the snail is pulled forward by columellar and longitudinal pedal muscle fibers (Fig. 12b,c) Slow stepping with the alternative application of penetration and terminal anchors is essentially an adaptation of normal gastropod locomotion for an infaunal life. This must have occurred in evolution on a number of separate occasions, for example, in the Nassaridae, Naticidae, and Terebridae.

In *Bullia digitalis* the foot is extremely flattened and has an enlarged pedal sinus containing blood which functions as a hydraulic organ (Trueman and Brown, 1976) (Fig. 12a). By contrast, the foot of *Polinices josephinus* is plow-shaped and contains both blood and water within an aquiferous pedal system (Trueman, 1968b). Details of pedal expansion with the taking in of water has also been demonstrated in other naticid species (Russell-Hunter and Russell-Hunter, 1968). The foot of members of both these groups contrasts to that of other gastropods which move over surfaces, e.g., *Patella,* and has no large, fluid-filled cavity. In *Bullia* there are two pressure pulses of up to 3 kPa during each step of digging cycle. One affects propodial extension and is generated by intrinsic pedal musculature, and the other occurs as the shell is drawn forward by contraction of the columellar muscle (Fig. 12). It is notable that the level of pressure in the pedal sinus rapidly increases, presumably due to muscle tension in the foot, for this also occurs in response to water currents when the snail is lying on the sand surface. Brown (1961) has frequently observed species of *Bullia* raising their feet and holding them turgid like a sail to allow waves to bring them up the beach. *Bullia digitalis* has also been observed to eject a small volume of water from the mantle cavity as the shell is drawn forward. This is done for the same function, but less powerfully and in a different manner, as the ejection of a jet of water from the mantle cavity of bivalves.

IV. Boring

Boring entails piercing a hole into a solid substance and may involve rotation about the longitudinal axis of the hole. Predatory gastropods, e.g., *Thais, Murex,* and *Natica,* obtain food by drilling small holes through calcareous shells (Owen, 1966). The radula abrades the shell, whereas the secretion of an accessory boring organ serves to dissolve and soften the shell (Carricker, 1961). It is,

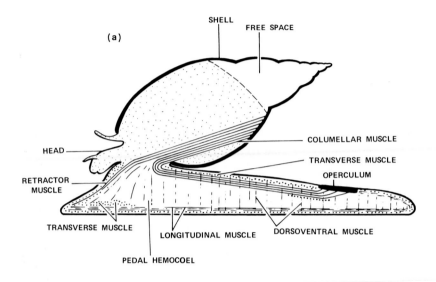

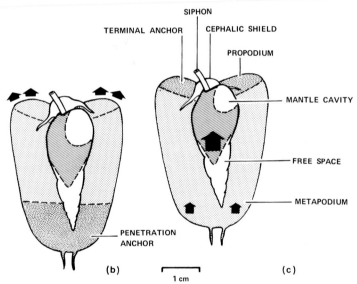

Fig. 12. (a) Diagrammatic sagittal section of *Bullia digitalis* with the foot extended, as in locomotion; (**b,c**) diagrams of *B. digitalis,* from ciné film, showing movement of the foot and shell. The propodium extends in "breast stroke" manner (**b,** arrows) with the cephalic shield being forced upward and the posterior of the foot anchored. The shell and metapodium are then drawn forward (**c,** arrows) and the head downward. (After Trueman and Brown, 1976.)

however, in the bivalves that the ability to bore into rock and wood is best studied.

Rock boring has been adopted independently in seven superfamilies of the Bivalvia, namely, the Mytilacea, Veneracea, Cardiacea (Tridacnidae), Gastrochaenacea, Saxicavacea, Myacea, and Adesmacea, but only in the Mytilacea is mechanical boring assisted by chemical means (Ansell and Nair, 1969). Rock boring has been arrived at by two different routes: (i) from an infaunal habitat by adaptation to burrowing in progressively stiffer muds, as by members of the Myacea (*Platyodon*) and Adesmacea (*Zirphaea*); and (ii) from animals attached byssally to a hard substratum where contraction of pedal retractors leads to abrasion of the rock, as in members of the Mytilidae (*Botula*) (Pojeta and Palmer, 1976; Yonge, 1955, 1963). The movements involved in boring in both of these categories are best presented diagrammatically (Fig. 13). Nair and Ansell (1968b) have analyzed the mechanism of boring in *Zirphaea* and demonstrated that maximum pressure pulses of 0.6 kPa occurred during the extrusion of pseudofeces. They concluded that a high-pressure hydraulic system was not a feature of boring in pholads, in contrast to those observed in many burrowing species. The ligament has been lost, allowing the umbones of the valves to act as a ball and socket joint, so that the hydraulic system of deep burrowers (Trueman 1968d) has been at least partially replaced by the shell acting as a jointed exoskeleton. The full development of this type of movement is reached in the wood-boring genus *Teredo* (Board, 1970; Nair and Ansell, 1968b).

V. Swimming

A. Introduction

Members of the Gastropoda and Bivalvia that swim have adopted their body form for this purpose in different ways; but in the Cephalopoda this is the characteristic mode of locomotion to which they are all adapted. Near neutral buoyancy is advantageous, and in all groups there is a tendency to reduce the weight of the shell and viscera, for example, in Heteropoda (Morton, 1964) or to develop buoyancy devices, such as air chambers in the cephalopod shell or substitution of heavy cations with ammonium ions in cranchid squids (Denton, 1961). Swimming in adult molluscs is achieved either by the rhythmic contraction of fins or paddle-like parapodial appendages or by jet propulsion. Molluscs swimming by these mechanisms have a Reynolds number in excess of 1000, so that stresses due to the viscosity of the medium are minimal and the animal relies on the inertia of water to generate forward thrust (Alexander, 1968, 1978).

Pelagic larval gastropods and bivalves characteristically swim by means of cilia. Little is known of the dynamics of their locomotion, but their Reynolds

(a)

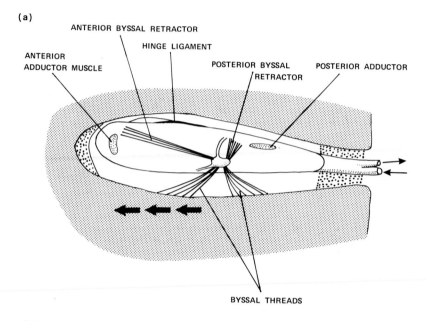

(b)

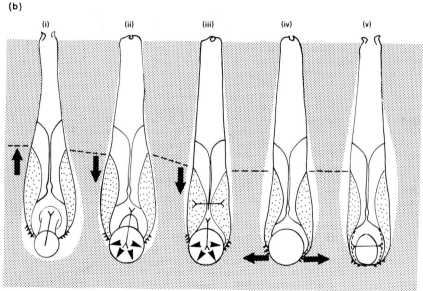

Fig. 13. Boring movements of bivalves. Large arrows indicate movement of the valves and main abrasive action; fine stippling shows rock. (**a**) *Botula falcata* main abrasive action by posterior byssal retractors; (**b**) successive stages of the boring cycle of *Zirphaea crispata* viewed ventrally. (**i**) foot extending, with siphons open; (**ii**) siphons closed, tip of the foot dilated with the margin extended, retraction starting; (**iii**) retraction continues, ventral margin of the valves

numbers must lie in the range $10^{-2}-10^{-3}$, where viscous forces are of major importance (Lighthill, 1969). An additional means of passive movement, drifting my means of byssal threads, has been described in young postlarval bivalves (Sigurdsson et al., 1976).

B. Undulatory Waves and Paddling

1. Undulatory Waves

Undulatory swimming is found throughout the animal kingdom. Apart from errant polychetes, all animals with elongate bodies swim by means of retrograde undulations that pass along the body, exerting backthrust against the adjacent water to provide motive force (Fig. 14a) (Gray, 1968; Trueman, 1975). Species that habitually swim tend to increase the area of the propulsive surface either by lateral compression, as in *Phyllirhoe,* or by enlarged parapodial lobes of the foot.

A few opisthobranchs, e.g., *Phyllirhoe,* swim by a rhythmic wave which passes over the whole laterally compressed body (Morton, 1979). Members of this genus have no foot but instead a sharp ventral margin. Other genera, such as *Dendronotus,* which are only temporarily pelagic, retain a narrow foot for creeping but swim effectively, with a complex retrograde muscular wave at a rate of about 5 cm/sec (Morton, 1960). The larger dorid, *Hexabranchus* (Fig. 14a,b), swims intermittently, with parallel undulatory waves passing down both sides of the body along a wide notal margin, each in about 4 sec. This illustrates the principles of undulatory propulsion clearly, but there is in addition a simultaneous slower bending of the body that forces water downward in the manner of a jet, moving the animal upward (Edmunds, 1968; Morton, 1964).

Undulatory contraction of enlarged parapodia is the most important swimming mechanism in the Opisthobranchia (for discussion, see Bebbington and Hughes, 1973). In *Pleurobranchus* (Thompson and Slinn, 1959) the body is short compared with the wavelength of the undulations and propulsion is achieved by distinct but successive power and recovery strokes (Fig. 14c). Perhaps the best developed parapodial swimming is seen in the aplysiid *Akera* (Morton, 1964), in which wide, cloaklike but unfused parapodia envelop the shell and visceral mass (Fig. 14d). Contraction closes the skirtlike parapodia, ejecting water downward and resulting in upward motion, although the heavy body soon falls. *Aplysia*

---

drawn together with the action of the accessory ventral adductor muscle; (iv) contraction of the posterior adductor muscle, anterior areas of the valves abrading the burrow as they diverge and siphons reopening; (v) siphons open, with contraction of the anterior adductor muscle and divergence of the posterior margins of the shell. >———<, contraction of adductor or retractor muscles; ———<, pressure in the hemocoel; extension of foot. (a: After Yonge, 1955; b: after Nair and Ansell, 1968b.)

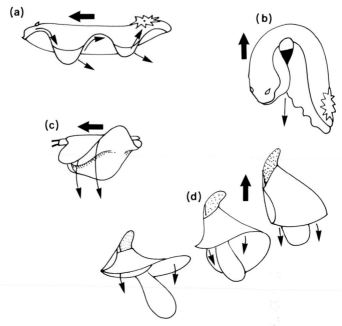

Fig. 14. Diagrams of opistobranchs swimming; large arrows indicate the movement of the animal and small arrows the contraction of parapodia or the motion of water. (**a, b**) *Hexabranchus marginalis*: (**a**) in the lateral aspect with retrograde locomotory wave in notal margin, (**b**) upward swimming by rapid flexing of the body. (**c**) *Pleurobranchus membranaceus* during the power stroke; (**d**) stages in the execution of the power stroke by *Akera bullata*. (**a, b, d**: After Morton, 1964; **c**: after Thompson and Slinn, 1959.)

fasciata swims in a rather similar manner, a specimen 20 cm in body length covering distances of 6–12 m, with some 27–75 beats/min (Bebbington and Hughes, 1973).

In *Notarchus* (Martin, 1966), a further modification of parapodial swimming results in the ejection of water through a small, funnel-shaped slit in a simple form of jet propulsion (Section V,C). These two latter types of swimming both occur in animals which are characteristically bottom dwellers, moving by pedal muscular waves, possibly as escape mechanisms.

In the permanently pelagic Heteropoda the foot forms a thin, muscular median fin (Morton, 1964). The least modified family, the Atlantidae, retains a small, spiral, compressed shell with a complete foot and operculum. In more specialized families, such as the Pterotracheidae, the shell and viscera are much reduced and the fin is the only remaining part of the foot. The swimming position is upside down, with the body flexing vigorously while the foot continually exhibits retrograde undulatory waves (Fig. 15a).

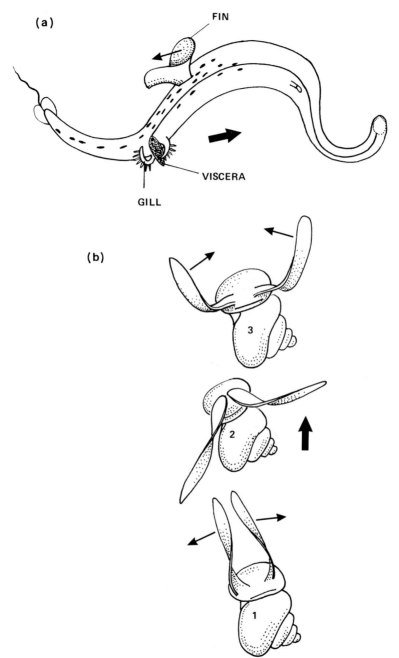

Fig. 15. Diagrams of swimming in (**a**) *Pterotrachea* by the median pedal fin, and (**b**) *Limacina* showing successive positions of the epipodial wings during upward swimming. 1, 2, power stroke; 3, recovery. Large arrows indicate swimming movement and small arrows movement of fins. (After Morton, 1954.)

The final example of undulatory propulsion is observed in the cuttlefish *Sepia*, in which movements of the marginal fins can move the animal gently backward or forward according to the direction of the undulatory waves (Boycott, 1958). The fins are more localized in squid, in which contraction resembles the motion shown in Fig. 14c for *Pleurobranchus*, and are probably of importance for maneuvering at low speeds (Lighthill, 1969). A similar but more rapid contraction of the fins of squid occurs during fast jet swimming a single beat being coincident with the expulsion of a powerful water jet in *Alloteuthis* (M. R. Clarke and E. R. Trueman, unpublished observations).

2. Paddling

In the Pteropoda, Thecosomata, the parapodia are specialized, forming long "wings" for permanent swimming. In *Limacina* (Morton, 1964) (Fig. 15b) the wings are used together in the manner of oars, each downward power stroke imparting upward motion to the animal on a spiral course. When outspread, the wings give a degree of flotation, but as a relatively heavy body and shell are retained, *Limacina* drops rapidly when the wings are held together above the body. *Limacina* remains pelagic by incessant swimming, but its small size (<2 mm) implies a Reynolds number of about 20 when viscous forces must play an important role in its maintenance of position.

Further evolution of the Thecosomata, with larger size, has led to the reduction of the shell, as in *Cavolinia*. The adaptations to swimming found within the Cavoliniidae and in the larger thecosomes of the Cymbuliidae are reviewed by Morton (1964). The Gymnosomata, which have converted the parapodial wings into two-way sculling organs and lost their shell, contain the fastest gastropod swimmers, e.g., *Clione* (Morton, 1964).

C. Jet Propulsion

1. Introduction

Jet propulsion is utilized by cephalopods, e.g., *Loligo* (Trueman and Packard, 1968); *Illex* (Bradbury and Aldrich, 1969), as the normal means of locomotion; and by the opisthobranch *Notarchus*, scallops, and most cephalopods for escape movements. The principle that water ejected in one direction results in motion in the opposite direction applies to all animals using jet propulsion. To accomplish this successfully, a body form is required which allows the expulsion of a maximal amount of water on each power stroke of the jet cycle, with the generation of a pulse of high pressure in a water-filled chamber. The mantle cavity fulfills this role in all molluscs except *Notarchus*, in which water is ejected from between parapodial folds (Martin, 1966). During recovery, the mantle cavity is refilled with water for a further power stroke, the pressure in the chamber falling below ambient. The effectiveness of jet propulsion depends on (i) a large Reynolds number, so that inertia is dominant over viscosity (Lighthill, 1969), and (ii) a large mass of ejected water in proportion to the mass of the animal. The velocity of the jet is related to the strength of the muscles producing a fluid pressure in the mantle chamber and the size of the jet aperture (Trueman, 1980). The relatively great capacity and powerful muscles of the mantle of the squid, *Loligo vulgaris,* a powerful swimmer, have been contrasted to the relatively smaller cavity of weaker muscles of *Octopus* (Trueman and Packard, 1968). Maximum jet aperture and chamber pressure produce the greatest thrust, but at the consequent high velocity the animal is retarded by drag forces even between rapidly executed, successive power strokes. *Loligo vulgaris* 20 cm in length swimming rapidly at 2 m/sec would have a Reynolds number of 4×10^5 and at this velocity an unstreamlined shape would result in high drag (Alexander, 1978). It should also be noted that in addition to normal profile drag, there will be a drag component due to the refilling of the chamber during the recovery stroke (Q. Bone and E. R. Trueman, unpublished observations). Packard (1969) has shown by stroboscopic and ciné photography that *L. vulgaris* with a body weight of 100 g has a maximum acceleration of 3.3 *g.* and that deceleration occurs less than 150 msec after acceleration commences. Jet swimming also involves consideration of the dynamics of swimming in a fluid and the effect of fins or buoyancy organs.

The presence of a web between the arms of members of the Octopoda facilitates jet swimming. This web is particularly well developed in deep benthic-living octopods of the superfamily Cirroteuthacea, such as *Cirrothauma,* in which the mantle and funnel are reduced and a large web hangs down like a skirt (Morton, 1979). This animal swims slowly by opening and closing the web.

2. Coleoid Cephalopods

a. Swimming. Coleoids have the ability to swim differently by means of jet of different velocities. Maximal contraction of the mantle muscles results in fast escape swimming; deep respiration, as in *Loligo* (Trueman and Packard, 1968), produces a cruising motion; and the gentle expulsion of water from respiratory movements of the mantle through a downwardly directed funnel, as in *Illex* (Bradbury and Aldrich, 1969), leads to motionless hovering. Associated rhythmic undulations of the fins have also been noted by these authors and by Zuev (1965a, 1965b).

b. Mantle Structure. The mantle cavity approximates a cylindrical shape in many coleoids, having slits between the mantle margin and the funnel and head through which water is drawn in during inhalation. Water flow is unidirectional. Water jets pass through the funnel, being prevented from passing through the slits by dilatation of the funnel base when positive pressures occur in the mantle cavity. Inhalation through the funnel is prevented by a flap valve. The mantle consists of muscle fibers sandwiched between outer and inner tunics containing collagen fibers whose axes in successive layers run alternatively in left- and right-handed helices (Wainwright et al., 1976). Recently, Bone et al. (1981) have described the presence of a more elaborate framework of connective tissue fibers running in different planes in the mantle of *Alloteuthis, Sepia,* and *Loligo* (Fig. 16). Some of these appear elastic, but in any event they must restrict the manner in which the mantle can be deformed during muscle contraction. The mantle muscles are principally circular fibers lying in the transverse plane of the mantle. These contract to effect the power stroke (Fig. 16, 17) when the mantle must either thicken or lengthen so that the total volume of muscle remains constant. Little change of length has been observed (Packard and Trueman, 1974; Ward and Wainwright, 1972), but thickening of the mantle is marked. The circular muscles are antagonized directly by thin sheets of radial fibers (Fig. 17), and contraction of the radials causes the mantle to become thinner, again with no perceptible change in mantle length (Packard and Trueman, 1974). Contraction of the radial muscles to give the mantle an almost balloon-like appearance, which

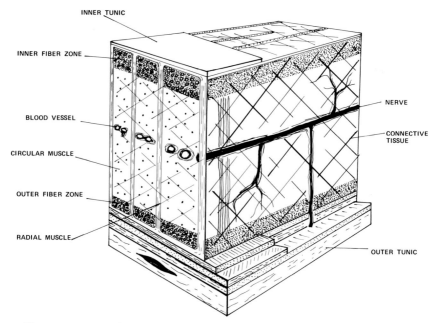

Fig. 16. Diagram of squid mantle showing a relatively thick outer tunic with the chromatophore, muscle, nerve, and blood vessels. The disposition of the outer and inner fiber zones and connective tissue are also indicated. (After Bone et al., 1981.)

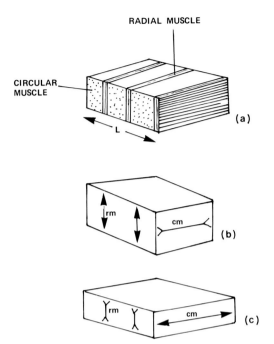

Fig. 17. Diagrams illustrating the changes of shape in a block of mantle muscle in which the length (L) remains constant. (**a**) Resting condition; (**b**) thickening resulting from contraction of the circular muscles (cm) and stretching of the radials (rm); (**c**) thinning by contraction of radial muscles, as occurs during hyperinflation of the mantle.

may be sustained for several seconds, has been observed by the latter authors in *Sepia*. It is likely that the radial muscles are not the sole means of recovery of the mantle after the power stroke, for pressure records (Fig. 18d) suggest that recovery is very rapid, possibly being effected by an elastic storage mechanism.

c. Power Stroke. Reduction of the mantle circumference causes a decrease in volume proportional to the square of the linear shortening of the circular mantle muscles. A reduction in the length of the mantle cavity would only yield a volume directly proportional to the shortening of the cavity (Ward and Wainwright, 1972). In *L. vulgaris* contraction of the circular muscles by about 30% of their length at a tensile stress of about $1.5 \times 10^5 N/m^2$ produces a pressure pulse of 30 kPa, with the expulsion of most of the water contained within the mantle cavity (Trueman and Packard, 1968). The maximal pressure pulse produced by an all-or-nothing contraction of the circular mantle muscles is simple in form, having a short rise time and duration (Fig. 18a). The ommastrephid squids are very powerful swimmers and have been shown to generate pressures in excess of

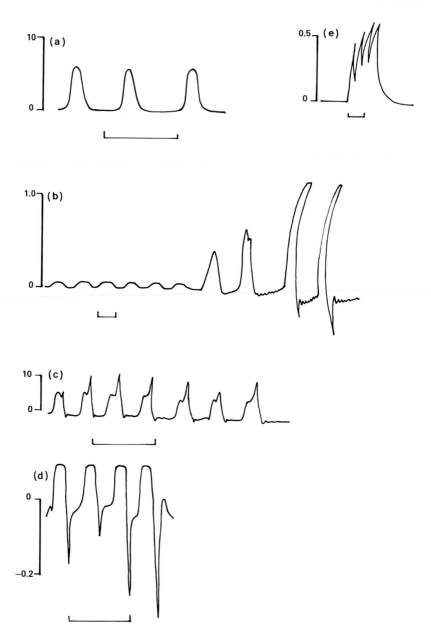

Fig. 18. Extracts of traces of pressure (kPa) from the mantle cavity during jetting, inhalation, and respiration of (**a–d**) *Sepia officinalis* and (**e**) *Chlamys opercularis*. (**a**) Succession of pulses due to contractions of the mantle muscles; (**b**) initial respiratory rhythm broken by jets of increasing amplitude, shorter rise time and duration; (**c**) rapid volley of pulses, with the super-

50 kPa of 150–200 msec duration (Trueman, 1980). Similar powerful pulses must also occur in the oceanic squid *Onycoteuthis* when it takes flight into the air during escape movements. Flight is due to expulsion of water from the mantle cavity during and immediately after takeoff, for their fins and arm membranes probably generate little lift. The giant squid *Dodiscus* can apparently attain a momentary velocity of about 7 m/sec in air (Morton, 1979).

Low-pressure pulses of about 0.2 kPa in *S. officinalis* and 0.5–1 kPa in *L. vulgaris* are produced by the same muscle fibers contracting less forcibly during respiration. In *Loligo*, which without positive buoyancy must continually swim, relatively deep respiratory pulses also have a significant locomotory function. The circular muscle fibers are of two types: (i) outer and inner mantle zones consisting of well-vascularized, mitochondrial-rich fibers, and (ii) mitochondrial-poor fibers in a sparse vascular bed (Fig. 16) (Bone et al., 1981). These authors suggest that the central fibers are used only in escape jetting, whereas the more peripheral fibers participate in rhythmic respiratory contractions. Maximal pressure pulses have a shorter duration and rise time than lower-amplitude, graded-pressure pulses (Fig. 18b). The association of shorter duration with higher pressures is probably due to the finite volume contained within the mantle cavity. When all available water has been expelled, the pressure pulse must terminate. This occurs most rapidly at maximal pressures provided the jet aperture is the same.

Recent observations using a catheter tip pressure transducer inserted into the mantle cavity of *S. officinalis* and the squid *Alloteuthis subulata* have, however, shown that the form of the pressure pulse is not always simple, *Sepia* (Fig. 18c) producing a spike superimposed on a more conventional pulse. Both of these vary independently, and the absence of evidence of a second contraction of the circular mantle muscles, together with retraction of the head when observed on ciné films of jetting, suggest that second system, the head and funnel retractor muscles, may be involved (Trueman, 1980). It may be noted that the head retractor muscles are served by second-order giant nerve fibers from the brain, as are the stellate ganglia (Young, 1938).

d. Recovery Stroke. Refilling of the mantle cavity is brought about by the radial muscles and elastic fibers of the mantle after jetting, but during respiration membranes at the base of the funnel in *Sepia* beat in a regular rhythm to draw water in (Trueman, 1980). During inhalation the mantle cavity of *Sepia* shows pressures of 1.5 kPa below ambient and of 50 msec duration while jetting at high pressure and frequency (Fig. 18d). These values correspond reasonably well with

---

imposed second pulse indicative of contractions of the head retractor muscles; (**d**) series of inhalant pulses, negative to ambient pressure, between jet pulses which saturate the recorder; (**e**) volley of jet pulses. (After Moore and Trueman, 1971.)

theoretical estimations based on the size of the inhalant and jet apertures, which unfortunately are impossible to measure accurately (Trueman, 1980).

e. Jet Thrust. This may be determined by attaching a squid to a force transducer with a thread so as to record the tension exerted simultaneously with pressure. This technique has the advantage that comparisons of performance may be made without consideration of the effect of drag and of carried surface water (Table III). The motion of squid has been discussed with particular reference to the relation between jet thrust and pressure (P) (Johnson et al., 1972). Jet thrust $= 2\,C_{Dn} \cdot a \cdot P$ where C_{Dn} is the coefficient of discharge of the funnel and a is the area of its aperture. The relation between thrust measured experimentally and estimated is reasonably good for *Sepia* and *Loligo* but not for *Octopus*. This is probably due to the difficulties of obtaining accurate measurements of the funnel aperture (a) during jetting and the possibility of the octopus constricting the funnel during jetting (Trueman, 1980). A poor performance for *Octopus*, a bottom dweller, compared with a squid which continually swims, is not, however-er, unexpected from the relatively poorly developed mantle musculature (Table III). Generation of a jet pulse of lower pressure and jet velocity for a longer duration may well result in more efficient swimming than a maximal jet thrust. The swimming velocity would be less, as would be the drag, enabling the animal to travel farther with the same output of water from the mantle cavity (Johnson et al., 1972).

TABLE III

Summary of Swimming Performance by Jet Propulsion[a]

Animal	% of body vol. expelled	Mantle muscle as % of body wt.	Mantle pressure resp.-max. jet (kPa)	Frequency of jetting (Hz)	Velocity (m/sec)
Notarchus punctatus	—	—	—	0.3	0.08
Chlamys opercularis	50	—	−0.6	3	0.25
Nautilus macromphalus (687 g)	20	—	0.1–3	1	0.2
Loligo vulgaris (350 g)	50	35	0.5–30	2	2
Sepia officinalis (250 g)	25	30	0.2–20	3	0.8
Octopus vulgaris (220 g)	15	<10	0.02–20	—	—

[a] Wet body weight is given where known. Velocity refers to maximum instantaneous velocity (after Trueman, 1980, with the exception of *Nautilus;* Packard et al., 1980).

3. Nautilus

Many coleoid cephalopods are well adapted for fast jet swimming because of their streamlined shape and nearly cylindrical mantle cavity. Loss of an external shell has allowed this development, for in *Nautilus* the external shell precludes fast movement by jetting and maximal pressures of only 4 kPa occur in the funnel (Packard et al., 1980). The range of maximum swimming speeds estimated by these authors fell within the range of 0.12–0.2 m/sec for animals observed by divers (Ward et al., 1977). Packard et al. (1980) discuss the mechanism for the production of jet pulses and suggest that the shell (or retractor) muscles contract to produce them. Low-amplitude respiratory pulses seem to involve only the funnel and its flaps in *Nautilus* in a manner similar to the quiet breathing of cuttlefish, which is carried out solely by the anterior mantle musculature (Packard and Trueman, 1974). The shell muscle of *Nautilus* consists of longitudinal and transverse fibers, the latter being anchored in the surrounding connective tissue. Contraction of the transverse fibers would antagonize the longitudinal, extend the body from the shell, and enlarge the mantle cavity, a mechanism similar to that encountered in the mantle of coleoids. Similar sheets of transverse fibers are found in the head retractor muscles of coleoids (Young, 1938) and must undoubtedly have a similar function.

4. Scallops

Infaunal bivalves (Section III) exploit a specialized form of jet propulsion for burial, and a similar method of producing a jet of water is used in swimming by *Pecten* and *Chlamys* (Moore and Trueman, 1971).

In *C. opercularis* single adductions of the valves produce pressure pulses of 0.6 kPa in the mantle cavity while rapidly repeated adductions lead to a buildup of pressure (Fig. 18e) (Table III). Adduction of the valves is rapidly executed by the specialized rapid adductor muscle fibers (Thayer, 1972), and the jet may be directed in most directions to effect locomotion in the opposite direction away from any potential predator, such as a starfish (Fig. 19) (Moore and Trueman, 1971; Thomas and Gruffydd, 1971). Inhalation occurs as the valves open by means of the elastic energy stored in the ligament. This ligament is mechanically more efficient than that found in other bivalves and, coupled with relatively thin valves with appropriate hydrodynamic properties (Thorburn and Gruffydd, 1979), allows swimming with minimum energy cost (Trueman, 1980).

5. Notarchus

In this opisthobranch the parapodia form a cavity with a single opening through which water is both expelled and inhaled. The result is jet propulsion in a peculiar looping manner (Martin, 1966) (Fig. 20), enough force being generated by the jet to lift the animal off the bottom, where it crawls away from predators.

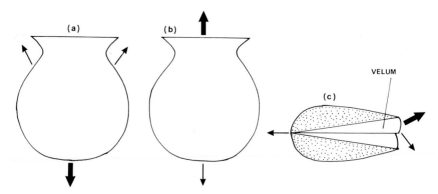

Fig. 19. Diagram of swimming movements of a scallop showing the direction of water jets (small arrows) and movement (large arrows). (a) Normal swimming from above; (b) "escape" movement with the ventral jet; (c) normal passing out between the mantle folds.

The looping motion consists of a single backward rotation during inhalation until the opening of the cavity faces downward when a further jet may recur, lifting *Notarchus.*

VI. Energetics of Locomotion

Quantitative comparisons between the different methods of locomotion in the Mollusca or between molluscs and members of other phyla may be made by determination of the mechanical energy required for the motion of unit body mass (J/kg) (Trueman and Jones, 1977). Currently, however, few determinations have been made. Energy cost may be estimated from oxygen consumption determinations during locomotion. However, it is difficult to achieve satisfactory results in a respirometer, and account must also be taken of anaerobic metabolism such as occurs during fast jetting of squid (Bone et al., 1981). An alternative method, used in *Donax* (Ansell and Trueman, 1973) and *Bullia* (Brown, 1979a, 1979b, 1981; Trueman and Brown, 1976), involves analysis to determine the mechanical energy used in movement. For burial, the rate of movement (*U*) may be measured directly, whereas the drag (*D*) can be ascertained by determining the maximum force exerted by attaching a thread from the animal to a force transducer. This force must normally overcome drag and represents the maximal drag experienced. *DU* represents the power required for burial, and from this the

Fig. 20. (a) Sagittal section through *Notarchus punctatus;* (b) reconstruction of swimming from ciné film; time interval, 0.4 sec. Propulsive (1–5, 12–13) and rolling (inhalation, 6–10) phases shown. (After Martin, 1966.)

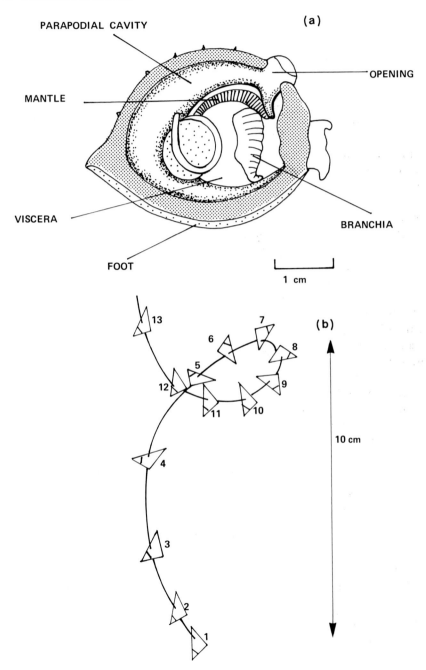

PARAPODIAL CAVITY   (a)

OPENING

MANTLE

VISCERA

BRANCHIA

FOOT

1 cm

13

7   (b)

6

8

5

12

9

11   10

4

10 cm

3

2

1

energy cost of burial may be calculated. For *Bullia digitalis* the energy cost of burial is 6.6 J/kg (Trueman and Brown, 1976). The energy cost of burial by the tropical clam *Donax incarnatus,* of 22 J/kg (Ansell and Trueman, 1973) is much greater than for *Bullia* and is probably related to its speed of burial, for the digging period is only 4.5 sec in duration. The European species *D. vittatus* burrows about 10 times more slowly and at less energy cost than *Bullia*.

A comparison may usefully be made between the amounts of mechanical energy required for burial in different animals (Table IV). The lower cost of burial of *Bullia* may be due to its relatively shallow penetration of the sand, whereas the two tropical species, *D. incarnatus* and the mole crab *Emerita,* burrow perpendicularly to the sand surface. The lower cost of burial of the latter, with their jointed exoskeleton, is significant. Trevor (1978) determined the cost of burrowing for some invertebrates and assumed a conversion efficiency of chemical to mechanical energy of 20%, comparing his results with data for locomotion obtained by respirometry. This showed the high cost of burrowing compared with running, flying, or swimming (Elder, 1980). Brown (1979b) has determined the energy cost of burial by *Bullia* both mechanically and by respirometry and suggests that a conversion factor of 6% would be more realistic. This further emphasizes the high cost of burial by molluscs. In an additional series of experiments, Brown (1979a, 1981) has determined the oxygen uptake during different activities (Table V). The high cost of transportation in the surf with the foot held turgid, like a sail, is significant in comparison to burrowing or crawling over sand; however, no account was taken of energy derived from anaerobic respiration, and the figures in Table V should be regarded as minimal values. This suggests that the cost of migratory movements up or down tropical beaches by *Donax* spp. may be much higher than previously assumed by Ansell and Trueman (1973).

TABLE IV

Energy Requirement for Burial Expressed as Mechanical Energy in relation to Dry Weight and Over Unit Distance (Mechanical Locomotory Cost, MLC)

	Mechanical energy (J/kg)	MLC (J/kg/m)
Bullia digitalis[a]	6.6	220
Donax incamatus[b]	22	1000
Nereis diversicolor[c]	100	1380
Emerita portoricensis[b]	3	200

[a] Trueman and Brown (1976).
[b] Ansell and Trueman (1973).
[c] Trevor (1978).

TABLE V

Energy Expenditure of *Bullia digitalis* during Single Tidal Cycles of Activity (Average of Eight Females for Standard-Sized Animal, 750 mg Dry Tissue Weight)[a]

Activity	Duration (min)	O_2 uptake per minute (μg/min)	Energetic equivalent (cal)
Transport in surf	14.5	20.8	1.03
Crawling	23.0	11.3	0.88
Burrowing	7.5	18.8	0.48
Emerging	3.8	18.8	0.25
Buried	695	9.95	24.8

[a] After Brown (1981).

The only evidence of the cost of crawling by pedal adhesion is presented by Denny (1981) in his work on the role of mucus in the crawling of the slug *Ariolimax*. He showed that the metabolic cost of crawling of 215 cal/kg/m was considerably greater than for burrowing (60) or crawling (36) by *Bullia* (Brown, 1979b). This may be due to the cost of producing mucus, by means of which the slug adheres to the substratum, rather than motion using the "breast-stroke" action of *Bullia*.

There are similarly few data available on the energy cost of swimming. Observations on the swimming of *Loligo* (Packard, 1969) and *Alloteuthis* (M. R. Clarke and E. R. Trueman, unpublished observations) indicate that the high velocity derived from a single jet cycle is of short duration and may be sustained only by frequent powerful jet cycles which rapidly exhaust the squid. This suggests that the cost of fast jet swimming is high. Alexander (1979) compares the Froude efficiency (useful power/useful power + power lost to the fluid) of jet propulsion in squid with the swiming of trout of the same size. At approximately similar swimming speeds (0.5 m/sec) the squid accelerates in its jets only one-seventh of the amount of water, but to a higher velocity than the trout to produce the same forward thrusts. The Froude efficiency of the squid is thus lower during fast swimming. However, it is likely that swimming at lower velocities with more gentle jets is a more economical means of locomotion. In rapid escape movements, avoidance of predation is more important than high energy cost (Trueman, 1980). The same principle also applies to burrowing, rapid reentry of the substratum after dislodgment being more important than the cost for active bivalves, such as *Donax*, and *Ensis*. In contrast, those that burrow slowly commonly have heavy shells, for example, *Mercenaria* and *Glycymeris*, and tend not to make rapid escape movements.

VII. Future Research

In the past 50 years there have been many articles describing the method of locomotion of molluscs. Although there may be opportunities to extend such qualitative work both in detail and in breadth of coverage in future years, the greatest need is for further quantitative investigations.

These should pay particular attention both to the analysis of mechanisms of locomotion, such as the role of mucus and the mechanisms involved in retrograde pedal waves, and to determining the energy cost of locomotion in different molluscs in the various modes of locomotion. Such work should lead to interesting comparisons between molluscs and between members of this phyla and other taxa.

References

Alexander, R. McN. (1968). "Animal mechanics." Sidgwick and Jackson, London.
Alexander, R. McN. (1978). Swimming. In "Mechanics and Energetics of Animal Locomotion" (R. McN. Alexander and G. Goldspink, eds.), pp. 222–248. Chapman and Hall, London.
Alexander, R. McN. (1979). "The invertebrates." University Press, Cambridge, England.
Ansell, A. D. (1967). Leaping and other movements in some cardiid bivalves. Anim. Behav. 15, 421–426.
Ansell, A. D. (1969). Leaping movements in the Bivalvia. Proc. Malac, Soc. London 38, 387–399.
Ansell, A. D., and Nair, N. B. (1969). A comparative study of bivalves which bore mainly by mechanical means. Am. Zool. 9, 857–868.
Ansell, A. D., and Trueman, E. R. (1967). Burrowing in Mercenaria mercenaria (L.) (Bivalvia, Veneridae). J. Exp. Biol. 46, 105–115.
Ansell, A. D., and Trueman, E. R. (1973). The energy cost of the migration of the bivalve Donax on tropical sandy beaches. Mar. Behav. Physiol. 2, 21–32.
Bebbington, A., and Hughes, G. M. (1973). Locomotion in Aplysia (Gastropoda, Opisthobranchia). Proc. Malac. Soc. London 40, 399–405.
Board, P. A. (1970). Some observations on the tunelling of shipworms. J. Zool. London 161, 193–201.
Bone, Q., Pulsford, A., and Chubb, A. D. (1981). Squid mantle muscle. J. Mar. Biol. Assoc. U. K. 61, 327–342.
Boycott, B. B. (1958). The cuttlefish - Sepia. New Biology 25, 98.
Bradbury, H. E., and Aldrich, F. A. (1969). Observations on locomotion of the short-finned squid, Illex illecebrosus illecebrosus (Lesueur, 1821), in captivity. Can. J. Zool. 47, 741–744.
Branch, G. M., and Marsh, A. C. (1978). Tenacity and shell shape in six Patella species: Adaptive features. J. Exp. mar. Biol. Ecol. 34, 111–130.
Breum, O. (1970). Stimulation of burrowing activity by wave action in some marine bivalves. Ophelia 8, 197–207.
Brown, A. C. (1961). Physiological-ecological studies on two sandy-beach Gastropoda from South Africa: Bullia digitalis Meuschen and Bullia laevissima (Gmelin). Z. Morph. Ökol. Tiere 49, 629–657.
Brown, A. C. (1964). Blood volumes, blood distribution and sea-water spaces in relation to expansion and retraction of the foot in Bullia (Gastropoda) J. Exp. Biol. 41, 837–854.

Brown, A. C. (1979a). Oxygen consumption of the sandy-beach whelk *Bullia digitalis* Meuschen at different levels of activity. *Comp. Biochem. Physiol. A* **62**, 673–675.

Brown, A. C. (1979b). The energy cost and efficiency of burrowing in the sandy-beach whelk *Bullia digitalis* (Dillwyn) (Nassariidae). *J. Exp. Mar. Biol. Ecol.* **40**, 149–154.

Brown, A. C. (1981). An estimate of the cost of free existence in the sandy-beach whelk *Bullia digitalis* (Dillwyn) on the west coast of South Africa. *J. Exp. Mar. Biol. Ecol.* **49**, 51–56.

Carricker, M. R. (1961). Comparative functional morphology of boring mechanisms in gastropods. *Am. Zool.* **1**, 263–266.

Clark, R. B. (1964). "Dynamics in metazoan evolution." Clarendon Press, Oxford, England.

Copeland, M. (1919). Locomotion in two species of the gastropod genus *Alectrion* with observations on the behaviour of pedal cilia. *Biol. Bull. Mar. Biol. Lab. Woods Hole* **37**, 126–138.

Copeland, M. (1922). Ciliary and muscular locomotion in the gastropod genus *Polinices*. *Biol. Bull. Mar. Biol. Lab. Woods Hole* **42**, 132–142.

Dale, B. (1974). Extrusion, retraction and respiratory movements in *Helix pomatia* in relation to distribution and circulation of the blood. *J. Zool. London* **173**, 427–439.

Denny, M. W. (1980). The role of gastropod pedal mucus in locomotion. *Nature Lond.* **285**, 160–161.

Denny, M. W. (1981). A quantitative model for the adhesive locomotion of the terrestrial slug, *Ariolomax columbianus*. *J. Exp. Biol.* **91**, 195–218.

Denton, E. J. (1961). The buoyancy of fish and cephalopods. *Prog. Biophys.*, **11**, 178–234.

Edmunds, M. (1968). On the swimming and defensive response of *Hexabranchus marginatus* (Mollusca, Nudibranchia). *J. Linn. Soc. London Zool.* **47**, 425–429.

Elder, H. Y. (1980). Peristaltic mechanisms. *In* "Aspects of Animal Movement" (H. Y. Elder and E. R. Trueman, eds.), pp. 71–92. University Press, Cambridge, England.

Elves, M. W. (1961). The histology of the foot of *Discus rotundatus,* and the locomotion of gastropod Mollusca. *Proc. Malac. Soc. London,* **34**, 346–355.

Gray, J. (1968). "Animal locomotion." Weidenfeld and Nicolson, London.

Haefelfinger, H. R. (1968). Die Lokomotion von *Aporrhais pes-pelicani* (L.) (Mollusca, Gastropoda, Prosobranchia). *Rev. Suisse Zool.* **75**, 569–574.

Hoggarth, K. R., and Trueman, E. R. (1967). Techniques for recording the activity of aquatic invertebrates. *Nature London* **213**, 1050–1051.

Johnson, W., Soden, P. D., and Trueman, E. R. (1972). A study in jet propulsion: an analysis of the motion of the squid, *Loligo vulgaris*. *J. Exp. Biol.* **56**, 155–165.

Jones, H. D. (1973). The mechanism of locomotion of *Agriolimax reticulatus* (Mollusca: Gastropoda). *J. Zool. London* **171**, 489–498.

Jones, H. D. (1975). Locomotion. *In* "Pulmonates" (V. Fretter and J. Peake, eds.), Vol. 1, pp. 1–32. Academic Press, London.

Jones, H. D. (1978). Fluid skeletons in aquatic and terrestial animals. *In* "Comparative Physiology: Water, Ions and Fluid Mechanics" (K. Schmidt-Nielsen, L. Bolis, and S. H. P. Maddrell, eds.), pp. 267–282. University Press, Cambridge, England.

Jones, H. D., and Trueman, E. R. (1970). Locomotion of the limpet, *Patella vulgata* L. *J. Exp. Biol.* **52**, 201–216.

Lighthill, M. J. (1969). Hydromechanics of aquatic animal propulsion. *Annu. Rev. Fluid Mech.* **1**, 413–446.

Lissmann H. W. (1945a). The mechanism of locomotion in gastropod molluscs, I. Kinematics. *J. Exp. Biol.* **21**, 58–69.

Lissmann, H. W. (1945b). The mechanism of locomotion in gastropod molluscs, II. Kinetics. *J. Exp. Biol.* **22**, 37–50.

Martin, R. (1966). On the swimming behaviour and biology of *Notarchus punctatus* Phillipi (Gastropoda, Opisthobranchia). *Pubbl. Stn. Zool. Napoli* **35**, 61–75.

Miller, S. L. (1974a). Adaptive design of locomotion and foot form in prosobranch gastropods. *J. Exp. Mar. Biol. Ecol.* **14**, 99–156.

Miller, S. L. (1974b). The classification, taxonomic distribution, and evolution of locomotor types among prosobranch gastropods. *Proc. Malac, Soc. London* **41**, 233–272.

Moore, J. D., and Trueman, E. R. (1971). Swimming of the scallop, *Chlamys opercularis* (L.). *J. Exp. Mar. Biol. Ecol.* **6**, 179–185.

Morton, J. E. (1954). The biology of *Limacina retroversa. J. Mar. Biol. Assoc. U. K.* **33**, 297–312.

Morton, J. E. (1960). The responses and orientation of the bivalve *Lasaea rubra* Montagu. *J. Mar. Biol. Assoc. U. K.* **39**, 5–26.

Morton, J. E. (1964). Locomotion. *In* "Physiology of Mollusca" (K. M. Wilbur and C. M. Yonge, eds.), Vol. 1 pp. 383–423. Academic Press, New York.

Morton, J. E. (1979). "Molluscs." Hutchinson, London. (5th edition.)

Mosher, J. I. (1972). The responses of *Macoma balthica* (Bivalvia) to vibrations. *Proc. Malac. Soc. London* **40**, 125–131.

Nair, N. B., and Ansell, A. D. (1968a). Characteristics of penetration of the substratum by some marine bivalve molluscs. *Proc. Malac. Soc. London* **38**, 179–197.

Nair, N. B., and Ansell, A. D. (1968b). The mechanism of boring in *Zirphaea crispata* (L.) (Bivalvia: Pholadidae). *Proc. R. Soc. B* **170**, 155–173.

Owen, G. (1966). Feeding. *In* "Physiology of Mollusca" (K. M. Wilbur and C. M. Yonge, eds.), Vol. 2, pp. 1–51. Academic Press, New York.

Packard, A. (1969). Jet propulsion and the giant fibre response of *Loligo. Nature London* **221**, 875–877.

Packard, A., and Trueman, E. R. (1974). Muscular activity of the mantle of *Sepia* and *Loligo* (Cephalopoda) during respiration and jetting and its physiological interpretation. *J. Exp. Biol.* **61**, 411–419.

Packard, A., Bone, Q., and Highnette, M. (1980). Breathing and swimming movements in a captive *Nautilus. J. Mar. Biol. Assoc. U. K.* **60**, 313–327.

Parker, G. H. (1911). The mechanism of locomotion in gastropods. *J. Morphol.* **22**, 155–169.

Pelseneer, P. (1935). Essai d'ethologie zoologique d'apres L'etude des mollusques. *Acad. r. Belg. Cl. Sci. Publ. Fond. Agathon Potter* **1**, 1–662.

Pojeta, J. Jr., and Palmer, T. J. (1976). The origin of rock boring in mytilacean pelecypods. *Alcheringa* **1**, 167–179.

Russell-Hunter, W. D., and Russell-Hunter, M. (1968). Pedal expansion in the naticid snails. 1. Introduction and weighing experiments. *Biol. Bull. Mar. Biol. Lab. Woods Hole,* **135**, 548–562.

Sigurdsson, J. B., Titman, C. W., and Davies, P. A. (1976). The dispersal of young post-larval bivalve molluscs by byssus threads. *Nature London* **262**, 386–387.

Stanley, S. M. (1969). Bivalve mollusk burrowing aided by discordant shell ornamentation. *Science* **166**, 634–635.

Stanley, S. M. (1970). Relation of shell form to life habits in the Bivalvia. *Mem. Geol. Soc. Am.* **125**, 1–296.

Stanley, S. M. (1975). Why clams have the shape they have: An experimental analysis of burrowing. *Paleobiology* **1**, 48–58.

Stanley, S. M. (1977). Coadaptation in the Trigoniidae, a remarkable family of burrowing bivalves. *Palaeontology* **20**, 869–899.

Thayer, C. W. (1972). Adaptive features of swimming monomyarian bivalves (Mollusca). *Forma Functio* **5**, 1–32.

Thayer, C. W. (1975). Morphologic adaptations of benthic invertebrates to soft substrata. *J. Mar. Res.* **33**, 177–189.

Thomas, G. E., and Gruffydd, Ll. D. (1971). The types of escape reactions elicited in the scallop *Pecten maximus* by selected sea star species. *Mar. Biol.* **10,** 87–93.

Thompson, J. E., and Slinn, D. J. (1959). On the biology of the opisthobranch *Pleurobranchus membranaceus. J. Mar. Biol. Assoc. U. K.* **38,** 507–524.

Thorburn, I. W., and Gruffydd, Ll. D. (1979). Studies of the behaviour of the scallop *Chlamys opercularis* (L.) and its shell in flowing sea water. *J. Mar. Biol. Assoc. U. K.* **59,** 1003–1023.

Trevor, J. H. (1978). The dynamics and mechanical energy expenditure of the polychaetea *Nephtys cirrosa, Nereis diversicolor* and *Arenicola marina* during burrowing. *Estuarine Coastal Mar. Sci.* **6,** 605–619.

Trueman, E. R. (1967). The dynamics of burrowing in *Ensis* (Bivalvia). *Proc. R. Soc. B* **166,** 459–476.

Trueman, E. R. (1968a). The burrowing activities of bivalves. *Symp. Zool. Soc. London* **22,** 167–186.

Trueman, E. R. (1968b). The mechanism of burrowing of some naticid gastropods in comparison with that of other molluscs. *J. Exp. Biol.* **48,** 663–678.

Trueman, E. R. (1968c). The burrowing process of *Dentalium* (Scaphopoda). *J. Zool.* London **154,** 19–27.

Trueman, E. R. (1968d). The locomotion of the freshwater clam *Margaritifera margaritifera* (Unionacea:Margaritanidae). *Malacologia* **6,** 401–410.

Trueman, E. R. (1971). The control of burrowing and the migratory behaviour of *Donax denticulatus.* (Bivalvia: Tellinacea). *J. Zool. London* **165,** 453–469.

Trueman, E. R. (1975). "The locomotion of soft-bodied animals," Arnold, London.

Trueman, E. R. (1976). Locomotion and the origins of Mollusca. *In* "Perspectives in Experimental Biology" (P. Spencer-Davies, ed.), Vol. 1, pp. 455–465. Pergamon, Oxford, England.

Trueman, E. R. (1980). Swimming by jet propulsion. *In* "Aspects of Animal Movement" (H. Y. Elder and E. R. Trueman, eds.), pp. 93–105. Cambridge Univ. Press, Cambridge, England.

Trueman, E. R., and Ansell, A. D. (1969). The mechanism of burrowing into soft substrata by marine animals. *Oceanogr. Mar. Biol.* **7,** 315–366.

Trueman, E. R., and Brown, A. C. (1976). Locomotion, pedal retraction and extension, and the hydraulic systems of *Bullia* (Gastropoda: Nassaridae). *J. Zool. London* **178,** 365–384.

Trueman, E. R., and Jones, H. D. (1977). Crawling and burrowing. *In* "Mechanics and Energetics of Animal Locomotion." (R. McN. Alexander and G. Goldspink, eds.), pp. 204–221. Chapman and Hall, London.

Trueman, E. R., and Packard, A. (1968). Motor performances of some cephalopods. *J. Exp. Biol.* **49,** 495–507.

Vlès, F. (1907). Sur les ondes pedieuses des mollusques repateurs. *C. R. Hebd. Séances Acad. Sci. Paris* **145,** 276–278.

Wainwright, S. A., Biggs, W. D., Currey, J. D., and J. M. Gosline (1976). "Mechanical design in organisms." Arnold, London.

Warburton, K. (1976) Shell form, behaviour and tolerance to water movement in the limpet *Patina pellucida* (L.) (Gastropoda: Prosobranchia). *J. Exp. Mar. Biol. Ecol.* **23,** 307–325.

Ward, D. F., and Wainwright, S. A. (1972). Locomotory aspects of squid mantle structure. *J. Zool. London* **167,** 437–449.

Ward, P., Stone, R., Westermann, G., and Martin, A. (1977). Notes on animal weight, cameral fluids, swimming speed and colour polymorphism of the cephalopod, *Nautilus pompilius* in the Fiji Islands. *Paleobiology* **3,** 377–388.

Webb, J. E. (1969). Biologically significant properties of submerged marine sands. *Proc. R. Soc. B.* **174,** 355–402.

Yonge, C. M. (1936). Mode of life, feeding, digestion and symbiosis with zooxanthellae in the Tridacnidae. *Scientific Reports Great Barrier Reef Exped.* **1**, 283–321.

Yonge, C. M. (1955). Adaptation to rock boring in *Botula* and *Lithophaga* (Lamellibranchia, Mytilidae) with a discussion on the evolution of this habit. *Q. J. Microsc. Sci.* **96**, 383–410.

Yonge, C. M. (1963). Rock boring organisms. *In* "Mechanisms of Hard Tissue Destruction" (R. F. Sognnaes, ed.). *Publ. Am. Assoc. Ad. Sci.* **75**, 1–24.

Yonge, C. M. (1972). On the primitive significance of the byssus in the Bivalvia and its effects in evolution. *J. Mar. Biol. Assoc. U. K.* **42**, 113–125.

Young, J. Z. (1938). The functioning of the giant nerve fibres of the squid. *J. Exp. Biol.* **15**, 170–185.

Zuev, G. V. (1965a). The functional basis of the structure of the locomotory apparatus in the phylogeny of cephalopods. *Zh. Obshch. Biol.* **26**, 616–619.

Zuev, G. V. (1965b). Concerning the mechanism involved in the creation of a lifting force by the bodies of cephalopod molluscs. *Biofizika Acad. Nauk SSSR* **10**, 360–361.

5

The Mode of Formation and the Structure of the Periostracum

A. S. M. SALEUDDIN HENRI P. PETIT

Department of Biology
York University
Toronto, Ontario, Canada

Baylor College of Dentistry
and
Dallas Museum of Natural History
Dallas, Texas

I. Introduction

The molluscan shell is covered externally by a thin, pliable, fibrous layer called the *periostracum*. This layer is thought to protect the shell and to serve as a matrix for the deposition of calcium carbonate crystals. New periostracum is extruded from the mantle edge and is subsequently insolubilized, hardened, and darkened. The biochemical process effecting the periostracal protein in this manner is called *sclerotization*. Fibrous proteins are cross-linked by reactive quinones, a process known as *quinone tanning*. In addition to tanned proteins, the periostracum contains some carbohydrate and lipid (Brown, 1952; Beedham, 1958; Hillman, 1961; Meenakshi et al., 1969; Bubel, 1973a; Waite, 1977; Hunt and Oates, 1978).

THE MOLLUSCA, VOL. 4
Physiology, Part 1

The periostracum of *Neopilina,* the only living member of Monoplacophora, is also tanned, but differs only slightly in chemical composition from the shell matrix (Meenakshi et al., 1970). Chitons (Polyplacophora) do not have a periostracum but a cuticle which consists in part of proteins. The girdle cuticle is not tanned, but the cuticle secreted at the shell-plate margin is quinone tanned (Beedham and Trueman, 1967; Haas, 1972). The periostracum of members of Aplacophora is not quinone tanned (Beedham and Trueman, 1968; Carter and Aller, 1975). However, with the exception of *Nautilus macromphalus* (Meenakshi et al., 1974a), the shells of all bivalves and gastropods studied so far have an outer proteinaceous periostracal layer and an inner calcified layer.

In many molluscs the function of the periostracum is unclear. It does protect the shell from acid dissolution, and it has been suggested that sclerotins give the periostracum its resistance to chemicals and its durability. The quinone tanning of the periostracum has been postulated to be an essential prerequisite for calcification, providing a suitable substratum for an orderly deposition of calcium carbonate (Beedham and Trueman, 1968; Taylor and Kennedy, 1969; Chan and Saleuddin, 1974; Carter and Aller, 1975; Petit et al., 1980b). Fully formed periostracum appears structurally simple under the light microscope; however, its ultrastructure varies immensely among species. In most molluscs two to three distinct layers are discernible, but this can vary from one distinct layer in *Helix* (Saleuddin, 1976a) to four in *Solemya parkinsoni* (Beedham and Owen, 1965) and many more in *Amblema* (Petit, 1977).

In all molluscs the periostracum is secreted by the mantle edge, but the mode of formation varies from species to species. In *Helix* it is formed by the tubular periostracal gland embedded in the connective tissue of the mantle edge. In *Lymnaea, Helisoma,* and *Physa* the periostracum is secreted by periostracal cells, and subsequently is thickened by the mantle edge gland (Kniprath, 1972; Saleuddin, 1975; Jones and Saleuddin, 1978). In bivalves the periostracum originates from epithelial cells lining the inner surface of the outer fold of the periostracal groove and sometimes from a row of basal cells at the bottom of the groove (Kawaguti and Ikemoto, 1962b; Dunachie, 1963; Neff, 1972; Bubel, 1973a,c,d; Saleuddin, 1974; Petit et al., 1979). Hunt and Oates (1978) studied the fine structure and molecular organization of the periostracum of *Buccinum undatum,* which is made up of alternate light and dark sheets of protein fibers. Each sheet is composed of protein subunits measuring 32 nm in length and 6.5 nm in width. They reported the presence of lectin-type proteins with globular as well as α-helical regions. Alternate light and dark sheets have also been reported in *Littorina littorea* (Bevelander and Nakahara, 1970). The cellular mechanisms of formation and the structures of periostracum have been reviewed by Saleuddin (1979). Several authors have reported on the chemistry of the periostracum (Beedham, 1958; Waite, 1977; Waite et al., 1979; Waite and Andersen, 1980).

A review by Waite on chemistry will appear in Chapter 11 of Volume 1 of this series.

Due to space limitation, this chapter will not attempt to document all the work done on periostracum structure, but it will make the reader aware of the diversity of structures by examining specific studies on cellular mechanisms of periostracum formation. The discussion of periostracum formation and structure will focus on a series of questions, which may lead to further work in the field.

II. Periostracum Formation in Pulmonates

Periostracum formation has been studied in detail in *Lymnaea stagnalis* (Kniprath, 1972), *Helisoma duryi* (Saleuddin, 1975, 1976b), *Helix aspersa* (Saleuddin, 1976a), and in *Physa* spp. (Jones and Saleuddin, 1978). A thin membrane-like lamella, overlying the main periostracal layer, has been reported in the aquatic pulmonates studied so far. This lamella is absent from the periostracum of terrestrial pulmonates (e.g., *Helix*) and of other gastropods (e.g., *L. littorea*) (Bevelander and Nakahara, 1970; Saleuddin, 1976a). The chemical nature of this layer is unknown. A morphologically similar layer has been reported in the cuticle of insects (Locke, 1966), and nematodes (Bird, 1971). The data presently available indicate that two modes of periostracum formation exist in pulmonates. *Lymnaea*, *Helisoma*, and *Physa* exhibit one mechanism, with only minor variations among the species. A lamellar layer is present in these three genera. The cellular mechanism of periostracum formation in *Helix* and *Otala* is quite different. These genera do not have a lamellar layer over the periostracum. These two modes of periostracum formation will be described now in detail.

A. *Helisoma*

Periostracum formation in *H. duryi* will be used to illustrate the mechanisms employed by aquatic pulmonates. The mantle edge of *Helisoma* is compartmentally specialized, allowing the periostracum to be secreted and formed in sequential steps. At the base of the periostracal groove, deep within the connective tissue, is a group of periostracal cells. At the ultrastructural level dark and light periostracal cell types can be distinguished. During development these cells originate in the epithelium, then recede into the underlying connective tissue, maintaining connections with the periostracal groove by means of long cell processes. The epithelial cells closer to the opening of the periostracal groove become tall and columnar, reaching a maximum height of 50 μm. They eventually become the mantle edge gland (Timmermans, 1969). Proximal to the mantle edge gland lie the dorsal epithelial cells. These low columnar cells are

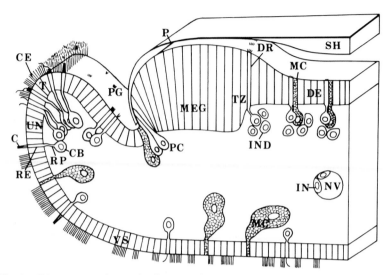

Fig. 1.   Diagrammatic longitudinal section through the mantle edge of *Helisoma* showing the general histological features and epithelial regions involved in shell formation. C, Ciliary tuft; CB, receptor cell body; CE, ciliated epithelium; DE, dorsal epithelium; DR, dorsal receptor; IN, intrinsic neuron; IND, dorsal plexus intrinsic neuron; MC, mucous cell; MEG, mantle edge gland; NV, longitudinal branch of the pallial nerve; P, periostracum; PC, periostracal cells; PG, periostracal groove; RE, receptor ending; RP, receptor process; SH, shell; T, tip of ventral mantle edge; TZ, transitional zone; UN, unciliated zone; VS, ventral surface. (Not to scale.) (Jones and Saleuddin, 1978.)

responsible for the deposition of the mineral part of the shell (Fig. 1). The transition zone is situated at the junction of the mantle edge gland and the dorsal epithelium (Chan and Saleuddin, 1974). Similar regional specialization of the mantle edge has been reported for *Lymnaea* (Kniprath, 1972) and for *Physa* (Jones and Saleuddin, 1978).

Periostracum formation begins at the base of the periostracal groove. The light periostracal cells are responsible for the production of periostracal units. In *Helisoma* these units measure 0.1–0.2 μm in length, and 10–13 nm in width (Fig. 2). The periostracal units in *Physa* are different from those in *Helisoma* (Fig. 3). The light cells have prominent Golgi bodies and smooth endoplasmic reticulum. Many isolated membrane-bound periostracal units are found in the vicinity of the Golgi bodies and are scattered throughout the cytoplasm. This indicates that the units are synthesized and packaged by the Golgi bodies. Large inclusions containing many periostracal units are also found in the cytoplasm of the light cells. It is unclear if these are formed by the fusion of individual membrane bound units, or by the formation of many units within a single large membrane-bound inclusion. The membrane-bound periostracal units migrate

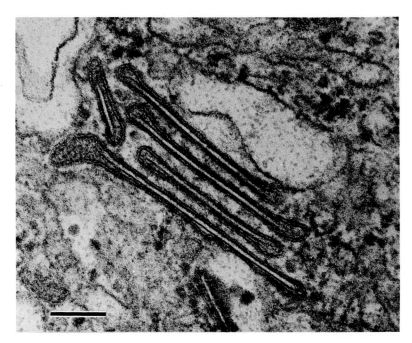

Fig. 2. *Helisoma.* Secretory vesicles containing periostracal units. Bar, 0.1 μm. (Saleuddin, 1975.)

along the cell processes and are eventually extruded into the base of the periostracal groove (Figs. 4 and 5). The units become aligned forming the lamellar layer (Fig. 5). The exact mechanism is unknown, but it is thought that the microvilli of the dark periostracal cells may participate in alignment of the units. A similar mode of alignment is found in *Physa* (Fig. 6).

Fully formed periostracum consists of an outer lamellar layer and an inner, dark homogeneous layer (Fig. 7). Once the lamellar layer has formed, the dark periostracal cells initiate the formation of the inner homogeneous layer. Occasionally sublayers can be seen within the homogeneous layer. Kniprath (1972) found three layers beneath the lamellar layer of the periostracum of *Lymnaea.* However, these sublayers are not consistently visible; therefore, their existence is debatable.

Newly formed periostracum gradually moves up and out of the periostracal groove. On the way it is thickened by secretions of the tall epithelial cells lying adjacent to the dark periostracal cells. Further thickening and maturation occur as secretions from the mantle edge gland are added to the newly formed periostracum (Fig. 8). Similar processes of periostracum secretion, thickening, and

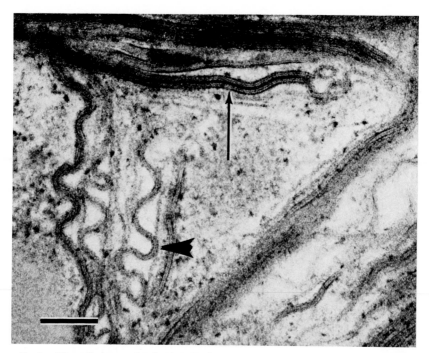

Fig. 3.   *Physa.* Periostracal units showing five-layered structure (arrow). Some units have β-helix configuration (arrowhead). Bar, 0.1 μm. (Jones and Saleuddin, 1978.)

maturation are observed in *Physa* (Jones and Saleuddin, 1978), and in *Lymnaea* (A. S. M. Saleuddin, personal observation).

B. *Helix*

Studies on *H. aspersa* will be used to illustrate the second mode of periostracum formation used by terrestrial pulmonates. In *Helix* the periostracum is secreted by the periostracal gland (Fig. 9). This is a tubular gland that runs parallel to and along the length of the mantle edge, opening at the base of the periostracal groove. The periostracal gland is comprised of one cell type (Saleuddin, 1976a). These tall columnar cells have prominent rough endoplasmic reticulum and Golgi bodies. Free ribosomes, lysosomes, and mitochondria are moderate in number and are scattered throughout the cytoplasm (Fig. 10). Periostracal units originate inside Golgi cisternae. These units appear as three electron-dense lines with two electron-lucent areas sandwiched between them (Fig. 11). Each unit is 0.4–0.6 μm in length and 9–10 nm wide. Mature cisternae with periostracal inclusions separate from the Golgi system and are moved through the cytoplasm by microtubules, becoming nascent secretory vesicles (Fig. 12). Mature secretory inclusions are spherical and contain from one to many periostracal units. They migrate to the apical part of the cell where they fuse with lysosomes. The significance of this process is not understood. Lysosomal enzymes may digest nonperiostracal components, or they may modify the periostracal units so that they are later able to cross-link to form periostracal sheets (Saleuddin,

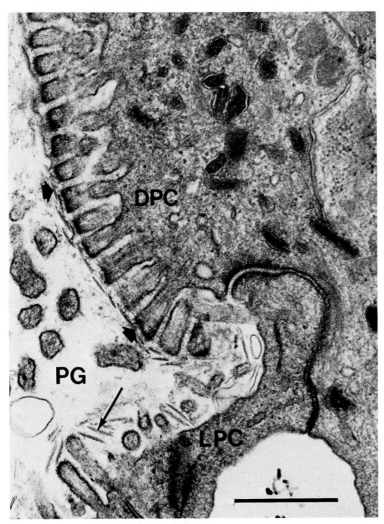

Fig. 4. *Helisoma.* Parts of light (LPC) and dark (DPC) periostracal cells. Note periostracal units (thin arrow) in the bottom of the periostracal groove (PG). Also note the alignment of periostracal units (thick arrows) to form the lamellar layer of the periostracum. Bar, 0.5 μm.

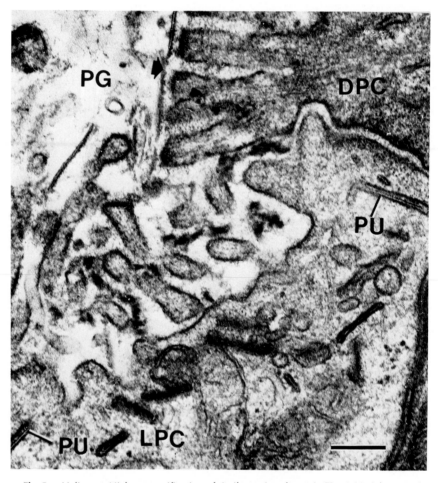

Fig. 5.   *Helisoma.* Higher magnification of similar region shown in Fig. 4. Vesicles containing a periostracal unit (PU). Note two periostracal units (arrow) aligned in the periostracal groove (PG). DPC, Dark periostracal cell; LPC, light periostracal cell. Bar, 0.1 μm. (Saleuddin, 1975.)

1976a). Lysosomal enzymes have also been found in secretory inclusions in *Physa* periostracal cells (Jones and Saleuddin, 1978).

The periostracal units are extruded into the lumen of the gland where they disperse and presumably become cross-linked to form periostracal sheets. In the lumen of the gland the periostracal units are still detectable by their periodicity; however, as the immature periostracum moves into the groove it becomes fibrous and individual units are no longer detectable (Fig. 13). When the mature periostracum leaves the groove it is a sheet of one homogeneously textured layer. The stages of development of the periostracum are diagrammatically shown in Fig. 14. The periostracal units in the secretory inclusions of *Littorina littorea* periostracal gland also show periodicity, and these can be traced to the fully formed periostracum (Bevelander and Nakahara, 1970). The outer lamellar layer in the periostracum of *Helisoma* (Saleuddin, 1975), *Lymnaea* (Kniprath, 1972), and *Physa* (Jones and Saleuddin, 1978) is not present in the periostracum of *Helix* (Saleuddin, 1976a, 1979).

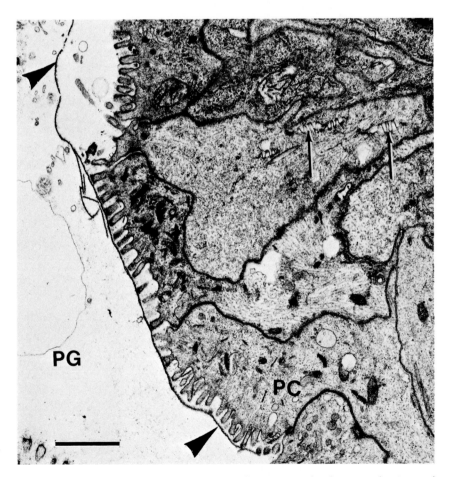

Fig. 6.   *Physa.* Apical regions of periostracal cells (PC). Note the alignment of periostracal units (arrowheads) in the periostracal groove (PG). Arrows point to membrane-bound periostracal units. Bar, 1.0 μm. (Jones and Saleuddin, 1978.)

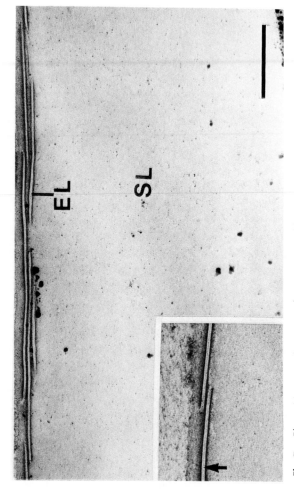

Fig. 7. *Physa.* Mature periostracum showing external lamella (EL). Note the homogenous sublamellar layer (SL). The external layer is not continuous but can form overlapping sheets in places. Bar, 0.1 μm. Inset: five-layered periostracal units. (Jones and Saleuddin, 1978.)

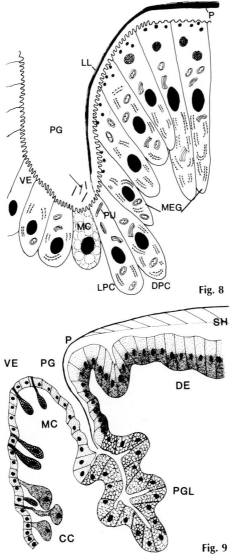

Fig. 8. *Helisoma.* Diagrammatic representation of epithelial cells lining the periostracal groove (PG) to show their spatial relationships and the formation of the periostracum. Note periostracal units (arrow) at the base of the groove. Periostracum formation is initiated first by the alignment of periostracal units forming the lamellar layer (LL). Dark periostracal cells (DPC) and cells of the mantle edge gland secrete the dark layer (P) underneath the lamellar layer. Only light periostracal cells (LPC) produce periostracal units (PU). MC, Mucous cell; VE, ventral epithelium. (Not to scale.)

Fig. 9. *Helix.* Diagrammatic representation of longitudinal section of the mantle edge. Periostracum (P), which is secreted by the periostracal gland (PG), protrudes from the periostracal groove (PGL). CC, Calcium cell; DE, dorsal epithelium; MC, mucous cell; SH, shell; VE, ventral epithelium. (Not to scale.) (Saleuddin, 1976a.)

209

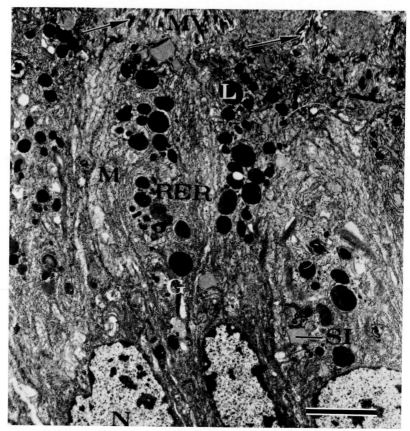

Fig. 10.   *Helix.* A few periostracal gland cells. Note prominent Golgi (G), lysosomes (L), and secretory inclusions (SI). Extended secretory materials (arrows) can be seen between microvilli (MV). M, Mitochondria; N, nucleus; RER, rough endoplasmic reticulum. Bar, 1.0 μm. (Saleuddin, 1976a.)

III. Periostracum Formation in Bivalves

As in other molluscan species, in bivalves the periostracum is a complex layer secreted in the periostracal groove of the mantle edge (Fig. 15). The processes of formation and maturation occur within the groove. The newly formed periostracum is extruded distally as a multilayered entity. Its innermost layers become continuous with the outer covering of the valve at the forming edge of the shell (Fig. 16).

The periostracum is also longitudinally nonhomogeneous. Amino acid analyses (Petit, 1981) show clearly that the composition of the periostracum is differ-

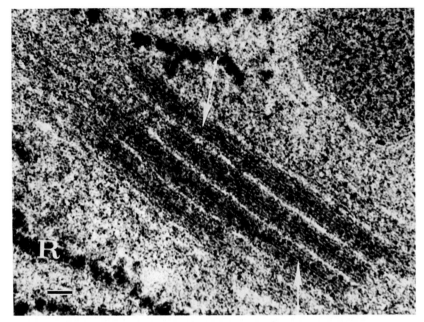

Fig. 11. *Helix.* Periostracal units (arrows) within Golgi cisternae. R, Ribosomes. Bar, 0.1 μm. (Saleuddin, 1976a.)

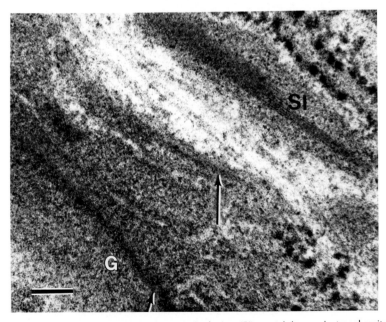

Fig. 12. *Helix.* Note a nascent secretory inclusion (SI) containing periostracal units (arrowhead). Periostracal units (arrows) are also seen in Golgi (G). Bar, 0.1 μm.

ent from proximal to distal parts. The periostracum may be considered an active structure since some chemical modifications are observed even after it loses all cellular contact with the periostracal groove. Several functionally different areas are arranged linearly within the different layers of this complex structure (Fig. 16).

Inside the periostracal groove is the *forming periostracum*, which has at least three layers. After extrusion from the distal edge, the *free periostracum* serves as a compartmental wall between the outer medium and the extrapallial fluid. Although it may be attached mesially to the maturing periostracum and distally to the forming edge of the shell, this area of the periostracum is "free" of cellular contact.

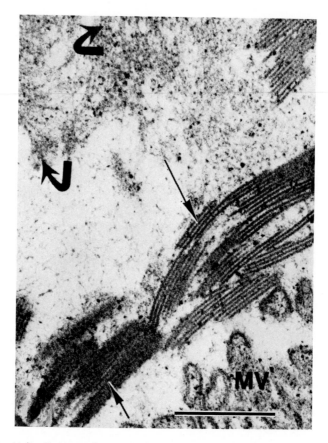

Fig. 13. *Helix.* Periostracal units (thin arrows) in the lumen of the periostracal gland. Curved arrows point to sites where dispersion of units has occurred. MV, Microvilli. Bar, 0.5 μm. (Saleuddin, 1976a.)

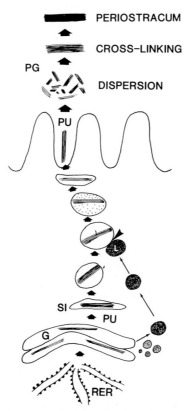

Fig. 14. *Helix.* Schematic presentation to show the major steps in the formation of the periostracum. Periostracal units (PU) are first recognized in Golgi (G). Lysosome (L) fuses with secretory inclusion (arrowhead). Units are then released into the lumen of the gland (PG) where dispersion and cross-linking of units occur to form the periostracum. RER, Rough endoplasmic reticulum. Lysosomes are Golgi-mediated (thin arrows). (Not to scale.) (Modified according to Saleuddin, 1976a.)

At the edge of the forming shell, the *outer periostracum* is an external layer, covering the outer prismatic layer (Figs. 16 and 17). This outer periostracum cleaves easily from the underlying layers.

This division into "forming," "free," and "outer" periostracum is based on macroscopic and microscopic observations and helps in understanding otherwise conflicting data in the literature. This division is not an artificial one, but represents three main stages in the transformation of the periostracum as they appear in *Amblema* and other unionid mussels.

The periostracum is initiated continuously by basal cells located deep in the periostracal groove, then formed and matured by epithelial cells (forming per-

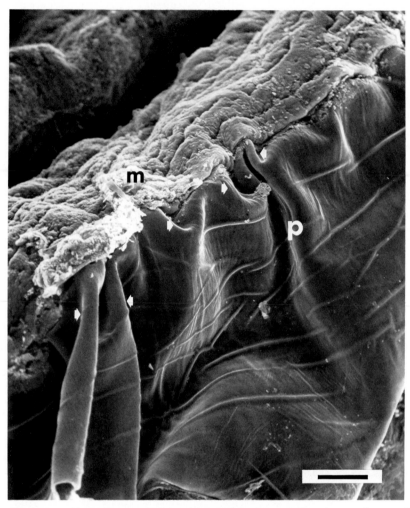

Fig. 15. *Amblema.* Scanning electron micrograph of the mantle edge (m) showing the periostracum (p) extruding from the periostracal groove. Note the "curtain-like" appearance (arrows) of the newly formed periostracum which will unfold and stretch when reaching the shell-forming edge. Note the incremental, or growth, lines. Bar, 100 μm. (Petit et al., 1978.)

Fig. 16. *Amblema.* Diagrammatic representation of the mode of formation of the periostracum. The periostracum is elaborated in a cornucopia-like periostracal groove (1). The groove is lined with long epithelial cells on its outer aspect and with short cuboidal epithelial cells on its inner aspect (A). Both layers participate to the formation and maturation of the periostracum by coating the initial pellicle. When in the groove, the periostracum is called "forming periostracum" (1). At the exit of the groove, the periostracum is a supple membrane which undulates and forms a cul-de-sac in order to join the tip of the mineralized shell. This is the "free periostracum" which limits distally the extrapallial space. Within the "free per-

214

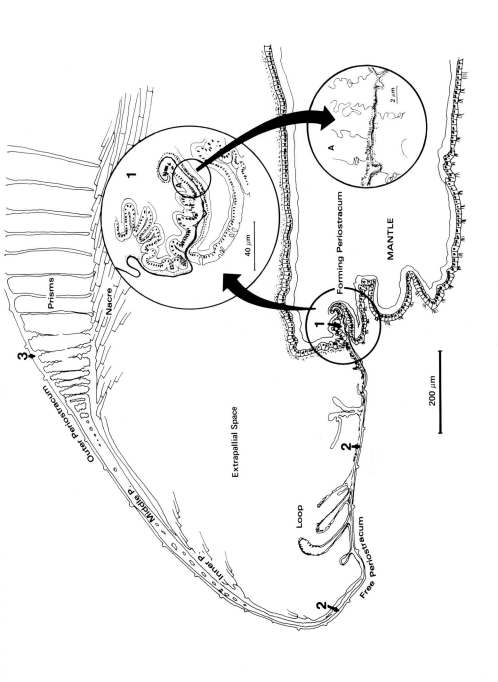

Prisms

Nacre

3

Outer Periostracum

Middle P.

Inner P.

Extrapallial Space

Loop

Free Periostracum

2

2

Forming Periostracum

MANTLE

1

1

A

2 μm

40 μm

200 μm

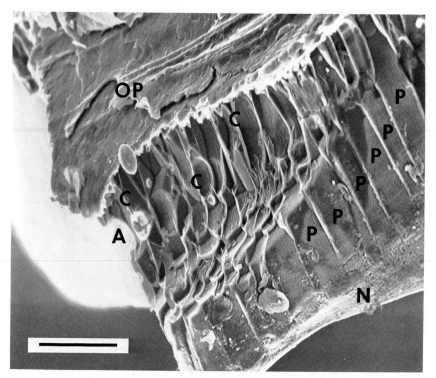

Fig. 17. *Amblema.* Scanning electron micrograph of the forming edge of the shell. The outer periostracum (OP) with one incremental line protects the forming prisms (P). Note that the bottom of the prism is formed prior to the elaboration of the matrices. The antrum (A) and the partitions, or columns (C), can be seen. Note the initial nacre (N) and the angle of the nacreous deposition over the bases of the prisms (prism/nacre angle, or P/N angle). Bar, 50 μm.

iostracum) (Figs. 16 and 18). This is a dynamic process within the periostracal groove.

From the tip of the mantle to the edge of the forming shell, the free periostracum undergoes transformations within each of its layers. Some harden to become the future outer periostracum, whereas other portions are selectively

iostracum," essential mineralization processes take place (2). The free periostracum can be cleaved in three series of layers: The innermost layer of the "forming periostracum" becomes the outer periostracum (3) covering the formed shell; the middle layers are involved with prism formation; and the outermost layers after making loops in the extrapallial space become the forming part of the innermost area of the mineralizing edge, the initial nacre.

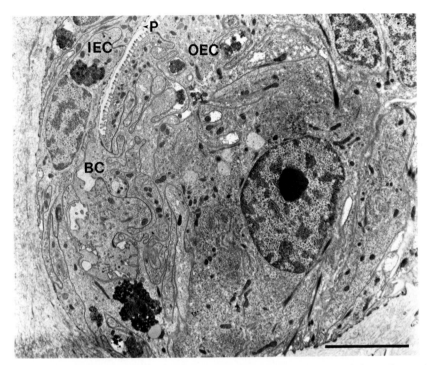

Fig. 18. *Amblema.* Low-power electron micrograph of the bottom of the periostracal groove. The three types of cells are evidenced: basal cell (BC), short inner epithelial cells (IEC), long outer epithelial cells (OEC). Note the secretion of the initial pellicle (P) at the tip of the protruding tongue of the basal cell. The pellicle is then seen undulating between the two walls of the epithelial cells. Bar, 7 μm. (Petit et al., 1979.)

modified for a specific mineral organization, giving rise to either prisms or nacre. At this stage mineralization occurs and crystal units can be seen.

One should keep in mind this dynamic evolution to understand fully and interpret the meaning of previous and current ultrastructural, histochemical, and crystallographic observations.

A. Periostracum Formation and Maturation

1. Periostracum Formation in the Periostracal Groove

Haas (1935) pointed out that periostracum formation varies considerably with species. Since that time, several electron microscopic studies have been conducted: Kawaguti and Ikemoto (1962a), Kawakami and Yasuzumi (1964), Wada (1968), Bevelander and Nakahara (1967, 1970), Neff (1972), Bubel (1973a,b), Saleuddin (1974), Petit (1977, 1978, 1981), and Petit et al. (1978, 1979,

1980b). These studies show some unique and some common features of periostracum formation. The periostracal groove is located at the distal edge of the mantle between the outer and middle folds, and in section is shaped like a cornucopia with its concavity turned toward the outer edge of the mantle (Fig. 17). Two layers of epithelial cells face each other within the groove. In some species, only one layer appears secretory; in others, both layers secrete a glycocalyx which coats the periostracal pellicle. The pellicle is formed at the bottom of the groove and differs structurally among species. In many cases, a row of basal cells is said to be responsible for the initiation of periostracum. These basal cells (Kawaguti and Ikemoto, 1962a), or intercalated cells (Petit, 1977, 1978), are squeezed between the last epithelial cells at the base of the groove (Figs. 16 and 18). Bevelander and Nakahara (1967) describe the basal cells as having extensive and modified microvilli, prominent tegumental fibers, abundant glycogen, and numerous, large mitochondria. Saleuddin (1974) describes the basal cells of *Astarte* as the largest epithelial cells of the outer fold, devoid of microvilli, but containing channels, or plicae. The cytoplasm has electron-dense inclusions, which are membrane-bound vesicles. These vesicles are thought to be precursors of periostracum. In *Astarte* the origin of the periostracum can be traced in the intercellular space between the basal cell and the cell of the middle fold. Petit (1977) describes in *Amblema* an intercalated row of cells (homologous to the basal cells). Only one aspect of the protruding tongue of the basal cells is exposed to the long microvilli of the first epithelial cells of the outer wall of the groove. At the tip of the tongue, a thin membrane-like layer is produced (Fig. 19). It receives secretory products from the two lateral walls of the groove. This layer is comparable with the pellicle described by other authors. It is a membranous prolongation reinforced by one or two glycocalytic coatings. In some cases it appears fibrillar; in others, a homogeneous electron-dense substance.

Further along the groove (Fig. 20), both outer and inner epithelial layers secrete a coating onto both sides of the pellicle. In *Amblema,* two types of cells are responsible for this coating: long epithelial cells with long microvilli on the outer aspect of the groove, and cuboidal cells with short microvilli on the inner aspect of the groove (Figs. 16 and 18).

The tall outer epithelial cells are tightly juxtaposed. The apex of each cell is covered by long microvilli and heavily coated with an osmium–pyroantimoniate-positive glycocalyx (Fig. 19). The membranes of the lumenal third of the cells are extremely intricate and are markedly labeled by osmium–pyroantimoniate fixation, suggesting the presence of calcium ions. The apical cytoplasm contains many vesicles and mitochondria. The membranes of the middle third are also osmium–pyroantimoniate-positive, but are straight and parallel. The cytoplasm in this area is rich in calcium-loaded lysosomes, often associated with glycogen. Numerous Golgi complexes are present. The basal third again has extremely

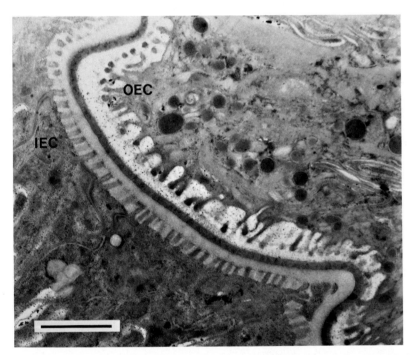

Fig. 19. *Amblema.* Periostracum within the periostracal groove is coated on both sides. Note the dramatic difference between the glycocalyx deposited by the long microvilli of the outer epithelial cells (OEC) and that associated with the short microvilli of the inner epithelial cells (IEC). Bar, 3 μm.

intricate membranes, and the infoldings are embedded in a thick basement membrane. The nucleus is basally located.

The short epithelial cells also rest on a thick basement membrane, and are bounded by osmium–pyroantimoniate-positive membranes. The nucleus is basally located, and the cytoplasm contains numerous vacuoles. Fewer lysosomes are present. The microvilli are very short and are coated with a thick glycocalytic material.

Once the periostracum is initiated, it is pushed distally by the continuous secretion of the pellicle. As this initial periostracum moves toward the exit of the groove, it receives supplementary coatings and treatments causing its maturation.

2. Maturation of the Periostracum in the Periostracal Groove

The process of maturation is not yet fully understood. Several events occurring within the groove are thought to contribute to periostracum maturation.

The tanning process, which polymerizes the glycocalytic material coating the pellicle, is still under investigation. Waite and Wilbur (1976), working on *Modiolus demissus* Dillwyn, demonstrated a phenoloxidase activity 200 times greater in the periostracum than in the mantle of this bivalve. They pointed out that phenoloxidase catalyzes the formation of *o*-quinones involved in cross-linking protein subunits in the periostracum.

Waite (1977) showed a structural difference between layers within the periostracum: a dense, amorphous upper layer and a loosely fibrous lower layer. Tanning may occur in both layers provided there are three ingredients: *o*-diphenols, fibrous proteins, and an oxidative enzyme such as phenoloxidase. DOPA is an *o*-diphenol, and was identified by Degens et al. (1967b) and Degens (1976) in the periostracum of *Mytilus edulis*. Waite et al. (1979) also identified in *M. edulis* a basic hydrophobic protein located only in the periostracal groove and marginal periostracum. They called this periostracin. Periostracin has *o*-diphenolic residues (DOPA) which may be oxidized during the polymerization process and is so similar in amino acid composition to the fully sclerotized periostracum that they suggest that it is the precursor of periostracum.

The precise location of the tanning process has not been confirmed histo-

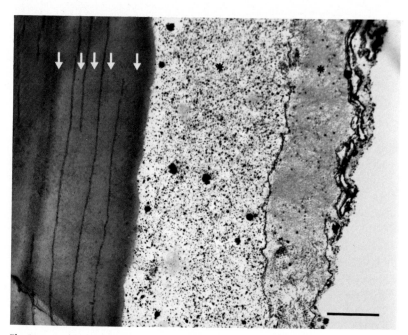

Fig. 20. *Amblema.* Note the layering in the periostracum. Four to six mature layers (arrows) contain homogenous and fine granular materials, and three newly deposited calcium-loaded glycocalytic layers (asterisks) have coarse granules. Bar, 1 μm.

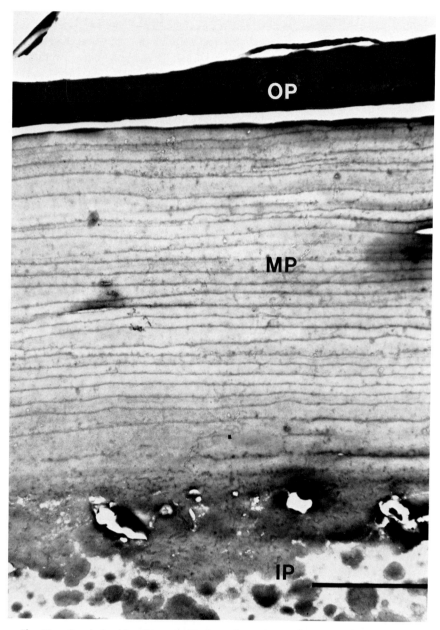

Fig. 21. *Quadrula.* After the extrusion from the periostracal groove, the three series of layers are easily identified: an outer dense homogenous periostracum (OP); a series of layers composing the middle periostracum (MP); and the maturing layer of the inner periostracum (IP). Bar, 1 μm.

chemically. Specific cells located either at the inner edge or along the whole periostracal groove are capable of tanning the outer part of the periostracum. Our own data has led us to propose that the cells of the periostracal groove in *Amblema* are a homogeneous population of cells which may have several consecutive roles: They may alternately produce the glycocalytic coating as a soluble, gellike material, and then polymerize it superficially in a second step.

Morphological studies demonstrate that the homogeneous initial deposition soon becomes stratified into layers clearly delineated by osmium–pyroantimoniate fixation (Figs. 20 and 21). This suggests a remodeling of the initial structure which receives superficial tanning, then a new glycocalytic deposition followed by tanning, with the last two processes being repeated several times.

3. Extrusion from the Periostracal Groove

The extrusion of the newly formed periostracum from the periostracal groove is the last event controlled by the cells of the mantle edge. At the exit of the groove, the last cells usually pinch the periostracum, and at the same time in *Amblema* a new internal coating is added. This layer originates in fact from the glycocalyx of the outer epithelial cells of the *outer mantle*. The free periostracum then makes a turn to reach the forming edge of the valve, causing the external face to become internalized. So the new periostracal coat becomes the "internal periostracum" in continuity with the bases of the shell prisms. The previously strongly polymerized inner layer will be in continuity with the top of the prisms, forming the "outer periostracum." In this process the middle layer will enlarge considerably and join the full length of the prisms (Figs. 16 and 22).

The periostracum is now a more complex structure, very different from the forming periostracum in the groove (Fig. 21).

B. Fine Structure of the Periostracum

1. Inside the Groove (Forming Periostracum)

Here the pellicle is secreted and receives successive coats of glycocalyces. This creates a multilayered structure (Fig. 20). The layers are beveled due to the dynamic process of extrusion. They reach a total average thickness of 0.5–4 μm. When the periostracum is extruded, the pellicle is often difficult to identify, and the periostracum is very thick: 25–50 μm. Probe analysis demonstrates the presence of calcium ions. Differences occur among species, and some authors report a certain number of layers for each species. The number of visible layers also depends upon the location of the sample taken. In *Amblema,* the periostracum begins as a one-layered pellicle, increasing to three layers, and at the distal end of the groove increasing to as many as 50 layers. All of these layers contain calcium.

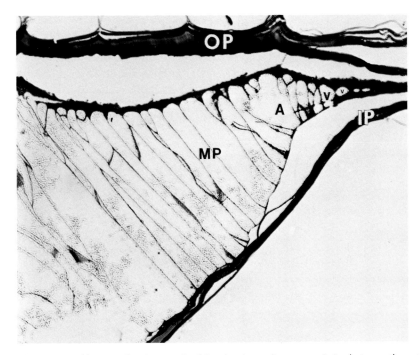

Fig. 22. *Amblema*. Light micrograph of the cleaving periostracum. Lying between the outer (OP) and inner (IP) periostracal layers is the distended middle layer (MP) which is continuous with the matricial top and bottom of the prisms. The total distention is the height of one prism. The vacuoles (V) and antrum (A) formation precede prism formation.

In some species the periostracal layers appear to be homogeneous, whereas in others they appear fibrillar. The mineral granules are coarser on the outer side, and finer, more evenly distributed on the inner side (Fig. 20).

2. Outside the Groove (Free Periostracum)

The outer layer of the periostracum is denser and separates somewhat from the middle layer (Fig. 22). Fixation with osmium–pyroantimoniate, together with microprobe analyses, demonstrate the presence of calcium in every underlying layer of the free periostracum. A reorganization occurs and morphological differences can be observed as one moves toward the forming edge of the shell. There are three major layers:

a. The outer layer. This layer is denser, and on its outer surface bears the same increments that were seen inside the groove. When these eventually reach the outer surface of the shell they are called *growth lines*. This outer layer appears homogeneous and bears few small osmium–pyroantimoniate-positive granules. It also cleaves slightly from the middle, underlying layers.

b. The middle layers. These layers of the periostracum comprise all the periostracum emerging from the periostracal groove with the exception of the outer layer, and they are strongly osmium–pyroantimoniate-positive. In *Amblema*, a series of sequential events is evident at the electron microscope level and appears to be related to the formation and organization of the prisms.

c. The inner layer. This layer, added at the mantle edge, also undergoes morphological changes. These are related to the formation of layers of initial nacre.

3. At the Forming Edge (Outer Periostracum)

The ultrastructure of the three periostracal layers again changes (Fig. 16). It is possible to follow the continuity of the outer periostracum which forms the outer covering of the shell. Externally it is pierced irregularly by perforations; internally it represents the top of the prisms.

At the forming edge, the middle layer appears greatly distended (Figs. 16 and 22). Its outer layer joins the top of the prisms, and its inner layer joins the base of the prisms. There is a continuity between the middle layers and the prismatic matrices.

The inner layer of the periostracum is in continuity with the initial nacre matrices.

IV. Histochemistry of the Mantle and the Periostracum Formation

Several enzymes have been localized in the mantle of various molluscs by both light and electron microscopy. Two enzymes have received special attention, namely acid phosphatase, because of its possible role in the modification of the periostracum, and phenoloxidase, because of its role in the tanning of periostracal proteins.

Acid phosphatase activity has been localized in the general epithelium of the mantle edge. The level of activity was found to vary cyclically after shell damage, and during periostracum and shell regeneration. After 12 h of regeneration the enzyme activity had decreased, returning to normal levels after 72 h of regeneration. Chan and Saleuddin (1974) suggested that the enzyme's function may be to modify the inner layer of the periostracum, enabling it to be calcified, so forming new shell.

The secretion and extrusion of new periostracum from the periostracal groove is accompanied by sclerotization of the periostracal protein. This process has been attributed to quinone tanning of proteins in many molluscs (Brown, 1952; Beedham, 1958; Hillman, 1961; Meenakshi et al., 1969; Bubel, 1973c; Jones and Saleuddin, 1978). This process requires three components to be present in a system: o-diphenols, fibrous proteins, and phenoloxidases. Both chemical and histochemical tests have shown all three components to be present in the periostracum. Phenoloxidase was extracted from the periostracum of M. demissus and partially characterized by gel filtration and SDS polyacrylamide gel electrophoresis (Waite and Wilbur, 1976). Waite and Wilbur (1976) suggest that the enzyme found in the mantle, with a molecular weight of 80,000, is in a latent form. The active form has a molecular weight of 70,000. Phenoloxidase has been histochemically localized in the mantle edge gland cells of Physa (Jones and Saleuddin, 1978); however, the preparative procedure for electron microscopy may have activated the enzyme.

An immediate precursor to quinone is 3,4-dihydroxyphenylalanine (DOPA). This has been found in the periostracum of various bivalve molluscs (Degens et al., 1967b; Waite, 1977; Waite and Andersen, 1978). Waite and Andersen (1980) suggest that DOPA plays a direct role in sclerotization of the periostracum of Mytilus. In support of their hypothesis they found that levels of DOPA, not tyrosine, which is also a precursor for quinones, decrease during sclerotization of the periostracum. Bubel (1973c) found that DOPA is abundant in the periostracum secreting cells of the mantle. Waite et al. (1979) found DOPA in proportions of up to 4% of formic acid-extracted periostracum. Waite and Andersen (1980) postulate that in Mytilus periostracal proteins are secreted from the mantle with tightly bound DOPA residues. The oxidized DOPA quinones then react with and cross-link the periostracal proteins during the process of tanning.

In Physa, polyphenols have been histochemically localized in secretory granules in mantle edge gland cells. These same granules were shown to contain periostracal proteins (Jones and Saleuddin, 1978). However, the polyphenols were not found in the Golgi system or in the rough endoplasmic reticulum.

V. Influence of the Periostracum on Calcification

The calcareous layer of the bivalve shell consists typically of two layers: an outer prismatic layer and an inner nacreous layer. Both are believed to be deposited by the extrapallial fluid secreted by the mantle epithelium. It has been suggested that the periostracum influences the deposition of calcium carbonate crystals in the prismatic layer (Taylor and Kennedy, 1969; Degens, 1976). Taylor and Kennedy (1969) suggest that calcium carbonate crystals may begin to form before the periostracum has completely polymerized. As a result, the un-polymerized periostracal material is pushed to the edge of the growing calcium carbonate crystals, preventing their lateral growth. The prismatic crystals grow perpendicular to the periostracum forming columnar structures until the periostracal material is no longer present. The calcium carbonate then forms a crystal mosaic, the nacreous layer.

Since the work of Taylor and Kennedy (1969), which suggested that the periostracum plays more than a passive role in the calcification process, it has been accepted that there is a structural continuity between the outer periostracum and the interprismatic matrices. This close relationship between the prisms and the periostracum demonstrates a role of the periostracum in organizing prism formation.

In 1969 Bevelander and Nakahara looked at the ligaments of *M. edulis* and *Pinctada radiata* and showed a close contact between the "conchiolin" of the ligament and needles of aragonite crystals embedded in the "matricial" compound. This was confirmed by Marsh et al. (1976) in *Spisula solidissima.*

In 1974 Meenakshi et al. (1974b) designed an experiment using the regeneration capacities of *Otala lactea,* a land snail. They introduced "substrata," which proved to be modifiers of the crystal arrangement pattern. The substrata were the individual periostraca of different species, such as *Mytilus, Elliptio complanatus, Megalonias gigantea,* and *L. stagnalis.* When no periostracum was added, *Otala* did not repair its shell "ad initium," but elaborated a repair mineral different in structural organization from that of its own shell. When a specific periostracum was added, *Otala* regenerated a new shell similar in organization to the shell from which the periostracum was taken.

Degens (1976) showed that calcium carbonates are associated with the proteinaceous matrix and from recent studies on ancestral forms (coral and algae), he suggests the possibility that, in addition to proteins and glycoproteins, carbohydrates may also function as a template in carbonate deposition.

From 1977 to 1980, Petit and his co-workers investigated the role of the periostracum in organizing mineral formation in the shell of a freshwater mussel, *Amblema plicata perplicata.* From a series of electron microscopic observations, x-ray diffraction, and microprobe analyses, with a careful preservation of hard and soft tissue relationships, it was evident that the periostracum was involved in several sequential processes associated with mineralization. All stages of these processes are clearly observable at the mantle edge of *Amblema.* A detailed discussion of the formation of the edge of the mineralized shell follows:

1. Inside the periostracal groove, the initial pellicle of the periostracum is coated with the glycocalyces of the epithelial cells lining the groove. The osmium–pyroantimoniate fixation and microprobe analysis show that these glycocalices contain calcium. The maturing periostracum appears gellike, con-

taining a dust of osmium–pyroantimoniate-positive granules. Within the groove, the coated pellicle receives more layers of glycocalyx material loaded with calcium. Events such as the polymerization or "tanning process" are very likely initiated within the groove (Waite, 1977). These structural modifications cause a layering of the gel, clearly visible as dark lines at the ultrastructural level.

The pellicle, with its attached glycocalytic products, is finally extruded from the tip of the mantle groove. Before losing any further cell contact, on the outer side it undergoes an ultimate tanning, and on the inner side it receives a further glycocalytic layer from the distal cells of the mantle fold. These cells are in continuity with the outer surface of the mantle, which is responsible for the general deposition of the nacre onto the inner surface of the valve (Fig. 16).

2. The periostracum is freed from the mantle groove and appears as three series of superimposed layers (Fig. 21): the nearly tanned protective outer layer, the middle layer, and the newly added inner layer.

The middle layer appears to be extremely active in the formation of prisms. From proximal to distal portions of the loop we observe the following: mineral segregation where the randomly distributed osmium–pyroantimoniate-positive granules appear in small clumps; vacuolization and cleavage of internal layers with further segregation and packing of amorphous material within the vacuoles; nucleation of individual needles at the internal gel–vacuole interphase; fusion of the vacuoles delineating a cavernous structure, the *antrum*. The rupture of the intervacuolar partitions forms columns, producing a honeycomb pattern (Fig. 23). Nucleation of initial needles proceeds on the roof, floor, and columns of the antrum. The first order of organization is seen as initial needles with a parallel orientation due to the specific angulation of their attachment to the matricial compound. They then associate into "flowers." This seems to unbalance their attachment to the gel, and eventually they become independent of the substrate, forming spherical associations of needles. This is the second order of organization. The balls of aragonite needles stack up in columns, eventually forming the prisms (Fig. 24), which are the third order of organization. In *Amblema* all of the active middle layer originates from the *periostracal groove* and serves exclusively in prism formation.

As it leaves the groove, the inner layer associates with the two other layers and probably undergoes polymerization and cleavage. It then initiates loops which extend deeply within the distal extrapallial fluid. As in the middle layer, we observe several sequential processes: mineral segregation within the loops; vacuolization and cleavage of some internal layers; continuation of nucleation and needle formation as the loops intimately reassociate with the middle periostracal layer; lamination of parallel sheaths (Fig. 25). Aragonite crystals pack together in bricks within the parallel sheaths. These rows of bricks constitute a third order of mineral organization called the *initial nacre* (Fig. 16).

The periostracum, then, appears to play a major organizing role in the com-

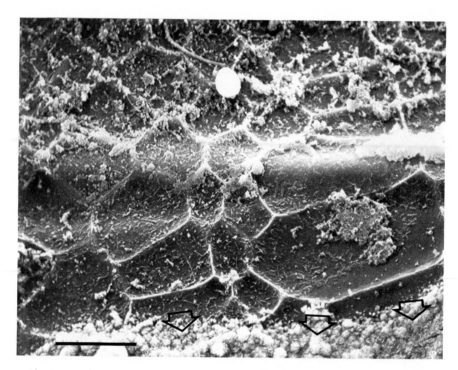

Fig. 23. *Amblema.* Microdissection of the inner aspect of the floor of an antrum, showing the honeycomb pattern. Note the mineralization front (arrows) starting the process of filling the honeycomb with mineral in the form of fuzzy balls. This will give the future prisms. Bar, 20 μm. (Petit et al., 1980a.)

plex series of events occurring at the forming edge of the shell of *Amblema.* The middle layer receives all its "information" during periostracum maturation from the cells lining the periostracal groove. This information helps in building prisms. The inner layer receives its information from the epithelial cells of the outer edge of the mantle groove, and this information helps in building the initial nacre.

3. *Distal to the mantle groove,* the outer covering of the shell remains unchanged as the tanned outer layer of the periostracum. The middle layer persists and remains associated with the prisms as prismatic bags, or matrices. The inner layer, which also persists, is associated with the initial nacre as nacreous matrices. This continuity is demonstrated morphologically at the electron microscopic level. It shows another example of close crystal-matrix relationships in mollusc shells and may be added to the observations of Watabe (1965) and Meenakshi et al. (1970). It is also in agreement with the theory of Degens (1976)

Fig. 24. *Amblema.* Newly formed prisms at the forming edge of the shell. Note the several orders of mineral organization. Bar, 0.5 μm. (Petit et al., 1980b.)

who described a protein fraction polymerized in the form of sheets, serving as a carrier and binding agent for an acidic polypeptide fraction which has a strong affinity for calcium ions.

Our amino acid analysis of periostracal material shows that the periostracum within the mantle groove is different from the free or the outer periostracum.

Fig. 25. *Amblema.* The lamination of parallel sheath starts from steplike structures within the inner periostracum. The fibrils initiate crystallization and also provide boundary walls between nacreous laminae. Bar, 2 μm. (Petit et al., 1980b.)

Periostracum within the groove has, in residues per thousand, a lower ratio of glycine (445/507), but higher ratios of alanine (38/19), aspartic acid (42/32), glutamic acid (31/21), and arginine (27/7).

We also find that the free periostracum, the outer shell edge periostracum, and the outer surface of the umbo are different in the amino acid ratios. This suggests that the periostracum is a dynamic structure. During each process related to biomineralization, some compounds, including amino acids, are altered with the forming structure. We propose that the amino acid compositions change prior to each order of mineral organization. Following the organization of the third order of mineral structure (prisms or nacre), the residual matrix becomes "relic structures that are frozen in mineral matter" (Degens, 1976).

VI. Future Research

The periostracum is formed sequentially as it passes out of the periostracal groove, undergoing tanning and finally a modification of the inner surface allowing calcification to begin. Attempts to induce the mantle edge to secrete periostracum *in vitro* have been unsuccessful in the past. Recently, however, S. C. Kunigelis and A. S. M. Saleuddin (unpublished data) cultured mantle edge tissue of *Helisoma* for 120 h, during which time fully tanned periostracum was produced. This periostracum was structurally different from normal periostracum. Studies are underway on the chemistry of periostracum formed *in vitro* and *in vivo,* and on the cellular formation of periostracum *in vitro.* Kunigelis and Saleuddin also observed that culturing mantle edge in the presence of brain, with or without lateral lobes, influences the incorporation of some labeled amino acids into the periostracum. They also found some evidence of deposition of calcium carbonate crystals on the inner surface of the periostracum after 56 hr of culture with L-3,4-dihydroxy-^{3}H-phenylalanine (^{3}H-DOPA).

The culture of mantle edge *in vitro* and the study of specific periostracal protein(s) *in vitro* and *in vivo* will be used to further examine the process and regulation of periostracum formation in molluscs.

Acknowledgments

This work was supported by the NSERC of Canada and Baylor College of Dentistry, Dallas, Texas. We thank Ms. Sharon Boothby and Ms. Judy Beeson for their help with the manuscript.

References

Beedham, G. E. (1958). Observation on the non-calcareous components of the shell of lamellibranchs. *Q. J. Microsc. Sci.* **99,** 341–357.

Beedham, G. E., and Owen, G. (1965). The mantle and the shell of *Solemya parkinsoni. Proc. Zool. Soc. London* **145,** 405–430.

Beedham, G. E., and Trueman, E. R. (1967). Relationship of the mantle and shell of the Polyplacophora in comparison with that of other Mollusca. *J. Zool.* **151,** 215–231.

Beedham, G. E., and Trueman, E. R. (1968). The cuticle of the Aplacophora and its evolutionary significance in Mollusca. *J. Zool.* **154,** 443–451.

Bevelander, G., and Nakahara, H. (1967). An electron microscope study of the formation of the periostracum of *Macrocallista maculata. Calcif. Tissue Res.* **1,** 55–67.

Bevelander, G., and Nakahara, H. (1969). An electron microscope study of the ligament of *Mytilus edulis* and *Pinctada radiata. Calcif. Tissue Res.* **4,** 101–112.

Bevelander, G., and Nakahara, H. (1970). An electron microscope study of the formation and structure of the periostracum of a gastropod, *Littorina littorea. Calcif. Tissue Res.* **5,** 1–12.

Bird, A. F. (1971). ''The Structure of Nematodes.'' Academic Press, New York.

Brown, C. H. (1952). Some structural protein of *Mytilus edulis. Q. J. Microsc. Sci.* **93,** 487–502.

Bubel, A. (1973a). An electron microscopic study of the periostracum formation in some marine bivalves. I. The origin of the periostracum. *Mar. Biol.* **20**, 213–221.

Bubel, A. (1973b). An electron microscopic study of the periostracum formation in some marine bivalves. II. The cell lining in the periostracal groove. *Mar. Biol.* **20**, 222–234.

Bubel, A. (1973c). An electron microscope study of periostracum repair in *Mytilus edulis. Mar. Biol.* **20**, 235–244.

Bubel, A. (1973d). An electron microscope investigation into the distribution of polyphenols in the periostracum and cells of the inner face of the outer fold of *Mytilus edulis. Mar. Biol.* **23**, 2–15.

Carter, J. G., and Aller, R. C. (1975). Calcification in the bivalve periostracum. *Lethaia* **8**, 315–320.

Chan, J. F. Y., and Saleuddin, A. S. M. (1974). Acid phosphatase in the mantle edge of the shell-regenerating snail *Helisoma duryi duryi. Calcif. Tissue Res.* **15**, 213–220.

Degens, E. T. (1976). Molecular mechanisms on carbonate, phosphate and silica deposition in the living cell. *Top. Curr. Chem.* **64**, 1–112.

Degens, E. T., Johannesson, B. W., and Meyer, R. W. (1967a). Mineralization processes in molluscs and their paleontological significance. *Naturwissenschaften* **54**, 638–640.

Degens, E. T., Spencer, D. W., and Parker, R. H. (1967b). Paleobiochemistry of molluscan shell proteins. *Comp Biochem. Physiol.* **20**, 553–579.

Dunachie, J. F. (1963). The periostracum of *Mytilus edulis. Trans. R. Soc. Edinburgh* **65**, 383–411.

Haas, F. (1935). Die Schale. *In* "Bronns Klassen und Ordung des Tier-Reichs" (H. G. Brown, ed.), Bd. iii, Abt. 3, Bivalvia. Tier 1.

Haas, W. (1972). Untersuchungen über die Miko-und Ultrastruktur der Polyplacophorenschale. *Biomineralisation* **6**, 1.

Hillman, R. E. (1961). Formation of the periostracum in *Mercenaria mercenaria. Science (Washington, D.C.)* **134**, 1754–1755.

Hunt, S., and Oates, K. (1978). Fine structure and molecular organization of the periostracum in a gastropod mollusc *Buccinum undatum* and its relation to similar structural protein systems in other invertebrates. *Phil. Trans. R. Soc. London Ser. B* **283**, 417–463.

Jones, G. M., and Saleuddin, A. S. M. (1978). Cellular mechanisms of periostracum formation in *Physa* spp. (Mollusca: Pulmonata). *Can. J. Zool.* **56**, 2299–2311.

Kawaguti, S., and Ikemoto, N. (1962a). Electron microscopy on the mantle of a bivalve, *Fabulina nitidula. Biol. J. Okayama Univ.* **8**, 1–2, 21–30.

Kawaguti, S., and Ikemoto, N. (1962b). Electron microscopy on the mantle of the bivalved gastropod. *Biol. J. Okayama Univ.* **8**, 1–2, 1–20.

Kawakami, I. K., and Yasuzumi, G. (1964). Electron microscope studies on the mantle of the pearl oyster *Pinctada martensii* Dunker, Prelim. report: The fine structure of the periostracum fixed with permanganate. *J. Electr. Microsc.* **13**, 119–123.

Kniprath, E. (1972). Feinstruktur der Periostrakumgrube von *Lymnaea stagnalis. Biomineralisation* **2**, 24–37.

Locke, M. (1966). The structure and formation of the cuticulin layer in the epicuticle of an insect *Calpodes ethlius* (Lepidoptera, Hesperiidae). *J. Morphol.* **118**, 461–494.

Marsh, M., Hopkins, G., Fisher, F., and Sass, R. L. (1976). Structure of the molluscan bivalve hinge ligament, a unique calcified elastic tissue. *J. Ultrastruct. Res.* **54**, 445–450.

Meenakshi, V. R., Martin, A. W., and Wilbur, K. M. (1974a). Shell repair in *Nautilus macrophalus. Mar. Biol. (Berlin)* **27**, 27–35.

Meenakshi, V. R., Donnay, G., Blackwelder, P. L., and Wilbur, K. M. (1974b). The influence of substrata on calcification patterns in molluscan shell. *Calcif. Tissue Res.* **15**, 31–44.

Meenakshi, V. R., Hare, P. E., Watabe, N., and Wilbur, K. M. (1969). The chemical composition of the periostracum of the molluscan shell. *Comp. Biochem. Physiol.* **29**, 611–620.

Meenakshi, V. R., Hare, P. E., Watabe, N., Wilbur, K. M., and Menzies, R. J. (1970). Ultrastructure, histochemistry and amino acid composition of the shell of *Neopilina*. *Anton Bruun Rep.* No. 2.

Neff, J. M. (1972). Ultrastructural studies of periostracum formation in the hard shelled clam *Mercenaria mercenaria* (L.) *Tissue & Cell* **4**, 311–326.

Petit, H. (1977). "The Mantle-Shell in the Fresh-Water Mussel *Amblema plicata perplicata*." Ph.D. Dissertation, Baylor Univ., Waco, Texas.

Petit, H. (1978). "Recherches sur des séquences d'événements périostracaux lors de l'elaboration de la coquille d'*Amblema plicata perplicata* Conrad 1834." Ph.D. Dissertation, Univ. Bretagne Occidentale, France.

Petit, H. (1981). Survol de la minéralisation chez les Unionidae. *Haliotis* **11**, 181–195.

Petit, H., Davis, W., and Jones, R. (1978). Morphological studies on the mantle of the fresh-water mussel *Amblema* (Unionidae). Scanning electron microscopy. *Tissue & Cell* **10**, 619–627.

Petit, H., Davis, W., and Jones, R. (1979). Morphological studies on the periostracum of the fresh-water mussel *Amblema* (Unionidae): L. M., T. E. M., S. E. M. *Tissue & Cell* **11**, 633–642.

Petit, H., Davis, W., and Jones, R. (1980a). A scanning electron microscopic study of the inorganic and organic matrices comprising the mature shell of *Amblema*, a fresh-water mollusc. *Tissue & Cell* **12**, 581–593.

Petit, H., Davis, W., Jones, R., and Hagler, H. (1980b). Morphological studies on the calcification process in the fresh-water mussel *Amblema*. *Tissue & Cell* **12**, 13–26.

Petit, H., Fullington, R., Matthews, J. L., Roa, R., and Skalnik, P. (1980c). Morphometry in a molluscan shell. *Proc. Int. Bone Histomorphometry Workshop 1980*, pp. 127–133.

Saleuddin, A. S. M. (1974). An electron microscopic study of the formation and structure of the periostracum in *Astarte* (Bivalvia). *Can. J. Zool.* **52**, 1463–1471.

Saleuddin, A. S. M. (1975). An electron microscopic study on the formation of the periostracum in *Helisoma* (Mollusca). *Calcif. Tissue Res.* **18**, 297–310.

Saleuddin, A. S. M. (1976a). Ultrastructural studies on the formation of the periostracum in *Helix aspersa* (Mollusca). *Calcif. Tissue Res.* **22**, 49–65.

Saleuddin, A. S. M. (1976b). Ultrastructural studies on the structure and formation of the periostracum in *Helisoma* (Mollusca). *In* "The Mechanisms of Mineralization in the Invertebrates and Plants" (N. Watabe, and K. M. Wilbur, eds.), pp. 309–337. Univ. of South Carolina Press, Columbia.

Saleuddin, A. S. M. (1979). Shell formation in molluscs with special reference to periostracum formation and shell regeneration. *In* "Pathways in Malacology" (S. Van der Spoel, A. C., Van Bruggen, and J. Lever, eds.), pp. 47–81. Bohn, Scheltema & Holkema, Utrecht, Netherlands.

Taylor, J. D., and Kennedy, W. J. (1969). The influence of the periostracum on the shell structure of bivalve molluscs. *Calcif. Tissue Res.* **3**, 274–283.

Timmermans, L. P. M. (1969). Studies on shell formation in molluscs. *Neth. J. Zool.* **19**, 417–523.

Wada, K. (1968). Electron microscopic observations of the formation of the periostracum in *Pinctada fucata*. *Kokuritsu Shinju Kenkyusho Hokoku (Bull. Natl. Pearl Res. Lab. [Jpn.])* **13**, 1540–1560.

Waite, J. H. (1977). Evidence for the mode of sclerotization in a molluscan periostracum. *Comp. Biochem. Physiol.* B **58**, 157–162.

Waite, J. H., and Andersen, S. O. (1978). 3,4-Dihydroxyphenylalanine in an insoluble shell protein of *Mytilus edulis*. *Biochem. Biophys. Acta* **541**, 107–114.

Waite, J. H., and Andersen, S. O. (1980). 3,4-Dihydroxyphenylalanine and sclerotization of periostracum in *Mytilus edulis*. *Biol. Bull (Woods Hole, Mass.)* **158**, 164–173.

Waite, J. H., and Wilbur, K. M. (1976). Phenoloxidase in the periostracum of the marine bivalve *Modiolus demissus* Dillwyn. *J. Exp. Zool.* **195**, 358–368.

Waite, J. H., Saleuddin, A. S. M., and Andersen, S. O. (1979). Periostracin—a soluble precursor of sclerotized periostracum in *Mytilus edulis* L. *J. Comp. Physiol.* **130,** 301–307.

Watabe, N. (1965). Studies on shell formation. XI. Crystal matrix relationships in the inner layer of mollusk shells. *J. Ultrastruct. Res.* **12,** 351–370.

6

Shell Formation

KARL M. WILBUR   **A. S. M. SALEUDDIN**

Department of Zoology   Department of Biology
Duke University   York University
Durham, North Carolina   Toronto, Ontario, Canada

Shells and the Reason of Their Shape

The creature that resides within the shell constructs its dwelling with joints and seams and roofing and the other various parts, just as man does in the house in which he dwells; and this creature

235

THE MOLLUSCA, VOL. 4
Physiology, Part 1

expands the house and roof gradually in proportion as its body increases and as it is attached to the sides of these shells.

Leonardo da Vinci
Manuscript F. 80r.

I. Summary and Perspective

The formation of shell can be described in terms of two major phases: (1) cellular processes of ion transport, protein synthesis, and secretion and (2) a series of physicochemical processes in which crystals of $CaCO_3$ are nucleated, oriented, and grow in intimate association with a secreted organic matrix.

In providing the mineral of shell, Ca^{2+} and HCO_3^- are first transported across epithelia at the body surface and the mantle epithelium facing the inner shell surface. Movement of these ions is incompletely understood but almost certainly involves active transport of Ca^{2+}. The physicochemical phase takes place in fluid between the outer mantle epithelium and the inner shell surface and on this shell surface. For crystals to be deposited, the following conditions are required (1) concentrations of Ca^{2+} and CO_3^{2-} exceeding the solubility product, (2) conditions favoring crystal nucleation, and (3) the elimination of H^+ resulting from $CaCO_3$ formation. Following crystal nucleation oriented crystal growth leads to a shell constructed of a variety of crystal patterns. In the process protein secreted by the mantle surrounds the individual crystals and becomes the cement which binds them together as a shell.

Our understanding of the processes of the crystal depositional phase of shell formation is fragmentary. However, we can expect that through analyses of the fluid at the site of crystallization and study of effects of organic fractions of the fluid and shell on crystallization, the conditions governing the initiation and control of crystal growth will be further defined. Substances that initiate crystallization have been proposed but have not been tested. Preliminary studies indicate that the compounds at the crystallization site include inhibitors of $CaCO_3$ deposition. A major aspect of shell construction of great interest yet to receive experimental study is the relation of particular segments of the mantle epithelium to the structure and orientation of strikingly different types of crystal patterns. A promising approach to these crystallographic problems appears to be the application of *in vitro* methods of physical biochemistry to crystal development and orientation in association with fractions of the organic phase of shell.

Studies indicating involvement of the nervous system and hormonal and neurosecretory changes associated with shell formation will almost certainly be expanded and contribute to the understanding of mechanisms of mineralization and their control.

II. Introduction

The study of shell formation comprises the processes involved in forming a marvelously complex structure from a myriad of crystal units assembled in a large variety of patterns. The crystals are calcium carbonate bound together in a matrix that is primarily protein. The shell is usually covered externally by a tanned protein sheet, the periostracum.

Our discussion of shell formation will be concerned in large measure with the physiological and biochemical mechanisms associated with the formation of crystals of $CaCO_3$ from Ca[1] and HCO_3[1] and deposition within the matrix. We shall also give attention to the assembly of some of the major crystal patterns. These mechanisms of shell formation can be discussed only within the context of the morphological system in which the crystals grow and form the shell, and so we shall be concerned with the general features of the shell-forming system present in all molluscs.

A cursory look at the form and structure of shells of the phylum Mollusca will show at once the great diversity that has evolved. In view of this diversity, it is evident that analysis of shell formation can come within reasonable and manageable limits if, and only if, common mechanisms can be found within and among taxa. This is not to say that differences in mineralizing mechanisms will not be found. Indeed, differences can be anticipated, but a search for similarities should be the first priority. In this chapter we shall make the assumption that similarities in mechanisms will cross taxonomic lines. There are sound reasons for that assumption, of course, the principal ones being the basic similarities in the morphological system of shell formation and the ultrastructure of shell. Our discussion of shell formation will begin with a brief description of that system. Following that, major attention will be directed to the complex of processes leading to shell formation. The first of these is the passage of ions from the medium through the tissues to the site at which they are converted to $CaCO_3$. Then the conditions required for $CaCO_3$ deposition and crystal growth will be examined. Because an organic matrix is associated with the crystalline phase of shell, attention will be given to the matrix and its possible functions in shell formation. Finally, neurohumoral mechanisms that may play a role in determining the shell deposition rate will be considered as well as the energy sources of shell deposition.

Within the past decade the subject of molluscan mineralization in its various phases has been discussed in two symposia (Watabe and Wilbur, 1976; Omori and Watabe, 1980) and in at least three reviews (Wilbur, 1972; Crenshaw, 1980; Wilbur and Manyak, 1983). These sources give a reasonably complete coverage of recent literature and provide background for the present chapter as well as details beyond those included here.

[1]The symbols Ca and HCO_3 hereafter will be understood to refer to Ca^{2+} and HCO_3^-.

The topic of shell growth, though intimately related to our discussion of mineralization and shell formation, will not be treated in the present chapter. Information on changes in shell form during growth are discussed by Wilbur and Owen (1964) and Seed (1980). Tevesz and Carter (1980) have considered environmental influences on shell form and structure, and Rhoads and Lutz (1980) have given an overview of microscopic shell patterns in relation to growth.

III. The System of Shell Formation

The mantle, a thin organ lining the inner shell surface, is directly responsible for the deposition of the crystals and the secretion of the organic matrix of shell. A layer of epithelium covers the mantle surfaces and the mantle interior, which is supplied by hemolymph, includes muscle, connective tissue, and nerve fibers. The outer mantle epithelium brings about the deposition of more than one type of shell structure, each with its characteristic crystal form and arrangement. The cells responsible for these different structural types also differ in form (Fig. 1). The total system of shell formation comprises four compartments: (1) the exter-

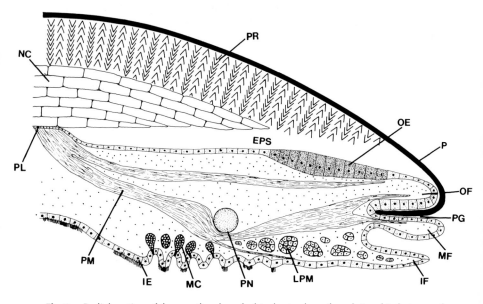

Fig. 1. Radial section of the mantle edge of a bivalve to show the relationship between the shell and mantle. (Not to scale.) EPS, Extrapallial space; IE, inner epithelium; IF, inner fold; LPM, longitudinal pallial muscle; MC, mucous cell; MF, middle fold; NC, nacreous shell layer; OE, outer epithelium; OF, outer fold; P, periostracum; PG, periostracal groove; PL, pallial line; PM, pallial muscle; PN, pallial nerve; PR, prismatic shell layer.

nal medium, (2) the hemolymph and body tissues, (3) the extrapallial fluid compartment between the mantle and the inner shell surface, and (4) the shell (Fig. 2). Two epithelia, each a single cell in thickness, limit the body compartments except for the shell. One epithelium covers the body surfaces and admits ions from the outer medium to the hemolymph and permits ion efflux (Simkiss and Wilbur, 1977). The second, the outer mantle epithelium, transfers Ca, HCO_3, and other ions from the hemolymph to the extrapallial fluid compartment and secretes organic components into the compartment. The body tissues provide a portion of the HCO_3 and may well be involved in other metabolic activities related to the synthesis of the organic compounds of the shell. It is the extrapallial fluid with its inorganic and organic substances contributed by the mantle that is the microenvironment of shell deposition.

Ion fluxes through the system are bidirectional. During shell deposition the net transport of Ca and HCO_3 will be inward toward the mineralizing shell surface, but under some conditions, the inner shell surface as a result of the solubilization of $CaCO_3$ is a source of Ca and HCO_3, which go to other parts of the body and to the outer medium (e.g., Greenaway, 1971a). These fluxes and their relation to shell formation will now be discussed.

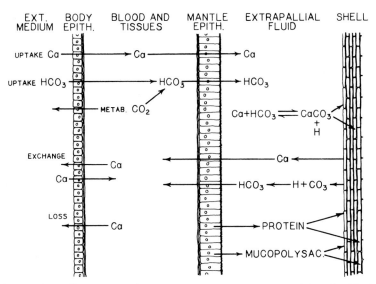

Fig. 2. The molluscan mineralizing system. Note: (1) movement of Ca and HCO_3 from the medium toward the shell as occurs during shell deposition; (2) source of HCO_3 from the external medium and tissue metabolism; (3) movement of Ca and HCO_3 outward as occurs from shell dissolution; (4) secretion by the mantle of organic material of shell; and (5) ion exchange at the body epithelium. (After Greenaway, 1971b.)

IV. Ion Uptake by the Organism

Various body surfaces of molluscs admit ions. In freshwater snails the external body surface is almost certainly a major site of ion uptake (van der Borght and van Puymbroeck 1964; van der Borght, 1963; Greenaway, 1971b; Thomas et al., 1974). The mantle surface facing the mantle cavity of the oyster *Crassostrea virginica* and the bivalve *Hyriopsis schlegeli* appears to be another area of intake (Jodrey, 1953; Horiguchi, 1958). Ions can also enter the hemolymph from the gut. *Lymnaea stagnalis* maintained on a diet of lettuce was found to obtain about 20% of its calcium by this route (van der Borght et al., 1966). The gills have also been shown to be important in Ca uptake in *H. schlegeli* (Horiguchi, 1958).

Ca uptake from the medium has been measured in *L. stagnalis* (van der Borght and van Puymbroek, 1964; Greenaway, 1971a; Schlichter, 1981) and *Biomphalaria glabrata* (Thomas and Lough, 1974). Van der Borght and van Puymbroeck (1964) found that *L. stagnalis* takes up Ca by active transport. Active transport of Ca against a small gradient is required in this species below 0.5 mM but not above that concentration (Greenaway, 1971a). Schlichter (1981), from comparison of the Nernst potential with the transepithelial potential difference in *L. stagnalis*, has also obtained evidence for active transport. One can expect that active transport across the outer epithelium will also occur in other freshwater molluscs from media of very low Ca concentrations. Active transport of Ca across the epithelium probably involves two stages: diffusion down an electrochemical gradient into the cytosol and active pumping by the cell membrane facing the hemolymph.

Ion uptake by marine and brackish water molluscs would appear to be somewhat different from freshwater molluscs in that the two former groups are osmoconformers without substantial ion gradients between the medium and the hemolymph. Moreover, in marine species, the medium is saturated with $CaCO_3$ and, as the same appears to be true of the hemolymph and extrapallial fluid, there is virtually no gradient among these three compartments. However, a Ca pump will still be required in marine as in freshwater and brackish water species provided that Ca does not gain entrance through intercellular spaces in the body epithelia. This follows from the adverse electrochemical gradient between the cytosol of the epithelial cells and the hemolymph, as we suggested for freshwater species.

Ca pumps in vertebrate red cells (Schatzmann, 1973) and in sarcoplasmic reticulum (de Meis and Vianna, 1979) are Ca-activated ATPases, and it is likely that the same is true in molluscan epithelia. The role of these enzymes in Ca transport deserves to be investigated not only in molluscs but also in other invertebrates that deposit calcareous skeletons.

If the rate of Ca transport is governed by a Ca-activated enzyme pump, the rate can be expressed approximately by the equation

$$v = \frac{V_{max}\ (s)}{K_m + (s)} \tag{1}$$

in which v is the velocity, V_{max} is the maximum velocity when the enzyme system is saturated, (s) is the Ca concentration, and K_m is a constant which determines the concentration at which v is one-half V_{max}. K_m expresses the affinity of the transportation system for Ca ions. As the Ca concentration of the medium is increased, v will increase to a maximum representing the saturation of the system with Ca. Values for the saturation concentration K_m, and V_{max} for freshwater gastropods are shown in Table I. The parameters in Eq. (1) indicate factors that will affect the rate of $CaCO_3$ deposition provided transport of Ca from the medium is limiting. For example, the rate of deposition will increase with Ca concentration of the medium up to the level at which the transport enzyme is saturated. Further, the smaller the value of K_m, the lower the concentration of external Ca at which V_{max} will be reached.

If the Ca concentration of the external medium is sufficiently low, a net loss to the medium will occur (Greenaway, 1971a). The minimum equilibrium concentration was found to lie between 0.012 and 0.025 mM at 26 ± 1° for *Biomphalaria glabrata* (Thomas and Lough, 1974); 0.062 mM for *L. stagnalis* at 10 ± 1° (Greenaway, 1971a); and 0.79 mM for *L. stagnalis appressa* at 21 ± 1°. One might expect that if uptake from the medium were the limiting factor in the rate of $CaCO_3$ deposition, then within certain limits of environmental Ca concentration, shell deposition might increase with Ca concentration. Thomas et al. (1974) have shown this to be the case in *B. glabrata*. The amount of shell mineral was also related to Ca concentration in *Lymnaea peregra* (Russell-Hunter, 1978) but not in *L. palustris* (Hunter, 1975) or *Ferrissia rivularis* (Rus-

TABLE I

Kinetics of Ca Uptake

Species	K_m (mM)	Saturation concentration (mM)	V_{max} (μM/g/h)	Temperature (°C)	Reference
Lymnaea stagnalis	0.3	1.0–1.5	0.4	10	Greenaway, 1971a
Ancylastrum fluviatilis					
hard water	0.175	0.75	0.6	20	Chaisemartin and Videaud, 1971
soft water	0.065	0.25	0.6		
Biomphalaria glabrata	0.27	1–2	10.2[a]	26	Thomas and Lough, 1974

[a] Net uptake value.

sell-Hunter et al., 1981). It is understandable that all species do not respond similarly since many factors other than Ca concentration of the medium affect deposition rate (Thomas et al., 1974). Whether an increase in shell growth will occur will depend ultimately on conditions within the extrapallial fluid and at the site of $CaCO_3$ deposition on the inner shell surface.

V. Ion Movement across the Mantle

We have mentioned that Ca and HCO_3 move from the hemolymph within the mantle across the outer mantle epithelium to the extrapallial fluid (Fig. 2). Mechanisms by which ions may move across membranes in this and other mineralization systems have been discussed by Simkiss (1976). They include (1) passive movement down an electrochemical gradient, (2) active transport, and (3) movement of Ca coupled to the movement of other ions. We should say immediately that the mechanism of transepithelial ion movement in molluscan mantles remains an open question. At the same time certain specific findings deserve mention. Neff (1972) suggested that the intercellular route is used in the central mantle area of the clam *Mercenaria mercenaria* as indicated by continuity of intercellular spaces with connective tissue spaces containing hemolymph and by the presence of Ca within intercellular spaces. Also, Crenshaw and Travis (Crenshaw, 1980), using La^{3+} as a tracer for Ca, found that this ion penetrated the length of the intercellular channels in the same area but not in the marginal zone where Ca deposition is normally more rapid. However, in *Helix aspersa* Ca can be localized in intercellular spaces of the mantle edge (A. S. M. Saleuddin, unpublished observations). Until further evidence is at hand, one should be cautious about attributing great importance to intercellular transport as a factor in shell deposition. That hemolymph fluid passing through intercellular channels cannot be the sole source of Ca and extrapallial fluid is shown by differences in composition of the hemolymph and extrapallial fluid (Table II; Florkin and Besson, 1935; Crenshaw, 1972a). Other evidence also demonstrates clearly that the mantle cells contribute to the composition of the extrapallial fluid.

Ion movement as indicated by potentials across mantle membranes has been investigated in the freshwater bivalves *Anodonta* (Istin and Kirschner, 1968; Istin and Fossat, 1972; Sorenson et al., 1980) and *Amblema costata* (Istin and Kirschner, 1968), the marine bivalve *Anomalocardia brasiliana* (Sorenson et al., 1980), and the terrestrial snails *Helix pomatia* and *H. aspersa* (Enyikwola and Burton, 1983). A resting potential across the entire mantle exists, the main electrogenic step being across the outer cell membrane of the epithelial layer facing the shell. The resting potential in isolated mantle preparations varies with species from a few millivolts in *Helix* to more than 60 mV in *Anomalocardia*, the shell side positive. With potentials of 47 mV and 60 mV present in *Anodonta* and *Anomalocardia*, respectively, and a Ca concentration of the cytosol of the epithelial cells of the order of other cells (10^{-7}–10^{-8} *M*) (Dipolo et al., 1976; Borle and Snowdowne, 1982), the electrochemical gradient between the cells and the extrapallial fluid will be strongly against a movement of Ca out of the mantle epithelium toward the shell (Sorenson et al., 1980). Moreover, Ca was relatively impermeant in *Anodonta* and *Anomalocardia* (Sorenson et al., 1980). The mantles were mainly permeable to K and Cl, and the resting potentials were consistent with a K diffusion potential. The authors concluded that the effects of Ca on the membrane potential were primarily due to effects on K permeability.

An examination of mantle potentials in *Helix pomatia* and *H. aspersa*, which are lower than in *Anodonta* and *Anomalocardia*, has indicated that electrogenesis is probably due to active transport of Cl from the shell side toward the hemolymph (Enyikwola and Burton, 1983). Ca, K, Na, Mg, and HCO_3 were not required for electrogenesis.

Results obtained thus far on ion movements in mantle epithelium indicate clearly the desirability of comparative studies on transport mechanisms in marine, freshwater, and terrestrial species. However, in carrying out such studies, we should recognize the unhappy possibility that information on normal ion movements in the intact animal may prove difficult to obtain in isolated mantle preparations. Since handling of bivalves is known to change shell growth, dissection of the mantle may well affect ion movements. The interruption of nerve conduction in isolated mantles may or may not affect epithelial cell functioning (see Dillaman et al., 1976). Also, the central mantle region used in the studies is an area of low deposition rate as compared with the mantle edge (Wilbur and Jodrey, 1952; Zischke et al., 1970), and as Sorenson et al. (1980) point out, Ca is relatively impermeant in this region.

The rate of movement of ions into the mantle and their incorporation into the shell can be followed by exposing molluscs to a medium containing ^{45}Ca or $H^{14}CO_3$. For short-term measurements of rates of shell deposition, one must ascertain that the mantle has reached isotopic equilibrium with the external medium. Once equilibrium has been attained, the rate of deposition of calcium and carbonate can be calculated from the specific activity of the medium and the rate of uptake by the shell using the relation

$$D = \frac{A_s}{A_w} \times C \tag{2}$$

in which D is milligrams of calcium or carbonate deposited per square centimeter or milligram of shell; A_s is counts per minute per square centimeter or milligram of shell; A_w is counts per minute per liter medium; and C is milligrams of calcium or bicarbonate per liter medium (Wilbur and Jodrey, 1952; Wheeler et al., 1975). By treating the shell samples of animals exposed to $H^{14}CO_3$ with acid to remove carbonate, the rate of carbon fixation into the organic compounds of shell can also be measured. The determination of the rate of deposition by this method presumes that the animals are not respiring anaerobically. Figure 3 shows an example of the rate of uptake of ^{45}Ca by the mantle and the rate of its deposition in the shell of the scallop *Argopecten irradians*. Equilibrium was reached in the mantle after about 2 h, and the rate of calcium deposition could accordingly be measured after this period. However, in *Mercenaria mercenaria*, equilibrium between the medium and the extrapallial fluid was not reached over several hours, as shown by sampling the extrapallial fluid with a catheter through a hole in the shell (C. K. Goddard, personal communication) and indicated by isotope experiments (Dillaman and Ford, 1982).

In the shipworm *Bankia gouldi*, two pools of Ca have been identified in the mantle underlying the region in which a calcified tube is deposited (Manyak,

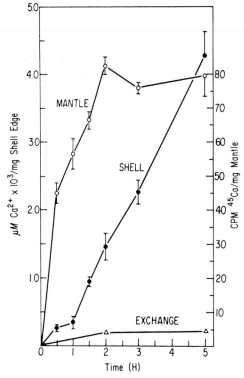

Fig. 3.   Uptake of ^{45}Ca by the mantle and shell edge of *A. irradians* as a function of time. After 2 h, the mantle was in equilibrium with the radioactive medium. The rate of ^{45}Ca deposition was approximately linear after 1 h in ^{45}Ca. Vertical bars show standard deviations. (After Wheeler et al., 1975.)

1982). Freely exchangeable and presumably ionic Ca reaches equilibrium with the external medium within about 2 h. Slowly exchangeable and presumably bound or complexed Ca does not reach equilibrium even after 120 h of exposure to ^{45}Ca. If bound or complexed Ca of the mantle that had not equilibrated with the medium were deposited in the calcified tube, measurements of $CaCO_3$ deposition rates based on incorporation of ^{45}Ca would not be valid. However, it appears that most or all of the Ca deposited in the tube is derived from turnover of ionic Ca of the mantle.

VI. Carbonate and Carbonic Anhydrase

The carbonate of shell is derived from two sources: the bicarbonate of the external medium and the bicarbonate from metabolic CO_2 (Hammen and Wilbur, 1959; Hammen, 1966; Campbell and Speeg, 1969; Dillaman and Ford, 1982) (Eqs. 3 and 4; Fig. 2). The conversion of CO_2 to HCO^- is a reversible reaction catalyzed by carbonic anhydrase (Eq. 3(1)). Ca and HCO_3 then combine to form $CaCO_3$ with the release of one proton for each molecule of $CaCO_3$ formed (Eq. 4).

We now turn to a discussion of the role of carbonic anhydrase in shell formation. This will be followed by an account of two methods proposed for the control of extrapallial fluid pH necessitated by the release of protons in $CaCO_3$ deposition.

$$CO_2 + H_2O \underset{(1)}{\rightleftharpoons} H_2CO_3 \rightleftharpoons H^+ + HCO_3^- \rightleftharpoons H^+ + CO_3^{2-} \tag{3}$$

$$Ca^{2+} + HCO_3^- \rightleftharpoons CaCO_3 + H^+ \tag{4}$$

For well over three decades, carbonic anhydrase has been given attention in molluscs. In addition to the catalysis of the reversible hydration of CO_2 by carbonic anhydrase, the H^+, HCO_3^-, and CO_3^{2-} in the associated reactions (Eq. 3) are involved in electrolyte balance, acid–base regulation, and mineralization. However, investigations in molluscs have dealt largely with shell mineralization. Because the reactions in Eq. 3 proceed without carbonic anhydrase, the enzyme cannot be considered a prerequisite of $CaCO_3$ deposition but only a factor affecting its rate (Freeman and Wilbur, 1948; Stolkowski, 1951; Freeman, 1960). Inhibitors, including Diamox, benezenesulfonilamide, p-toluenesulfonamide, sulfonamide, and ethoxyzolamide, have been employed to support the role of carbonic anhydrase in shell deposition (Wilbur and Jodrey, 1955; Abolins-Krogis, 1958; Freeman, 1960). The commonly observed effect of inhibitors has been a decrease, but not a cessation, of calcium deposition or shell growth. Freeman (1960) has examined the effects of various concentrations of inhibitors on *Physa heterostropha* and found an inhibition of the rate of increase in shell length of 32 to 44%. This partial inhibition is to be expected because the reaction involving HCO_3 will occur without catalysis, as we mentioned. There was no inhibition in snails growing at a slow rate as a result of a difference in diet. This lack of inhibition indicates that catalysis is not taking place, an effect which deserves further examination. Carbonic anhydrase activity of the mantle edge has been observed to increase during shell regeneration in *Helisoma duryi*, the activity being greatest in the mantle edge underlying the area of shell removal. Mantle tissue carbonic anhydrase specific activity has also been found to increase with increasing shell deposition rate (S. C. Kunigelis and A. S. M. Saleuddin, in preparation). Wilbur and Anderson (1950) found a 47-fold increase in activity of carbonic anhydrase at the outset of shell formation in larvae of *Busycon carica* whereas in older animals, which deposit shell at a lower rate, the enzyme activity was lower. The results with *Physa, Helisoma,* and *Busycon* indicate that the activity of carbonic anhydrase is not constant but may be altered with physiological changes of the organism.

In the early work the basic assumption for carbonic anhydrase action was that it functioned in the catalysis of the conversion of metabolic CO_2 to HCO_3 (Eq. 3(1)), which then entered into the formation of $CaCO_3$ (Wilbur, 1964; see also Loest, 1979a). In this way $CaCO_3$ deposition would be accelerated provided that HCO_3 was in short supply and CO_2 was limiting in its formation. A current view is that carbonic anhydrase is involved in the catalysis of the reverse reaction $HCO_3 \rightarrow CO_2$ within the extrapallial fluid with a consequent reduction of acidity resulting from $CaCO_3$ formation (Wheeler, 1975) (see Section VII for details). A third proposed function attributes to carbonic anhydrase the acceleration of the conversion of spherules of $CaCO_3$ to Ca. This is hypothesized to take place in those bivalves in which spherules are present in the interstitial cells of the mantle (Istin and Maetz, 1964; Istin and Masoni, 1973). The spherules have carbonic anhydrase in their periphery (Istin and Girard, 1970; Rionel et al., 1973). The enzyme is presumed to catalyze the hydration of CO_2 within the cells, increasing the acidity with a consequent increased rate of dissolution of the spherules, thus providing an increase in Ca ions. It is to be noted that the free Ca would be in the interstitial cells rather than the epithelial cells bordering the extrapallial space. An increased turnover rate of $CaCO_3$ of the spherules could be advantageous as a source of Ca and HCO_3 for shell mineralization, especially in freshwater molluscs with reduced Ca availability, in shell repair (Watabe et al., 1976), and in buffering under anaerobic conditions in which organic acids accumulate.

VII. Control of Extrapallial Fluid pH

Since the formation of $CaCO_3$ from Ca and HCO_3 results in the release of protons (Eq. 4), it will be apparent that the removal of protons is mandatory for the continued deposition of shell crystals. Two methods proposed for accomplishing proton removal will be summarized.

A. Proton Removal by Carbonic Anhydrase Catalysis

Wheeler (1975) has proposed that the protons react with bicarbonate forming CO_2 within the extrapallial fluid (Eqs. 3 and 5).

$$H^+ + HCO_3 \rightarrow CO_2 + H_2O \tag{5}$$

CO_2, in turn, must be removed from the extrapallial fluid. It would be expected to diffuse down the gradient from the extrapallial fluid into the mantle cells and so to the hemolymph. As CO_2 diffuses out of the extrapallial fluid, Eq. 5 will continue to the right. Carbonic anhydrase, known to be present in the mantle, could function in two ways to increase the rate of proton and CO_2 removal (Wheeler, 1975): by catalyzing the dehydration of HCO_3 (Eq. 5) and by favoring the diffusion of CO_2 through the unstirred layers immediately external and internal to the outer plasma membranes of the mantle cells. The action of carbonic anhydrase in increasing movement of bicarbonate has been demonstrated for artificial membranes (Gutknecht et al., 1977). In coccolithophorid cells, Sikes and Wheeler (1982) have suggested that carbonic anhydrase applied externally may maintain CO_2 equilibrium and so increase the CO_2 level at the external surface. The authors consider that facilitation of diffusion of CO_2 is an important function of carbonic anhydrase (Sikes and Wheeler, 1983).

Fixation of CO_2 can take place in mantle tissue (Hammen, 1966), and this process will also assist in removal of CO_2 from the extrapallial fluid.

B. Proton Removal by Ammonia

A second method proposed for the removal of protons resulting from $CaCO_3$ deposition is by ammonia formation (Eq. 6) (Campbell and Boyan, 1976).

$$Ca^{2+} + HCO_3^- + NH_3 \rightarrow CaCO_3 + NH_4^+ \tag{6}$$

The ammonia would be formed from urea by the action of urease, an enzyme present in certain, but not all, pulmonate snails. However, species lacking urease may still produce ammonia through deamination of purine, purine nucleotides, and purine nucleosides (Campbell and Boyan, 1976). Loest (1979b), in an examination of 14 species of pulmonates, found that adenosine deaminase or urease was present in every species and that adenosine deaminase was the predominant enzyme in 10 species. Various reactions involving the mantle, extrapallial fluid, carbonic anhydrase, and the transfer of NH_3, NH_4^+, H^+, and other ions are hypothesized by Campbell and Boyan to account for proton removal from the site of mineralization. Periodicity of ammonia release of approximately 24 h has been observed in the snail *Otala lactea* and is presumed to be accompanied by CO_2 release (see also Loest, 1979a). The mechanism involved is considered consistent with periodic shell deposition. However, the frequency of increment formation of the shell was not examined. In order for the NH_3 mechanisms to be effective, an adequate rate of production of NH_3 and the elimination of NH_4^+ from the extrapallial fluid at a rate which will prevent its accumulation will obviously be required. From measurements of the activities of adenosine de-aminase and urease of the 14 species examined, Loest (1979b) calculated that the rate of ammonia formation was more than adequate to react with protons released in shell growth. Thus far, the ammonia formation has been examined in ter-restrial species. Extension of the studies to marine and freshwater molluscs would be of interest.

Simkiss (1976) has pointed out that ammonia production is a poor mechanism for removing protons resulting from the formation of $CaCO_3$ (Eq. 4). Although NH_3 lacking a charge would be expected to pass readily across plasma mem-branes, the charged NH_4^+ that is formed (Eq. 6) would penetrate plasma mem-branes with difficulty and so would accumulate in the extrapallial fluid. More-over, NH_4^+ would be expected to accumulate in the extrapallial fluid if the site of calcification has a pH slightly below that of the hemolymph (Simkiss, 1976).

Another possible mechanism favoring the elimination of protons from the extrapallial fluid and maintenance of pH would be a Ca/H exchange across the outer cell membrane of the outer mantle epithelium.

C. Anaerobiosis and pH of Extrapallial Fluid

On shell closure when there is no access to the external medium and the animal becomes anaerobic, the pH of the extrapallial fluid decreases due to an increase in acids. In the case of molluscs exposed to air during the tidal cycle, the pH change may be cyclic, as in the bivalve *Cerastoderma edule* which shows a variation from approximately pH 7.1 to 7.7 (Richardson et al., 1981). Because pH at surfaces may be below that of bulk solution (Plummer et al., 1979), the crystal surfaces may be exposed to a pH lower than that measured. At a suffi-ciently low pH, the rate of $CaCO_3$ deposition will be decreased and recently deposited $CaCO_3$ may be dissolved. Whether dissolution occurs will depend upon the crystal type, the pH equilibrium for that type, and the pH at the crystal surface. Because the pH equilibrium of aragonite in seawater occurs at pH 8.2 ($P_{CO_2} \sim 0.2$ Torr), *Cerastoderma* with an aragonitic shell may well experience periodic shell deposition and dissolution with each tidal cycle even though the pH of the extrapallial fluid is on the alkaline side of neutrality (Richardson et al., 1981; C. S. Sikes, personal communication).

The decrease in pH in the extrapallial fluid under anaerobic conditions results from an increase in succinic, lactic, and propionic acids (Simpson and Awapara, 1966; Stokes and Awapara, 1968; Crenshaw and Neff, 1969; Zurburg and Kluytmans, 1980). In each of the these studies, succinic acid was found in high concentration as compared with lactic and propionic acids. The predominance of succinic acid is advantageous in that in equivalent concentrations it will give a smaller increase in H^+ concentration than lactic and propionic acids, as evident from their pKs. Dissolution of crystals on the inner shell surface will accordingly be less. The mechanisms of shell dissolution have been discussed by Crenshaw (1980).

VIII. Extrapallial Fluid: The Environment of $CaCO_3$ Deposition

The mineral and organic composition of shell will be governed by the constituents and state of the extrapallial fluid enclosed between the outer mantle epithelium and the inner shell surface. The volume of the extrapallial fluid may be a few milliliters or only a thin layer as at the shell margin. In the latter case, the transfer of materials from epithelial cells to shell surface is essentially direct. The extrapallial fluid is a complex mixture of a large number of inorganic and organic substances. The inorganic ions are derived primarily from the hemolymph in the mantle after diffusion or active transport across the outer mantle epithelium (see Section V). The organic compounds are secreted from the cells of the epithelium. Whether the compounds are synthesized chiefly in the mantle cells or in other tissues and then transported to the mantle is unknown. Substances in the external medium could also enter the extrapallial fluid directly if the thin sheet of periostracum connecting the mantle edge and the shell edge becomes temporarily ruptured through withdrawal of the mantle. If the medium is seawater, this would supply a saturated $CaCO_3$ solution but in freshwater medium, the entering fluid would be very low in inorganic ions (Table II).

Information on the main constituents of the extrapallial fluid will be summarized briefly in the sections which follow.

A. Ionic Constituents

Analyses of the extrapallial fluid of several species, both marine and freshwater, all show the major cations to be Na, K, Ca, and Mg (Crenshaw, 1972a, Wada and Fujinuki, 1976; Table II). The Ca fractions in the extrapallial fluid of *Mytilus edulis* are shown in Table III. The major anions are HCO_3, Cl, and SO_4 (Table II). In addition to the major cations, trace quantities of a number of metals are present (Wada and Fujinuki, 1976).

A comparison of concentrations of inorganic ions of external media and extrapallial fluids shows that in marine bivalves the differences in Ca, Mg, Na, and K are small (Crenshaw, 1972a; Wada and Fujinuki, 1976; Table II). However, in freshwater bivalves, these ions and HCO_3 and Cl are greatly increased in the extrapallial fluid over concentrations in the external medium (Table II). Of

TABLE II

Inorganic Composition of Extrapallial Fluids[a]

	Na	K	Ca	Mg	HCO$_3$	Total CO$_2$ (mM)	Cl	SO$_4$	P
Extrapallial fluids—marine species									
Mercenaria mercenaria	444 ± 9[b]	9.6 ± 0.8	11.8 ± 1.0	60 ± 5	—	5.2 ± 1.9	472 ± 8	46.1 ± 5.1	—
Crassostrea virginica	441 ± 9	9.4 ± 0.5	10.8 ± 0.9	57 ± 3	—	5.0 ± 0.8	480 ± 0.9	48.3 ± 2.3	—
Mytilus edulis	442 ± 10	9.5 ± 0.5	10.7 ± 0.6	58 ± 3	—	4.2 ± 0.5	477 ± 8	47.3 ± 2.3	—
seawater	427 ± 9	9.0 ± 0.1	9.3 ± 0.2	53 ± 3	—	2.5 ± 0.1	496 ± 6	51.1 ± 2.6	—
Pinctada fucata	431.5[c]	12.7	9.7	50.7	3.7		524.0	28.0	1.54
Pinna attenuata	422.8	9.6	9.7	48.6	2.4		521.0	26.4	0.20
Crassostrea gigas	429.8	10.8	9.5	49.2	5.2		540.8	28.5	0.29
Chlamys nobilis	425.4	10.9	9.9	48.7	3.7		520.2	26.2	0.53
seawater	452.8	9.0	10.2	51.2	2.2		533.1	27.4	0.002
Extrapallial fluids—freshwater species									
Hyriopsis schlegeli	22.1	0.6	4.1	0.6	10.5		15.0	5.2	0.12
Cristaria plicata	22.8	0.6	3.9	0.7	11.5		14.9	5.7	0.13
fresh water	0.4	0.1	0.3	0.2	0.7		0.4	0.2	0.001

[a] Values for the first three species are taken from Crenshaw (1972a). Values for other species are calculated from Wada and Fujinuki (1976).
[b] Mean ± S.D. All values are in mM.
[c] All values except P are rounded to the first decimal place.

TABLE III

Calcium in Extrapallial Fluid[a]

Calcium species	Percentage total
Free	15.3 ± 0.5
Bound to small chelates	74.3 ± 0.8
Bound to insoluble carbohydrate	9.2 ± 0.6
Bound to soluble macromolecular components	0.88 ± 0.04

[a] Misogianes and Chasteen, 1979.

necessity this must be true if concentrations are to be adequate for $CaCO_3$ deposition (see Section IX). In both marine and freshwater bivalves, the concentrations of heavy metals are much higher in the extrapallial fluid than in the external medium. The various differences between the medium and the extrapallial fluid could be brought about as ions pass through cells of two major partitions, the first between the external medium and the hemolymph, and the second between the hemolymph and the extrapallial fluid. The concentrations will also be influenced by the binding properties of the organic compounds of the extrapallial fluid.

The concentrations of Ca and HCO_3 in the extrapallial fluid are of importance in meeting the requirement that the solubility product must be exceeded for $CaCO_3$ crystals to form (see Section IX). Measurements indicated that the total Ca concentration in marine species exceeded that of two freshwater species examined by more than twofold whereas the bicarbonate concentration in freshwater species was two- to fourfold that in the marine species examined (Table II; Wada and Fujinuki, 1976). These differences merit further analyses in other species with attention to the concentration of free Ca since it is this, rather than total Ca, that is critical for crystal formation.

The pH of the extrapallial fluid has been measured in many bivalve and gastropod species from marine and freshwater environments. Most values lie between pH 7.4 and 8.3 (Wilbur, 1964; Crenshaw, 1972a; Pietrzak et al., 1973; Wada and Fujinuki, 1976; Misogianes and Chasteen, 1979; Richardson et al., 1981). Freshwater species are in the upper part of this range. The higher pH and greater HCO_3 concentration of freshwater molluscs could be advantageous for $CaCO_3$ crystal formation in view of their lower total Ca as compared with marine species. However, a generalization is not warranted in the absence of information on the concentration of free Ca and several other factors which may affect rates of $CaCO_3$ deposition.

Control of the pH of the extrapallial fluid is discussed in Section VII,A,B.

B. Organic Compounds

The organic compounds of extrapallial fluids include amino acids, proteins, mucopolysaccharides, organic acids, and probably lipids, since lipids are found in shell (Wilbur and Simkiss, 1968; Ravindranath and Rajeswari Ravindranath, 1974).

1. Proteins and Amino Acids

Using paper and cellulose acetate electrophoresis, Kobayashi (1964a,b) examined the extrapallial fluids of 14 species of marine and freshwater molluscs and found one or more proteins depending upon the species. Eight species exhibiting three or more protein bands had shells of aragonite or aragonite and calcite whereas species showing a single protein were calcitic. Six other genera examined more recently by acrylamide gel electrophoresis were found to have 5 to 10 proteins in the extrapallial fluid (Pietrzak et al., 1973; Misogianes and Chasteen, 1979). The number of bands in a given species was not constant. Such an inconstancy of protein composition might result from the secretion of specific proteins at different times. Another possible explanation for the variability could be sampling of the extrapallial fluid at different stages of self-assembly of the shell matrix.

Peptides and free amino acids of extrapallial fluids have been given relatively little detailed attention. Wada (1982) has analyzed the total amino acid content of the extrapallial fluid of seven bivalve species as well as the amino acid components of the dialysate (membrane cut off > 10,000 MW). The latter fraction may include small peptides. Differences between species were marked and generalizations of relative amounts of free amino acids cannot be made. Taurine was in highest concentration in four species, with glutamic acid, aspartic acid, glycine, and alanine being prominent in two species but not in the others. An ultrafiltrate (cutoff 10,000 MW) from *Mytilus edulis* extrapallial fluid was found to contain 15 amino acids (Misogianes and Chasteen, 1979). Glycine accounted for more than half of the residues, and some of the amino acids were thought to be in free form.

2. Carbohydrates

Extrapallial fluid contains a substantial amount of carbohydrate (up to 40% of the dialyzed dry weight) (Misogianes and Chasteen, 1979), including acid mucopolysaccharide (Crenshaw, 1972a). Acid and neutral polysaccharides have also been demonstrated by color reactions following electrophoretic separation (Kobayashi, 1964b). A number of carbohydrates bound to protein have also been identified in the shell layers of the bivalve *Lamellidans marginalis* by chromatography, indicating the presence of carbohydrates in the extrapallial fluid of this species (Ravindranath and Rajeswari Ravindranath, 1974).

IX. Crystal Formation

In order for crystals to be formed within the extrapallial fluid, two requirements must be met: The fluid must be supersaturated with $CaCO_3$, and ion clusters of critical size must form. We now consider both conditions.

A solution in which a strong electrolyte is in equilibrium with crystals of the same cation (C^+) and anion (A^-) is said to be saturated. The activity of the ions in the liquid phase and at the crystal surface will be the same. The relation can be expressed in the following way:

$$K_{SP} = [C^+] [A^-] f_{C^+} \cdot f_{A^-} \tag{7}$$

in which K_{SP} is the solubility product or solubility product constant, C^+ and A^- are the molar concentrations, and f_{C^+} and f_{A^-} are activity coefficients. It will be apparent that K_{SP} will be constant only at a given temperature and if activities at the crystal surface remain unchanged (see Robertson, 1982 for a discussion of solubility).

The extrapallial fluid, with its large number of organic compounds and various ion species, both bound and free, cannot in any sense by considered a solution in which rules of simple solutions apply (Robertson, 1982). Nonetheless, Misogianes and Chasteen (1979) have undertaken an estimate of the *apparent* solubility product of the extrapallial fluid of *Mytilus edulis*. The *apparent* solubility product K'_{SP} is defined as

$$K'_{SP} = m_{Ca^{2+}} \, m_{CO_3^{2-}} \tag{8}$$

where m is the molal concentration of the two ions. The *apparent* solubility product constants were determined for seawater matrices (Ingle et al., 1973); $m_{Ca^{2+}}$ of free Ca was measured and $m_{CO_3^{2-}}$ was assumed. Only 15% of the Ca in the extrapallial fluid was present as free ions as determined with a Ca electrode, the greatest amount being bound to small chelates (Table III). Macromolecules accounted for only a small fraction of the Ca, as Crenshaw (1972a) had also found in other bivalve species. The calculated ion product K'_{SP} for the extrapallial fluid was 0.62×10^{-7} mol²/kg². The K'_{SP} required for saturation of $CaCO_3$ was calculated to be 4.5×10^{-7} mol²/kg². That is, the extrapallial fluid was estimated to be undersaturated, and crystals would not be expected to form. However, the saturation level could be reached by sufficient increase in transport by the outer mantle epithelium and/or by sufficient decrease in chelation of Ca (Table III). Actually, a calculated level of undersaturation is not surprising in that the extrapallial fluid was taken from the central area of the shell, a region in which $CaCO_3$ deposition is commonly low or negligible (Wilbur and Jodrey, 1952). Comparable information on extrapallial fluid at the shell periphery, where the deposition rate is much higher, is not available because of the difficulty of obtaining sufficient fluid for analyses.

The general behavior of crystals as a function of concentration and particle size has been described by Neuman and Neuman (1958) and Garside (1982). Once the solubility product is exceeded, ion clusters may reach a critical size and form nuclei termed *critical nuclei*. If the size of the nuclei increases slightly by the addition of ions, growth will continue and the nuclei may become small crystals. As Neuman and Neuman (1958) point out, ions on the surface lattice of small crystals may not be restrained from escaping into solution to the same degree as from large crystals and, as a result, small crystals may go into solution. This would be expected in biological systems if the concentration of ions at the site of mineralization were to decrease below the level of supersaturation due to an inadequate supply during mineral deposition or as a result of increase in H^+.

As the concentrations of the cation and an ion which form the crystals are increased to or beyond the K_{SP}, solid particles may be present but limited to a particular size range and, for one reason or another, may not grow into crystals. This represents a metastable condition. When a metastable solution fails to form crystals, one can consider that this condition is the result of an energy barrier and that energy of activation is required to overcome it (Neuman and Neuman, 1958; Wilbur and Simkiss, 1979). Conditions and substances which then initiate crystal formation in metastable solutions can be viewed as reducing the barrier sufficiently for crystallization to proceed. With crystal seeding of a metastable solution, precipitation will normally occur since the added crystals, being larger and less soluble than the small particles, will grow at the expense of the small particles. In molluscan shell when mineral deposition is in progress, a metastable state obviously will not occur if exposed crystal surfaces are available. If a metastable state does not block growth of nuclei and there is a continuing addition of Ca^{2+} and CO_3^{2-} to the crystal lattice, crystal seeds will form and, through further growth, may coalesce to become crystals of the shell. At the growing shell edge where deposition may be rapid, the layer of extrapallial fluid between mantle and shell is thin. Ion transport will need to be relatively rapid to maintain supersaturation, and the rate of crystal growth could vary directly with the rate of ion transport.

Various factors in addition to concentrations of Ca and HCO_3 may come into play in initiating, inhibiting, and controlling the rate of $CaCO_3$ deposition. During crystal growth foreign ions, either inorganic or organic, attaching to the lattice and affecting lattice forces may bring about increased or decreased solubility (Neuman and Neuman, 1958). Also, organic compounds of high molecular weight which cover the lattice surface may inhibit crystal growth.

As one approach to crystallization studies in molluscs, the influence of substances on the time required for initiation of $CaCO_3$ crystal formation in solutions supersaturated with respect to $CaCO_3$ has been determined (Wheeler et al., 1981; Sikes and Wheeler, 1983; Wilbur and Bernhardt, 1982). Polyaspartic and poly-glutamic acids with many COO^- groups on each molecule were strongly inhibi-tory whereas positively charged free and polyamino acids, neutral free and polyamino acids, and free negatively charged amino acids were without inhibito-ry action. In *in vitro* systems inhibition may be brought about by the binding of multiple COO^- groups in a polypeptide chain to the crystal lattice surface (Sikes and Wheeler, 1983; see also Pytkowicz, 1969; Crenshaw and Ristedt, 1976) and interfere with further additions to the lattice.

Soluble shell matrix of *Crassostrea virginica* was found inhibitory in the initiation of $CaCO_3$ crystal formation and growth *in vitro* (Wheeler et al., 1981). The same was true of extrapallial fluid from *C. virginica* and *Mercenaria mer-cenaria* (K. M. Wilbur and A. M. Bernhardt, unpublished observations). Inhibi-tion in these cases may be produced by the same mechanism that is suggested for acidic polyamino acids because both the soluble matrix and the extrapallial fluid would be expected to have polypeptide chains with COO^- groups (Weiner and Hood, 1975; Weiner, 1979). This is supported by the finding that the inhibitory action of extrapallial fluid can be reversibly removed by passing it through a Sephadex column, which binds negative groups (K. Wilbur and A. M. Bernhardt, unpublished observations). These authors also observed that per-iostracum and insoluble shell matrix have little or no inhibitory effect on crystall-ization nor do they accelerate crystal formation.

Magnesium is present in the extrapallial fluid of marine bivalves in a total concentration of 49–60 mM (Crenshaw, 1972a; Wada and Fujinuki, 1976). The inhibitory action of Mg on initiation of crystallization was demonstrated in a solution containing the ionic components and concentrations reported for the extrapallial fluid of marine bivalves (Crenshaw, 1972a; Wada and Fujinuki, 1976). As the Mg concentration was increased beyond 5 mM, the delay in crystallization became increasingly greater (K. M. Wilbur and A. M. Bernhardt, unpublished observations).

At first sight, the *in vitro* findings represent a curious situation. The extra-pallial fluid and the soluble matrix, the media in which crystallization takes place, contain substances inhibitory to crystal nucleation and growth. Conceiv-ably, the reduced rate of crystal growth of the central region of the shell (Wilbur and Jodrey, 1952; Zischke et al., 1970) from which the extrapallial fluid was taken for the *in vitro* experiments may, in fact, be the result of inhibition by the extrapallial fluid. Perhaps at the shell edge, where growth is more rapid, the extrapallial fluid is different. In any case, it will be readily apparent that the *in vitro* system, even with the same ionic composition and pH as the extrapallial fluid (Wilbur and Bernhardt, 1982), is markedly different in lacking protein and other organic compounds present *in vivo* (see Section VIII,B). Another difference is the alternation of phases of initiation and cessation of crystal growth as individual layers of crystals are deposited in nacreous and foliated shell. In forming a layer, nucleation of crystals first takes place, the crystals grow to a given thickness, and growth in thickness ceases. The sequence then repeats. A third difference between the *in vitro* and *in vivo* systems is the formation of insoluble matrix in the latter.

At this point in our discussion, it is appropriate to summarize the present information on extrapallial fluid as the environment of crystallization of $CaCO_3$.

1. Samples of extrapallial fluid central to the pallial attachment where growth is slow or absent have been analyzed, but analyses from the shell edge region where growth is rapid have not been carried out.
2. Extrapallial fluid contains the common cations and anions of body fluids and a number of proteins and carbohydrates. Further information is needed on Ca binding, free amino acids, peptides, and possible self-assembly within the extrapallial fluid.
3. Differences occur among species in both inorganic and organic constituents of extrapallial fluid, indicating that $CaCO_3$ of molluscan shells may be deposited in quite different milieux. This is evident in Ca and HCO_3 concentrations which differ markedly between freshwater and marine bivalves.
4. The pH of extrapallial fluids is commonly between 7.4 and 8.3.
5. Extrapallial fluid in low concentration inhibits the initiation of $CaCO_3$ crystals *in vitro* at normal pH. Soluble shell matrix derived from extrapallial fluid is also inhibitory at low concentrations *in vitro*.
6. Magnesium present in relatively high concentrations in extrapallial fluid inhibits initiation of $CaCO_3$ crystal formation at low concentrations *in vitro*.
7. Extracts of extrapallial fluid and soluble shell matrix that may act as crystal nucleators have not as yet been examined.

X. Shell Construction

A. Basic Structure

In this section we shall be concerned with the morphological systems in which crystals by their manner of growth provide various types of shell structures. The architectural patterns of shells, even of a single class, may be complex but not infinitely so. In fact, the individual patterns of crystal microstructures that molluscs have employed in the assembly of shell can be classified into a few broad types (Taylor et al., 1969; Taylor, 1973; Carter, 1980; Watabe, 1981). The common basic types are the following:

1. *Nacreous:* tabular crystals arranged in laminae or columns (Fig. 4)
2. *Foliated:* elongate flattened crystals arranged side-by-side or in overlapping sheets
3. *Prismatic:* parallel polycrystalline elongate prisms enclosed in an organic sheath (Fig. 5)
4. *Cross lamellar:* tablet-shaped crystals, each composed of parallel rodlike elements which, in adjacent crystals, run in different directions and which, in alternate crystals, run in the same direction (Fig. 6)
5. *Spherulitic:* spherical structures of elongate crystallites radiating from a center (Fig. 7)
6. *Homogeneous and granular:* Taylor et al. (1969) designate homogeneous as a term of convenience for any fine-grained structure. The grains may be irregular or rounded and are usually less than 5 μm across. Aggregates of larger size are called *granular*.

For details of the above types, other crystalline structures, and subcategories of prismatic, cross lamellar, and spherulitic structures, Taylor et al. (1969), Taylor (1973), Carter (1980), and Watabe (1981) should be consulted.

B. Control of Crystal Form

Considering the expected relative uniformity of the microenvironment of crystal deposition adjoining the mantle, the number of crystal forms and layer types of molluscan shells is surprising at first sight. Within a single extrapallial space, microstructures as different as a vertical prismatic layer and a horizontal tabular nacreous layer are commonly contiguous. The question as to the causes of differences in crystal morphologies has not commonly been asked, and experimental studies that might give insight have been few.

Based upon growth characteristics of nonbiogenic minerals under various conditions, we can assume that the predominant factors influencing crystal form in shell will be soluble substances (both inorganic and organic) in the extrapallial fluid, the substrata, and the rate of crystal growth as influenced by environmental conditions. For example, Wada (1957) found that the size and shape of nacreous crystals were different in seasons of fast and slow growth.

Different shell layers with distinct crystal types are directly related to different cell types in the mantle epithelium which overlies them (Fig. 1). These cells

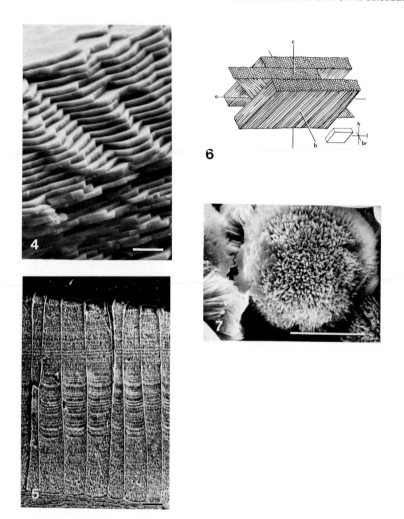

Fig. 4. Nacreous shell of *Geukensia demissa*. Fracture surface. Bar, 10 μm. (Micrograph by M. A. Crenshaw.)

Fig. 5. Prismatic shell of *Anodonta woodiana*. Vertical section of simple prisms. Brief etching after polishing. Bar, 50 μm. (Suzuki and Uozumi, 1981.)

Fig. 6. Diagram of cross lamellar structure in polyplacophorans. The three large units are first-order lamellae (lamels). The smaller units which make up the first-order lamellae are called *second-order lamellae* (lamels). a, b, c: Crystallographic axes; h, vertical axis of first-order lamella; l, long horizontal axis of a first-order lamella; br, short horizontal axis of a first-order lamella. (Haas, 1972.)

Fig. 7. Aragonite spherulite formed at an early stage of shell regeneration in the freshwater snail *Pomacea paludosa*. Bar, 10 μm. (Blackwelder and Watabe, 1977.)

clearly secrete different organic compounds, as shown by the differences in the composition of the organic matrix of the layers (e.g., Wilbur and Simkiss, 1968; Ravindranath and Rajeswari Ravindranath, 1974). It follows that the physical nature of the secretions of the different cell types is such that the secretions remain localized within the extrapallial fluid or that the epithelium is applied to the shell surface during crystal deposition. By either means, specific epithelial regions control crystal form.

The substrata on which crystals of the shell grow will undoubtedly be an important factor in governing their form. There are three distinctly different substrata within the extrapallial microenvironment that may have an influence: (a) the inner surface of the periostracum at the outer shell surface, (b) the surfaces of crystals previously deposited, and (c) the organic matrix, which is largely protein with carbohydrate, covering the surfaces of the periostracum and the previously deposited crystals. The influence of these three substrata on crystal form could perhaps be investigated by placing them in supersaturated $CaCO_3$ solutions and observing the crystals deposited on each by scanning electron microscopy.

C. Shell Types and Their Formation

In discussing shell formation, attention will be given to the manner in which crystals grow and are assembled to form nacreous, cross lamellar, and prismatic shell. Because growth and assembly are closely integrated with secretion and alteration of organic matrix, both the mineral and organic phases are considered together.

Before beginning, however, some statements pertaining to common crystallographic features and differences of shells may be useful.

1. Molluscan shell is constructed of polycrystalline aggregates of $CaCO_3$ of preferred orientation and usually with c-axes having a common direction (Watabe, 1981).
2. Crystals grow within the extrapallial fluid from which the organic matrix of shell develops.
3. The organic matrix of shell serves to separate individual crystals and to bind the crystals and crystal layers into a unified structure.
4. A shell is made up of more than one major layer which may differ in crystal type.
5. Shell crystals are usually aragonite or calcite, rarely vaterite. During shell repair, vaterite may be present (see Chapter 7, this volume).

1. Formation of Nacreous Shell.

The discussion of nacreous shell formation will be limited to the gastropods and bivalves with primary attention given to the processes involved. For further details, the papers of Wise and Hay (1968a,b), Wise (1970), Erben (1972), Mutvei (1978), and Watabe (1981) will be found informative. The study by Erben (1972) was carried out on representatives of 11 families of gastropods and 8 families of bivalves. It is from these sources that the descriptive accounts of layer formation are largely drawn. At the same time, we have not hesitated to include speculative ideas as well. Shell structure per se will be included for orientation only and will not be discussed in detail. Information on this aspect can be found in Schmidt (1923), Taylor et al. (1969), Grégoire (1972), and Carter (1980).

a. Gastropod nacreous layer formation. The formation of a mineralized lamella of gastropod nacre is initiated with the secretion of matrix by the mantle and the deposition of mineralized granules (seeds) distributed randomly on its surface. Through growth, the granules come to form rounded, flattened crystals, each covered by a delicate organic membrane. As the crystals are deposited, conical stacks are formed (Fig. 8; also Mutvei, 1978). The conical form results from nucleation of crystals in the center of the underlying crystal and an increased period of lateral growth of the crystals formed earlier in the stack. The most recently formed crystal will have the smallest diameter. During lateral growth the crystal may become somewhat elongate, but the axis of elongation and the crystallographic axis are not uniform from stack to stack (Wise, 1970; Erben, 1972). Sheets of matrix connect the stacks, bridging the intervening spaces (Fig. 8). The sheets will become the interlamellar matrix of completed layers. Erben (1972) has reported that the matrix sheets separating crystal layers are formed only every three to eight layers as the crystals of a single layer become contiguous. This vertical spacing of the sheets is at variance with the accounts of Wise (1970), Watabe (1981), and Nakahara (1981), who indicate that all lamellae are separated by a sheet of interlamellar matrix. Since matrix separates the crystals in a stack, the vertical crystal alignment could be due either to continuity of crystal growth through matrix interstices (Watabe, 1981), by provision of a central nucleation site in the matrix which covers the underlying crystal (Crenshaw and Ristedt, 1976), or in the matrix in the central part of the tablets (Mutvei, 1978).

Continued lateral growth brings crystals of neighboring stacks into contact, giving polygonal shape to the previously rounded crystals. Crystals are in register in neighboring stacks. This could be brought about by the secretion of a sheet of matrix covering a considerable area, which would limit growth in crystal thickness. Because crystals in neighboring stacks are in register, they form lamellae extending over many stacks. The continuity of lamellae can be observed in fracture surfaces (J. G. Carter, personal communication) and to a limited degree from observations of growing surfaces with scanning electron microscopy. Because the crystals differ in area and extend somewhat irregularly at the

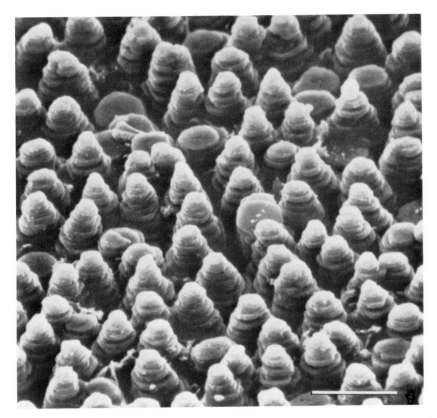

Fig. 8. Developing nacreous shell of the gastropod *Turbo castanea* showing formation of stacks of crystals. Crystal size is a function of age, the topmost crystals being the most recently deposited. Bar, 10 μm. (Wise, 1970.)

edge of the stacks, the crystals of neighboring stacks interdigitate when the stacks grow together, which gives increased resistance of the shell to fracture (Fig. 9).

The conical stacks are reported by Erben (1972) to become more cylindrical before they grow into contact with neighboring stacks. This could result only from a relatively slower growth of older crystals lower in the stack. A differential growth rate of this kind might result from a decreasing rate of diffusion within a gelled matrix with increasing distance from the source of Ca and HCO_3, that is, the mantle epithelium, and increasing distance required for diffusion of H^+ from the site of $CaCO_3$ deposition (Eq. 4). It will be evident that with crystal growth taking place beneath sheets of matrix, the sheets as well as other matrix material are permeable to ions.

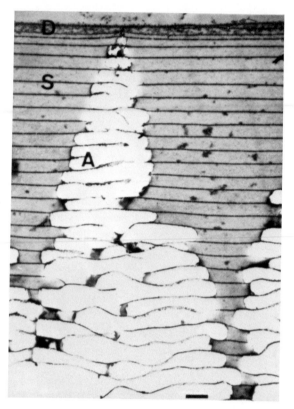

Fig. 9. Growing nacreous layer of the gastropod *Calliostoma unicum* showing matrix sheets prior to the completion of crystal growth. Crystals, A, were decalcified during staining and are represented by white areas. S, Matrix sheets; D, surface sheets. Note interdigitation of crystals in lower part of micrograph. Bar, 1 μm. (Nakahara, 1981.)

Wise (1970) has directed attention to the method of nacreous formation in gastropods in relation to deposition rate. A crystal stack may have 15 to 25 crystals, all undergoing growth. Moreover, 20,000 to 30,000 stacks may be growing within 1 mm² of shell surface. The result is that $CaCO_3$ deposition proceeds three-dimensionally in gastropods as compared with bivalves, in which crystals are growing on crystal laminae largely in two-dimensional fashion (see the following section). Once the crystals of the stacks attain maximal thickness, deposition occurs on the lateral crystal surfaces. Within a stack these surfaces can be roughly considered to constitute a right cone. The increase in the surface offered by this construction as compared with a surface area such as the base of the cone can be estimated roughly. A cone in which the height is 1.5 times the diameter of the base will have a surface area 3 times the basal area. If the height is twice the basal diameter, the surface is 4 times the basal area. Because of the irregularity of the surfaces of the stacks of crystals, the actual increase in area will be greater than the geometric estimate. Although the increased crystal surface area would seem to give gastropods an advantage over bivalves in deposition rate, it will be readily apparent that crystal surface is only one of several factors determining the rate.

b. Bivalve nacreous layer formation. The nacreous layer in bivalves, like that in gastropods and cephalopods, commonly (but not invariably) consists of lamellae of tabular crystals one crystal thick separated by a layer of organic matrix. The deposition of a mineral lamella begins in the same manner as in gastropods with formation of mineral granules randomly distributed on a layer of matrix. Single crystals may develop in various ways.

1. Small crystals may grow within a matrix layer on the matrix sheet. As they grow, some come into contact with others, coalesce, and become larger. Other small crystals are dissolved in the process.
2. Minute crystals may grow on the broad face of larger crystals (Fig. 10). The superimposed crystals are likely to have their origin near crystal edges in that these sites are more favorable energetically for growth (Grégoire, 1972; Wada, 1972; Fig. 11).
3. Spiral growth may occur around screw dislocations on the broad face of crystals (Fig. 10, insert) (Wada, 1961).
4. Crystals may grow by forming dendrites as described for foliated and nacreous layer formation (Watabe, 1954; Watabe and Wilbur, 1961; Taylor, 1973). The dendrites are thought to represent the parallel portions of the crystals seen in thin sections (Watabe, 1965).

As the crystals become larger, they take various shapes including rounded flattened forms and plaques with polygonal borders (Fig. 12). Finally, as a result of increasing size, the crystals become contiguous and form a mosaic one layer thick as an outward extension of a previously formed lamella (Fig. 12b). At any one time crystals may be added to more than one layer at their periphery. Because each layer is recessed behind the edge of the underlying layer, the overall growth front has the form of a stepped terrance (Fig. 12). In any restricted area the crystals all have the same crystallographic axes. Over the shell inner surface, however, some variation in axis orientation may occur (Wada, 1961).

In summary, there are differences in the manner of lamellar formation in most gastropod and bivalve nacreous shells:

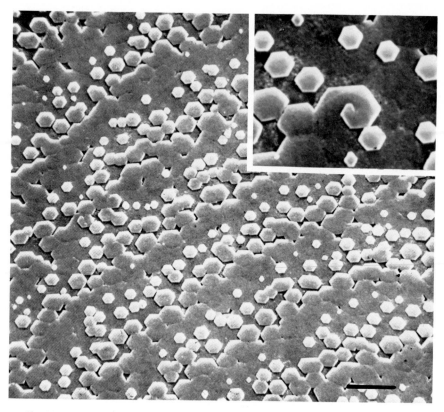

Fig. 10. Deposition of crystals in step arrangement in growing nacreous shell of *Pinctada radiata*. Shell edge at upper left. Bar, 10 μm. Inset: Screw dislocation in newly deposited crystal of nacreous shell of same species. (Wise, 1970.)

1. Initial growth of noncontiguous stacks characterizes gastropods, whereas crystal growth in bivalves occurs on previously deposited lamellae at the edges of completed areas.

2. During any one period, formation of many lamellae may take place in gastropods but in only a few lamellae in bivalves. That is, the relative frequency of secretion of matrix sheets is greater in gastropods (see Watabe, 1981).

3. Crystal orientation is nonuniform in gastropods but uniform in any one shell area in bivalves.

Bevelander and Nakahara (1969, 1980) and Nakahara (1979) have contended that crystals of nacreous layers of bivalves and gastropods form in preformed

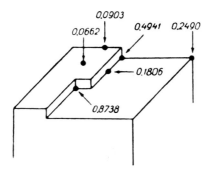

Fig. 11. Relative energy values released by an ion on being attached in various positions on a crystal, for example, halite. (Grigor'ev, 1965.)

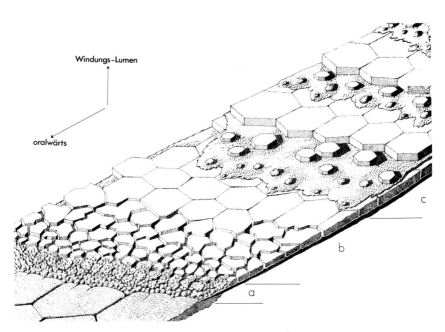

Fig. 12. Diagram of the formation of nacreous layers near the growing edge of a bivalve shell. The shell edge is to the left. Four layers are in the process of crystal deposition and growth. a, Early crystal formation; b, crystals increase in size and form a complete lamella; c, development of a lamella on a sheet of matrix. (Erben, 1972.)

rectangular organic chambers. This method of crystal development has been contested by Erben (1972, 1974) and Erben and Watabe (1974), who were unable to observe such preformed chambers. These authors and Crenshaw (1980) pointed out that chambers without crystals or with incomplete crystals could appear in sections as a result of decalcification during preparation. An example put forward by Nakahara (1981) as evidence for preformed chambers is a section showing a pyramid of developing crystals, all layers separated by matrix sheets with unfilled chambers lateral to each crystal (Fig. 9). Watabe (1981) has offered an alternative interpretation of the development of the layering of the pyramid of crystals similar to the account for deposition of nacreous shell of gastropods.

2. Cross Lamellar Shell Formation

a. Structure. Cross lamellar structure characteristically consists of thin rectangular blocks termed *first-order lamels,* which are made up of parallel laths designated *second-order lamels.* A distinctive feature of this type of structure is an alternation in angle of the second-order lamels in adjacent first-order lamels (Fig. 6). The result is that alternate first-order lamels will have parallel second-order lamels. Bandel (1979) has measured the angles of adjacent lamels in 84 species of gastropods including Mesogastropoda, Neogastropoda, Basommatophora, and Stylommatophora and found that most angles are between 90 and 130° but with a wide spread on both sides of these values. Even in single species the variation may be great. Cross lamellar shell often departs from the regularity shown in Fig. 6 and exhibits branching, lens formation, and a variety of patterns (Taylor et al., 1969). Decalcified sections of cross lamellar shells show the presence of organic material enveloping the second-order lamels. *Third-order lamels* surrounded by organic material may also be present (Uozumi et al., 1972).

In the freshwater snail *Ampullarius glaucus,* sections of the first-order lamels show that the crystals (presumably of the second-order lamels) consist of rows of crystallites 50–60 nm in thickness (Zischke et al., 1970). A first-order lamel 12 μm in width would comprise more than 200 rows of crystallites.

b. Formation. The formation of cross lamellar shell presents a deeply perplexing problem, and we agree fully with Taylor (1973) who states ". . . no sensible explanation can be provided with present knowledge." Not only does the deposition of first-order lamels involve a formation of parallel laths made up of crystallite subunits (Zischke et al., 1970; Uozumi et al., 1972) but also, as we have mentioned, the angle of the laths alternates as each succeeding first-order lamel is formed. In a fast-growing *A. glaucus,* the alternation may occur every 40 min on the average (Zischke et al., 1970).

Nakahara et al. (1981) from sections of *Strombus gigas* shell described the formation of cross lamellar shell as beginning with threadlike or fine granular particles that appear to develop into single-layered envelopes. Rod-shaped crystals form within the envelopes. Adjacent envelopes of similar orientation produce second-order lamels. Adjacent *groups* of envelopes have a different angle or orientation as in neighboring first-order lamels of shell. This account would seem to suggest that molecular organization within the matrix brings about the orientation of envelopes, which governs the orientation of crystals within the envelopes of the prisms after decalcification (Watabe and Wada, 1956; Taylor et al., 1969; Grégoire, 1972; Manyak et al., 1980; Suzuki and Uozumi, 1981). Whether crystal growth is partially or completely interrupted by these horizontal organic layers is uncertain.

In some species, we pointed out that the angle may vary, indicating that control by the mantle epithelium may not be rigidly fixed (Bandel, 1979). A still greater epithelial lability is evident from observations that cross lamellar shell may develop following the deposition of a simple precursor shell type and be followed by another type of shell structure (Bandel, 1979).

3. Prismatic Shell Formation

In its structure, prismatic shell contrasts strikingly with nacreous and cross lamellar shell. Its elements are long crystal prisms consisting of elongate spicular elements. The prisms are kept separate, yet bound together, by substantial amounts of organic matrix (Fig. 5). Surrounded by extrapallial fluid, the prisms extend from the periostracum at the outer shell surface, and as the prisms elongate the prismatic layer becomes thicker.

The early stages in the development of the prismatic layer can be seen at the growing shell edge of a bivalve. Here mineral bodies are randomly distributed in organic material on the inner shell surface of the newly formed sheet of periostracum. The bodies are spherulites and appear to be the precursors of the mineral prisms. Evidence for the origin of the prisms comes from scanning electron microscopic examination of developing prismatic shell of *Pecten maximus* (Taylor, 1973), *Amblema plicta perplicata* (see Chapter 5, this volume), the burrow lining of the boring mollusc *Bankia gouldi* (Manyak et al., 1980), and the shell of *Nautilus macromphalus* undergoing repair (Meenakshi et al., 1974). Taylor et al. (1969) in describing the development of simple prismatic shell point out that the growth of the spherulites follows a course known for inorganic systems. The following sequence is taken from their account (Fig. 13).

1. The spherulites grow into juxtaposition, forming polygonal blocks at the outer shell surface. This union together with the limitation imposed by the periostracum at their outer surface would limit growth of the spherulites to one general direction, namely inward. Through continuing growth, elongate crystal structures would begin to form.

2. If the spherulites are at different stages of development, the growth of some

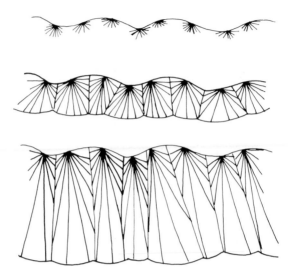

Fig. 13. Possible method of formation of prismatic crystals from spherulites. *Upper:* Spherulites formed in contact with inner surface of periostracum. *Middle:* Prisms developing from spherulites. *Lower:* Layer of fully formed prisms. The shorter diagonal lines represent prisms whose growth has been terminated as a result of contact with other prisms. Long vertical and diagonal lines indicate single prisms with continuing growth. Prismatic shell is commonly constructed of prisms considerably more uniform in width than shown in the figure (cf. Fig. 5). (After Grigor'ev, 1965, and Taylor et al., 1969.)

will be blocked for lack of space, producing a structure somewhat like that in Fig. 13. Such structures are seen in vertical sections of bivalve shell.

3. As the crystals continue their growth and become larger in cross section, the organic material in which they are growing will be displaced and form intercrystalline matrix as in nacreous shell formation. This is clearly evident in shell sections (Fig. 5). At the same time, matrix will also be included around the spicular elements of the prisms (Nakahara et al., 1980; Suzuki and Uozumi, 1981), which presumably develop from the crystallites of the spherulites. In those genera in which the intercrystalline matrix takes the form of firm honeycomb structures, one may reasonably suppose that sclerotization of the matrix involving phenoloxidase has occurred.

From the length of the prisms, it is apparent that crystal growth in prismatic shell is not interrupted in the same way as in nacreous shell. However, there are irregularly spaced horizontal organic layers in the prisms and in the organic envelopes of the prisms after decalcification (Watabe and Wada, 1956; Taylor et al., 1969; Grégoire, 1972; Manyak et al., 1980; Suzuki and Uozumi, 1981). Whether crystal growth is partially or completely interrupted by these horizontal organic layers is uncertain.

XI. Organic Matrix and Shell Crystals

The organic matrix in which the crystals of the shell develop has long been considered a possible determinant of the following: (1) nucleation, (2) crystal orientation, (3) crystal size, and (4) crystal type. None of the functions is firmly established, yet they merit inclusion in our discussion of shell formation. We shall consider some of the mechanisms proposed. Summaries of crystal-matrix relations have been given by Grégoire (1972), Wilbur (1972), Degens (1976), Crenshaw (1980), and Wilbur and Manyak (1983).

A. Crystal Nucleation

Crystal nucleation and crystal growth require that the extrapallial fluid be supersaturated with $CaCO_3$ and that Ca and HCO_3 continue to be supplied by the mantle (see Section IX). However, these conditions do not ensure crystal nucleation, and it is a commonly held view that compounds present in the shell matrix are required for the initiation of crystal formation on the inner shell surface. The evidence rests largely on the association of crystals with organic structures. The relation is illustrated in the deposition of aragonite in the larval shell of the archeogastropod *Haliotis discus hannai* (Iwata, 1980). Prior to mineralization, organic spherules are formed on the periostracum. Within these spherules birefringent nuclei first appear, followed by the formation of crystallites, which develop into spherulites that finally constitute a complete shell layer. A second type of evidence is the formation of crystals in structures staining for acid mucopolysaccharides and commonly for protein on coverslips inserted between mantle and shell in *Pinctada fucata* and *Pinna attenuata* (Wada, 1980). Also, acid mucopolysaccharide and probably protein were present at the site of crystal formation during the initial stages of shell repair in *Helix pomatia* (Saleuddin and Chan, 1969). During calcified tube formation by the wood-borer *B. gouldi*, formation of an organic sheath precedes the appearance of deposited crystals by at least 12 h. Initial $CaCO_3$ deposition is in the form of individual crystallites in intimate association with strands of organic material (Manyak et al., 1980).

Possible involvement of skeletal matrix in crystal nucleation is also indicated by the remineralization of decalcified and fixed matrix of the crab *Carcinus maenas* (Roer, 1979) and the tube of the marine borer *B. gouldi* (Manyak, 1982) in physiological electrolyte solutions.

As an approach to the determination of specific organic substances that may initiate nucleation of $CaCO_3$ crystals, Ca-binding compounds have been isolated from the soluble matrix of shells. Several Ca-binding polypeptides considered as possible nucleators of $CaCO_3$ have been isolated from the soluble shell matrix of bivalves, gastropods, and a cephalopod. A glycoprotein isolated from *Mercenaria* shell was suggested as a nucleator of $CaCO_3$ through concentration of Ca by ester sulfate groups associated with hexosamine residues (Crenshaw, 1972b). This view was supported by histochemical findings on *Nautilus* that demonstrated the presence of sulfated acid polysaccharide groups and Ca sites on the membrane overlying nacreous crystals (Crenshaw and Ristedt, 1976). Ca-binding proteins differing from that in *Mercenaria* have been separated from the soluble matrix of several species (Krampitz et al., 1976). Subfractions of soluble matrix with high aspartic acid content have been isolated from *Nautilus, Plicatula,* and *Strombus* and are thought to be important in mineralization because of Ca binding by COO⁻ groups (Weiner, 1979). Soluble proteins rich in aspartic acid residues are postulated to align with insoluble matrix and to make a β-pleated sheet conformation. To bring about nucleation, these proteins would have appropriate spacing of COO⁻ groups to bind Ca at distances corresponding to a multiple of the distance separating Ca atoms along the b-axis of aragonite. With different spacing of COO⁻ groups, inhibition of crystal growth could result from secretion of the polypeptide on the crystal surface (Weiner and Traub, 1981). Polypeptide chains from *Crassostrea, Mercenaria,* and *Nautilus* with repeating sequences of aspartic acid separated by glycine or serine were also considered as possible nucleators (Weiner and Hood, 1975), but these chains are now suggested to be possible inhibitors of crystal growth because of the spacing of their Ca-binding residues (Weiner and Traub, 1981). A Ca-binding shell protein isolated by Samata et al. (1980) from *Crassostrea gigas* was associated with an isoprenoid hydrocarbon and differed from proteins isolated by Crenshaw (1972b), Weiner and Hood (1975), and Weiner (1979) in having no sulfate and low concentrations of aspartic and glutamic residues. Ca binding was attributed to the amino acid sequence in the protein chain.

Degens (1976) proposed matrix structure and functions somewhat parallel to those later advanced by Weiner (1979) and Weiner and Traub (1981). Two protein fractions were suggested: a carrier protein of pleated sheets to which an acidic polypeptide fraction called the *mineralizing matrix* with a strong affinity for calcium ions was bound. The mineralizing matrix is presumed to nucleate the $CaCO_3$ crystals when activated by attachment with the carrier protein. This would be followed by the epitaxial deposition of crystals. During the course of crystal growth, the secretion of a layer of mineralizing matrix on the crystal surface would inhibit growth along the c-axis. Lateral growth of crystals along the b-axis would continue. A model of crystals and matrix fractions is shown in Fig. 14. The double function of the matrix in initiating and inhibiting crystal growth has also been proposed by Crenshaw and Ristedt (1976) and Crenshaw (1982), and is thought to be brought about in the following way: (1) Glycoprotein is adsorbed on the center of the exposed crystal surface where the concentration of free Ca is highest. The effect is to inhibit further growth on this surface. (2) The free portions of the adsorbed glycoprotein concentrate Ca ions (and secondarily, carbonate ions), initiating a new crystal through ionotropy.

To summarize, the long-held belief that the matrix plays a role in crystal

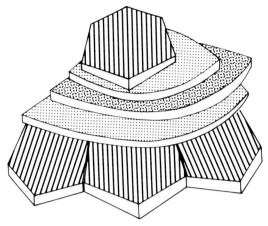

Fig. 14. Model of hypothetical structural relationship of crystals, *mineralization matrix* (dotted), and *carrier protein* (banded). (After Degens, 1976.)

nucleation finds indirect support in (1) Ca binding by fractions of the soluble matrix, (2) the presence of protein and acid mucopolysaccharide at sites of crystal formation, and (3) recalcification of decalcified skeletal matrix. A number of different matrix fractions have the capacity to bind Ca and whether one or more is critical for nucleation is not known. Indeed, Ca binding per se cannot be considered evidence of a nucleating function. More persuasive would be the finding that a fraction or fractions of the extrapallial fluid or matrix would initiate crystallization in metastable solutions of $CaCO_3$ and produce crystals corresponding in type to the shell from which the fraction had been extracted. However, none of the Ca-binding substances has been examined in this manner except for the Ca-binding soluble matrix of the shell of *Crassatrea virginica*, which is inhibitory in very low concentration (Wheeler et al., 1981) (see also Section IX).

B. Crystal Orientation

In the following three sections on crystal orientation, control of crystal size, and control of crystal type, we shall be concerned with chemical influences on growth of crystals differing in form and type in different layers of a single shell. As we indicated previously, the composition of the organic material differs between layers (Grégoire, 1972), and these differences derive directly from the secretions of the mantle epithelium facing the inner shell surface (Fig. 1).

Crystal orientation of one shell type in a limited shell region may be relatively uniform or show variability. We have pointed out the uniformity in bivalve nacreous shells and variation in gastropod nacreous and cross lamellar shell. More than one factor may play a role in crystal orientation. The periostracum, covered by extrapallial fluid, is the substratum of the crystals first deposited at the growing shell edge. This may have an initial influence on crystal orientation. These outermost crystals may, in turn, orient successive crystal layers if there is crystal continuity between layers as Wada (1972) observed in nacreous shell. In prismatic shell, it appears likely that the orientation of prisms developing at the periostracum is determined by the direction of growth of the crystallites of spherulites from which the prisms originate. Orientation in later growth of prisms is simply due to continuity of individual prisms (see page 267). The orientation of crystals of nacreous and foliate shell has been attributed to the matrix on or in which the crystals grow (see the following discussion). There is a double problem here in that the orientation of each layer of matrix requires explanation as well as the orientation of the crystals. If, through binding, the surface lattice of the outermost layer of tabular crystals could orient matrix molecules, the matrix in turn could conceivably determine the orientation of the next formed crystal layer (Wilbur, 1976). Such a mechanism, like crystal continuity from layer to layer, could maintain uniformity of orientation of both matrix and crystals as successive layers are formed. Experimental data are lacking on such a mechanism.

By means of X-ray diffraction, Weiner and Traub (1980, 1981) have examined the matrix–crystal relationship of shell in six species by comparing the orientation of molecules of the insoluble matrix and the crystallographic axes of associated $CaCO_3$ crystals. The nacreous layer of the septum of *Nautilus repertus* gave a well-oriented β-chitin pattern with the fiber axis parallel to the interlamellar matrix. Other spacing indicated a β-pleated sheet structure perpendicular to the fiber axis with side chains perpendicular to the interlamellar matrix. The results showed very good alignment between matrix components and crystallographic axes of the nacreous crystals. The prismatic layer of *Mytilus californianus* also gave a β-pleated sheet structure with random orientation in the plane of the specimen and with side chains perpendicular to the long axis of the prisms. In other species, alignment of crystals and the β-pleated sheet structure of the proteinaceous insoluble matrix was good in some and poor in others.

X-ray diffraction is clearly a powerful method in the analysis of shell, and future studies will be of particular interest. As yet, the alignment of soluble matrix polypeptides and their β-sheet conformation on which crystals are presumed to nucleate have not been demonstrated. Weiner et al. (S. Weiner, personal communication) using electron diffraction have reported that in the nacreous layers of the bivalve *Pinctada* and the gastropod *Tectus* the matrix macromolecules were oriented in a preferred direction within the plane of the specimen. The b-axis of the aragonite crystals was aligned with the polypeptide chains and the a-axis with the chitin. This is the same relationship observed in *Nautilus* shell using X-ray diffraction. The results of the X ray and electron diffraction have been considered strong support for epitaxial crystal growth on a matrix template.

C. Crystal Size

One of the characteristic features of shell growth is the deposition of crystals of relatively uniform thickness within the individual shell layers. This uniformity is often strikingly evident in nacreous, foliate, and cross lamellar shell as seen in sections (Fig. 4), but it is also true of the thickness of crystals in prismatic shell (Fig. 5).

The presence of matrix overlying crystals has been considered a factor in limiting crystal thickness. This is illustrated in Fig. 9 showing matrix sheets thought to act as barriers. Sclerotization of the sheets involving phenoloxidase would provide rigidity (Gordon and Carriker, 1980). As a further suggestion of the manner in which overlying matrix may limit crystal thicknes, Crenshaw and Ristedt (1976) and Weiner and Traub (1981) have proposed that protein of the soluble matrix containing negatively charged groups could inhibit by reacting with Ca of the lattice of the upper crystal surfaces and so prevent further additions to the lattice. Polyglutamic and polyaspartic acids having multiple COO^- groups apparently inhibit crystal growth *in vitro* in this way (see p. 255). Another method of control of crystal thickness could conceivably be the transport of fixed amounts of Ca and HCO_3 by the mantle epithelium, resulting in crystals of uniform thickness. However, there is no evidence of such a means of control. Also, adjustments of pH relating to protons formed as $CaCO_3$ is deposited (see Section VII) has been suggested as a possible mechanism that results in periodic deposition (Campbell and Boyan, 1976). The four hypothetical mechanisms of providing uniformity of crystal thickness—secretion of matrix sheets that act as barriers, secretion of compounds having negatively charged groups, units of ion transport, and alteration of pH—would clearly require well-controlled physiological processes.

Crystal length and breadth in nacreous and foliate shell will depend upon the number of spacing of growing crystals per unit area, a wider spacing resulting in larger crystals because of greater growth in area prior to inhibition by contact with surrounding crystals. In prismatic shell, because the elongation of the crystallites and the prisms do not appear to be permanently inhibited by the presence of matrix, the prisms become greatly elongate.

D. Crystal Type

Calcium carbonate crystals of molluscan shell occur as aragonite in some species, calcite in others, and still others have aragonite in one layer and calcite in another. Vaterite, a third polymorph, is not commonly present. The mechanism of control of crystal type in molluscan shell is not well understood. However, it is known that various inorganic and organic substances affect crystal type *in vitro* (Kitano et al., 1969, 1976), and some of these may be present in the extrapallial fluid.

The influence of the shell matrix on crystal type has been examined *in vivo* and *in vitro*. The *in vivo* experiments were carried out by inserting shell matrix from a species differing in crystal type between the mantle and shell and determining the crystal type deposited on the inserted matrix. For the *in vitro* tests, matrix from an aragonitic shell was placed in a solution in which $CaCO_3$ precipitated as calcite (Wilbur and Watabe, 1963). Here too, the type of the crystals forming on the matrix was determined. The insertion of aragonitic matrix into the calcitic *Crassostrea* resulted in aragonitic crystals on the inserted matrix in 4 out of 17 cases (Watabe and Wilbur, 1960). An influence of aragonitic matrix in inducing aragonite crystals *in vitro* occurred in four out of eight cases. For various reasons matrix inserted within a mollusc differing in species could scarcely be anticipated to have its normal influence on crystal type. The decalcification of the matrix prior to insertion would remove the soluble matrix leaving the insoluble matrix. Further, conditions between mantle and shell might well be expected to predominate over the influence of the inserted foreign matrix. In view of these considerations, the positive cases in the *in vivo* and *in vitro* experiments, few as they are, and the lack of effect of inert materials in both experiments would seem to carry some weight and perhaps point to the influence of the insoluble matrix still present after decalcification. However, a similar experiment in *Elliptio complanatus* failed to demonstrate an effect of inserted matrix over that of inert material. Foreign matrix inserted in the shell repair area of *Cepaea nemoralis* showed an influence on crystal type in 1 case out of 10 (Watabe et al., 1979).

A possible influence of insoluble matrix on crystal type was suggested in shell regeneration experiments in which X-ray diffraction spacings of the matrix were determined (Wilbur and Watabe, 1963). *C. virginica*, a calcitic species, and *E. complanatus*, an aragonitic species, both showed a change in crystal type in regenerated shell. The X-ray diffraction spacing of the matrix of the regenerated shell was also changed as compared with normal matrix. On the other hand, *M. mercenaria*, an aragonitic species, maintains its usual crystal type and also normal X-ray diffraction spacing of its insoluble matrix during shell regeneration.

The experiments cited can scarcely be considered more than suggestive that insoluble matrix controls crystal type *in situ*. The possible roles of the soluble matrix and of the nonprotein soluble phase of the extrapallial fluid on crystal type remain to be examined. It would appear possible that the effects of both fractions on crystal type could be studied using solutions resembling extrapallial fluid in their inorganic composition.

XII. Endocrine Control of Shell Formation

The possibility of endocrine control of shell formation and shell regeneration has been suggested on several occasions. However, experimental evidence was not forthcoming until 1976 when Dillaman et al. showed that a group of neurosecretory cells present in the visceral ganglion undergo a change in activity during shell regeneration. The light microscopic stain paraldehyde fuchsin (PAF) was applied to localize neurosecretory cells in the brain of *Helisoma duryi*. The intensity of the stain was used as an indication of the amount of neurosecretory material present in the cells. The staining intensity of the neurosecretory cells decreased during the first 24 h of shell regeneration, returning to control levels at 72 h. Other neurosecretory cells present in the brain did not show this change. The authors postulated that decrease in staining corresponds to the release of hormone(s) in response to shell damage followed by the buildup of newly synthesized hormone(s). At the ultrastructural level the picture appeared more complex. In the visceral ganglion three types of neurosecretory cells were recognized (Khan and Saleuddin, 1979). Of these only type 1 cells showed ultrastructural alterations during shell regeneration. In control snails, two populations of neurosecretory cells were found:

1. *Dark cells.* These cells are active as evidenced by abundant rough endoplasmic reticulum and Golgi complexes. Dark neurosecretory granules are also abundant.
2. *Light cells.* These cells are inactive and electron lucent granules are abundant.

During shell regeneration, the ratio of dark:light cells changes. At 12 and 24 h the ratio is approximately 2 dark:1 light. By 72 h, the ratio returns to control levels 4 dark:1 light. This behavior of type 1 cells may be related either to a functional subdivision of cellular activity or an asynchrony in the timing of their response during shell regeneration.

Another line of evidence for the involvement of neurosecretory cells in shell formation comes from the Dutch workers. Geraerts (1976a) showed that cauterization of light green neurosecretory cells (LGC) resulted in retardation of body growth in *Lymnaea stagnalis*. Removal of other neurosecretory cells in the cerebral ganglia or the dorsal body cells (a putative endocrine gland) did not affect body growth. Implantation of cerebral ganglia containing LGC into snails without LGC restored normal growth. Geraerts therefore suggested that the LGC secrete a growth hormone which influences body growth as well as shell growth. He postulated that the growth hormone acts primarily on the mantle edge, increasing shell height and weight. Geraerts (1976a,b) also showed that the removal of the paired lateral lobes (LL), located on either side of the cerebral ganglia, results in giant growth, including shell. Removal of both LGC and LL prevents giant growth. Therefore, he postulated that for the LL to be effective in affecting body growth LGC are required.

Dogterom et al. (1979) in other studies of *L. stagnalis* found that removal of the LGC resulted in a marked decrease in the concentration of Ca in the mantle edge and a decrease in the rate of incorporation of ^{47}Ca into the shell edge. They suggested that the growth hormone produced by the LGC of the cerebral ganglia is involved in increasing shell growth. In a subsequent investigation of *L. stagnalis*, Dogterom and Doderer (1981) reported a Ca-binding protein specific for the mantle edge and not present in the buccal mass. This protein fraction contained some carbohydrate and a small amount of lipid and has not as yet been purified. Snails from which the LGC had been removed showed a decreased amount of the protein. The Ca-binding protein in the mantle edge would provide a high Ca content in that area of the mantle. The proposed function of the Ca-binding protein is the transport of Ca from mantle to shell. Accordingly, a reduction in the amount of this protein resulting from removal of the LGC would decrease the rate of Ca deposition in the shell edge, as had been observed (Dogterom et al., 1979).

Burton (1977) found ecdysone to have no effect on hemolymph calcium in *Helix pomatia* whereas the injection of subesophageal ganglia extracts resulted in elevated hemolymph Ca. Injection of ecdysone and other steroids into *Biomphalaria glabrata* caused increased incorporation of ^{45}Ca into regenerated shell (Whitehead, 1974, 1977). Whitehead and Saleuddin (1978) reported that ecdysone stimulates the incorporation of ^{45}Ca into regenerating shell at the expense of the remaining shell. They suggested that steriod enhancement of shell regeneration may mimic a factor or factors normally derived from vegetation.

Further evidence of the involvement of the brain in shell formation comes from the laboratory of Saleuddin and his co-workers in Canada. When daily linear shell deposition rates (SDR) were monitored in *H. duryi*, snails could be distinguished as being either slow (SG), normal, or fast growing (FG) (Kunigelis and Saleuddin, 1978). The effect of brain on shell growth rates (SDR) has been evaluated. Whole brains or parts were homogenized, freeze-dried, and then reconstituted in saline to give a final concentration of 1 brain or fraction per 2 μl. Injections of 2 μl per snail were administered. The whole brain extracts or parts (supra- or subesophageal ganglion) from fast-growing snails stimulated the shell growth rate in slow-growing snails. The fraction of supraesophageal ganglia from regenerating snails had a greater stimulatory effect than the subesophageal ganglia. Whole brain fractions were less effective in stimulating SDR than either supraesophageal or subesophageal fractions, suggesting that there may be an antagonistic interaction present in the intact brain (Table IV).

XIII. Energy of Shell Formation

The energetics of shell formation can be approached only in a general way because of the processes involved and the paucity of data for computations. Nonetheless it is of interest to consider the nature of the energy requirements for shell deposition. Three major processes must be operative during crystal formation and shell growth: (1) ion movement to the site of mineralization, (2) the

TABLE IV

Effects of Brain Homogenates on Linear Shell Growth in *Helisoma duryi*

Donor and treatment[b,c]	Recipient[b,c]	SDR 7 days preinjection[a]	SDR 7 days postinjection[a]	SDR % change[a]
Whole brain, FG	SG	0.83 ± 0.08	1.44 ± 0.16	+73.49
Supraesophageal fraction, FG	SG	0.63 ± 0.08	1.55 ± 0.54	+144.03
Subesophageal fraction, FG	SG	0.63 ± 0.08	1.31 ± 0.16	+107.94
Supraesophageal fraction, FG, shell-repairing	SG	0.46 ± 0.04	1.24 ± 0.14	+169.57
Subesophageal fraction, FG, shell-repairing	SG	0.63 ± 0.08	0.87 ± 0.13	+38.10
Foot tissue	FG	1.50 ± 0.08	2.06 ± 0.22	+37.33
Fish brain	FG	1.71 ± 0.13	2.21 ± 0.67	+29.34
Untreated	FG	1.92 ± 0.13	2.37 ± 0.14	+23.44

[a] SDR, Mean linear shell deposition rate (mm × 10^{-2}/h ± S.E.).
[b] FG, Fast growing.
[c] SG, Slow growing.

synthesis of organic compounds of the periostracum and shell matrix, and (3) the secretion of these compounds by the mantle epithelium.

With respect to ion movement, energy will be required for active transport at body surfaces and the mantle epithelium facing the shell. Based on data for active Ca transport in mammalian mitochondria (Carafoli and Crompton, 1976) and sacroplasmic reticulum (de Meis and Vianna, 1979), the expected requirement is one ATP molecule for each two Ca ions actively transported at each surface (but see Schatzmann, 1973). If HCO_3 is actively transported, as Wheeler (1975) has proposed for the outer mantle epithelium, additional metabolic energy will be required for this ion, which becomes shell carbonate. Because information is lacking on the relative amounts of active transport and intercellular diffusion, energy estimates of ion movement are not possible. Otherwise, the energy cost of ion transport could be calculated from isotope measurements of the rate of $CaCO_3$ deposition.

An approximation of the energy of synthesis of shell protein can be made from the shell protein content and the number of peptide bonds. For each bond formed, four ATP molecules will be required (Lehninger, 1975). A measure of the energy represented by shell protein can also be calculated from the protein content assuming a caloric value for protein (Morowitz, 1968; Dame, 1972). Shell matrix also contains carbohydrate as the second most abundant compound, and matrix carbohydrate differs from the carbohydrates of soft tissue. Because of this and the corresponding energy differences, analyses of compounds present should be carried out in calculating energy equivalents of shell and tissue. Sufficient attention has not been given to such analyses. Further, in relating the energy of shell formation to the total energy of growth of the organism, the energy represented by the gametes released should be taken into account because this may amount to the energy invested in a year's growth of soft tissue (Jørgensen, 1976).

The organic matter of shell is less than 10% by weight with rare exceptions (Hare and Abelson, 1965; Price et al., 1976). However, the shell organic matter, small on a weight basis, may constitute between one-third and one-half the total body organic matter. This has been taken to indicate that one-quarter to one-third of the total energy of growth is required for shell deposition (Dame, 1976; Jørgensen, 1976; Griffiths and King, 1979). This range may be conservative in that it does not include the energy of active ion transport, metabolic steps in the synthesis of shell organic matrix, or secretion of organic compounds by the mantle epithelium. Yet from the estimates, though incomplete and rough, it is clear that shell represents a very considerable investment of the total energy required for growth in some molluscs. There are differences among species to be recognized, as the range of values mentioned indicates. Burnell and Rodhouse (P. G. Rodhouse, personal communication) from determinations of the energy represented by somatic tissue, gonad, and shell organics in the scallop *Chlamys varia,* ages 2 to 10 yr, found that a considerably smaller proportion of the total energy of growth was required in depositing the shell of this light-shelled swimming species as compared with the heavier-shelled oyster *Ostrea edulis* (Rodhouse, 1978).

A technical point in these various studies should be mentioned. Because the energy of shell deposition is computed on the basis of shell organics, the method of determining organic content becomes important. In the commonly used combustion method, the organic content of the shell is determined from the loss in weight on heating at a temperature which removes the organic fraction leaving the inorganic fraction. If the combustion temperature is too high or too long-extended, a portion of the shell carbonate may be volatilized as well. The organic fraction will then be calculated as erroneously high, leading to an overestimation of the energy of shell formation in the energy budget of the organism. P. G. Rodhouse (personal communication) has pointed out that some published values of shell organic content may be too high because of an error due to loss of carbonate during combustion.

Acknowledgments

We thank A. M. Bernhardt, S. E. Ford, P. Greenaway, D. M. Manyak, P. G. Rodhouse, and N. Watabe for their assistance in the preparation of the manuscript; L. M. Gebo for typing the manuscript; M. A. Crenshaw for an unpublished electron micrograph; P. G. Rodhouse for unpublished data; and the National Science Foundation, the Office of Naval Research, and Natural Sciences and Engineering Research Council of Canada for support of several studies reported in the chapter.

References

Abolins–Krogis, A. (1958). The morphological and chemical characteristics of organic crystals in the regenerating shell of *Helix pomatia* L. *Acta Zool.* **39**, 19–38.

Bandel, K. (1979). Ubergänge von einfacheren Strukturtypen zur Kreuzlamellenstruktur bei Gastropodenschalen. *Biomineralisation* **10**, 9–38.

Beedham, G. E. (1958). Observations on the mantle of the Lamellibranchia. *Q. J. Microsc. Sci.* **99**, 181–197.

Bevelander, G., and Nakahara, H. (1969). An electron microscope study of the formation of the nacreous layer in the shell of certain bivalve molluscs. *Calcif. Tissue Res.* **3**, 84–92.

Bevelander, G. B., and Nakahara, H. (1980). Compartment and envelope formation in the process of biological mineralization. *In* "The Mechanisms of Biomineralization in Animals and Plants" (M. Omori, and N. Watabe, eds.), pp. 19–27. Tokai Univ. Press, Tokyo, Japan.

Blackwelder, P. L., and Watabe, N. (1977). Studies on shell regeneration. II. The fine structure of normal and regenerated shell of the freshwater snail *Pomacea peludosa*. *Biomineralisation* **9**, 1–10.

Borle, A. B. (1981). Control, modulation, and regulation of cell calcium. *Rev. Physiol. Biochem. Pharmacol.* **90**, 13–153.

Borle, A. B., and Snowdowne, K. W. (1982). Measurement of intracellular free calcium in monkey kidney cells with aequorin. *Science (Washington, D.C.)* **217**, 252–254.

Burton, R. F. (1977). Hemolymph calcium in *Helix pomatia:* effects of EGTA, ganglion extracts, ecdysterone, cyclic AMP and ionophore A23187. *Comp. Biochem. Physiol. C* **57**, 135–137.

Campbell, J. W., and Boyan, B. D. (1976). On the acid-base balance of gastropod molluscs. *In* "The Mechanisms of Mineralization in the Invertebrates and Plants" (N. Watabe, and K. M. Wilbur, eds.), pp. 109–133. Univ. of South Carolina Press, Columbia.

Campbell, J. W., and Speeg, K. V. (1969). Ammonia and the biological deposition of calcium carbonate. *Nature (London)* **224**, 725–726.

Carafoli, E., and Crompton, M. (1976). Calcium ions and mitochondria. *Symp. Soc. Exp. Biol.* **30**, 89–115.

Carter, J. G. (1980). Guide to bivalve shell microstructures. *In* "Skeletal Growth of Aquatic Organisms" (D. C. Rhoads, and R. A. Lutz, eds.), pp. 645–673. Plenum, New York.

Chaisemartin, C., and Videaud, A. (1971). Seuils calciques de l'eau et economie du calcium chez *Ancylastrum fluviatilis* (Gastropodes Pulmones). *C. R. Seances Soc. Biol. Fil.* **165**, 2401–2404.

Crenshaw, M. A. (1972a). The inorganic composition of molluscan extrapallial fluid. *Biol. Bull. (Woods Hole, Mass.)* **143**, 506–512.

Crenshaw, M. A. (1972b). The soluble matrix from *Mercenaria mercenaria* shell. *Biomineralisation* **6**, 6–11.

Crenshaw, M. A. (1980). Mechanisms of shell formation and dissolution. *In* "Skeletal Growth of Aquatic Organisms" (D. C. Rhoads, and R. A. Lutz, eds.), pp. 115–132. Plenum, New York.

Crenshaw, M. A. (1982). Mechanisms of normal biological mineralization of calcium carbonates. *In* "Biological Mineralization and Demineralization" (G. H. Nancollas, ed.), pp. 243–257. Springer–Verlag, Berlin.

Crenshaw, M. A., and Neff, J. M. (1969). Decalcification at the mantle-shell interface in molluscs. *Am. Zool.* **9**, 881–885.

Crenshaw, M. A., and Ristedt, H. (1976). Histochemical localization of reactive groups in septal nacre from *Nautilus pompilius* L. *In* "The Mechanisms of Mineralization in the Invertebrates and Plants" (N. Watabe, and K. M. Wilbur, eds.), pp. 355–367. Univ. of South Carolina Press, Columbia.

Dame, R. F. (1972). The ecological energies of growth, respiration and assimilation in the intertidal American oyster *Crassostrea virginica. Mar. Biol. (Berlin)* **17**, 243–250.

Dame, R. F. (1976). Energy flow in an intertidal oyster population. *Estuarine and Coastal Mar. Sci.* **4**, 243–253.

Degens, E. T. (1976). Molecular mechanisms on carbonate, phosphate, and silica deposition in the living cell. *Top. Curr. Chem.* **64**, 1–112.

de Meis, L., and Vianna, A. L. (1979). Energy interconversion by the Ca^{2+}-dependent ATPase of the sarcoplasmic reticulum. *Ann. Rev. Biochem.* **48**, 275–292.

Dillaman, R. M., and Ford, S. E. (1982). Measurement of calcium carbonate deposition in molluscs by controlled etching of radioactively labeled shells. *Mar. Biol. (Berlin)* **66**, 133–143.

Dillaman, R. M., Saleuddin, A. S. M., and Jones, G. M. (1976). Neurosecretion and shell regeneration in *Helisoma duryi* (Mollusca: Pulmonata) *Can. J. Zool.* **54**, 1771–1778.

Dipolo, R., Requena, J., Brinley, F. J., Jr., Mullins, L. J., Scarpa, A., and Tiffert, T. (1976). Ionized calcium concentrations in squid axons. *J. Gen. Physiol.* **67**, 433–467.

Dogterom, A. A., and Doderer, A. (1981). A hormone dependent calcium-binding protein in the mantle edge of freshwater snail *Lymnaea stagnalis. Calcif. Tissue Res.* **33**, 505–508.

Dogterom, A. A., and van der Schors, R. C. (1980). The effect of the growth hormone of *Lymnaea stagnalis* on (Bi) carbonate movements, especially with regard to shell formation. *Gen. Comp. Endocrinol.* **41**, 334–339.

Dogterom, A. A., van Loenhout, H., and van der Schors, R. C. (1979). The effect of the growth hormone of *Lymnaea stagnalis* on shell calcification. *Gen. Comp. Endocrinol.* **39**, 63–68.

Enyikwola, O., and Burton, R. F. (1983). Chloride-dependent electrical potentials across the mantle epithelium of *Helix*. *Comp. Biochem. Physiol.* **74A**, 161–164.

Erben, H. K. (1972). Über die Bildung und das Wachstum von Perlmutt. *Biomineralisation* **4**, 16–46.

Erben, H. K. (1974). On the structure and growth of the nacreous tablets in gastropods. *Biomineralisation* **7**, 14–27.

Erben, H. K., and Watabe, N. (1974). Crystal formation and growth in bivalve nacre. *Nature (London)* **248**, 128–130.

Florkin, M., and Besson, G. (1935). Le liquide extrapalléal de l'anodonte n'est pas identique au sang de cet animal. *Soc. Belge. Biol.* **18**, 1222–1223.

Freeman, J. A. (1960). Influence of carbonic anhydrase inhibitors on shell growth of a freshwater snail, *Physa heterostrapha*. *Biol. Bull. (Woods Hole, Mass.)* **118**, 412–418.

Freeman, J. A., and Wilbur, K. M. (1948). Carbonic anhydrase in molluscs. *Biol. Bull. (Woods Hole, Mass.)* **94**, 55–59.

Garside, J. (1982). Nucleation. *In* "Biological Mineralization and Demineralization" (G. H. Nancollas, ed.), pp. 23–35. Springer–Verlag, Berlin.

Geraerts, W. P. M. (1976a). Control of growth by the neurosecretory hormone of the light green cells in the freshwater snail *Lymnaea stagnalis*. *Gen. Comp. Endocrinol.* **29**, 61–71.

Geraerts, W. P. M. (1976b). The role of the lateral lobes in the control of growth and reproduction in the hermaphrodite freshwater snail *Lymnaea stagnalis*. *Gen. Comp. Endocrinol.* **29**, 97–108.

Gordon, J., and Carriker, M. R. (1980). Sclerotized protein in the shell matrix of a bivalve mollusc. *Mar. Biol. (Berlin)* **57**, 251–260.

Greenaway, P. (1971a). Calcium regulation in the freshwater mollusc, *Limnaea stagnalis* (L.) (Gastropoda: Pulmonata). I. The effect of internal and external calcium concentration. *J. Exp. Biol.* **54**, 199–214.

Greenaway, P. (1971b). Calcium regulation in the freshwater mollusc *Limnaea stagnalis* (L.) (Gastropoda: Pulmonata). II. Calcium movements between internal calcium compartments. *J. Exp. Biol.* **54**, 609–620.

Grégoire, C. (1972). Structure of the molluscan shell. *In* "Chemical Zoology" (M. Florkin, and B. T. Scheer, eds.), Vol. 7, pp. 45–102. Academic Press, New York.

Griffiths, C. L., and King, J. A. (1979). Energy expended on growth and gonad output in the ribbed mussel *Aulacomya ater*. *Mar. Biol. (Berlin)* **53**, 217–222.

Grigor'ev, D. P. (1965). "Ontogeny of Minerals." Israel Program for Scientific Translations, Jerusalem.

Gutknecht, J., Bisson, M. A., and Tosteson, D. C. (1977). Diffusion of carbon dioxide through lipid bilayer membranes. *J. Gen. Physiol.* **69**, 779–794.

Haas, W. (1972). Untersuchungen über die Mikro und Ultrastruktur der Polyplacophorenschale. *Biomineralisation* **5**, 1–52.

Hammen, C. S. (1966). Carbon dioxide fixation in marine invertebrates. V. Rate and pathway in the oyster. *Comp. Biochem. Physiol.* **17**, 289–296.

Hammen, C. S., and Wilbur, K. M. (1959). Carbon dioxide fixation in marine invertebrates. I. The main pathway in the oyster. *J. Biol. Chem.* **234**, 1268–1271.

Hare, P. E., and Abelson, P. H. (1965). Amino acid composition of some calcified proteins. *Carnegie Inst. Wash. Year Book* **64**, 223–231.

Horiguchi, Y. (1958). Biochemical studies on *Pteria (Pinctada) martensii* (Dunker) and *Hyriopsis schlegeli* (Martens). IV. Absorption and transference of ^{45}Ca in *Hyriopsis schlegeli* (Martens). *Nippon Suisan Gakkaishi (Bull. Jpn. Soc. Sci. Fish.)* **23**, 710–715.

Hunter, R. D. (1975). Variation in population of *Lymnaea palustris* in upstate New York. *Amer. Midl. Nat.* **94**, 401–420.

Ingle, S. E., Culberson, C. H., Hawley, J. E., and Pytkowicz, R. M. (1973). The solubility of calcite in seawater at atmospheric pressure and 35/00 salinity. *Mar. Chem.* **1**, 295–307.

Istin, M., and Fossat, B. (1972). Etude du profil du potentiel éléctrique de la partie central du manteau du moule d'eau douce. *C. R. Hebd. Seances Acad. Sci. Ser. D* **274**, 119–121.

Istin, M., and Girard, J. P. (1970). Dynamic state of calcium reserves in freshwater clam mantle. *Calcif. Tissue Res.* **5**, 196–205.

Istin, M., and Kirschner, L. B. (1968). On the origin of the bioelectrical potential generated by the freshwater clam mantle. *J. Gen. Physiol.* **51**, 478–496.

Istin, M., and Maetz, J. (1964). Perméabilité au calcium de manteau de lamellibranches d'eau douce etudieé à l'aide des isotopes ^{45}Ca et ^{47}Ca. *Biochim. Biophys. Acta* **88**, 227–230.

Istin, M., and Masoni, A. (1973). Absorption et redistribution du calcium dans le manteau des lamellibranches en relation avec le structure. *Calcif. Tissue Res.* **11**, 151–162.

Iwata, K. (1980). Mineralization and architecture of the larval shell of *Haliotis discus hannai* Ino, (Archaeo-gastropoda). *J. Fac. Sci., Hokkaido Univ. Ser. 6* **19**, 305–320.

Jodrey, L. H. (1953). Studies on shell formation. III. Measurement of calcium deposition in shell and calcium turnover in mantle tissue using the mantle-shell preparation and Ca^{45}. *Biol. Bull. (Woods Hole, Mass.)* **104**, 398–407.

Jørgensen, C. B. (1976). Growth efficiencies and factors controlling size in some Mytilid bivalves, especially *Mytilus edulis L.*: review and interpretation. *Ophelia* **15**, 175–192.

Khan, H. R., and Saleuddin, A. S. M. (1979). Osmotic regulation and osmotically induced changes in the neurosecretory cells of the pulmonate snail *Helisoma*. *Can. J. Zool.* **57**, 1371–1383.

Kitano, Y., Kanamori, N., and Tokuyama, A. (1969). Effects of organic matter on solubilities and crystal form of carbonates. *Am. Zool.* **9**, 681–688.

Kitano, Y., Kanamori, N., and Yoshioka, S. (1976). Influence of chemical species on the crystal type of calcium carbonate. *In* "The Mechanisms of Mineralization in the Invertebrates and Plants" (N. Watabe, and K. M. Wilbur, eds.), pp. 191–202. Univ. of South Carolina Press, Columbia.

Kobayshi, S. (1964a). Studies in shell formation. X. A study of the proteins of the extrapallial fluid in some molluscan species. *Biol. Bull. (Woods Hole, Mass.)* **126**, 414–422.

Kobayashi, S. (1964b). Calcification in fish and shellfish. II. A paper electrophoretic study on the acid mucopolysaccharides and PAS-positive materials of the extrapallial fluid in some molluscan species. *Nippon Suisan Gakkaishi (Bull. Jpn. Soc. Sci. Fish.)* **30**, 893–907.

Krampitz, G., Engels, J., and Cazaux, C. (1976). Biochemical studies on water-soluble proteins and related components of gastropod shells. *In* "The Mechanisms of Mineralization in the Invertebrates and Plants" (N. Watabe, and K. M. Wilbur, eds.), pp. 155–173. Univ. of South Carolina Press, Columbia.

Kunigelis, S. C., and Saleuddin, A. S. M. (1978). Regulation of shell growth in the pulmonate gastropod *Helisoma duryi*. *Can. J. Zool.* **56**, 1975–1980.

Kunigelis, S. C., and Saleuddin, A. S. M. (1982). Shell repair rates and carbonic anhydrase activity during shell repair in *Helisoma duryi* (Mollusca). *Can. J. Zool.* **61**, 597–602.

Lehninger, A. L. (1975). "Biochemistry," 2nd ed. Worth, New York.

Loest, R. A. (1979a). Ammonia volatilization and absorption by terrestrial gastropods: A comparison between shelled and shell-less species. *Physiol. Zool.* **52**, 461–469.

Loest, R. A. (1979b). Ammonia-forming enzymes and calcium carbonate deposition in terrestrial pulmonates. *Physiol. Zool.* **52**, 470–483.

Manyak, D. M. (1982). "Calcified Tube Formation by the Shipworm *Bankia gouldi*." Ph.D. Thesis, Duke Univ., Durham, North Carolina.

Manyak, D. M., Dillaman, R. M., and Wilbur, K. M. (1980). Structure of the burrow lining and calcified tube of the shipworm *Bankia gouldi* (Mollusca, Teredinidae). *Scanning Electron Microsc.* **3**, 549–554.

Meenakshi, V. R., Martin, A. W., and Wilbur, K. M. (1974). Shell repair in *Nautilus macromphalus. Mar. Biol.* **27,** 27–35.

Misogianes, M. J., and Chasteen, N. D. (1979). Extrapallial fluid: a chemical and spectral characterization of the extrapallial fluid of *Mytilus edulis. Anal. Biochem.* **100,** 324–334.

Morowitz, H. S. (1968). "Energy Flow in Biology." Academic Press, New York.

Mutvei, H. (1978). Ultrastructural characteristics of the nacre of some gastropods. *Zool. Ser.* **7,** 287–296.

Nakahara, H. (1979). An electron microscope study of the growing surface of nacre in two gastropod species, *Turbo cornutus* and *Tegula pfeifferi. Venus* **38,** 205–211.

Nakahara, H. (1981). The formation and fine structure of the organic phase of the nacreous layer in mollusc shell. *In* "Study of Molluscan Paleobiology," pp. 225–230. Niigata Univ., Japan.

Nakahara, H., Kakei, M., and Bevelander, G. (1980). Fine structure and amino acid composition of the organic "envelope" in the prismatic layer of some bivalve shells. *Venus* **39,** 167–177.

Nakahara, H., Kakei, M., and Bevelander, G. (1981). Studies on the formation of the crossed lamellar structure in the shell of *Strombus gigas. Veliger* **23,** 207–211.

Neff, J. M. (1972). Ultrastructure of the outer epithelium of the mantle in the clam *Mercenaria mercenaria* in relation to calcification of the shell. *Tissue & Cell* **4,** 591–600.

Neuman, W. F., and Neuman, M. W. (1958). "The Chemical Dynamics of Bone Mineral." Univ. of Chicago Press, Chicago, Illinois.

Omori, M., and Watabe, N. (1980). "The Mechanisms of Biomineralization in Animals and Plants." Tokai Univ. Press, Tokyo, Japan.

Pietrzak, J. E., Bates, J. M., and Scott, R. M. (1973). Constituents of unionid extrapallial fluid. I. Electrophoretic and immunological studies of protein components. *Biol. Bull. (Woods Hole, Mass.)* **144,** 391–399.

Plummer, L. N., Parkhurst, D. L., and Wigley, T. M. L. (1979). Critical review of the kinetics of calcite dissolution and precipitation. *In* "Chemical Modeling in Aqueous Systems" (E. A. Jenne, ed.), pp. 537–573. American Chemical Society, Washington, D.C.

Price, T. J., Thayer, G. W., LaCroix, M. W., and Montgomery, G. P. (1976). The organic content of shells and soft tissues of selected estuarine gastropods and pelecypods. *Proc. Nat. Shellfish Assoc.* **65,** 26–31.

Pytkowicz, R. M. (1969). Chemical solution of calcium carbonate in sea water. *Am. Zool.* **9,** 673–679.

Ravindranath, M. H., and Rajeswari Ravindranath, M. H. (1974). The chemical nature of the shell of molluscs. I. Prismatic and nacreous layers of a bivalve *Lamellidans marginalis* (Unionidae). *Acta Histochem.* **48,** 26–41.

Rhoads, D. C., and Lutz, R. A. (1980). "Skeletal Growth in Aquatic Organisms." Plenum, New York.

Richardson, C. A., Crisp, D. J., and Runham, N. W. (1981). Factors influencing shell deposition during a tidal cycle in the intertidal bivalve *Cerastoderma edule. J. Mar. Biol. Assoc. U.K.* **61,** 465–476.

Rionel, N., Morel, F., and Istin, M. (1973). Etude des granules calcifiés du manteau des lamellibranches à l'aide de la microsonde électronique. *Calcif. Tissue Res.* **11,** 163–170.

Robertson, W. G. (1982). The solubility concept. *In* "Biological Mineralization and Demineralization" (G. H. Nancollas, ed.), pp. 5–21. Springer–Verlag, Berlin.

Rodhouse, P. G. (1978). Energy transformations by the oyster *Ostrea edulis* L. in a temperate estuary. *J. Exp. Mar. Biol. Ecol.* **34,** 1–22.

Roer, R. D. (1979). "Mechanisms of Deposition and Resorption of Calcium in the Carapace of the Green Crab, *Carcinus maenas.*" Ph.D. Thesis, Duke Univ., Durham, North Carolina.

Russell-Hunter, W. D. (1978). Ecology of freshwater pulmonates. *In* "Pulmonates 2A" (V. Fretter and J. Peake, eds.), pp. 335–383. Academic Press, London.

Russell-Hunter, W. D., Burky, A. J., and Hunter, R. D. (1981). Interpopulation variation in calcareous and proteinaceous shell components in the stream limpet, *Ferrissia rivularis*. *Malacol.* **20**, 255–266.

Saleuddin, A. S. M., and Chan, W. (1969). Shell regeneration in *Helix*: shell matrix composition and crystal formation. *Can. J. Zool.* **47**, 1107–1112.

Samata, T., Sanguansri, P., Cazaux, C., Hamm, M., Engels, J., and Krampitz, G. (1980). Biochemical studies on components of mollusc shells. In "The Mechanisms of Biomineralization in Animals and Plants" (M. Omari, and N. Watabe, eds.), pp. 37–47. Tokai Univ. Press, Tokyo, Japan.

Schatzmann, H. J. (1973). Dependence on calcium concentration and stoichiometry of the calcium pump in human red cells. *J. Physiol. (London)* **235**, 551–569.

Schlichter, L. C. (1981). Ion relations of hemolymph, pallial fluid, and mucus of *Lymaea stagnalis*. *Can. J. Zool.* **59**, 605–613.

Schmidt, W. J. (1923). Bau und Bildung der Perlmuttermasse. *Zool. Jahrb. Abt. Anat. Ontog. Tiere* **45**, 1–148.

Seed, R. (1980). Shell growth and form in the bivalvia. In "Skeletal Growth in Aquatic Organisms" (D. C. Rhoads, and R. A. Lutz, eds.), pp. 23–68. Plenum, New York.

Sikes, C. S., and Wheeler, A. P. (1982). Carbonic anhydrase and carbon fixation in coccolithophorids. *J. Phycol.* **18**, 423–426.

Sikes, C. S., and Wheeler, A. P. (1983). A systematic approach to some fundamental questions of carbonate calcification. In "Biomineralization and Biological Metal Accumulation" (P. Westbroek and E. W. de Jong, eds.), pp. 285–289. Reidel, Dordrecht, Holland.

Simkiss, K. (1976). Cellular aspects of calcification. In "The Mechanisms of Mineralization in the Invertebrates and Plants" (N. Watabe, and K. M. Wilbur, eds.), pp. 1–31. Univ. of South Carolina Press, Columbia.

Simkiss, K., and Wilbur, K. M. (1977). The molluscan epidermis and its secretions. *Symp. zool. Soc. London* **39**, 35–76.

Simpson, J. W., and Awapara, J. (1966). The pathway of glucose degradation in some invertebrates. *Comp. Biochem. Physiol.* **18**, 537–548.

Sorenson, A. L., Wood, D. S., and Kirschner, L. B. (1980). Electrophysiological properties of resting secretory membranes in lamellibranch mantles. *J. Gen. Physiol.* **75**, 21–37.

Stokes, T. M., and Awapara, J. (1968). Alanine and succinate as end-products of glucose degradation in the clam *Rangia cuneata*. *Comp. Biochem. Physiol.* **25**, 883–892.

Stolkowski, J. (1951). Essai sur le déterminisme des formes minéralogiques du calcaire chez les êtres vivants (calcaires coquiller). *Ann. Inst. Oceanogr. (Paris)* **26**, 1–113.

Suzuki, S., and Uozumi, S. (1981). Organic components of prismatic layers in molluscan shells. *J. Fac. Sci. Hokkaido Univ. Ser. 4* **20**, 7–20.

Taylor, J. D. (1973). The structural evolution of the bivalve shell. *Paleontol.* **16**, 519–534.

Taylor, J. D., Kennedy, W. J., and Hall, A. (1969). The shell structure and mineralogy of the bivalvia. Introduction. Nuculacea-Trigonacea. *Bull. Br. Mus. (Nat. Hist.)* **3**, 1–125.

Tevesz, M. J. S., and Carter, J. G. (1980). Environmental relationships of shell form and structure of Unionacean bivalves. In "Skeletal Growth in Aquatic Organisms" (D. C. Rhoads, and R. A. Lutz, eds.), pp. 295–322. Plenum, New York.

Thomas, J. D., and Lough, A. (1974). The effects of external calcium concentration on the rate of uptake of this ion of *Biomphalaria glabrata* (Say). *J. Anim. Ecol.* **43**, 861–872.

Thomas, J. D., Benjamin, M., Lough, A., and Aram, R. H. (1974). The effects of calcium in the external environment on the growth and natality rates of *Biomphalaria glabrata*. *J. Anim. Ecol.* **43**, 839–860.

Uozumi, S., Iwata, K., and Togo, Y. (1972). The ultrastructure of the mineral in and the construction of the crossed-lamellar layer in molluscan shell. *J. Fac. Sci. Hokkaido Univ. Ser. 4* **15**, 447–478.

van der Borght, O. (1963). In- and outfluxes of calcium ions in freshwater gastropods. *Arch. Int. Physiol. Biochim.* **71**, 46–50.

van der Borght, O., and van Puymbroeck, S. (1966). Calcium metabolism in a freshwater mollusc: Quantitative importance of water and food as supply for calcium during growth. *Nature (London)* **210**, 791.

van der Borght, O., and van Pymbroak, S. (1966). Calcium metabolism in a freshwater mollusc: Quantitative importance of water and food as supply for calcium during growth. *Nature (London)* **210**, 791.

Wada, K. (1957). Electron-microscopic observations of the shell structures of pearl oyster (*Pinctada martennsii*). II. Observations of the aragonite crystals on the surface of nacreous layers. *Kokuritsu Shinju Kenkyusho Hokoku (Bull. Natl. Pearl Res. Lab. [Jpn.])* **2**, 74–85.

Wada, K. (1961). Crystal growth of molluscan shells. *Kokuritsu Shinju Kenkyusho Hokoku (Bull. Natl. Pearl Res. Lab. [Jpn.])* **7**, 703–828.

Wada, K. (1972). Nucleation and growth of aragonite crystals in the nacre of bivalve molluscs. *Biomineralisation* **6**, 141–159.

Wada, K. (1980). Initiation of mineralization in bivalve molluscs. *In* "The Mechanisms of Biomineralization in Animals and Plants" (M. Omori and N. Watabe, eds.), pp. 79–92. Tokai Univ. Press, Tokyo, Japan.

Wada, K. (1982). Composition of amino acids in the extrapallial fluid of bivalved molluscs. *10th Ann. U. J. N. R. Meetings*, Rehobeth, New Jersey.

Wada, K., and Fujinuki, T. (1976). Biomineralization in bivalve molluscs with emphasis on the chemical composition of the extrapallial fluid. *In* "The Mechanisms of Mineralization in the Invertebrates and Plants" (N. Watabe, and K. M. Wilbur, eds.), pp. 175–190. Univ. of South Carolina Press, Columbia.

Watabe, N. (1954). Electron microscopic observations of the aragonite crystals on the surface of cultured pearls. I. *Rep. Fac. Fish. Prefect. Univ. Mie* **1**, 440–454.

Watabe, N. (1965). Studies on shell formation. XI. Crystal-matrix relationships in the inner layers of mollusk shells. *Ultrastructure Research* **12**, 351–370.

Watabe, N. (1981). Crystal growth of calcium carbonate in the invertebrates. *Prog. Cryst. Growth Charact.* **4**, 99–147.

Watabe, N., and Dunkelberger, D. G. (1979). Ultrastructural studies on calcification in various organisms. *Scanning Electron Microscopy* **II**, 403–416.

Watabe, N., and Wada, K. (1956). On the shell structure of the Japanese pearl oyster, *Pinctada martensii*. I. Prismatic layer. *I. Rep. Fac. Fish. Prefect. Univ. Mie* **2**, 227–232.

Watabe, N., and Wilbur, K. M. (1960). Influence of the organic matrix on crystal type in molluscs. *Nature (London)* **188**, 334.

Watabe, N., and Wilbur, K. M. (1961). Studies on shell formation. IX. An electron microscope study of crystal layer formation in the oyster. *J. Biophys. Bioch. Cytol.* **9**, 761–772.

Watabe, N., and Wilbur, K. M. (1976). "The Mechanisms of Mineralization in the Invertebrates and Plants." Univ. of South Carolina Press, Columbia.

Watabe, N., Meenakshi, V. R., Blackwelder, P. L., Kurtz, E. M., and Dunkelberger, D. G. (1976). Calcareous spherules in the gastropod, *Pomacea paludosa*. *In* "The Mechanisms of Mineralization in the Invertebrates and Plants" (N. Watabe, and K. M. Wilbur, eds.), pp. 283–308. Univ. of South Carolina Press, Columbia.

Weiner, S. (1979). Aspartic acid-rich proteins: major components of the soluble organic matrix of mollusk shells. *Calcif. Tissue Int.* **29**, 163–167.

Weiner, S., and Hood, L. (1975). Soluble proteins of the organic matrix of mollusc shells: a potential template for shell formation. *Science (Washington, D.C.)* **190**, 987–989.

Weiner, S., and Traub, W. (1980). X-ray diffraction study of the insoluble organic matrix of mollusk shells. *FEBS Lett.* **111**, 311–316.

Weiner, S., and Traub, W. (1981). Organic-matrix-mineral relationships in mollusc shell nacreous layers. *In* "Structural Aspects of Recognition and Assembly in Biological Macromolecules" (M. Balaban, J. L. Sussman, W. Traub, and A. Yonth, eds.)¸ pp. 467–482. Balaban ISS, Rehovot and Philadelphia.

Wheeler, A. P. (1975). "Oyster Mantle Carbonic Anhydrase: Evidence for Plasma Membrane-Bound Activity and for a Role in Bicarbonate Transport." Ph.D. Thesis, Duke Univ., Durham, North Carolina.

Wheeler, A. P., Blackwelder, P. L., and Wilbur, K. M. (1975). Shell growth in the scallop *Argopecten irradians*. I. Isotope incorporation with reference to diurnal growth. *Biol. Bull. (Woods Hole, Mass.)* **148**, 472–482.

Wheeler, A. P., George, J. W., and Evans, C. A. (1981). Control of calcium carbonate nucleation and crystal growth by soluble matrix of oyster shell. *Science (Washington, D.C.)* **212**, 1397–1398.

Whitehead, D. L. (1974). Steroid involvement in calcification of the invertebrate exoskeleton. *J. Endocrinol.* **61**, 77–78.

Whitehead, D. L. (1977). Steroids enhance shell regeneration in an aquatic gastropod (*Biomphalaria glabrata*). *Comp. Biochem. Physiol. C* **58**, 137–141.

Whitehead, D. L., and Saleuddin, A. S. M. (1978). Steroids promote shell regeneration in *Helix aspersa* (Mollusca: Pulmonata). *Comp. Biochem. Physiol. C* **59**, 5–10.

Wilbur, K. M. (1964). Shell formation and regeneration. *In* "Physiology of Mollusca" (K. M. Wilbur, and C. M. Yonge, eds.), Vol. 1, pp. 243–282. Academic Press, New York.

Wilbur, K. M. (1972). Shell formation in molluscs. *In* "Chemical Zoology" (M. Florkin, and B. T. Scheer, eds.), Vol. 7, pp. 103–145. Academic Press, New York.

Wilbur, K. M. (1976). Recent studies of invertebrate mineralization. *In* "The Mechanisms of Mineralization in the Invertebrates and Plants" (N. Watabe and K. M. Wilbur, eds.), pp. 79–108. Univ. of South Carolina Press, Columbia.

Wilbur, K. M., and Anderson, N. G. (1950). Carbonic anhydrase and growth in the oyster and *Busycon*. *Biol. Bull. (Woods Hole, Mass.)* **98**, 19–24.

Wilbur, K. M., and Bernhardt, A. M. (1982). Mineralization of molluscan shell: Effects of free and polyamino acids on crystal growth rate *in vitro*. *Am. Zoologist* **22**, 952.

Wilbur, K. M., and Jodrey, L. H. (1952). Studies on shell formation. I. Measurement of the rate of shell formation using Ca⁴⁵. *Biol. Bull. (Woods Hole, Mass.)* **103**, 269–276.

Wilbur, K. M., and Jodrey, L. H. (1955). Studies on shell formation. V. The inhibition of shell formation by carbonic anhydrase inhibitors. *Biol. Bull. (Woods Hole, Mass.)* **108**, 359–365.

Wilbur, K. M., and Manyak, D. M. (1983). Biochemical aspects of molluscan shell mineralization. Symposium on Marine Biodeterioration. (In press.)

Wilbur, K. M., and Owen, G. (1964). Growth. *In* "Physiology of Mollusca" (K. M. Wilbur, and C. M. Yonge, eds.), pp. 211–242. Academic Press, New York.

Wilbur, K. M., and Simkiss, K. (1968). Calcified shells. *In* "Comprehensive Biochemistry" (M. Florkin, and E. H. Stotz, eds.), Vol. 26A, pp. 229–295. Elsevier, New York.

Wilbur, K. M., and Simkiss, K. (1979). Carbonate turnover and deposition by metazoa. *In* "Biogeochemical Cycling of Mineral-forming Elements" (P. A. Trudinger, and D. J. Swaine, eds.), pp. 69–106. Elsevier, Amsterdam.

Wilbur, K. M., and Watabe, N. (1963). Experimental studies in molluscs and the alga *Coccolithus huxleyi*. *Ann. N.Y. Acad. Sci.* **109**, 82–112.

Wise, S. W., Jr. (1970). Microarchitecture and mode of formation of nacre (mother-of-pearl) in pelecypods, gastropods, and cephalopods. *Eclogae Geol. Helv.* **63,** 775–797.

Wise, S. W., Jr., and Hay, W. W. (1968a). Scanning electron microscopy of molluscan shell ultrastructure. I. Techniques for polished and etched sections. *Trans. Am. Microsc. Soc.* **87,** 411–418.

Wise, S. W., Jr., and Hay, W. W. (1968b). Scanning electron microscopy of molluscan shell ultrastructures. II. Observations of growth surfaces. *Trans. Am. Microsc. Soc.* **87,** 419–430.

Wood, D. (1973). "Calcium Movement and Electrogenesis across the Isolated Clam Mantle." Ph.D. Thesis, Washington State Univ. Pullman.

Zischke, J. A., Watabe, N., and Wilbur, K. M. (1970). Studies on shell formation: measurement of growth in the gastropod *Ampullarius glaucus. Malacologia* **10,** 423–439.

Zurburg, W., and Kluytmans, J. H. (1980). Organ specific changes in energy metabolism due to anaerobiosis in the sea mussel *Mytilus edulis* (L.). *Comp. Biochem. Physiol. B* **67,** 317–322.

7

Shell Repair

NORIMITSU WATABE

Electron Microscopy Center
Department of Biology
and
Belle W. Baruch Institute for Marine Biology
and Coastal Research
University of South Carolina
Columbia, South Carolina

I. Summary and Perspective

Molluscs are capable of repairing damage inflicted to the shell. The repair, often called *shell regeneration,* is accomplished by a deposition of new shell material in the damaged area. The process of repair may or may not be similar to that of normal shell formation, and the characteristics of the repaired shell vary with the organisms and the region of the shell damage. This chapter discusses (1) the processes of shell repair, (2) structure and mineralogy of repaired shell, (3)

THE MOLLUSCA, VOL. 4
Physiology, Part 1

chemical characteristics of organic matrix and its role on crystal initiation, (4) histology and histochemical aspects of the mantle epithelium during shell repair, and (5) factors controlling shell repair.

Shell repair is a highly regulated process, and attention is increasingly being paid to its on–off control mechanisms. Recent physiological and ultrastructural investigations have revealed that the mechanisms are intimately associated with neurosecretory or hormonal actions of the organisms. It will probably not be long before the control substances and their targets, that is, ion concentration and transport, synthesis and secretion of organic matrices for shell repair, and inhibition of normal shell formation, are identified and their mode of action and counteraction established.

Characterization of organic matrix in repaired shell is an important topic, which deserves concentrated research efforts. In many cases, repaired shell consists of a number of layers, each composed of crystals, the type and/or morphology of which differ from the normal, and very often a bimineralic layer is present as well. The organic matrix has also been shown to differ from that of the normal shell in its composition and/or structure. It is hoped that chemical and structural characteristics of insoluble and soluble (if any) protein and carbohydrate fractions, and lipid components of the matrix will be investigated extensively and their roles in nucleation, orientation, morphology, polymorphisms of crystals, and inhibition of crystal growth clarified.

Shell repair is an excellent experimental system in which to investigate various phases of molluscan shell formation, and the information obtained not only will contribute to the knowledge of shell repair but also will be very useful in understanding the mechanisms of biological calcification.

II. The Repair Process

In general, initiation and completion of shell repair are much more rapid in terrestrial molluscs than in aquatic species (Kessel, 1933; Wagge and Mittler, 1953; Beedham, 1965; Saleuddin, 1969; Meenakshi et al., 1973; Wilbur, 1973). In terrestrial species, exposure of the mantle to air by shell damage may lead to a difference in gas exchange or cause desiccation (Wilbur, 1973) so that immediate repair of exposed area of the mantle surface is essential, but it is not so critical in aquatic species (Beedham, 1965).

In the terrestrial pulmonates *Helix* and *Otala,* the repair is much more rapid when the shell damage occurs in the mid-shell region than at the edge (Saleuddin and Wilbur, 1969). Generally, the repair is faster at the edges than in the central region where the mantle edge cannot retract (Timmermans, 1973). In the freshwater pulmonates such as *Limnaea* and *Planorbis,* and the prosobranchs such as *Viviparus* as well as marine forms, a hole made in the central region of a shell will not be repaired or will be repaired slowly if it is not covered with some material (Kessel, 1933; Wilbur, 1964). However, Beedham (1965) reported in the freshwater bivalve *Anodonta* that the cover did not necessarily hasten the process of the repair. In many cases, the secretion of a new shell material was said to be actively stimulated by the exposed mantle coming into direct contact with the environment.

III. Structure and Mineralogy of Repaired Shell

A. Shell Structure

When the repair takes place at the shell edge, the structure of repaired shell is essentially the same as the normal. In the freshwater pulmonate *Helisoma duryi,* it is similar to the normal shell structure except for the absence of the periostracum (Wong and Saleuddin, 1972). In the case of the terrestrial pulmonate *Pleurodonte rostrata,* a "cuticle" (or periostracum) is formed followed by the formation of normal shell (Andrews, 1934; Saleuddin, 1980). In the marine prosobranch *Littorina irrorata* (N. Watabe, unpublished observations) and the marine bivalve *Pinctada fucata martensii* (N. Watabe, unpublished observations), the regenerated shells are also similar to the normal type.

If the damage occurs away from the shell edges where the mantle cannot retract, the repaired shell is often different from the normal in structure, composition of organic matrix, mineralogy, and morphology of crystals. Table I shows structure and mineralogy of normal and repaired shell in several molluscan species.

The first material observed after the shell removal is the fluid in the damaged area. This is seen, for example, in *Helix pomatia* (Wilbur, 1964) and the marine bivalve *Mytilus edulis* (Uozumi and Ohta, 1977; Uozumi and Suzuki, 1978, 1979). Formation of organic membranes follows. The membranes are either periostracum-type or non-periostracum-type material. According to von Levetzow (1932) and Kessel (1933), the periostracum is not produced when the damaged area is not in contact with the mantle border and the mantle groove. On the contrary, periostracum or the material equivalent to the periostracum are reported to be formed in *Lymnaea stagnalis* (Timmermans, 1973), the freshwater bivalves *Musculus senhousia* (Kawaguti and Ikemoto, 1962) and *Anodonta* (Tsujii, 1960, 1968; Beedham, 1965), and the marine bivalve *M. edulis* (Meenakshi et al., 1973). Non-periostracum-type membrane is formed in the terrestrial pulmonates *P. rostrata* (Andrews, 1934), *Helix aspersa* (Wagge, 1951), *H. pomatia* (Kessel, 1933; Abolins-Krogis, 1968; Saleuddin and Wilbur, 1969), *Otala lactea* (Meenakshi et al., 1974c), and *Cepaea nemoralis* (N. Watabe,

TABLE I

Structure and Mineralogy of Normal and Repaired Shell (Mid-Shell Region) in Several Molluscan Species

Species	Normal shell[a]	Repaired shell[a]		
		Early stage	Middle stage	Later stage
Helix pomatia[b,c]	Per + Pr(A) + Cl(A) + N(A)	Npm	Sph(A) + Poly or Rhom(C) + Sph(A)	Sph(A) (Also vaterite)
Otala lactea[d,e]	Per + Pr(A) + Cl(A) + N(A)	Npm	Sph(A) + Poly(C)	N(A)
Cepaea nemoralis[f]	Per + Pr(A) + Cl(A)	Npm	Sph(A) + Rhom(C)	Sph(A)
Viviparus viviparus[c]	Per + Cl(A)	Npm	Rhom(C) + Sph(A)	Sph(A) + Cl(A)
Pomacea paludosa[g]	Per + Pr(A) + Cl(A)	Npm	Fol(C) + Sph(A)	Pr(A) + Cl(A)
Anodonta sp.[h]	Per + Pr(A) + N(A)	Per	Pr(A)	N(A)
Mytilus edulis[i,j]	Per + Pr(C) + N(A)	Lm + Per + Conch	Sph(A) (Complex crossed lamellar)	Pr(C) + N(A); also Fol(A)
Nautilus pompilus[k]	Sph-pr(A) + N(A) + Semipr(A)	Npm	Sph(A) + Sph-pr(A); also Semipr(A)	N(A)

[a] Per, Periostracum; Pr, prismatic layer; Cl, crossed lamellar layer; N, nacreous layer; Npm, non-periostracum-type matrix; Sph, spherulites; Poly, polygonal crystals; Rhom, rhombohedral crystals; Fol, foliated layer; Lm, laminated membraneous layer; Conch, conchiolin membraneous layer; Sph-pr, spherulitic prismatic layer; Semipr, semiprismatic layer; (A), aragonite; (C), calcite.
[b] Abolins-Krogis, 1963a.
[c] Kessel, 1933.
[d] Wilbur, 1973.
[e] Meenakshi et al., 1974c.
[f] Watabe, 1981a.
[g] Blackwelder and Watabe, 1977.
[h] Tsujii, 1976.
[i] Meenakshi et al., 1974a.
[j] Uozumi and Suzuki, 1979.
[k] Meenakshi et al., 1974b.

unpublished observations), and in the freshwater prosobranchs *Viviparus viviparus, V. fasciatus* (Kessel, 1933), and *Pomacea paludosa* (Meenakshi et al., 1975; Blackwelder and Watabe, 1977); the membrane is believed not to be of the periostracum type in the terrestrial pulmonate *Euplecta indica* (Kapur and Gupta, 1970). The cephalopod *Nautilus macromphalus* (Meenakshi et al., 1974b) does not form a periostracum, which is also lacking in normal shells.

In addition to the periostracum, a membraneous layer structurally and histochemically similar to the organic matrix of "calcified layer" is found in *L. stagnalis* (Timmermans, 1973). In *Anodonta*, nacreous and prismatic layer-type matrix layers are deposited prior to the formation of periostracum (Beedham, 1965). In *M. edulis*, a membrane equivalent to the periostracum layer is formed after the laminated membraneous layer, followed by the deposition of a layer of conchiolin membranes (Fig. 1) (Uozumi and Suzuki, 1978, 1979).

Calcification takes place after the formation of the organic membranes. This occurs in two ways. One follows more or less the sequence of normal shell formation. For example, the prismatic and nacreous layers are successively deposited as in normal shell in *Anodonta* (Beedham, 1965; Tsujii, 1960, 1969, 1976). Also, the spherulitic–prismatic layer, nacreous layer, and semiprismatic layer are formed in sequence in *N. macromphalus* (Meenakshi et al., 1974b).

In the majority of cases of repair, abnormal types of crystals and/or shell layers are produced. Interestingly, a formation of aragonite spherulites is a common phenomenon in almost all the abnormal shell-type repairs so far reported, although the time of their formation may vary in different species. The crystal formation is initiated either in organic spherules developed on the organic membrane or directly on the membrane. Abolins-Krogis (1958) found in *H. pomatia* that small (0.6–5.0 μm in diameter) boat-shaped organic spherules containing *b-granules* were deposited on an organic membrane. These spherules came to be calcified, assuming the form of "dendrite-like" spherulties. The spherulites were said to be transformed into polygonal calcium carbonate crystals. However, the polygonal crystals could develop independently. Wada (1961, 1964, 1980) also reported calcification of organic granules deposited on coverslips inserted between the mantle and shell of marine bivalves. Although this is not shell repair, the formation must have been induced by the stimulus of the coverslip and could be regarded as similar to shell repair. Similarly, in the early stage of the repair in *M. edulis* organic crystals developed in the conchiolin membrane (the third organic layer) and grew into the spherulites (Uozumi and Suzuki, 1978, 1979). These authors called them complex crossed lamellar structures, but spherulite seems to be the proper term because of their radial structure. However, Meenakshi et al. (1973) reported that shell repair of the same species followed the normal sequence of periostracum–prismatic layer–nacreous layer formation (except for the frequent presence of a complex prismatic-type structure), and spherulite formation was not mentioned.

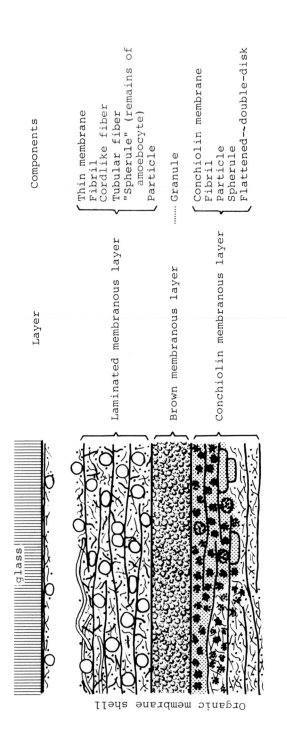

Fig. 1. A diagram of the structure of organic membrane shell formed at an early stage of shell regeneration in *Mytilus edulis*. (From Uozumi and Suzuki, 1979.)

Not all calcification in shell repair has been reported to be associated with organic spherules. In *O. lactea*, doubly pointed crystals (0.6–2.0 μm) developed on the matrix membrane within 2 to 3 h after the shell damage and grew into dumbbell-shaped aggregates. They were said to form large polycrystalline aggregates. In addition, tabular euhedral crystals with depressed central portions were also present. Through coalescence, those crystals developed into outer calcified shell layers (Wilbur, 1973; Meenakshi et al., 1974c). Similar configurations were reported in *H. pomatia* (Biedermann, 1902; Sioli, 1935), the marine prosobranchs *Oncomelania formosana* (Davis, 1964), *Neritina reclinata* (Andrews, 1935), and the terrestrial pulmonate *P. rostrata* (Andrews, 1934).

However, according to Kessel (1933), regularly arranged rhombic and hexagonal crystals (calcite) mixed with a small number of spherulites were the first to be formed in *H. pomatia*, followed by the formation of aggregates of needles which developed into spherulites. This was also the case in the repair of *V. viviparus* and *V. fasciatus* (Kessel, 1933). Similarly, a mixture of rhombohedral calcite and spherulitic crystals was the early form of shell repair in *C. nemoralis* (Fig. 2) (Watabe, 1981a,b). Aggregates of spherulites were formed at a later stage (Fig. 3) (Watabe, 1981a,b). Shell repair in *P. paludosa* followed a some-

Fig. 2. A mixture of rhombohedral calcite and spherulitic aragonite at an early stage of shell regeneration in *Cepaea nemoralis*. (2000×.) (From Watabe, 1981a.)

Fig. 3.   Aragonite spherulites in a regenerated shell of *Cepaea nemoralis*. (2700×.) (From Watabe, 1981a.)

what different process in that a foliated layer composed of an aggregate of polygonal crystals (calcite) was formed prior to the formation of a spherulitic layer (Blackwelder and Watabe, 1977). Those spherulites and rhombohedral crystals are not present in the normal shells in these species. After the spherulite aggregate formation, the normal shell layer formation resumes in some species. The examples are aragonitic prismatic layer and crossed lamellar layer in *P. paludosa* (Blackwelder and Watabe, 1977); "lower" layer in *P. rostrata* (Andrews, 1934) and the "third" layer in *V. viviparus* (Kessel, 1933)—both layers are probably equivalent to the crossed lamellar layer; and the prismatic and nacreous layer in *M. edulis* (Uozumi and Suzuki, 1979). However, in *M. edulis* a structure similar to the calcitic foliated layer, which is not a normal layer, was also formed (Uozumi and Suzuki, 1979).

In addition to the spherulite aggregate, some species produce other abnormal layer(s) before the deposition of normal shell layer(s). Polygonal and multi-layered groups of tabular crystals without uniform orientation, and compact and more planar crystals were formed prior to the normal nacreous layer development in *O. lactea* (Wilbur, 1973; Meenakshi et al., 1974c). The normal shell of this species (and *H. pomatia*) contains the prismatic layer and crossed lamellar layer in addition to the nacreous layer. It is not clear if and when these two layers were formed in the repaired shell.

Thus, depending on the species and the region of the shell damage, structure and mineralogy of repaired shell vary. However, abnormal types of shells are generally formed when the damage is away from the shell edges, and spherulitic crystals appear to be formed in many cases, although the time of their formation is different depending on the species.

B. Mineralogy

Calcium carbonate crystals in the repaired shells are vaterite, aragonite, or calcite. Not only does the repaired shell show abnormal types of shell layers, but also crystal types are often different from the normal (see Table I). However, in the freshwater prosobranch *Ampullarius glaucus*, both normal and repaired shells are aragonite (Saleuddin et al., 1970). Of considerable interest is the occurrence of vaterite in several species including *Helix* (Mayer, 1931; von Levetzow, 1932; Abolins-Krogis, 1968; Saleuddin and Wilbur, 1969), *Viviparus intertextus* and *Elliptio complanatus* (freshwater bivalve) (Wilbur and Watabe, 1963), and *Tegula funebralis* and *T. ligulata* (marine prosobranchs) (Reed-Miller et al., 1980). In contrast to the other polymorphic forms, vaterite is not present in normal shells. This mineral is the most unstable form of calcium carbonate crystals and its occurrence is reported in the mantle epithelium of *V. viviparus* (Kessel, 1933); egg capsules of *Ampullarius* species (Hall and Taylor, 1971); spherules and egg capsules of *Pomacea paludosa, P. urceus,* and *Pila vivens* (Meenakshi et al., 1974a); spherules and spicules within the tissues of cestodes, larval insects, and nudibranch molluscs (Prenant, 1927, 1928); otoliths of some fishes (Carlstrom, 1963); eggshells of some pelicans (Gould, 1972) and cuckoos (Board and Perott, 1979); pathological structures such as urinary and gallstones (Meier and Moenke, 1961; Lagergren, 1962; Sutor and Wooley, 1968). Morphology of molluscan vaterite is little known except in *Pomacea* (Meenakshi et al., 1974a).

In many repaired shells, calcite and aragonite or vaterite coexist (see Fig. 2; also Wilbur and Watabe, 1963), but in others, such as *P. paludosa,* calcite and aragonite are present in separate discrete layers.

IV. Organic Matrix of Crystalline Layers

Crystalline aggregates or layers of repaired shell contain organic matrix. Chemical or histochemical characteristics of the matrix of each individual crystalline layer are not very well known. However, several comparisons have been made between the normal and repaired shell matrices as a whole. Glycine content is much lower in repaired shell than in normal calcified layers in *Helix* (Saleuddin and Hare, 1970) and *Pinctada* (K. Wada, personal communication; Wilbur, 1973), whereas the opposite trend is seen in *Anodonta* (Wilbur, 1973). Arginine, proline, and cysteic acid are higher in repaired shell in *Helix* (Saleuddin and Hare, 1970). The significance of these differences is not clear.

Meenakshi et al. (1975) compared the amino acid composition of shell of *Pomacea paludosa* during the 30-day period of repair (Table II). The high acidic mucopolysaccharide content in the early stage of shell repair suggests an increased mucous secretion at the beginning due to the stimulus given to the mantle (see also Saleuddin and Hare, 1970). The early shell repair matrix is also characterized by lower levels of glycine and higher levels of acidic amino acids. This trend is also seen in *Helix* (Saleuddin and Hare, 1970). In *Pomacea*, the crystal

TABLE II

Amino Acid Composition of Normal and Regenerated Shell of *P. paludosa*[a,b]

	Normal shell		Regenerated shell (days)			
	Periostracum	Calcified layer	4	7	15	30
Aspartic	39	92	126	123	117	107
Threonine	19	48	78	69	74	58
Serine	26	70	65	63	65	65
Glutamic	38	98	126	129	122	101
Proline	32	84	58	61	59	68
Glycine	519	166	76	75	79	148
Alanine	36	76	79	79	76	70
Cystine	—	—	5	3	3	—
Valine	26	35	66	69	68	63
Methionine	—	74	13	14	14	14
Isoleucine	45	25	40	39	47	43
Leucine	59	73	85	89	86	75
Tyrosine	38	25	34	35	38	32
Phenylalanine	77	39	40	39	39	37
Histidine	12	31	42	42	43	38
Lysine	25	33	36	40	35	41
Arginine	11	32	32	31	35	40
Ammonia	47	86	88	87	86	80
Glucosamine	—	Traces	104	87	81	50
Ratio of acidic to basic residues	0.61	1.08	1.52	1.51	1.35	1.10

[a] The figures represent residues per 1000.
[b] From Meenakshi et al. (1975).

Mineral Composition of Regenerated Shell of *P. paludosa*[a,b]

	Time of regeneration (days)			
	4	7	15	30
Weight percentage of calcite	60	55	5	Traces
Weight percentage of aragonite	40	45	95	100

[a] Percentages of calcite and aragonite calculated from X-ray diffractogram of 20 pooled regenerated shells at each time interval.
[b] From Meenakshi et al. (1975).

type associated with this matrix is calcite, which is an abnormal type in this snail. The glycine content and levels of acidic amino acids gradually change during the repair, and at the 30th day when the shell mineral becomes almost 100% aragonite (Table III) they are at about the same levels as in the normal shell. Thus, the initial difference in amino acid composition appears to be related to the foliated layer (calcitic) formation, and as reviewed by Wilbur previously (1973), repaired shell matrix was different from the normal, and the difference was accompanied by the difference in shell mineralogy.

V. Histological and Histochemical Aspects of Shell Repair

A. Mantle Epithelium

The mantle epithelium undergoes structural and histochemical changes during shell repair. If the damage occurs at the shell edge, the changes are slight (Beedham, 1965). In *Anodonta*, RNA content decreased in the outer mantle epithelium immediately after damage but increased later. The increase may be correlated with the deposition of conchiolin. Alkaline and acid phosphatase activity also increased (Saleuddin, 1967) and probably played some role in the RNA synthesis (Saleuddin, 1967; Chan and Saleuddin, 1974a). In *Helix* the cells change their shape, accumulate secretory granules, and increase protein synthesis and the number of mitochondria (Saleuddin, 1970). As mentioned earlier, the repaired shell in this region is essentially similar to the normal shell in structure.

The changes of the mantle epithelium are more drastic in the mid-shell regions. The changes occur in a large area, and the whole epithelium from the mantle edge up to and including the repair area is affected (Timmermans, 1973).

In the freshwater bivalves *Anodonta* and *Musculus senhousia,* the epithelial cells, which normally secrete the nacreous layer, are sequentially transformed into periostracum- and prismatic layer-secreting-type cells before resuming their normal activities (Tsujii, 1960, 1968, 1969, 1976; Kawaguti and Ikemoto, 1962; Beedham, 1965). This indicates that the functions of the mantle epithelium are not permanently fixed, but can be altered by a stimulus such as shell damage.

As seen in the repair at the shell edges in *Anodonta* and *Helix* mentioned earlier, RNA and alkaline phosphatase activities increased in the outer epithelium of *L. stagnalis;* the cells also became taller (Timmermans, 1973). Timmermans (1973) related the alkaline phosphatase activity to calcium deposition. The amount of calcium in *Anodonta* mantle increased and reached a maximum 24–48 h after injury (Saleuddin, 1967; Tsujii, 1976). This was also observed in *H. pomatia* (Abolins-Krogis, 1963a) as the result of the increase in number of calcium cells and subsequent disintegration of their calcium spherules within the mantle tissue. Acid mucopolysaccharides and proteinaceous granules are also reported to be liberated by disruption of pigment granules and other intracellular bodies in the epithelium and transported onto the mantle surface (Abolins-Krogis, 1963a).

B. Amoebocytes

Besides mantle epithelium, amoebocytes, calcium cells, and hepatopancreas are believed to be directly involved in shell repair. In almost all cases, amoebocytes migrate onto the outer surface of the mantle epithelium at the early stage of shell repair (Wagge 1951, 1952; Wagge and Mittler, 1953; McGee-Russel, 1957; Durning, 1957; Abolins-Krogis, 1963a, 1968, 1973, 1976; Beedham, 1965; Kapur and Gupta, 1970; Watabe and Blackwelder, 1976; Uozumi and Suzuki, 1979). This differs from Kawaguti and Ikemoto (1962) and Tsujii (1976) who reported that amoebocytes adhered to the basal part of the outer epithelium during shell repair in *M. senhousia* and *Anodonta*. Also, Saleuddin (1970) rarely noticed amoebocytes in the mantle of *Helix*.

The reports on the function and the time of migration of amoebocytes are controversial (Saleuddin, 1980). According to Beedham (1965), amoebocytes pass through breaks in the damaged epithelium and form a protective barrier for the mantle before the formation of organic membrane in *Anodonta*. Abolins-Krogis (1963a) has made somewhat similar observations in *H. pomatia;* the amoebocytes moved out from the mantle along with acid mucopolysaccharides, proteins, and other materials. Barrier formation was not mentioned. Other observations made in *Euplecta indica* (Kapur and Gupta, 1970), in *P. paludosa* (Watabe and Blackwelder, 1980), and in *Brachydontes exustus* (Watabe and Blackwelder, 1976) showed that the migration of the amoebocytes occurred after the formation of organic membrane at the beginning of repair.

The sequence of structural changes of the mantle epithelium in relation to the amoebocytes and crystal formation in *Pomacea paludosa* is as follows (Watabe and Blackwelder, 1980): During the first 3 days, some epithelial cells showed progressively degenerative alteration with reduction of microvilli and mitochondria and increase in the number of vesicles, multivesicular bodies, and vacuoles. The organic membrane was formed, covering the epithelium. The intercellular spaces were widened and cell dissociation became evident. At a later stage, the detached cells degenerated and polygonal calcite crystals developed in association with the cells and cellular debris, but not intracellularly. After the cell detachment and calcite formation, amoebocytes migrated out through the opening in the epithelium and formed a protective barrier over the epithelium. Calcium cells with spherules were also present close to the epithelial surface. In about 7 days the amoebocytes degenerated, and the extrapallial space was filled with these degenerating cells and cellular debris. Aragonite spherulites developed in fibrous materials in the debris and eventually formed a spherulitic layer.

Amoebocytes have been reported to transport calcium and other repair material to the sites of shell repair (Wagge, 1951, 1952; Wagge and Mittler, 1953; Dunachie, 1963; Abolins-Krogis, 1963a, 1968; Kapur and Gupta, 1970; Tsujii, 1960, 1976). In contrast, McGee-Russel (1957) and Durning (1957) did not assign them any role in the repair process. Timmermans (1973) and Watabe and Blackwelder (1980) did not detect significant amounts of calcium in the amoebocytes. As Wilbur (1964) and Beedham (1965) assume, the amoebocytes may play a limited role in the early stage of shell repair, but the majority of the repair process is accomplished by the mantle epithelium (see also Uozumi and Suzuki, 1979). Abolins-Krogis (1976) also reported in *H. pomatia* that although amoebocytes released proteoglycan granules, lipid-containing vesicles, lipofuchsin pigment granules, and calcium in the repair membrane, the bulk of substances needed for the complete restoration of the damaged shell came from other sources in the body.

C. Calcium Sources for Shell Repair

Calcium cells are present in the connective tissue of the mantle, foot, hepatopancreas, and the epithelium of the mantle and contain calcium carbonate or calcium phosphate spherules (Watabe et al., 1976; Simkiss 1976a; Sminia et al., 1977). The calcium for shell repair is derived from the calcium cells of the mantle (Durning, 1957; Guardabassi and Piacenza, 1958; Tsujii, 1960; Abolins-Krogis, 1963a, 1968; Watabe et al., 1976) and foot (Watabe et al., 1976). At the onset of shell repair, calcium spherules in the mantle connective tissue were shown to dissolve, and calcium movement was followed histochemically from the connective tissue spaces to the basal and apical portion of the mantle epithelium through cells. Extrusion of the spherules through breakage of the epithelium into the extrapallial space is also conceivable (Watabe and Blackwelder, 1980). This intracellular pathway of calcium is different from the normal shell formation in which the intercellular space may be one route of calcium transport (Neff, 1972; Simkiss, 1974, 1976a,b; Watabe and Blackwelder, 1976; Simkiss and Wilbur, 1977). Calcium in the calcium cells in the hepatopancreas is also considered to be the source for shell repair (Sioli, 1935; Wagge, 1951, 1952; Abolins-Krogis, 1960, 1963a,b, 1968; Sioli, did not specifically refer to the "calcium cells.") However, Burton (1972), Campbell and Boyan (1976), and Whitehead and Saleuddin (1978) showed hepatopancreas calcium spherules were not greatly mobilized for shell formation or repair. Timmermans (1973) assumed that calcium for shell repair in *L. stagnalis* was obtained directly from the water or food and not much from the calcium cells in the mantle and digestive gland (hepatopancreas).

The mantle epithelium also contains calcium, which is utilized for shell repair (Wilbur, 1964). Another source of at least a portion of calcium for shell repair is the existing shell (Sioli, 1935; Wagge, 1951, 1952; Chan and Saleuddin, 1974b; Watabe et al., 1976; Whitehead and Saleuddin, 1978; Reed-Miller, 1982).

In addition to calcium, spherules of hepatopancreas have been reported to release materials such as proteinaceous granules (*a-granules* and *b-granules*) (Abolins-Krogis, 1960, 1963a,b, 1968, 1980), two electrophoretically identified protein components (Saleuddin et al., 1970) and other materials (Wagge, 1951, 1952; Wagge and Mittler, 1953), which are transported to the sites of shell repair. These materials are thought to contribute only a part of the total requirement for the components of the repaired shell (Whitehead and Saleuddin, 1978).

VI. Crystal Initiation in Shell Repair

The role of organic matrix in crystal nucleation has been discussed frequently in biological calcification (see Wilbur, 1976, 1980; Wilbur and Simkiss, 1979; Watabe, 1981a,b, 1983; Chapter 6, this volume). At the early stage of shell repair, specific organic granules are reported to induce crystal formation. These granules include the "metachromatic organic matter" described by Wada (1961, 1980) that is formed on the coverslip inserted between the mantle and shell of bivalves, the organic spherules in *H. pomatia* reported by Abolins-Krogis (1963b) and the organic spherules in the conchiolin layer in *M. edulis* observed by Uozumi and Suzuki (1979).

The metachromatic granules contain sulfated acid mucopolysaccharides, the sulfated groups of which are negatively charged and thought to bind calcium, thus inducing crystal nucleation. The surfaces of granules may also have exposed carboxyl groups of aspartic acid residues and provide negative charges (Wada, 1980). These concepts of nucleation have been proposed by Crenshaw and Ristedt (1976) and Weiner and Hood (1975). On the other hand, Abolins-Krogis (1976) thought that the lipid or phospholipid components of the b-granules within the cores of the organic spherules and in the vesicles of the repair membrane served as loci of calcification in *H. pomatia*. She showed that the demineralized shell-repair membrane of *H. pomatia* could be recalcified *in vitro* in *H. pomatia* saline containing $CaCl_2$ and $NaHCO_3$. The recalcification was inhibited when the membrane was inactivated or deprived of lipids (Abolins-Krogis, 1979a). Addition of phospholipids gave stimulatory (although slow) effects on recalcification (Abolins-Krogis, 1979b). She suggested that the cytoplasmic vesicles transported to the repair membrane were responsible for the accumulation of calcium and nucleation of crystals. She further speculated that lipids were involved in these processes, as in the case of vertebrate hard tissue mineralization (Abolins-Krogis, 1979a). However, it is yet to be seen whether the nucleator for the calcium carbonate in the invertebrates is a similar material as that for the calcium phosphate in the vertebrate system.

VII. Factors Controlling Shell Repair

A. Control of the Rate of Shell Repair

Much of our knowledge of the rate of shell repair comes from work with the repair at shell edges. The region of the damaged area, conditions of the animal, and environmental parameters (e.g., temperature, salinity, etc.) all affect the rate.

In *Crassostrea virginica* (Loosanoff and Nomejko, 1955) and *L. stagnalis* (Timmermans, 1973) repair at the shell edges was faster than the normal shell growth. Similarly, the rate of ^{45}Ca deposition in the repair of shell edge was higher than that in the normal shell growth in *Helisoma duryi* (F. Farley, personal communication). Kobayashi (1951) drilled holes at different areas of the shell edges of the Japanese pearl oyster *Pinctada martensii* (Fig. 4), and the oysters were checked every day for the period of 1 yr to see if the holes were filled. When filled, the regenerated areas were removed for new repairs to occur, and the number of repairs per year was recorded. As shown in Table IV, holes No. 3 and 4 were repaired most frequently, that is, the repair rate was highest. The lowest two were No. 1 and 2. The positions of the holes No. 3 and 4 coincide with the region of bivalve shells that generally show the highest rate of linear growth (see Wilbur and Owen, 1964) and calcium deposition (Wheeler et al., 1975).

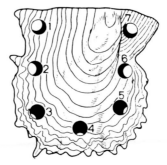

Fig. 4. Holes drilled at different portions of a shell of *Pinctada martensii* to measure the rate of shell regeneration. (From Kobayashi, 1951.)

Higher temperature regimes result in higher rates of shell repair in *Pinctada martensii* (Kobayashi, 1951), in *Helix* (Saleuddin and Chan, 1969), and in *Pomacea paludosa* (P. L. Blackwelder and N. Watabe, unpublished observations). No or almost no repair takes place in *P. martensii* when the water temperature falls below 13°C (Kobayashi, 1951), or 12–11°C (Miyauti, 1970).

Shell repair seems to show a circadian rhythm, and in *P. martensii* the rate was highest in the evening (7:30 P.M.) and lowest in the morning (7:30 A.M.) (Yuki, 1951). This trend has also been observed in the normal shell of *Helisoma duryi*, in which the linear rate of growth is higher in continuous darkness than in continuous light or on a 12L:12D photoperiod (Kunigelis and Saleuddin, 1978; Saleuddin et al., 1980).

Miyauti (1970) carried out an extensive series of experiments on the effects of environment and seasons on the shell repair rate in *P. martensii*. It was found

TABLE IV

The Frequency of Shell Repair per Year at Different Portions of the Shell in *Pinctada martensii*[a,b]

Animal number	Hole number						
	1	2	3	4	5	6	7
1	32	32	63	62	57	56	56
2	31	31	86	86	64	66	53
3	41	36	74	84	52	68	54
4	33	37	95	96	87	73	57
5	37	44	89	92	84	80	53
	34.8 ± 4.1[c]	36 ± 5.1[c]	81.4 ± 12.8[c]	84 ± 13.1[c]	68.8 ± 15.8[c]	68.6 ± 8.9[c]	54.4 ± 1.5

[a] See Fig. 4.
[b] From Kobayashi, 1951.
[c] Mean ± standard deviation.

that: (1) the cumulative shell repair in a year followed a sigmoid curve; (2) the rate was affected by the seasonal changes in water temperature, density, nutrients, turbidity, and other conditions of the environment; (3) the rate of shell repair was much more sensitive to environmental factors than the rate of normal shell growth; and (4) no difference was found in the rate among age groups of 2–5 yr.

In *Helisoma duryi* (S. Ford, personal communication), no difference was found in the rate of shell repair regardless of the rate of normal shell growth, or of the animals' age (up to 10 months) as long as they were fed. Interestingly, animals that had stopped growing before the injury frequently regenerated new shell beyond the point of complete repair, thus increasing the size of their shell over what had originally appeared to be its maximum size.

B. Control of the Morphology and Mineralogy of Repaired Shell

Very little is known about the factors determining the morphology and mineralogy of repaired shell. Meenakshi et al. (1974c) have experimentally shown in *O. lactea* that the morphology of the crystals in the repaired shell reflected the microtopography of the substrata surface put in contact with the mantle in the area of shell removal (Fig. 5a, b) (see also Wilbur, 1973). The fact that this mold-and-cast relationship was observed is remarkable in view of the fact that an organic membrane is usually deposited initially on the surface of the substrata, which would reduce the influence of the substrata. It may be that the matrix membrane was so thin that it reflected the substrata topography, which in turn controlled the repaired crystal morphology (Watabe and Dunkelberger, 1979). Judging from the micrographs of Meenakshi et al. (1974c), spherulites, which would have been the first crystals to be formed in the normal (i.e., without inserted substratum) shell repair, were absent from those repairs with substrata. Thus, the type of substrata may also affect the growth pattern of crystals.

Several factors which may affect crystal types in shell repair have been proposed. One is the chemical and structural characteristics of the organic matrix, which may induce epitaxial growth of aragonite or calcite crystals. It was shown *in vitro* and *in vivo* that the organic matrix from calcitic and aragonitic shell induced calcite and aragonite, respectively (Watabe and Wilbur, 1960; Wilbur and Watabe, 1963). In *Cepaea nemoralis,* the early shell repair consisted of a mixture of rhombohedral calcite and spherulitic aragonite (see Fig. 2). However, when the organic matrix from the nacreous layer (aragonitic) of *Haliotis* was used as the substratum for shell repair, only aragonite with the morphology typical of the nacreous layer was formed in the repaired shell (Fig. 6); likewise, matrix from the calcitic prismatic layer from *Pinna* species induced rhombohedral calcite (Fig. 7) (Watabe, 1981a,b). Although the effect occurred in only

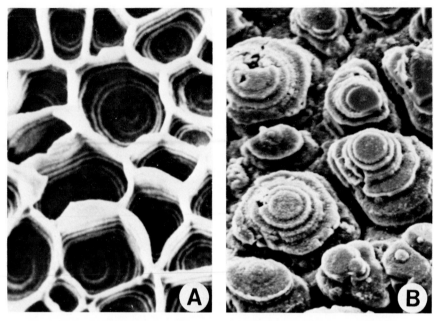

Fig. 5. (A) Honeycomb pattern of *Megalonais* periostracum. (2400×.) (B) Crystals deposited on the *Megalonais* periostracum inserted into regenerating *Otala*. Note the crystal morphology reflecting the honeycomb pattern of the periostracum (cf. Fig. 5A). (1600×.) (From Wilbur, 1973.)

one out of 10 animals in each case, this low percentage may be attributed to the formation of a new organic membrane on the substratum, thus blocking its possible effect on mineralogy. Meenakshi et al. (1974c) reported that the aragonite-to-calcite ratio in the shell repair of *O. lactea* varied among individuals and appeared to be independent of experimental substrata placed in the area of repair. However, this may also be attributed to intervention by the newly deposited organic membranes.

Temperature or inorganic ions are also considered to affect the mineralogy of repaired and normal shells (see Wilbur, 1964; Watabe, 1981a,b). It was shown in the shell repair of *Viviparus intertextus* that within the experimental temperature range of 13 to 25°C, the percentage of aragonite increased with temperature and the period of shell repair; calcite and vaterite percentages increased at low temperatures (Wilbur and Watabe, 1963). Similarly, the percentage of aragonite was higher (58%) at 27°C than at 18°C (45%) in *P. paludosa* at day 7 of the repair (P. L. Blackwelder and N. Watabe, unpublished observations). However, calcite and aragonite are present in separate layers in repaired shells of this species (see Table I), and the formation of aragonite spherulites, spherulitic

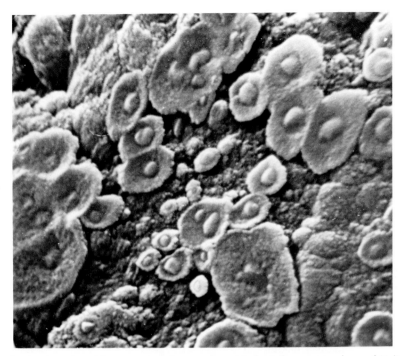

Fig. 6. Argonite tablets grown on the organic matrix from the nacreous layer of *Haliotis* during shell regeneration in *Cepaea nemoralis*. (8300×.) (From Watabe, 1981a.)

prismatic, and crossed lamellar layers was accelerated at higher temperature. Thus, the temperature effect was indirect in this case and seemed to enhance the recovery of the organism to normal conditions. On the other hand, Saleuddin and Chan (1969) reported that mostly aragonite crystals were formed at lower temperatures (5–15°C) while calcite crystals were predominant at higher (25–30°C) temperatures in the repaired shell of *Helix*. More detailed studies are needed to establish the effects of temperature on mineralogy during shell repair process.

Many *in vitro* studies exist on the effects of various inorganic ions on crystal types of calcium carbonate (see Wilbur, 1964; Kitano et al., 1976). Yet not much is known of these effects in *in vivo* systems. An experiment on *P. paludosa* (Watabe and Blackwelder, 1976, unpublished observations) showed that an increase of Sr^{2+} content by 0.25 to 0.50 mM or Mg^{2+} content by 2.6 to 5.0 mM in the water progressively increased the aragonite percentage from 45% in the control to 85% at day 7 of the shell repair. Increase of Sr^{2+} to 0.74 mM or Mg^{2+} to 8.0 mM did not cause a further increase of aragonite. Actually, this seemingly positive effect of Sr and Mg ions on the crystal type was found to be indirect, as in the case of the temperature effect. Scanning electron microscope

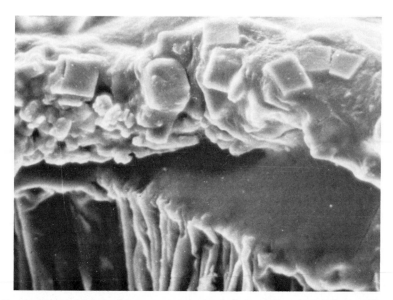

Fig. 7. Rhombohedral calcite grown on the matrix from calcitic prismatic layer of *Pinna* sp. during shell regeneration in *Cepaea nemoralis*. (2000×.) (From Watabe, 1981a.)

observation revealed that aragonite spherulites were already present at day 2 at higher Sr and Mg concentrations, whereas the repair shells contained only the calcitic foliated layer in the control. Thus, these ions apparently accelerated the repair process. Exactly how these ions acted on the animals' metabolism is unknown.

No information is available concerning the factors governing the formation and growth of vaterite crystals in shell repair.

C. On–Off Control Mechanisms

The ordered sequence of matrix and crystal layer formation, and the regularity of crystal type and morphology in each organism and each shell layer imply that shell repair as well as the normal shell formation are highly regulated processes. The cellular responses in the synthesis of organic matrix, transport and deposition of clacium must be well coordinated (Saleuddin, 1980) and also must be initiated and terminated by some precise control mechanisms. Wilbur (1976, 1980) suggested neurosecretory or hormonal control as one of them.

A number of investigations have recently been carried out on the controls for the repair at the shell edges in gastropods. The stimulus of shell damage, either mechanical or chemical (Sioli, 1935; Wilbur, 1964), is received by the sensory cells in the dorsal side of the mantle. Several types of receptors have been identified in gastropods (see Jones and Saleuddin, 1978 for references). In *H. duryi*, dorsal receptors are present at the transitional zone and could respond to the absence of shell, changed topography of the damaged surface, or loss of extrapallial fluid. They could initiate or monitor both shell repair and normal growth acting through the dorsal plexus, the peripheral nervous system, or the central nervous system (Jones and Saleuddin, 1978). Zylstra et al. (1978) also identified ciliated free nerve endings in the zone 4 (which corresponds to the transitional zone of Jones and Saleuddin, 1978) of the epithelium of *L. stagnalis* and *Biomphalaria pfeifferi* and assigned them similar functions, although chemoreception to detect the nature of extrapallial fluid is believed to be most likely.

Some neurosecretory responses during shell repair have been revealed in gastropods by several investigators. Geraerts (1976) suggested that the cerebral ganglia in *L. stagnalis* contain neurosecretory cells (light green cells) producing a shell growth hormone. Dillaman et al. (1976) assumed that in *H. duryi* one group of neurosecretory cells in the visceral ganglia was involved in the early stages of shell repair. They showed that paraldehyde–fuchsin-positive material was depleted from those cells between 24 and 48 h after the shell removal (from the edges), and reappeared by 72 h. The secretory material is considered to be transported in the pallial axon to the mantle edge gland, which is responsible for the shell repair and normal shell formation (Saleuddin and Dillaman, 1976). It was also shown that the growth hormone in *L. stagnalis* acted specifically on the mantle edge and stimulated the formation of the outer crystalline layer (Dogterom et al., 1979) and periostracum (Dogterom and Jentjens, 1980).

Injections of brain homogenates gave stimulatory effects in normal shell growth and shell repair, and the subesophageal fraction may contain materials to stimulate shell repair in *Helisoma duryi* (Kunigelis and Saleuddin, 1978) and *Helix pomatia* (Abolins-Krogis, 1980). Since injection of subesophageal brain fractions causes an increase in blood calcium level in *H. duryi*, one possible mechanism for control during shell repair may involve ionic regulation and transport of calcium to the damaged area (Saleuddin et al., 1980). Similar suggestions were made by Abolins-Krogis (1980) for *H. pomatia*. Release of the lipoprotein units serving for calcification "centra" was also considered by the same author (Abolins-Krogis, 1980).

Ecdysteroids have been proposed in the control of calcification and ecdysone has been found to increase deposition of ^{45}Ca in repaired shells in *Biomphalaria glabrata* (Whitehead, 1977). Ecdysterone injection into *H. aspersa* deprived of steroids from food causes an increasing incorporation of ^{45}Ca in the repaired shell at the expense of the intact shell (Whitehead and Saleuddin, 1978). It is suggested that these steroids mimic the action of a factor that is present after shell damage in normal circumstances. In normal shell formation in *L. stagnalis*, the shell growth hormone is considered to elevate the calcium concentration at the mantle edge (Dogterom et al., 1979) by means of specific calcium-binding proteins and to stimulate the formation of periostracum and outer calcified shell layer (Dogterom and Jentjens, 1980; Dogterom and van der Schors, 1980).

Inasmuch as the repair process at the shell edge is essentially similar to that of the normal shell growth, the control mechanism of the repair could simply be the release of the growth hormone or other neurosecretory substances to stimulate the mantle epithelium to form shell layers more or less similar to the normal. In contrast, the repair at the region where the mantle edge cannot retract involves substantial structural and histochemical alterations of the mantle epithelium and successive formation of several different shell layers, either normal or abnormal types. These layers are not normally formed by the epithelial cells covering that region of the shell. The control mechanisms to alter the epithelial structures and initiate and terminate the formation of different shell layers in orderly sequence are undoubtedly more complex than those for the shell edge repair. Several different groups of neurosecretory cells of endocrine gland(s) or neurohemal organ(s) releasing different substances may be responsible for the repair process. In most cases, the shell damage at this region in particular is accompanied by damage in the mantle epithelium (Watabe and Blackwelder, 1980). Accordingly, shell repair also involves mantle tissue repair. The extrusion of amoebocytes onto the mantle surface in fact may be related to this process as in the case of wound healing (Sminia et al., 1973). Probably these two types of repair are controlled by different, but rather well-coordinated, mechanisms. At present no studies exist concerning the control mechanisms relative to the repair in the mid-shell regions or in any region of bivalves.

Acknowledgment

I am most grateful to Dr. Karl M. Wilbur, Department of Zoology, Duke University, and Ms. Roni J. Kingsley, Marine Science Program, University of South Carolina, for their valuable suggestions in the preparation of the manuscript. Also, I thank the National Science Foundation and the Alexander von Humboldt Foundation (Federal Republic of Germany) for support of portions of the studies reported here.

References

Abolins-Krogis, A. (1958). The morphological and chemical characteristics of organic crystals in the regenerating shell of Helix pomatia L. Acta Zool. (Stockholm) **39**, 19–38.

Abolins-Krogis, A (1960). The histochemistry of the hepatopancreas of Helix pomatia (L.) in relation to the regeneration of the shell. Ark. Zool. **13**, 159–201.

Abolins, Krogis, A. (1963a). The histochemistry of the mantle of Helix pomatia (L.) in relation to the repair of the damaged shell. Ark. Zool. **15**, 461–474.

Abolins-Krogis, A. (1963b). On the protein stabilizing substances in the isolated b-granules and in the regenerating membranes of the shell of Helix pomatia (L.). Ark. Zool. **15**, 475–484.

Abolins-Krogis, A. (1968). Shell regeneration in *Helix pomatia* with special reference to the elementary calcifying particles. *Symp. Zool. Soc. London* **22**, 75–92.

Abolins-Krogis, A. (1973). Fluorescence and histochemical studies of the calcification-initiating lipofuscin type pigment granules in the shell-repair membrane of the snail, *Helix pomatia* L. *Z. Zellforsch. Mikrosk. Anat.* **142**, 205–221.

Abolins-Krogis, A. (1976). Ultrastructural study of the shell-repair membrane in the snail, *Helix pomatia* L. *Cell Tissue Res.* **172**, 455–476.

Abolins-Krogis, A. (1979a). *In vitro* recalcification of the demineralized shell-repair membrane of the snail, *Helix pomatia* L. *Cell Tissue Res.* **200**, 487–494.

Abolins-Krogis, A. (1979b). The effect of adenosine triphosphate, magnesium chloride and phospholipids on crystal formation in the demineralized shell-repair membrane of the snail, *Helix pomatia* L. *Cell Tissue Res.* **204**, 497–505.

Abolins-Krogis, A. (1980). The effects of ganglia extracts on shell regeneration in *Helix pomatia* L. *In* "The Mechanism of Biomineralization in Animals and Plants" (M. Omori, and N. Watabe, eds.), pp. 111–119. Tokai Univ. Press, Tokyo, Japan.

Andrews, E. A. (1934). Shell repair by the snail *Pleurodonte rostrata* Pfr. *Biol. Bull. (Woods Hole, Mass.)* **67**, 294–299.

Andrews, E. A. (1935). Shell repair by the snail *Neritina*. *J. Exp. Zool.* **70**, 75–107.

Beedham, G. E. (1965). Repair of the shell in species of *Anodonta*. *Proc. Zool. Soc. London* **145**, 107–125.

Biedermann, W. (1902). Über die Bedeutung von Kristallisationsprozessen bei der Bildung der Skelett wirbelloser Tiere, namentlich, der Molluskenschalen. *Z. Allg. Physiol.* **1**, 154–208.

Blackwelder, P. L., and Watabe, N. (1977). Studies on shell regeneration. II. The fine structure of normal and regenerated shell of the freshwater snail *Pomacea paludosa*. *Biomineralisation* **9**, 1–10.

Board, R. G., and Perott, H. R. (1979). Vaterite a constituent of the eggshells of the nonparasitic cuckoos, *Guira guira* and *Crotophagi ani*. *Calcif. Tissue Int.* **29**, 63–69.

Burton, R. F. (1972). The storage of calcium and magnesium phosphate and of calcite in the digestive glands of the Pulmonata. *Comp. Biochem. Physiol. A* **43**, 655–663.

Campbell, J. W., and Boyan, B. D. (1976). On the acid-base balance of gastropod molluscs. *In* "The Mechanisms of Mineralization in the Invertebrates and Plants" (N. Watabe, and K. M. Wilbur, eds.), pp. 109–133. Univ. of South Carolina Press, Columbia.

Carlstrom, D. (1963). Crystallographic study of vertebrate otoliths. *Biol. Bull. (Woods Hole, Mass.)* **125**, 441–463.

Chan, J. F. Y., and Saleuddin, A. S. M. (1974a). Acid phosphatase in the mantle of the shell regenerating snail *Helisoma duryi duryi*. *Calcif. Tissue Res.* **15**, 213–220.

Chan, W., and Saleuddin, A. S. M. (1974b). Evidence that *Otala lactea* (Müller) utilizes calcium from the shell. *Proc. Malacol. Soc. London* **41**, 195–200.

Crenshaw, M. A., and Ristedt, H. (1976). The histochemical localization of reaction group in septal nacre from *Nautilus pompilius* L. *In* "The Mechanism of Mineralization in the Invertebrates and Plants" (N. Watabe, and K. M. Wilbur, eds.), pp. 335–367. Univ. of South Carolina Press, Columbia.

Davis, G. M. (1964). Shell regeneration in *Oncomelania formosana*. *Malacologia* **2**, 145–158.

Dillaman, R. M., Saleuddin, A. S. M., and Jones, G. M. (1976). Neurosecretion and shell regeneration in *Helisoma duryi* (Mollusca: Pulmonata). *Can. J. Zool.* **54**, 1771–1778.

Dogterom, A. A., and Jentjens, T. (1980). The effect of the growth hormone of the pond snail *Lymnaea stagnalis* on periostracum formation. *Comp. Biochem. Physiol. A* **66**, 687–690.

Dogterom, A. A., and van der Schors, R. C. (1980). The effect of growth hormone of *Lymnaea stagnalis* on (Bi) carbonate movements, especially with regard to shell formation. *Gen. Comp. Endocrinol.* **41**, 334–339.

Norimitsu Watabe

Dogterom, A. A., van Loenhout, H., and van der Schors, R. C. (1979). The effect of the growth hormone of *Lymnaea stagnalis* on shell calcification. *Gen. Comp. Endocrinol.* **39**, 63–68.

Dunachie, J. F. (1963). The periostracum of *Mytilus edulis*. *Trans. R. Soc. Edinburgh* **65**, 383–410.

Durning, W. C. (1957). Repair of a defect in the shell of the snail *Helix aspersa*. *J. Bone Jt. Surg. Am. A* **39**, 377–393.

Geraerts, W. P. M. (1976). Control of growth by the neurosecretory hormone of the light green cells in the freshwater snail *Lymnaea stagnalis*. *Gen. Comp. Endocrinol.* **29**, 61–71.

Gould, R. W. (1972). Brown pelican eggshells: X-ray diffraction studies. *Bull. Environ. Contam. Toxicol.* **8**, 84–88.

Guardabassi, A., and Piacenza, M. L. (1958). Le manteau de l'escargot *Helix pomatia*. Étude cytologique et histochimique. *Arch. Anat. Microsc. Morphol. Exp.* **47**, 25–46.

Hall, A., and Taylor, J. D. (1971). The occurrence of vaterite in gastropod egg shells. *Mineral Mag. (London)* **38**, 521–522.

Jones, G. M., and Saleuddin, A. S. M. (1978). Ultrastructural observations on sensory cells and the peripheral nervous system in the mantle edge of *Helisoma duryi* (Mollusca:Pulmonata). *Can. J. Zool.* **56**, 1807–1821.

Kapur, S. P., and Gupta, A. S. (1970). The role of amoebocytes in the regeneration of shell in the land pulmonate, *Euplecta indica* (Pfieffer). *Biol. Bull. (Woods Hole, Mass.)* **139**, 502–509.

Kawaguti, S., and Ikemoto, N. (1962). Electron microscopy on the mantle of a bivalve *Musculus senhousia*, during regeneration of shell. *Biol. J. Okayama Univ.* **8**, 31–42.

Kessel, E. (1933). Über die Schale von *Viviparus viviparus* L. und *Viviparus fasciatus* Müll. Ein Beitrage zum Strukturproblem der Gastropodenschale. *Z. Morphol. Oekol. Tiere* **27**, 129–198.

Kitano, Y., Kanamori, N., and Yoshioka, S. (1976). Influence of chemical species on the crystal type of calcium carbonate. *In* "The Mechanisms of Mineralization in the Invertebrate and Plants" (N. Watabe, and K. M. Wilbur, eds.), pp. 191–202. Univ. of South Carolina Press, Columbia.

Kobayashi, S. (1951). "Shinju-no-Kenkyu"-Studies of Pearls (S. Kobayashi and N. Watabe), pp. 119–120. Gihodo Press, Tokyo.

Kunigelis, S. C., and Saleuddin, A. S. M. (1978). Regulation of shell growth in the pulmonate gastropod *Helisoma duryi*. *Can. J. Zool.* **56**, 1975–1980.

Lagergren, C. (1962). Calcium carbonate precipitation in the pancreas, gallstones, and urinary calculi. *Acta Chir. Scand.* **124**, 320–325.

Loosanoff, V. L., and Nomejko, C. A. (1955). Growth of oysters with damaged shell-edges. *Biol. Bull. (Woods Hole, Mass.)* **108**, 151–159.

Mayer, F. K. (1931). Röntogenographische Untersuchungen an Gastropoden Schalen. *Jena. Z. Naturwiss.* **65**, 487–512.

McGee-Russel, S. M. (1957). Tissue for assessing histochemical methods for calcium. *Q. J. Microsc. Sci.* **98**, 1–8.

Meenakshi, V. R., Blackwelder, P. L., and Wilbur, K. M. (1973). An ultrastructural study of shell regeneration in *Mytilus edulis* (Mollusca:Bivalvia). *J. Zool.* **171**, 475–484.

Meenakshi, V. R., Blackwelder, P. L., and Watabe, N. (1974a). Studies on the formation of calcified egg-capsules of ampullarid snails. I. Vaterite crystals in the reproductive system and egg-capsules of *Pomacea paludosa*. *Calcif. Tissue Res.* **16**, 283–291.

Meenakshi, V. R., Martin, A. W., and Wilbur, K. M. (1974b). Shell repair in *Nautilus macromphalus*. *Mar. Biol. (Berlin)* **27**, 27–35.

Meenakshi, V. R., Donnay, G., Blackwelder, P. L., and Wilbur, K. M. (1974c). The influence of substrata on calcification patterns in molluscan shell. *Calcif. Tissue Res.* **15**, 31–44.

Meenakshi, V. R., Blackwelder, P. L., Hare, P. E., Wilbur, K. M., and Watabe, N. (1975). Studies on shell regeneration. I. Matrix and mineral composition of the normal and regenerated shell of *Pomacea paludosa*. *Comp. Biochem. Physiol. A* **50**, 347–351.

Meier, W., and Moenke, H. (1961). Über die Natur der Kalziumcarbonate in Gallensteinen. *Natur-wissenschaften* **48,** 521.

Miyauti, T. (1970). Studies of the method to judge the vitality of the Japanese pearl oyster, *Pteria (Pinctada) martensii* (Dunker). *Shinju-Gijustsu-Kenkukaiho* **8** (special issue), p. 221.

Neff, J. M. (1972). Ultrastructure of the outer epithelium of the mantle in the clam *Mercenaria mercenaria* in relation to calcification of the shell. *Tissue & Cell* **4,** 591–600.

Prenant, M. (1927). Les formes minéralologiques du calcaire chez les êtres vivants et le problème de leur déterminisme. *Biol. Rev.* **2,** 365–393.

Prenant, M. (1928). Contributions a l'étude cytologique du calcaire IV. La vaterité chez les animaux. *Bull. Biol. Fr. Belg.* **62,** 21–50.

Reed-Miller, C. (1982). Crystalline and amorphous calcium in the tissues of shell-regenerating snails. *Am. Zool.* **22,** 982.

Reed-Miller, C., Wise, S. W., Jr., and Siegel, H. J. (1980). Correlated techniques for identifying mineral phases and morphologies in regenerated archaeogastropod shell: a progress report. *In* "The Mechanisms of Biomineralization in Animals and Plants" (M. Omori, and N. Watabe, eds.), pp. 73–77. Tokai Univ. Press, Tokyo, Japan.

Saleuddin, A. S. M. (1967). The histochemistry of the mantle during the early stages of shell repair. *Proc. Malacol. Soc. London* **37,** 371–380.

Saleuddin, A. S. M. (1969). Isoenzymes of alkaline phosphatase in *Anodonta grandis* (Bivalvia-Unionidae) during shell regeneration. *Malacologia* **9,** 501–508.

Saleuddin, A. S. M. (1970). Electron microscopic study of the mantle of normal and regenerating *Helix. Can. J. Zool.* **48,** 409–416.

Saleuddin, A. S. M. (1980). Shell formation in molluscs with special references to periostracum formation and shell regeneration. *In* "Pathway in Malacology" (S. van der Spoel, A. C. van Bruggen, and J. Lever, eds.), pp. 47–81. Schellema & Holkema, Utrecht, The Netherlands.

Saleuddin, A. S. M., and Chan, W. (1969). Shell regeneration in *Helix:* shell matrix composition and crystal formation. *Can. J. Zool.* **47,** 1107–1111.

Saleuddin, A. S. M., and Dillaman, R. M. (1976). Direct innervation of the mantle edge gland by the neurosecretory axons in *Helisoma duryi* (Mollusca:Pulmonata). *Cell Tissue Res.* **171,** 397–401.

Saleuddin, A. S. M., and Hare, P. E. (1970). Amino acid composition of normal and regenerated shell of *Helix pomatia. Can. J. Zool.* **48,** 886–888.

Saleuddin, A. S. M., and Wilbur, K. M. (1969). Shell regeneration in *Helix pomatia. Can. J. Zool.* **47,** 51–53.

Saleuddin, A. S. M., Kunigelis, S. C., Khan, H. R., and Jones, G. M. (1980). Possible control mechanisms in mineralization in *Helisoma duryi. In* "The Mechanisms of Biomineralization in Animals and Plants" (M. Omori, and N. Watabe, eds.), pp. 121–129. Tokai Univ. Press, Tokyo, Japan.

Saleuddin, A. S. M., Miranda, E., Losada, F., and Wilbur, K. M. (1970). Electrophoretic studies on blood and tissue proteins of normal and regenerating *Ampullarius glaucus* (Gastropoda). *Can. J. Zool.* **48,** 495–499.

Simkiss, K. (1974). Calcium translocation by cells. *Endeavour* **33,** 119–123.

Simkiss, K. (1976a). Intracellular and extracellular routes in biomineralization. *Symp. Soc. Exp. Biol.* **30,** 423–424.

Simkiss, K. (1976b). Cellular aspects of calcification. *In* "The Mechanisms of Mineralization in the Invertebrates and Plants" (N. Watabe, and K. M. Wilbur, eds.), pp. 1–32. Univ. of South Carolina Press, Columbia.

Simkiss, K., and Wilbur, K. M. (1977). The molluscan epidermis and its secretion. *Symp. Zool. Soc. London* **39,** 35–76.

Sioli, H. (1935). Über den Chemismus der Reparatur von Schalendefekten bei *Helix pomatia. Zool. Jahrb. Allg. Zool. Physiol. Tiere* **54**, 507–534.

Sminia, T., Pietersma, K., and Scheerboom, J. E. M. (1973). Histological and ultrastructural observations on wound healing in the freshwater pulmonate *Lymnaea stagnalis. Z. Zellforsch. Mikrosk. Anat.* **141**, 561–573.

Sminia, T., de With, N. D., Bos, J. L., van Nieuwmegen, M. E., Witter, M. P., and Wondergem, J. (1977). Structure and functions of the calcium cells of the freshwater pulmonate snail *Lymnaea stagnalis. Neth. J. Zool.* **27**, 195–208.

Sutor, J. D., and Wooley, S. E. (1968). Gallstones of unusual composition: Calcite, Aragonite, and Vaterite. *Science (Washington, D.C.)* **159**, 1113–1114.

Timmermans, L. P. (1973). Mantle activity following shell injury in the pond snail, *Lymnaea stagnalis* L. *Malacologia* **14**, 53–61.

Tsujii, T. (1960). Studies on the mechanisms of shell and pearl formation in mollusca. *J. Fac. Fish. Prefect. Univ. Mie* **5**, 1–70.

Tsujii, T. (1968). Studies on the mechanisms of shell and pearl formation. XI. The submicroscopical observation on the mechanism of formation of abnormal pearls and abnormal shells. *Rep. Fac. Fish. Prefect. Univ. Mie* **6**, 59–77.

Tsujii, T. (1969). Studies on the mechanisms of shell– and pearl–formation. Electron microscopic determination of calcium in pearl sac and mantle epithelium in the pearl oyster, *Pteria (Pinctada) martensii* (Dunker). *In* "Hard Tissue, with Specific Reference to Tooth" (S. Araya, S. Ijiri, T. Kirino, T. Mimura, S. Suga, S. Takuma, and K. Wada, eds.), pp. 431–442. Ishiyaku, Tokyo, Japan.

Tsujii, T. (1976). An electron microscopic study of the mantle epithelial cells of *Anodonta* sp. during shell regeneration. *In* "The Mechanisms of Mineralization in the Invertebrates and Plants" (N. Watabe, and K. M. Wilbur, eds.), pp. 339–353. Univ. of South Carolina Press, Columbia.

Uozumi, S., and Ohta, S. (1977). Scanning electron microscopy and electron probe microanalysis of initial mineralization in marine shell regeneration. *Chishitsugaku Zasshi (Geol. Mag. [Tokyo])* **83**, 425–432.

Uozumi, S., and Suzuki, S. (1978). 'Organic membrane shell' in the early stage of regeneration in *Mytilus edulis* (Bivalvia). *Chikyu Kagaku (Earth Sci. [Tokyo])* **32**, 113–119.

Uozumi, S., and Suzuki, S. (1979). 'Organic membrane shell' and initial calcification in shell regeneration. *J. Fac. Sci. Hokkaido Univ. Ser. 4* **19**, 37–74.

von Levetzow, K. G. (1932). Die Struktur einiger Schneckenschalen und ihre Entstehung durch typisches and atypisches Wachstum. *Jena. Z. Naturwiss.* **66**, 41–105.

Wada, K. (1961). Crystal growth on molluscan shells. *Kokuritsu Shinju Kenkyusho Hokoku (Bull. Natl. Pearl Res. Lab [Jpn.])* **7**, 703–828.

Wada, K. (1964). Studies on the mineralization of the calcified tissue in molluscs. VIII. Behavior of eosinophil granules and of organic crystals in the process of mineralization of secreted organic matrices in glass coverslip preparations. *Kokuritsu Shinju Kenkyusho Hokoku (Bull. Natl. Pearl Res. Lab. [Jpn.])* **9**, 1087–1098.

Wada, K. (1980). Initiation of mineralization in bivalve molluscs. *In* "The Mechanism of Bio-mineralization in Animals and Plants" (M. Omori, and N. Watabe, eds.), pp. 79–92. Tokai Univ. Press, Tokyo, Japan.

Wagge, L. E. (1951). The activity of amoebocytes and of alkaline phosphatases during the regeneration of the shell in the snail, *Helix aspersa. Q. J. Microsc. Sci.* **92**, 307–321.

Wagge, L. E. (1952). Quantitative studies of calcium metabolism in *Helix aspersa. J. Exp. Zool.* **120**, 311–342.

Wagge, L. E., and Mittler, T. (1953). Shell regeneration in some British molluscs. *Nature (London)* **171**, 528–529.

Watabe, N. (1981a). Crystal growth of calcium carbonate in the invertebrates. *Prog. Cryst. Growth Charact.* **4**, 99–147.

Watabe, N. (1981b). Some problems on the structure and formation of calcium carbonate crystals and their aggregates in the invertebrates. *In* "Study of Molluscan Paleobiology" (T. Habe and M. Omori, eds.), pp. 35–46. Prof. Masae Omori Mem. Vol. Publ. Comm., Niigata Univ., Niigata, Japan.

Watabe, N. (1983). Mollusca: Shell. *In* "Biology of the Integument," Vol. 1. Invertebrates. (K. S. Richards, ed.), in press. Springer-Verlag, New York/Berlin.

Watabe, N., and Blackwelder, P. L. (1976). Studies on shell regeneration in molluscs. III. Effects of inorganic ions on shell mineralogy; ultrastructural changes of mantle epithelium. *Am. Zool.* **16**, 249.

Watabe, N., and Blackwelder, P. L. (1980). Ultrastructural and calcium localization in the mantle epithelium of the freshwater gastropod *Pomacea paludosa* during shell regeneration. *In* "The Mechanisms of Biomineralization in Animals and Plants" (M. Omori, and N. Watabe, eds.), pp. 131–144. Tokai Univ. Press, Tokyo, Japan.

Watabe, N., and Dunkelberger, D. G. (1979). Ultrastructural studies on calcification in various organisms. *Scanning Electron Microscopy 1979 II*, 403–416.

Watabe, N., and Wilbur, K. M. (1960). Influence of organic matrix on crystal type in the molluscs. *Nature (London)* **188**, 334.

Watabe, N., Meenakshi, V. R., Blackwelder, P. L., Kurtz, E. M., and Dunkelberger, D. G. (1976). Calcareous spherules in the gastropod, *Pomacea paludosa. In* "The Mechanisms of Mineralization in the Invertebrates and Plants" (N. Watabe, and K. M. Wilbur, eds.), pp. 283–308, Univ. of South Carolina Press, Columbia.

Weiner, S., and Hood, L. (1975). Soluble protein of the organic matrix of mollusk shells: a potential template for shell formation. *Science (Washington, D.C.)* **190**, 987–989.

Wheeler, A. P., Blackwelder, P. L., and Wilbur, K. M. (1975). Shell growth in the scallop *Argopecten irradians*. I. Isotopic incorporation with reference to diurnal growth. *Biol. Bull. (Woods Hole, Mass.)* **148**, 172–183.

Whitehead, D. L. (1977). Steroid enhance shell regeneration in an aquatic gastropod (*Biomphalaria glabrata*). *Comp. Biochem. Physiol. C* **58**, 137–141.

Whitehead, D. L., and Saleuddin, A. S. M. (1978). Steroids promote shell regeneration in *Helix aspersa* (Mollusca: Pulmonata). *Comp. Biochem. Physiol. C* **59**, 5–10.

Wilbur, K. M. (1964). Shell formation and regeneration. *In* "Physiology of Mollusca" (K. M. Wilbur, and C. M. Yonge, eds.), Vol. 1, pp. 243–282. Academic Press, New York.

Wilbur, K. M. (1973). Mineral regeneration in echinoderms and molluscs. *Ciba Found. Symp.* **11** (new ser.), 7–33.

Wilbur, K. M. (1976). Recent studies on invertebrate mineralization. *In* "The Mechanisms of Mineralization in the Invertebrates and Plants" (N. Watabe, and K. M. Wilbur, eds.) pp. 79–108. Univ. of South Carolina Press, Columbia.

Wilbur, K. M. (1980). Cells, crystals and skeletons. *In* "The Mechanisms of Biomineralization in Animals and Plants" (M. Omori, and N. Watabe, eds.), pp. 3–11. Tokai Univ. Press, Tokyo, Japan.

Wilbur, K. M., and Owen, G. (1964). Growth. *In* "Physiology of Mollusca" (K. M. Wilbur, and C. M. Yonge, eds.), Vol. 1, pp. 211–242. Academic Press, New York.

Wilbur, K. M., and Simkiss, K. (1979). Carbonate turnover and deposition by metazoa. *In* "Biochemical Cycling of Mineral Forming Elements" (P. A. Trudinger, and D. J. Swaine, eds.), pp. 69–106. Elsevier, New York.

Wilbur, K. M., and Watabe, N. (1963). Experimental studies on calcification in molluscs and the alga *Coccolithus huxleyi. Ann. N.Y. Acad. Sci.* **109**, 82–112.

Wong, V., and Saleuddin, A. S. M. (1972). Fine structure of normal and regenerated shell of *Helisoma duryi duryi. Can. J. Zool.* **50,** 1563–1568.

Yuki, R. (1951). ''Shinju-no-Kenkyu''—Studies of Pearls (S. Kobayashi and N. Watabe). p. 130. Gihodo Press, Tokyo.

Zylstra, U., Boer, H. H., and Sminia, T. (1978). Ultrastructure, histology, and innervation of the mantle edge of the freshwater pulmonate snail *Lymnaea stagnalis* and *Biomphalaris pfeifferi. Calcif. Tissue Res.* **26,** 271–282.

8

Endocrinology

J. JOOSSE

Department of Biology
Free University
Amsterdam, The Netherlands

W. P. M. GERAERTS

Department of Biology
Free University
Amsterdam, The Netherlands

THE MOLLUSCA, VOL. 4
Physiology, Part 1

I. Introduction

Defined in classical terms, *hormones* are substances that are secreted in minute quantities by one part of the body and transported to other parts, the target organs, in which they evoke physiological responses. In vertebrates the production centers of hormones can be distinguished in groups of nerve cells (neurosecretory cells, or NSC) and endocrine organs. Many neurohormones released by NSC control the activity of endocrine organs, whereas others act directly on target tissues. In molluscs, the number of organs that are now known to produce hormones is much lower compared to the vertebrates. On the other hand, there is now good evidence to suggest that the diversity of the neuroendocrine system is not very different from that of the vertebrates. Consequently, the majority of the neurohormones of molluscs act diretly on target tissues.

The number of functions of hormones/neurohormones of molluscs known at the moment is still rather small. In contrast, the basic mechanisms of some types of neurosecretory cells are known in great detail. This is especially due to the fact that in some gastropods these neurons, like their conventional neurons, are giant cells. For this reason, they are highly suitable for neurophysiological studies. The great progress recently made in this field, and the importance of this work for neuroendocrinology in general, are the reasons much attention is paid to these results in this chapter.

It is impossible to present a complete survey of all the data in the literature on molluscan endocrinology. Only those that are necessary to establish a comprehensive view of the present situation will be mentioned.

As will become clear, very little is known of the occurrence or functions of hormones in molluscs other than gastropods and cephalopods. This is regrettable, particularly with regard to the Bivalvia, as so much is already known of the physiology of these animals. This absence of clear data is mainly due to the enormous technical problems encountered in applying the classical endocrinological techniques (operations and transplantations) to these animals.

For reviews particularly devoted to molluscan endocrinology, see Gabe (1966), Simpson et al. (1966), Durchon (1967), Martoja (1968), Joosse (1972, 1975, 1976, 1979a, 1979b), Golding (1974), and Boer and Joosse (1975). For those of the whole field of invertebrate endocrinology, see Highnam and Hill (1977), Durchon and Joly (1978), Maddrell and Nordmann (1979), and Goldsworthy et al. (1981).

II. Histology of Hormone-Producing Centers

A. Dorsal Bodies

These organs are found in all freshwater (Basommatophora) and terrestrial (Stylommatophora) pulmonate gastropods studied so far. Early reports on their presence are from de Lacaze Duthiers (1872). Lever (1957, 1958) was the first to suggest an endocrine role for the dorsal bodies (DB) in freshwater pulmonates.

DB are so named because of their dorsal position on the cerebral ganglia (Fig. 1). In the majority of basommatophoran species the paired mediodorsal bodies (MDB) are located dorsally on the origins of the intercerebral commissure. In planorbids the two bodies form one continuous organ. In a number of species, a pair of laterodorsal bodies (LDB) is also found. All possible evidence points to one common function for all DB (see below). For reports on DB in Basommatophora, see Lever (1957, 1958), Joosse (1964), Röhnisch (1964), Simpson et al. (1966), Boer et al. (1968), and Boer and Joosse (1975).

In the Stylommatophora the situation is quite different. Compact DB, similar to those of the Basommatophora, have been observed in only a few species, e.g., *Succinea putris* (Cook, 1966; Lever, 1958), *Strophocheilus oblongus* (Kuhlmann, 1966), and *Acavus phoenix* (Boer and Joosse, 1975). In all other species studied, the DB consist of groups of cells dispersed in the relatively thick sheath of connective tissue surrounding the cerebral ganglia (Cook, 1966; Kuhlmann, 1966; Nolte, 1965; Van Mol, 1967; Wijdenes, 1981).

Structures similar to the DB of the pulmonates have been reported for prosobranchs and opisthobranchs by Martoja (1965).

Dorsal body cells (DBC) are most probably mesodermal in origin (Boer et al., 1968). The DB of the Basommatophora are well-defined organs, delimited by a connective tissue capsule. Histologically, DB can be divided into a cortex, consisting of groups of cell bodies and a medulla, which is primarily composed of DBC processes (Boer et al., 1968, 1976; Joosse, 1964; Lever, 1958).

The dorsal body hormone (DBH) is most probably confined in small (70–90 nm), moderately electron-dense, membrane-limited granules, which are particularly numerous in the DBC processes, where the hormone is released by exocytosis (Boer et al., 1968, 1976; Roubos et al., 1980; Simpson, 1969). At the ultrastructural level, the DBC are characterized by the presence of great numbers of mitochondria, with prominent cristae (e.g., Boer et al., 1968). In *Lymnaea stagnalis* the DBC show a hyperplasmic stage just prior to the reproductive season (Joosse, 1964). Concurrent increase in the number of Golgi zones is also seen (J. Van Minnen, unpublished observations) which may indicate increased hormone production.

Great attention has been given to the possible morphological contact of the DB with the neuronal elements in the cerebral ganglia. In light microscopic investigations of some freshwater pulmonates, the projections of the DBC appear to be directed to a central area where they are continuous with projections of neurosecretory neurons. This is particularly clear in planorbid snails, in which the perineurium splits in an area centrally located under the MDB to form a small disk. Lever (1958) and Simpson et al. (1966) suggested that those contacts could indicate a physiological relationship of a neuroendocrine nature. This view was not shared by Joosse (1964), because in *L. stagnalis* the DB are clearly separated

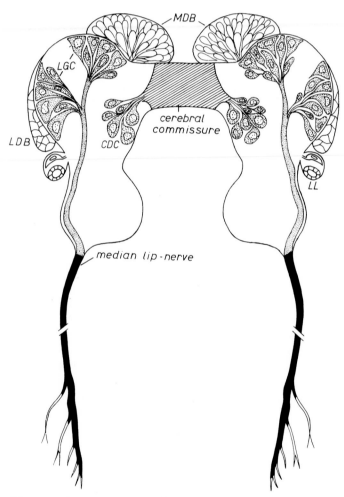

Fig. 1. Diagrammatic transverse section through the cerebral ganglia of *L. stagnalis*, showing the location of the various neuroendocrine and endocrine centers. CDC, Caudodorsal cells; LGC, light green cells; LL, lateral lobes; MDB and LDB, mediodorsal and laterodorsal bodies. The periphery of the cerebral commissure is the neurohemal area of the CDC. The peripheries of the median lip nerves are the neurohemal areas of the LGC.

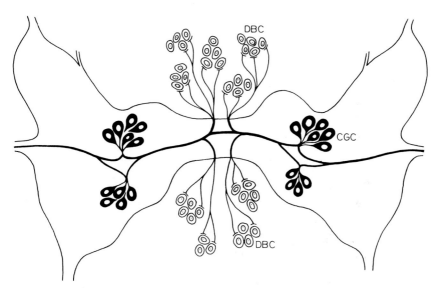

Fig. 2. Tentative schematic representation of the synapse-like innervation of dispersed groups of dorsal body cells (DBC), located in the connective tissue sheath around the cerebral ganglia of terrestrial pulmonates, by neurites of the neurosecretory cerebral green cells (CGC). The CGC have also projections running in the median lip nerves. (Based on Nolte, 1978, and Wijdenes and Vincent, 1981.)

from the adjacent NSC. Subsequent histochemical and ultrastructural investigations revealed that in planorbid as well as other basommatophoran snails the DB and the neuronal elements are separated at least by a basement membrane (Boer et al., 1968; Simpson, 1969; Simpson et al., 1966). Therefore, an innervation of the DBC could not be demonstrated in the freshwater pulmonates studied so far.

In contrast, early studies of terrestrial pulmonates carried out by Nolte (1965), Kuhlmann (1966), and Van Mol (1967) demonstrated the presence of small commissural nerves branching in the DB area. Nolte (1978) showed that in *Theba pisana* axons are in close contact with processes in DBC (Fig. 2), and in *Helix aspersa* Wijdenes and Vincent (1981) observed synapse-like structures between axons of commissural nerves and DBC (Fig. 3). The latter authors suggested that these axons originate from neurosecretory cerebral green cells. These observations leave little doubt that the DBC of Stylommatophora are at least partly under nervous control.

B. The Gonad

From experimental studies reported below, it is clear that in some groups of gastropods the gonad is the source of hormone(s) (see Section III,C,3). As in vertebrates, the function of these hormones may be control of gamete produc-

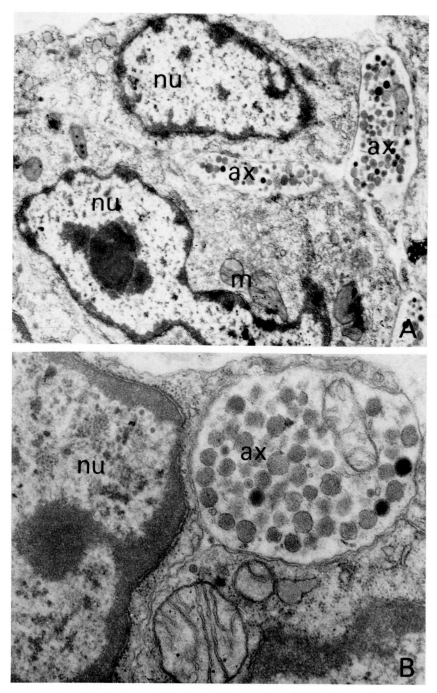

Fig. 3. (A) Synapse-like contacts of neuroendocrine cerebral green cells, characterized by the presence of electron-dense and less electron-dense elementary granules in their axons (ax) with dorsal body cells, characterized by their nuclei (nu) and mitochondria (m). (15,000X.) (B) Detail of the same type of contact. (45,000X.) (Courtesy of Dr. J. Wijdenes.)

tion, synchronization of gonadal and accessory sex gland activities, feedback control of the gonadotropic center(s), or control of reproductive behavior. Many authors have demonstrated in gonadal tissue the synthetic capacity for or the presence of steroids (e.g., in gastropods: Bardon et al., 1971; de Jong-Brink et al., 1981; Gottfried and Dorfman, 1969, 1970a,b,c; Lehoux and Williams, 1971; Lupo di Prisco and Dessi Fulgheri, 1975; Rohlack, 1959; Takeda, 1979; Teshima and Kamazawa, 1971; in bivalves: Idler et al., 1969; Longcamp et al., 1974; Saliot and Barbier, 1971; and in cephalopods: Carreau and Drosdowsky, 1977). Steroid-synthesizing capacity is not restricted to gonadal tissue. Krusch et al. (1979) report a steroid-synthesizing capacity in several other tissues (DB, ovotestes, and buccal ganglia) of *H. pomatia*.

The identification of specific hormone-producing cell types in the gonad of molluscs appears to be very difficult. Nonsexual elements such as the Sertoli cells and the follicle cells might be involved in hormone production (Fig. 19). A distinct correlation between biochemical and histological data could be shown by de Jong-Brink et al. (1981). They found simultaneous diurnal activity changes of steroid-synthesizing enzymes and the extension of the smooth endoplasmic reticulum (SER) in Sertoli cells carrying late male phase cells and in Sertoli cells after spermeation in the ovotestis of *L. stagnalis*. In this case, the steroids probably have only a local hormonal effect.

In conclusion, it is clear that further studies are needed to elucidate the endocrine role of the gonad and particularly to demonstrate the production centers of gonadal steroids in molluscs.

C. Optic Tentacles

Pelluet and Lane (1961) were the first to suggest an endocrine function of the optic tentacles of Stylommatophora. Since then, many controversial studies have been published in this field. However, the experiments of Wattez (1973, 1975, 1978, 1980) with *Arion subfuscus* clearly indicate that in this species, and thus most probably also in other terrestrial pulmonates, the optic tentacles contain and release a masculinizing substance.

Up to now, all efforts to identify the endocrine centers in the optic tentacles have failed. This is a great disadvantage for the reproductive endocrinology of pulmonate gastropods, as little is known about production centers of androgenic substances in these hermaphrodite snails. This contrasts clearly to the detailed knowledge of the control centers of female reproductive activity. To stimulate further research in this area, a survey of the morphological studies is presented.

The results of the investigations of Tuzet et al. (1957), Sanchez and Bord (1958), Lane (1962, 1964), Smith (1966), Bierbauer and Török (1968), Rogers (1969), and Bierbauer and Vigh-Teichmann (1970) suggest that in the optic tentacle a tentacular ganglion is located close to the eye. This ganglion is connected with the procerebrum of the cerebral ganglion, sending six nerve branches to the eye and the skin of the tip of the tentacle. Primarily on the basis of their reactions with classic neurosecretory stains, four cell types, located outside the ganglion, were assumed to be hormone-producing cells: the collar cells, two types of lateral cells, and the gland cells of the dermatomuscular layer. Further histochemical and electron microscopic studies indicate that all these cell types are subepidermal gland cells (e.g., Lane, 1964; Rogers, 1969; Röhlich and Bierbauer, 1966). Of these, the collar cells have an intimate relationship with the tentacular ganglion, which is clear from their fine cytoplasmic projections into the ganglion (Bierbauer and Török, 1968; Lane, 1964). However, the collar cells have a process, which runs between the epidermal cells to the surface of the tentacle where external secretion occurs, and their secretion granules differ greatly from neurosecretory elementary granules (Rogers, 1969). In the future, attention should be focused (as suggested by Boer and Joosse, 1975) on the tentacular ganglion, because surprisingly, no detailed studies have been devoted to this possibly neuroendocrine structure.

D. Optic Glands

From the elegant experiments of Wells and Wells (1959, 1969), we know that in the coleoid cephalopods the optic glands control reproductive activity. These glands are located upon the optic tracts, which connect the brain with the large optic lobes (Fig. 4). The glands are highly vascularized and are innervated by a nerve from the subpedunculate lobe of the supraesophageal part of the brain. The activity of the glands is controlled by an inhibitory nerve supply: Removing the subpedunculate lobe or cutting the nerves, which run to the glands, results in a rapid enlargement of the optic glands, which become bright orange. At the same time, sexual maturation starts (Wells and Wells, 1959). This can also be induced by implantation of optic glands in intact immature recipients (Wells and Wells (1975) or in organ cultures of optic glands with gonadal material (Durchon and Richard, 1967). These experimental data prove that the optic glands are endocrine organs involved in the control of reproduction. The structure and innervation of the optic glands have been studied in detail by various authors (Björkman, 1963; Boycott and Young, 1956; Defretin and Richard, 1967; Froesch, 1974, 1979; Mangold and Froesch, 1977; Nishioka et al., 1970; Wells and Wells, 1959). They contain only one type of granular cell, the stellate cell. In some species, these cells have long projections which run along the wall of blood capillaries but remain separated from the lumen by the pericytes of the capillaries (cf. Froesch, 1979). The stellate cells contain large numbers of tubular mitochondria and free ribosomes. Defretin and Richard (1967) gave a clear description of the presence of typical secretion granules (type α; ϕ, 100 nm), their origin from the Golgi zones, and their change in quantity in relation to the activity of

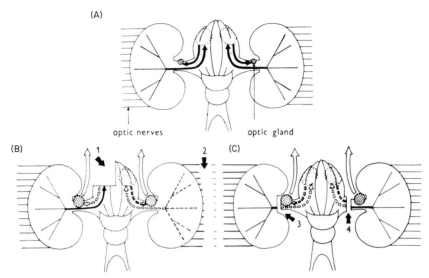

Fig. 4. Localization of the optic glands and the mechanism of hormonal control of gonad maturation in *Octopus*. (**A**) Situation in an immature, unoperated *Octopus*, in which secretion of the optic glands is held in check by an inhibitory nerve supply. (**B**) Two operations that cause the optic glands to secrete a product stimulating the gonad to enlarge: (1) removal of the source of the inhibitory nerve supply and (2) optic nerve section. (**C**) Further operations having the same effect on the gonads, thus eliminating the possibility that there is also excitatory innervation: (3) optic lobe removal and (4) optic tract section. Operation (3) produces enlargement at the same rate as optic nerve section and operation (4) at the same rate as subpedunculate lobe removal. (From Wells and Wells, 1977a.)

the gland in *Sepia officinalis*. On the other hand, Mangold and Froesch (1977) were unable to observe changes in activity of the cells in the optic glands of *Octopus vulgaris* and *Eledone moschata* at any moment in their life cycles. They suggest that the presence of lipofuscin and hemocyanin in the cells of the optic glands indicates that these organs have other functions as well. Froesch et al. (1978) showed that the stellate cells of *O. vulgaris* take up injected horse spleen ferritin. Froesch (1979) found that the stellate cells respond to a variety of foreign proteins by a mass production of particulated material. He suggested this to be the optic gland hormone. This hormone then would be a steroid, as it is not stored in granules but produced in cells rich in tubular mitochondria. The hormone would be involved not only in the control of sexual maturation but also in defense mechanisms against foreign proteins.

In conclusion, it can be said that the endocrine nature of the optic glands is undisputed, but that the opinions on the chemical structure of the hormone and its processing in the optic gland are still controversial. Moreover, the optic glands may also have nonendocrine functions.

E. Neuroendocrine Systems

1. General Aspects and New Developments

The synthesis and release of hormones by neurons is called *neurosecretion*. The neurons showing this phenomenon are *neurosecretory cells (NSC)*, and their products are called *neurohormones*. Initially, the demonstration of the presence of NSC was mainly the work of histologists. The first report on molluscs dates as far back as 1935 (B. Scharrer). The application of the Gomori methods, i.e., the chrome–hematoxylin and paraldehyde fuchsin staining techniques, based on the reactions with protein-bound cysteine, appeared to be rather specific for NSC. NSC were distinguished in Gomori-positive and Gomori-negative types. A great number of works on neurosecretion in molluscs, based on these techniques, were published between 1955 and 1965. Gabe's book, "Neurosecretion" (1966), presents many of these data. With these techniques, it was also possible to discover complete neurosecretory systems. A neurosecretory system includes the cell bodies, where synthesis of the neurohormone occurs; the neurosecretory tract, which is the bundle of axons transporting the neurosecretory material; and the neurohemal organ or area, consisting of the neurohemal endings from which the neurohormone is released into the blood.

The number of identified NSC types increased considerably after the introduction of the alcian blue–alcian yellow technique by Wendelaar Bonga (1970). Cell types differ by showing different shades of green and yellow. In *L. stagnalis,* 11 NSC types could be identifed. However, neurohemal areas of only some of these cell types could be identified.

Of great importance for the further development of neuroendocrinology has been the introduction of electron microscopic techniques. It appears that the NSC of all animal groups have basic characteristics in common: The secretion product is synthesized in rough endoplasmic reticulum (RER), packaged in membrane-bound elementary granules (generally with a size above 100 nm) by the Golgi apparatus, and finally released by exocytosis (Fig. 5) (e.g., Wendelaar Bonga, 1971a). Quantitative electron microscopy allows investigators to determine the rate of synthesis (Golgi apparatus volume) and release activity (number of exocytotic figures) of the NSC (Wendelaar Bonga, 1971a).

Further progress could be made by studying the chemistry of the neurohormones. From knowledge gained in other animal groups, it has become clear that many NSC produce neurohormones of a peptide nature. This is most probably also true for the molluscan neurohormones (see below). Gradually, this characteristic is also used in the reverse way: All "peptidergic" neurons are considered to be NSC.

Two important recent developments have changed the classical picture of neuroendocrinology in general, and also that of molluscs. First, techniques have been developed to raise antibodies against purified neurohormones and other

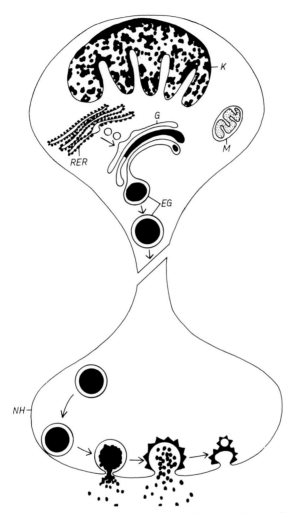

Fig. 5. Schematic representation of the ultrastructural aspects of the production, transport, and release of the secretory product of a neurosecretory cell. Synthesis of the neurohormone or a prohormone occurs in the rough endoplasmic reticulum (RER). This product is transported in small vesicles to the Golgi zones (G), where packaging in the elementary granules (EG) takes place. These are transported through the axon to the numerous neurohemal endings (NH), where release by exocytosis occurs. K, Nucleus; M, mitochondrion. (Courtesy of Dr. E. W. Roubos.)

biologically active peptides. These antibodies are now available, mostly for peptides from vertebrate origin. The antibodies are used to test immunoreactivity in neurons. When a positive reaction is shown in the cytoplasm or the axons of neurons, such cells are considered to be peptidergic. In fact, immunocytochemistry has become a new means for the identification of neurosecretory (peptidergic) cells (Boer et al., 1980; Grimm-Jørgensen, 1978, 1979; Grimm-Jørgensen and Jackson, 1975; Schot et al., 1981; Van Noorden et al., 1980). It is not yet clear whether a positive reaction of a neuron with a specific antibody means that the secretory product and the molecules to which the antibody was raised are identical. Therefore, such neurons are indicated by designations such as "ACTH-like" cells (Fig. 6). Often, antibodies against vertebrate peptides give positive

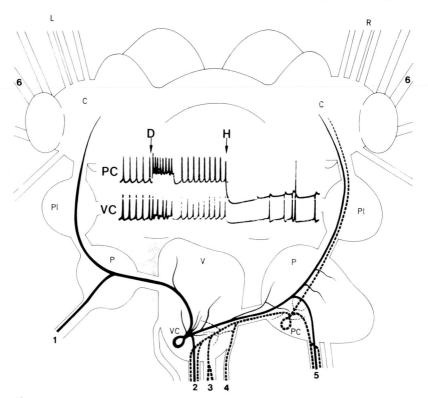

Fig. 6. Diagram of the position and fiber pattern of two peptidergic cells (VC and PC) identified immunocytochemically with anti-ACTH 1–39 and 1–24 in *L. stagnalis*. By means of the intracellular horseradish peroxidase injection technique, it was shown that the cells send fibers into the neuropils of various ganglia and into nerves. The cells are electrotonically coupled: VC follows PC when the latter is depolarized (D) or hyperpolarized (H) (inset; cf. Figs. 14 and 23). L, Left; R, right. C, cerebral; P, parietal. Pl, pleural; V, visceral ganglion. 1, left pallial; 2, anal; 3, intestinal; 4, genital; 5, right internal pallial; 6, median lip nerve. (Derived from Boer et al., 1979.)

TABLE I

Peptidergic Neurons in the Central Nervous System of *Lymnaea stagnalis*

Peptidergic neuron	Reference
Thyrotropin releasing hormone (TRH)	Grimm-Jørgensen, 1978
ACTH	Boer et al., 1979
FMRFamide	Boer et.al., 1980
α-MSH	Schot et al., 1981
Arginine vasopressin (AVP)	Schot et al., 1981
Arginine vasotocin (AVT)	Schot et al., 1981
Calcitonin	Schot et al., 1981
Gastrin	Schot et al., 1981
Gastrointestinal polypeptide (GIP)	Schot et al., 1981
Glucagon	Schot et al., 1981
Insulin	Schot et al., 1981
Met-enkephalin	Schot et al., 1981
Oxytocin	Schot et al., 1981
Pancreatic polypeptide (PP)	Schot et al., 1981
Secretin	Schot et al., 1981
Somatostatin	Schot et al., 1981
Substance P	Schot et al., 1981
Vasoactive intestinal polypeptide (VIP)	Schot et al., 1981

reactions in invertebrates, and vice versa (Boer et al., 1979). In this way, in *L. stagnalis* the number of identified peptidergic or neurosecretory cell types increased from 10 to about 25 (Schot et al., 1981), which at the moment is the maximum for any invertebrate (Table I).

The second development concerns the study of the function of NSC, which for many years has been in the hands of endocrinologists. Their extirpation, transplantation, and injection experiments were based on techniques derived from studies on endocrine organs, which are richly vascularized but poorly innervated. One must bear in mind, however, that important information is transduced to NSC by means of synaptic input from conventional neurons, because they are in fact also neurons. Studies on the integration of the input and on the electrical, biosynthetic, and release activities of these cells have only recently been made (see Sections III,B,1 and III,C,5). The need for integration of the work on neurosecretory neurons of neurophysiologists and endocrinologists has recently become urgent.

Evidence has become available from studies in mammals (e.g., Buijs and Swaab, 1980; De Wied and Gispen, 1977) that peptides in neurons may have a role in neurotransmission, i.e., neuropeptides may act as transmitter substances in synaptic contacts between neurons and perhaps also between neurons and cells of nonnervous target tissues.

Because a number of gastropods contain easily identifiable, large neuronal cell bodies, they are particularly suitable for the study of the electrical characteristics of both neurosecretory and conventional neurons (cf. Joosse et al., 1982).

2. General Morphology of Molluscan Neurosecretory Systems

The shape and size of the NSC of molluscs do not differ from those of conventional neurons. The cells are often unipolar and located at the periphery of the ganglia. They tend to be constant in number and position. Cells of the same type are often located together in a specific area of the ganglion, where the smaller cell bodies lie near the neuropil and the larger ones (up to 120 μm) at the periphery.

Only in gastropods and cephalopods have neurohemal areas been described. In molluscs the release sites are not concentrated in specific organs, such as the sinus glands of crustaceans, the corpora cardiaca of insects, or the median eminence and neurohypophysis of vertebrates. In gastropods the entire periphery of the nervous system, including nerves, connectives, and commissures, may be used for hormone release. For instance, the intercerebral commissure of *L. stagnalis* is used for the release of the ovulation hormone of the caudodorsal Cells (CDC) (Figs. 1 and 7) and the median lip nerve for that of the growth hormone of the light green cells (LGC) (Joosse, 1964; Wendelaar Bonga, 1970). Release occurs inside the perineurium. The hormones have to diffuse through the perineurium and the connective tissue sheath before entering the blood (Roubos et al., 1981b). However, in a number of cases, the axons penetrate the perineurium and ramify into the connective tissue around the nervous system (Fig. 7; cf. the release areas of the bag cells of *Aplysia californica* located around the cerebropleural connectives; Fig. 8). Alternatively in other cases, when axons enter peripheral nerves, release may even occur in target tissues. This latter phenomenon is called *peripheral neurosecretion* (cf. Maddrell, 1967). It is common in gastropods, particularly in pulmonate snails (Plesch, 1977; cf. Boer and Joosse, 1975). Wendelaar Bonga (1971b, 1972) reported it for the neurosecretory dark green cells (DGC) of *L. stagnalis,* which have projections into the kidney and probably are involved in the control of ion transport. Saleuddin and Dillaman (1976) found release of neurosecretory material near cells of the mantle edge gland of the planorbid *Helisoma duryi.* Swindale and Benjamin (1976a, 1976b) even found, in *L. stagnalis,* an innervation of the entire water-exposed body wall by branches of the DGC which react to osmotic stimuli (cf. Roubos and Moorer-van Delft, 1976). The most striking case was found in the cephalopod *E. cirrosa,* in which a neurohormone appears to be released into the orbit (Froesch and Mangold, 1976).

Another variation of the neurohemal area is the release of neurosecretory materials in the capsule of blood vessels. Early descriptions of this type of release refer to the cerebral arteries of *H. pomatia* (Nolte and Kuhlmann, 1964)

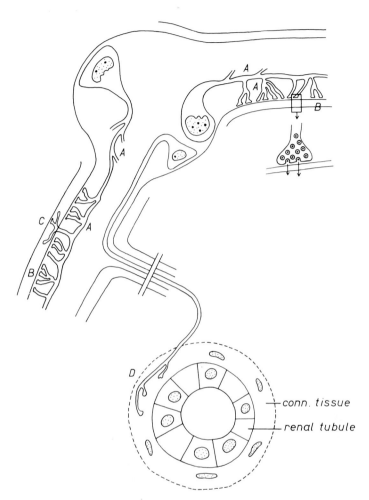

conn. tissue

renal tubule

Fig. 7. Various types of neurohemal areas in pulmonate gastropods. The axons show (**A**) frequent ramifications, which result in numerous neurohemal endings where release occurs by exocytosis. Release is found (**B**) underneath the perineurium of nerves, connectives, and commissures, (**C**) in the connective tissue sheath around the nervous system, in the wall of blood vessels, and (**D**) in the target tissues (peripheral neurosecretion). (After Joosse, 1976.)

and *Arion ater* (Van Mol, 1967), and to the vena cava of cephalopods (Alexandrowicz, 1965). The latter system consists of 2–3 million cells, the axons of which line the inner surface of the blood vessel (see Section VI; Blanchi et al., 1973).

In the large release areas, great numbers of neurohemal endings are present. The endings are formed by frequent branching of the axons of the NSC (Fig. 7).

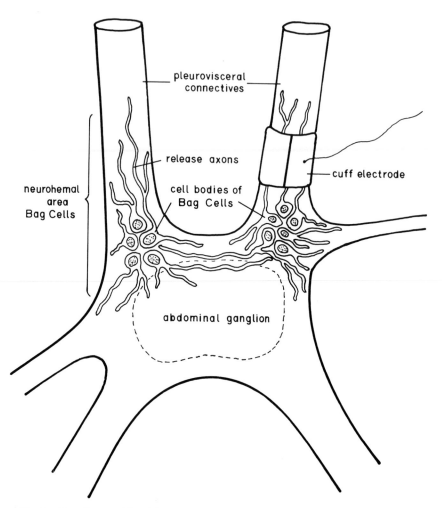

Fig. 8.   Dorsal view of the abdominal ganglion of *Aplysia brasiliana*. A small number of the numerous bag cells with their release axons is indicated. The position of the cuff electrode for *in vivo* extracellular recordings of the discharge activity from bag cell axons is indicated for the right connective. (Based on Dudek et al., 1979.)

Wendelaar Bonga (1971a) has calculated that the ovulation hormone cells (CDC) of *L. stagnalis* have a total number of 80,000 endings, that is, 800 per cell.

The enormous dimensions of the release areas and the absence of concentration contrast strongly to the situation in vertebrates, crustaceans, and insects. Maddrell and Nordmann (1979) have suggested that this might be related to the fact that molluscs lack a blood–brain barrier. Peptides do not easily pass blood–brain barriers. Therefore, in vertebrates, crustaceans, and insects, release of peptide hormones is concentrated in special organs outside the barrier. However, data on blood–brain barriers in molluscs are scarce. Therefore, the above suggestion, although interesting, needs further research.

Morphological data on specific neuroendocrine systems will be presented together with the data on their functional significance.

III. Hormonal Control of Reproduction

Experimental studies on the hormonal control of reproduction have been performed in only a small number of molluscan species. The main reason for this is the lack of suitable techniques for maintenance and (micro)surgery. At present, the control of reproduction of gastropods and cephalopods only is known in more detail. From a comparative endocrinological point of view, studies on molluscs are most important in respect to the basic problems of the control of simultaneous hermaphroditism and sex reversal, and the exceptional "terminal" reproduction of cephalopods.

A. Gastropoda Prosobranchia: The Control of Sex Reversal

The majority of prosobranchs are gonochoristic. However, a number of species (e.g., *Patella vulgata*, *Crepidula fornicata*, and *Calyptraea sinensis*) show protandric sex reversal: A male phase precedes a female phase, and these are linked by a hermaphroditic phase (cf. Choquet, 1971). Remarkably, most attention has been given to the endocrinology of these nongonochoristic species.

1. Control of Gonadal Activity

In the protandrous species, the juvenile gonad tends to be bisexual, but when the animals reach sexual maturity, the male line develops first. In the study of the endocrine control of sex cell differentiation, the organ culture technique developed by French schools has been of considerable value. In some cases, female cells were found to differentiate and develop in a culture medium which lacks hormones. This ovarian autodifferentiation was shown in *C. sinensis* by Streiff (1966) and in *C. fornicata* by Lubet and Streiff (1969), but it could not be demonstrated in *P. vulgata* by Choquet (1971) and in *Viviparus viviparus* by Griffond (1975). In *Calyptraea*, *Crepidula* and *Patella*, male cells differentiate and develop only when juvenile gonadal tissue is cultured in the presence of cerebral ganglia or of hemolymph of a specimen in the male phase. These investigators concluded that these ganglia produce an androgenic factor of a hormonal nature. In *P. vulgata* this factor is also produced by animals in the female phase. In this phase, it stimulates the mitotic multiplication of the oogonial cells. Choquet (1971), therefore, called this factor *gonadostimulin* or *mitogenic substance*. This effect on the female cells may be an exception, because oogonial multiplication is not often found in gastropods. These results have been further substantiated by the effects of extirpation and implantation of cerebral ganglia of *C. fornicata*, carried out in the male as well as in the female phase (Lubet and Silberzahn, 1971). Extirpation of these ganglia leads, in both phases, to inactivation of the gonad, whereas reimplantation restores gametogenesis. Thus, oogenesis is also dependent on a cerebral ganglion factor. This factor could also be demonstrated in organ culture conditions: In the gonads of *C. sinensis* and *P. vulgata* vitellogenesis occurred only in the presence of cerebral ganglia of active female phase specimens (Choquet, 1971; Streiff, 1967).

Apart from these effects of the cerebral ganglia, in *P. vulgata* the tentacles seem to produce a hormonal factor which suppresses the male activity, thus causing seasonal periods of rest in spermatogenesis and sex reversal (Choquet, 1971).

Up to now, the cellular origin of the factors mentioned has not been clearly demonstrated. Most probably, the active centers of the cerebral ganglia of *C. fornicata* are located near the cerebropleural connectives (P. Lubet, personal communication). There are no experimental indications for a role of the juxtaganglionar organ in the control of reproduction. After the first description by Martoja (1968), the presence of this organ in *Patella* was confirmed by Choquet and Lemaire (1969). Martoja could not identify this structure in other prosobranchs. An analogy of its function with that of the DB of the Pulmonata has been suggested (Boer and Joosse, 1975; Joosse, 1972).

2. Control of Accessory Sex Organs

During the sex reversal phase of, e.g., *C. fornicata* the male accessory sex organs (ASO), consisting of a sperm duct, seminal vesicles, an external sperm groove, and an elaborate penis, are replaced by an oviduct, a gonopericardial canal, a receptaculum seminis, a uterus, and a vagina (Fig. 9). The testis simultaneously changes into an ovarium. Streiff (1966, 1967) has shown in organ culture experiments that in *Calyptraea* the changes of the ASO are independent of the gonad, and that they occur under the direct control of neuroendocrine and tentacular centers, S. Le Gall (1974, 1978, 1981) has clarified the control of the differentiation and lysis of the penis of *Crepidula*. She showed that the differentiation of the penis in a morphogenetic territory located near the right tentacle is stimulated by a neurohormone produced by the pedal ganglia (Fig. 10). This hormone is released into the hemolymph and seems to accumulate in specific hemal lacunae of the right tentacle. Such a peripheral accumulation of a hormone has been hitherto unknown and is therefore poorly understood. It is, however, apparent from the stimulating effect of the isolated right tentacle on penis differ-

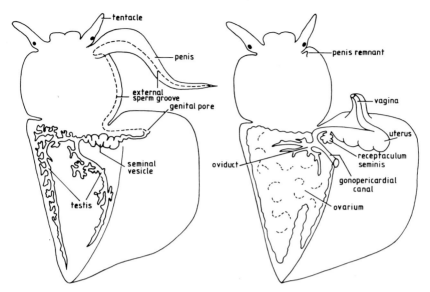

Fig. 9. Structure of the male and female reproductive systems of *Crepidula fornicata*. (After P. Le Gall, 1980.)

entiation in organ culture (Fig. 11) (Lubet and Streiff, 1969) and from the absence of this effect when these tentacles are taken from snails deprived of their pedal ganglia. This morphogenetic factor is not species specific, as was found in experiments with *Crepidula* and *Calyptraea* (Streiff et al., 1970). Moreover, pedal ganglia of males of gonochoristic species (*Littorina, Buccinum*) appeared to be active in organ cultures of the penis of *Crepidula* (S. Le Gall, 1981).

The release of the pedal ganglion neurohormones can be stimulated by a factor released by the cerebropleural ganglia. With this factor, inactive pedal ganglia of snails which are, after sex reversal, in the female phase can be reactivated in organ cultures (S. Le Gall, 1978).

In *Crepidula* the dedifferentation of the penis during the transition to the female phase is controlled by still another neurohormone produced by NSC located in the mediodorsal area of the pleural ganglia. These cells release their products via the cerebropleural connectives (Fig. 10). The necrotic action on the penis was demonstrated in organ cultures. Extirpation of the area *in vivo* resulted in the halting of regression of the penis in animals during sex reversal (S. Le Gall, 1974; S. Le Gall and Streiff, 1975; Lubet and Streiff, 1969; Streiff, 1966). Again, this factor is not species specific (S. Le Gall, 1974).

The uniqueness of the control system of *Crepidula* is apparent not only from the fact that two neurohormones are involved in the control of the exclusively morphogenetic processes in the penis, but also because the release of these

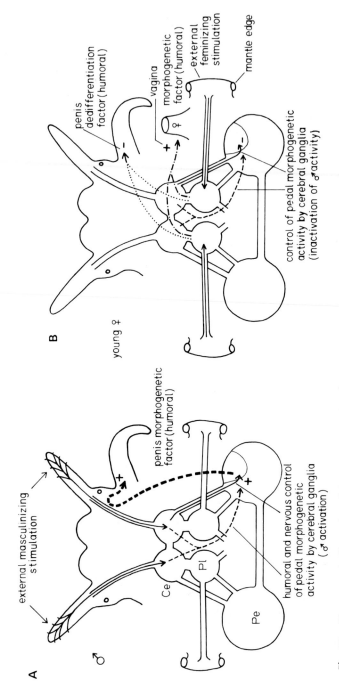

Fig. 10. Diagrammatic illustration of the environmental, nervous, and humoral factors involved in (**A**) the differentiation of the male external genital tract and (**B**) the dedifferentiation of the male and the differentiation of the female external genital tract in *C. fornicata*. Ce, Cerebral ganglia; Pe, pedal ganglia; Pl, pleural ganglia. (Derived from P. Le Gall, 1980.)

Labels in figure:

A
external masculinizing stimulation
penis morphogenetic factor (humoral)
humoral and nervous control of pedal morphogenetic activity by cerebral ganglia (♂ activation)
Ce
Pl
Pe
♂

B
young ♀
penis dedifferentiation factor (humoral)
vagina
♀ morphogenetic factor (humoral)
external feminizing stimulation
mantle edge
control of pedal morphogenetic activity by cerebral ganglia (inactivation of ♂ activity)

Fig. 11. Series of experiments to demonstrate the role of the pedal ganglia and the right ocular tentacle in the control of morphogenesis of the penis of C. *fornicata*. (**A**) The penis regenerates in the presence of the pedal ganglia. (**B**) No regeneration of the penis occurs after simultaneous removal of the pedal ganglia. (**C**) Implantation of pedal ganglia restores the capacity of penis regeneration. (**D**) After removal of the right tentacle and the penis, both regenerate in the presence of the pedal ganglia, and the new tentacle is able to induce penis morphogenesis from a territory in organ culture. (**E**) After removal of the right tentacle, the penis, and the pedal ganglia, only the tentacle regenerates. However, it is not able to induce penis morphogenesis. (Derived from S. Le Gall and Streiff, 1975.)

factors appears to depend on social conditions. *Crepidula* has the peculiar habit of living in colonies or chains of up to 15 individuals (Fig. 12). The snails at the top of the chain are small males. whereas those at the bottom are large females. The intermediate animals are in the hermaphroditic sex reversal phase. P. Le Gall (1980) has shown that placing a small male specimen in a chain, always at the top, induces rapid body growth and a gradual transition to the female sex of the specimen in the former top position. He observed that contacts between the members of the chain occur mainly by the tentacles and the pallial border, and he demonstrated that a masculinizing factor is released by the border, which is registered by the tentacles of the higher specimens. A feminizing factor is released by the tentacles of the higher specimens and registered by the pallial border of the lower specimens. In the chain, these effects are cumulative. Finally, a growth-accelerating factor affects the growth of the lower specimens, but this effect is not cumulative: It is a trigger mechanism. The chemical character of the external messengers is not yet known. They are not carried by the water. Therefore, the term *pheromone* is not applicable (cf. Joosse, 1979a). P. Le Gall (1980) has shown that the social contact in a chain is not a condition for the start

Fig. 12.   Colony of C. *fornicata,* with a small male specimen on top and large females at the bottom. (Courtesy of Drs. S. and P. Le Gall.)

of the change of sex. Isolated specimens can spontaneously start rapid growth and sex reversal. However, this occurs in older animals.

3. Induction of Ovulation

A new dimension was added to the endocrinology of prosobranchs by Ram's (1977) discovery of an egg-laying inducing substance in two species of the genus *Busycon*. The active factor was mainly found in the parietal ganglia and has been called *egg capsule-laying substance* (ECLS). It may be a peptide, because it is destroyed by pronase and is relatively heat stable. This fact, together with its nervous origin, makes it likely that the substance is a neurohormone. Because ECLS is produced in both sexes, the question arises as to whether it also has a function in males. It is hoped that the above-mentioned findings will stimulate further endocrinological research in these gonochoristic prosobranchs.

An interesting but isolated report from Morse et al. (1977) announces the induction of spawning in *H. rufescens* by externally applied hydrogen peroxide.

In conclusion, prosobranch snails appear to have several gonadotropic centers: (a) the cerebropleural connective areas, which have a non-sex-specific stimulating action on the gonad and the accessory sex organs; (b) the cerebral ganglia, which have an androgenic effect (and in *Patella* a mitogenic effect on the oogonia); (c) the tentacles, which have an inhibiting effect on spermatogenesis; and (d) the pedal ganglia and cerebropleural connectives, which have morphogenetic differentiating and dedifferentiating effects, respectively, on the penis. These gonadotropic factors, although discovered in hermaphroditic species, also occur in gonochorists, and they are not species specific. Synthesis and release of the gonadotropins in the neuroendocrine production centers are controlled via nervous pathways, whereas the production of the morphogenetic factor of the pedal ganglia is also influenced by a hormonal stimulus. In *Crepidula*, sensory information about the situation in the chain colony is important for this control, but it is not conditional.

There is no clear information about the control of the differentiation and activity of the ASO. The gonad does not seem to have an endocrine function. An egg-laying inducing substance is recorded in a gonochoristic species.

B. Gastropoda Opisthobranchia: Egg-Laying Hormones

Members of this subclass live in marine habitats and are hermaphrodites. The present status of reproductive endocrinology in this group is quite peculiar. On the one hand, hardly anything is known about the hormones controlling the production of gametes and the activity of the ASO; on the other hand, the egg-laying hormone of *A. californica* is the most thoroughly studied molluscan hormone.

Vicente (1969) has studied the effects of the extirpation of the cerebral ganglia and tentacles in *A. rosea*. He observed a stop in mating and egg-laying after extirpation of the cerebral ganglia. Extirpation of the tentacles seemed to stimulate reproductive activity, and at the same time the juxtaganglionar organ was hypertrophied. This organ is suggested to be an endocrine gland (see Section II,A), perhaps homologous to the DB of the Pulmonata, the latter being the production centers of a female gonadotropic hormone.

1. The Egg-Laying Hormones

For various opisthobranch species, egg-laying is controlled by a neurohormone called *egg-laying hormone* (ELH). For recent reviews of this subject, see Arch (1976), Strumwasser et al. (1980), and Blankenship (1980). Kupfermann (1967, 1970) was the first to discover that in *A. californica* homogenates of paired groups of neurosecretory bag cells, which are located around the pleuroabdominal connectives in the parietovisceral (abdominal) ganglion (cf. Fig. 8), could induce egg-laying in this species (Fig. 13). From then on, *Aplysia* was one of the most prominent animal species for the study of the basic electrophysiological characteristics of neuroendocrine cells and for the effects of neurohormones on behavior.

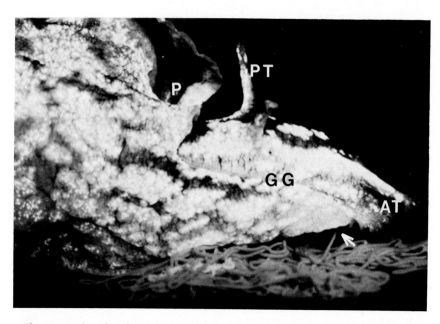

Fig. 13. *Aplysia brasiliana* photographed during egg laying, The long-winded egg string is visible at the bottom of the picture and emerges (arrow) from the common genital groove (GG) near the mouth at the base of the anterior tentacle (AT). P, Parapodium; PT, posterior tentacle. (Original picture courtesy of Dr. H. Pinsker; from Cobbs and Pinsker, in press.)

In *A. californica* the bag cells (about 50 μm) are represented by two groups of about 400 cells each (Coggeshall, 1967; Frazier et al., 1967). The cell clusters are located adjacent to the abdominal ganglion. Their axons ramify in the connective tissue around the pleuroabdominal connectives and the ganglion. This region is the neurohemal area of the bag cells. Other species of this genus also have bag cells (e.g., *A. brasiliana;* Dudek et al., 1979) (Fig. 13).

Davis et al. (1974) and Ram et al. (1977) also demonstrated the presence of an egg-laying hormone in the opisthobranch *Pleurobranchaea californica.* The NSC which produce this hormone lie in the medial lobes of the pedal ganglia, in two groups of about 100 cells each (70–90 μm).

A single injection of a homogenate of the bag cells induces the whole repertoire involved in egg-laying: ovulation of the oocytes; their transport, fertilization, and packaging in the egg string; and extrusion of the egg string and its fixation to a substrate (Fig. 13). (For a discussion of the structure and function of the reproductive system of aplysiids, see Beeman, 1977.) The action of ELH on the ovotestis is very rapid. Coggeshall (1970) found that muscle cells associated with the acini of the ovotestis contract, and oocytes are already present in the hermaphrodite duct within 1 min after ELH injection. The action of the hormone on the gonad is direct (see Fig. 17), because Dudek and Tobe (1978) could evoke the release of oocytes from fragments of ovotestis kept *in vitro,* in a dose-dependent manner, after addition of crude extracts of the entire parietovisceral ganglion. The same effect was obtained with the substance(s) released by *in vitro* activated bag cells (Dudek et al., 1980).

It is as yet not clear whether the release of the secretion products of the albumin gland and the formation of the egg capsules and the egg string are also under the direct control of ELH. The complexity of these procedures makes local nervous control highly probable (see Fig. 17). Observations on pulmonate snails suggest that a genital nervous plexus is activated by ELH (see Section III,C,5).

The latent period between injection of ELH-containing extracts and the onset of egg-laying in *A. californica* is 72 ± 2.5 min at 14°C (Strumwasser et al., 1969). With the injection technique, egg-laying can be induced by a sudden rise of the ELH content of the blood, which is followed by a gradual, for peptide hormones often rather quick (cf. Maddrell and Nordmann, 1979), decrease of the ELH titer. Therefore, the question arises as to how the ELH titer proceeds under normal circumstances. At the neurohemal endings, the release of neurohormones occurs by exocytosis after the arrival of an action potential. Bag cells are normally silent, but after electrical stimulation *in vitro* they show an afterdischarge (cf. Fig. 14), which is a series of action potentials lasting as long as 30 min (Kupfermann and Kandel, 1970). All cells within one cluster show a synchronous spiking, which has its morphological basis in electrotonic contacts between the cells (Kaczmarek et al., 1979). This simultaneous activity results in a rapid release of ELH during the afterdischarge. In this respect, the observations

Fig. 14. Recording of a spontaneous discharge of the bag cells of a freely behaving specimen of *A. brasiliana* obtained by using a cuff electrode (at 20°C) around one of the pleuro-

of Dudek et al. (1979) are important. They developed a technique for recording the electrical activity of the bag cells in freely behaving *A. brasiliana* (see Fig. 8). Egg-laying in this species always occurs simultaneously with spontaneous bag cell activity (Fig. 15). Thus, the bag cells behave similarly both *in vivo* and *in vitro*. The ELH titer, therefore, will normally rise very rapidly to high levels after the start of the release, and it will decrease again when the spike train has stopped. Dudek et al. (1979) did not find a correlation between the duration of the electrical activity and the number of eggs laid. There are as yet no indications that electrical activity of the bag cells or hormone release occurs beyond the egg-laying periods. For *Aplysia* species no environmental stimuli are known which induce egg laying, i.e., bag cell activity.

In *A. californica* egg-laying can also be induced by extracts of the atrial gland (Arch et al., 1978). This gland is connected to the most rostral part of the large hermaphroditic duct (Beeman, 1977; Thompson and Bebbington, 1969). After injection of this extract, the response latency for egg-laying is identical to that after ELH injection. Therefore, the question arises as to the chemical nature of the atrial gland product and of ELH. After initial studies by Toevs (1970) and Arch et al. (1976), Chiu et al. (1979) succeeded in determining the primary amino acid structure of ELH. This peptide molecule consists of 36 amino acids (Fig. 16). The pI is about 9.3, so that the molecule is highly basic (Arch et al., 1976). ELH shows hardly any resemblance to other vertebrate or invertebrate peptides the chemical structure of which is known. It appeared that at 20°C 2.5 nm of pure ELH suffice to induce egg-laying in *A. californica*.

It is important to notice that during *in vitro* afterdischarges of bag cells, at least three peptides other than ELH are released. One of these has a molecular weight of about 6000, but it has no egg-laying inducing activity (Arch et al., 1976; Strumwasser et al., 1980). The other two peptides are smaller. Their function is unknown.

Regarding the atrial gland factor, Heller et al. (1979) isolated two bioactive peptides from this gland. Both (A and B) consist of 34 amino acids, which are identical except for four amino acid positions (4, 7, 8, and 9) (Fig. 16). Although the number of amino acids in the atrial gland peptides nearly equals that of ELH,

25 | μV
100 sec

visceral connectives (cf. Fig. 8). (Original picture courtesy of Dr. H. Pinsker; from Pinsker and Dudek, 1977.)

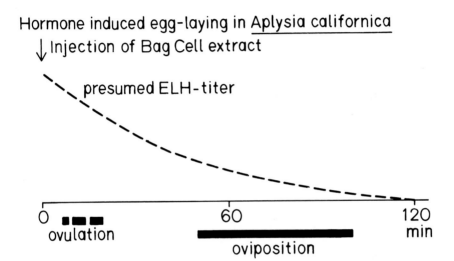

Hormone induced egg-laying in Aplysia californica
↓ Injection of Bag Cell extract

presumed ELH-titer

0 ▪▪▪   60   120
ovulation   min
oviposition

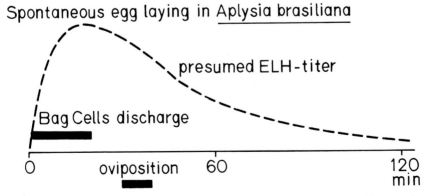

Spontaneous egg laying in Aplysia brasiliana

presumed ELH-titer

Bag Cells discharge

0   oviposition   60   120
min

Fig. 15. Time schedules of ovulation and egg-laying (oviposition) in *A. californica* after injection of the egg-laying hormone (ELH) (based on Strumwasser et al., 1980) and in *A. brasiliana* after spontaneous discharge of the bag cells, which was recorded *in situ* (based on Dudek et al., 1979).

Egg-Laying Hormone

10
NH$_2$-Ile-Ser-Ile-Asn-Gln-Asp-Leu-Lys-Ala-Ile-
Thr-Asp-Met-Leu-Leu-Thr-Glu-Gln-Ile-Arg-
Glu-Arg-Gln-Arg-Tyr-Leu-Ala-Asp-Leu-Arg-
Gln-Arg-Leu-Leu-Glu-Lys-OH
36

Atrial Gland Peptides

10
A   H-Ala-Val-Lys-Leu-Ser-Ser-Asp-Gly-Asn-Tyr-
B   H-Ala-Val-Lys-Ser-Ser-Ser-Tyr-Glu-Lys-Tyr-

A   Pro-Phe-Asp-Leu-Ser-Lys-Glu-Asp-Gly-Ala-
B   Pro-Phe-Asp-Leu-Ser-Lys-Glu-Asp-Gly-Ala-

A   Gln-Pro-Tyr-Phe-Met-Thr-Pro-Arg-Leu-Arg-
B   Gln-Pro-Tyr-Phe-Met-Thr-Pro-Arg-Leu-Arg-

A   Phe-Tyr-Pro-Ile
B   Phe-Tyr-Pro-Ile
34

Fig. 16. Amino acid composition of the egg-laying hormone and atrial gland peptides A and B of *A. californica*. The atrial gland peptides differ only at positions 4, 7, 8, and 9. (From Chiu et al., 1979, and Heller et al., 1980.)

their composition is quite different. The authors presented clear evidence that the atrial gland peptides cause egg-laying in an indirect way, because *in vitro* they induce an afterdischarge in bag cells. This means that *in vivo* the atrial gland peptides will activate the ELH system. The action is exerted on an unknown center in the central nervous system, which is not located in the parietovisceral ganglion (Fig. 17).

These results seem to conflict with those of Arch et al. (1980), who could not identify a cell type in the atrial gland showing the ultrastructural characteristics of peptide production. Moreover, they could induce egg-laying by injection of atrial gland extracts in specimens of *A. californica* from which the parieto-visceral ganglion and the majority of the neurohemal area of the bag cells had been removed. This suggests that the atrial gland peptides are also able to act directly on the reproductive system. Arch et al. (1980) consider the possibility of a minor specificity of the receptors in the targets the possible reason of the cross-reactivity of both hormones. However, structurally the molecules differ consid-erably. Further research is needed to show whether and when (in relation to egg laying) the atrial gland peptides are released into the blood. The fact that in some aplysiids copulation and egg-laying have some correlation is used as an argument in favor of the endocrine role of the atrial gland. The gland would be stimulated at copulation (cf. Strumwasser et al., 1980).

In view of the scarcity of data on the endocrinology of reproduction of opisthobranchs, it is difficult to establish whether the gonadotropic role of ELH is restricted to the induction of egg-laying. At the moment, nothing is known about other gonadotropic effects of this hormone.

In contrast, much is known about the effects of ELH on behavior. It induces egg-laying behavior. This consists of cessation of locomotion, extrusion of the egg string, egg-winding behavior, and inhibition of feeding. This behavior is shown before egg-laying. Experiments with purified ELH preparations have revealed that, in isolated ganglia, four neurons are directly activated by the hormone (Stuart and Strumwasser, 1980). It is suggested that these neurons might be involved in the control of food uptake. Previously numerous neurons in the parietovisceral ganglion (e.g., R15) were demonstrated to show a variety of indirect or possibly direct reactions to ELH, such as slow inhibition, transient excitation, prolonged excitation or a biphasic response (Mayeri et al., 1979a,b).

This survey of our present knowledge of bag cell and ELH function in aplysiids has demonstrated that at the moment this neuroendocrine system functions as a model system in neuroendocrinology (Strumwasser et al., 1980). It is expected that in the near future, when synthetic ELH has become available and radioimmunoassay studies are possible, the importance of this model will increase further. Hopefully, broader knowledge of other aspects of the reproductive endocrinology of these animals will then provide a better indication of the exact role of ELH in the survival of the aplysiid species.

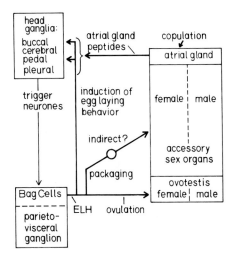

Fig. 17. Tentative schematic representation of the effects of ELH and the atrial gland peptides in *A. californica*. For details, see text. (Based on Strumwasser et al., 1980.)

C. Gastropoda Pulmonata: Control Systems for Male and Female Activity in Hermaphrodites

In recent years, our knowledge of the endocrinology of pulmonate reproduction has increased considerably. The number of species studied experimentally is rather small, but these include both the freshwater Basommatophora and the terrestrial Stylommatophora. There is considerable similarity in the control systems of these groups.

1. Structure of the Reproductive System

All pulmonates are hermaphrodites. The gametes of both sexes are produced in each of the numerous acini of the ovotestis (Fig. 18). In some species each acinus is compartmentalized: The female cells develop abluminally and the male cells in the acinar lumen. Both compartments are separated by a layer of Sertoli

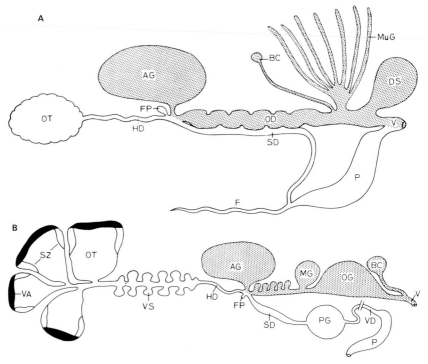

Fig. 18. Schematic representation of the reproductive systems of (**A**) a stylommatophoran (*H. aspersa*) and (**B**) a basommatophoran (*L. stagnalis*) pulmonate snail. The female ASO are hatched. AG, Albumin gland; BC, bursa copulatrix; DS, dart sac; F, flagellum; FP, fertilization pocket; HD, hermaphrodite duct; MG, muciparous gland; OD, oviduct; OG, oothecal gland; OT, ovotestis; P, penis; PG, prostate gland; SD, sperm duct; SZ, spermatogenetic zones; V, vagina; VA, vitellogenetic area; VD, vas deferens; VS, vesiculae seminales.

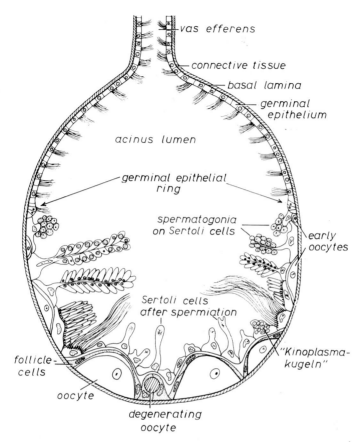

Fig. 19. Schematic representation of a longitudinal section through an acinus of the ovotestis of *L. stagnalis.* The continous layer of Sertoli cells separates the male from the female cells. (After Joosse and Reitz, 1969.)

cells to which the male cells are attached (Fig. 19) (cf. de Jong-Brink et al., 1976; Joosse and Reitz, 1969). This compartmentalization may also have consequences for the accessibility of hormones.

In all pulmonates there is one hermaphrodite duct arising from the ovotestis (see Fig. 18). At the *carrefour* this duct divides into a male and a female part or tract, each with accessory structures. The degree of separation of the tracts can differ: For example, in *L. stagnalis* the separation is complete; in *H. aspersa* ducts meet again at the common genital aperture.

2. Control of Differentiation of Sex Cells and Gametogenesis

There is clear evidence that the juvenile gonad of pulmonates contains only one type of gonadal stem cell (cf. Hogg and Wijdenes, 1979; for an alternative theory, see de Jong-Brink et al., in press, and Geraerts and Joosse, in press). Both the sex cells and their supporting cells (Sertoli cells and follicle cells) originate from this stem cell. From organ culture experiments with juvenile gonads of *H. aspersa* (Gomot, 1970; Guyard, 1971a) and *Arion subfuscus* (Wattez, 1980), we know that the female cells arise by autodifferentiation. Differentiation of the male cells is stimulated by a factor from the optic tentacles as well as the cerebral ganglia of young animals. This factor suppresses the development of female cells. Wattez (1980) has shown *in vivo*, by tentacle extirpation and injection experiments, that in *A. subfuscus* the masculinizing factor is found only in the optic tentacle of snails in the male phase. This factor stimulates the spermatogonial mitoses. From organ culture experiments, it is clear that this factor acts directly on the sex cells. The cellular origin of the masculinizing factor is still unknown.

The investigations of Wattez (1980) have given the clearest answer to the question in the literature of the exact influence of the optic tentacles on reproduction in Stylommatophora, which was raised by the tentacle extirpation experiments of Pelluet and Lane (1961) and Pelluet (1964). For a survey of the literature, see Boer and Joosse (1975) and Wattez (1980).

Further research in this area has been greatly facilitated by the introduction of *Limax maximus* by Sokolove and McCrone (1978) and McCrone and Sokolove (1979) as a model gastropod for studies concerning the start of puberty. These authors have shown that long-day photoperiods (16L–8D) induce sexual maturation in *Limax*. In this process, the cerebral ganglia play a decisive role, because transplantation of these ganglia from long-day donors to short-day acceptors induces maturation in the latter animals. Maturation includes growth of the gonad, which involves gametogenesis, and of the ASO. Most probably, a masculinizing as well as an oocyte-stimulating factor is released by these cerebral ganglia. The apparently irreversible activation of the cerebral ganglia will allow investigators to trace the cellular components involved.

Little is known about the control of differentiation of the sex cells in Basommatophora. The pulmonates derive their name from the fact that their eyes are located at the base of the structurally simple tentacles. Extirpation of neither the eyes nor the tentacles appeared to have an effect on the reproductive activity of *L. stagnalis* (unpublished results). Geraerts (1976b) has shown that the small lateral lobes (LL) of the cerebral ganglia of *L. stagnalis* (Fig. 1) have a stimulating effect on spermatogenesis in maturing young snails. It is not known, however, whether this includes the stimulation of male sex cell differentiation. The lobes contain several types of neurosecretory cells (Lever and Joosse, 1961; for further literature, see Van Minnen and Reichelt, 1980).

Cerebral ganglia taken from specimens of *H. aspersa* (Guyard, 1971b) or *A. subfuscus* (Wattez, 1980), which are in the female phase of their reproductive activity, stimulate the growth of the oocytes in ovotestes kept in organ cultures. This growth may be mainly due to vitellogenesis. On the basis of extirpation and implantation experiments, Geraerts and Joosse (1975) have shown that the DBH (see above) of *L. stagnalis* stimulates vitellogenesis in this species. No changes of spermatogenesis were observed. Wijdenes and Runham (1976) demonstrated that DBH has a similar effect on oogenesis in *Deroceras (Agriolimax) reticulatum:* in slugs deprived of their DB, the maturation of the oocytes was retarded.

Effects of brain extracts on oocyte maturation in the basommatophorans *Helisoma duryi* and *H. tribolis* have been demonstrated by Saleuddin et al. (1980). Stimulation was apparent from the appearance of endocytotic profiles on large oocytes and from the activation of oviposition within a few days. These effects were attributed to DBH, which was present in the ganglionic extracts.

The above results clearly indicate that in all pulmonates DBH stimulates vitellogenesis in the oocytes, the effects of total cerebral ganglia extracts being due to the presence of the DB on these ganglia (cf. Wattez, 1980).

In young specimens of *L. stagnalis* which are deprived of their LL, the maturation of the oocytes is delayed (Geraerts, 1976b). This effect can be restored by LL implantation. In adult snails, extirpation of the lobes results in a lower level of egg mass production (for the effect on growth, see Section IV,A, B). This effect is most probably exerted indirectly via the DB. In snails without LL, the growth of the DB is delayed. Moreover, the synthesis and release of the secretion product of the DB cells are reduced (Roubos et al., 1980). Thus, in juvenile snails, the LL affect the male and female lines, both of which are accelerated, and in the adults the ovipository activity, which is stimulated. Similar effects of the LL were observed in *Bulinus truncatus* (Geraerts and Mohamed, 1982).

As mentioned above (Section II,B) there is evidence that the Sertoli cells in the gonad of Basommatophora produce steroids. These steroids would act only intragonadally. A possible function could be the synchronization of spermatogenesis of the male cells attached to them. A negative feedback effect from the gonad on DB activity could not be demonstrated in castration experiments in *B. truncatus* (Boer et al., 1976). Thus, in Basommatophora, control of the DB occurs via neuroendocrine pathways, in which the LL might play an important role.

It is tempting to suggest that in the Basommatophora, where simultaneous hermaphroditism is preceded by a very short protandric period, the start of sexual activity is controlled by one center, the LL, which stimulates both sexes. On the other hand, in Stylommatophora, where hermaphroditism is more successive (male cells reach maturity first, followed later by female cells), such as *A. subfuscus* (Wattez, 1980), *D. reticulatum* (Runham and Laryea, 1968), and *H. aspersa* (Guyard, 1971a), the onset of reproductive activity is controlled by a factor which stimulates the male and inhibits the female line. The release of this factor decreases gradually after puberty. In all pulmonates, female cell maturation is controlled by the exclusively female DBH.

The mature oocytes of pulmonates are rather small (130 μm) and contain little yolk. It is therefore not surprising that signs of extragonadal vitellogenin production could not be observed. In *Lymnaea* there are indications that yolk granules also function as lysosomes, with a role in the digestion of the perivitellin fluid (albumin gland secretion) by the developing embryos (de Jong-Brink and Geraerts, 1982).

3. Control of Accessory Sex Organs

In Basommatophora, the possible endocrine-controlling effect of the gonad on the ASO has been studied nearly exclusively in planorbid species. In these animals, the gonad lies in the apical coils of the shell, posterior to the digestive gland. This makes castration a rather easy procedure, as contrasted to other groups, in which the lobes of the ovotestis are embedded between those of the digestive gland. Castration experiments in young but sexually mature snails were performed by Harry (1965) and Vianey-Liaud (1979) in *Biomphalaria glabrata* and by Brisson (1971), Boer et al. (1976), de Jong-Brink et al. (1979), and Geraerts and Mohamed (1981) in *B. truncatus*. Contrary to expectation, the castrated snails continued to produce egg masses, although at a lower rate. Such egg masses consist of the secretion product of the albumin gland (perivitellin fluid), relatively normal egg membranes, and an egg mass capsule secreted by the oothecal gland, but of course, no egg cells are present (Fig. 20). The female ASO grow normally and are apparently able to synthesize their products (de Jong-Brink et al., 1979). In view of these results, it is highly improbable that the ASO of freshwater pulmonates are controlled by the gonad.

On the other hand, in Stylommatophora, gonadal control of the ASO has been shown. Here again a particular group, the slugs, is exceptionally suitable for castration experiments, as their gonad has a less intimate contact with the digestive gland. Castration experiments were performed on *L. maximus* (Abeloos, 1943; Laviolette, 1954), *Arion* spp. (Laviolette, 1954), and *D. reticulatum* (Runham et al., 1973). The fate of juvenile ASO implanted into castrated snails was also studied. Our final conclusion is based on the experiments of Runham et al. (1973). They showed that in castrated *D. reticulatum* implanted juvenile ASO did not develop, whereas in intact animals the male part of implants developed when the acceptors were in the male phase and the female part in acceptors, which were in the female phase of the reproductive cycle. This points clearly to a sex-specific control by the gonad. For the male organs, this includes the penis (Wijdenes, 1981b). The gonadal hormones of stylommatophorans may be of a steroid nature, because Takeda (1979) could stimulate oocyte development and egg production in *D. reticulatum* and *L. flavus* by injections of steroids, particularly estrogens (cf. Gottfried and Dorfman, 1970c). However, the doses used (1 μg/g snail) were very high. Thus, in Stylommatophora, a dual endocrine control of the ASO by the gonad seems well established.

However, this is only part of the picture. Bailey (1973) obtained survival and growth of the prostate gland of *D. reticulatum* in an organ culture medium only if

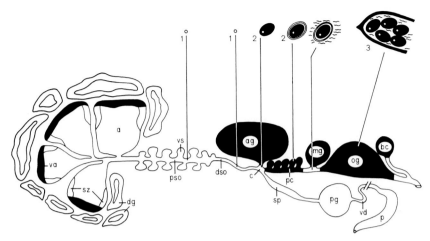

Fig. 20. The various stages of egg mass formation in *L. stagnalis*. (1) After the induction of ovulation, oocytes are transported from the acini (a) to the distal spermoviduct (dso), where fertilization occurs. (2) Eggs are formed in the pars contorta (pc) by the formation of an egg membrane around the egg cell and the perivitelline fluid secreted by the albumin gland. (3) Egg mass formation occurs in the oothecal gland (og). Between the eggs: secretion of the muciparous gland (mg). Around the egg mass, a capsule is formed by the oothecal gland. bc, Bursa copulatrix; c, carrefour; dg, digestive gland lobes; p, penis; pg, prostate gland; pso, proximal part of spermoviduct; sp, sperm duct; sz, spermatogenic zone; va, vitellogenic area; vd, vas deferens; vs, vesiculae seminales.

gonadal tissue, together with cerebral ganglia and optic tentacles, was present. In *L. stagnalis* the female ASO do not grow in juvenile animals deprived of the DB, whereas the male organs develop normally (Geraerts and Joosse, 1975). Juvenile tracts implanted in adult snails without DB show cellular differentiation and growth of the male part only. In intact snails, the male and female structures of the implants develop normally (Geraerts and Algera, 1976). Moreover, the synthesis of galactogen, an important component of the albumin gland secretion product, is very low in pond snails deprived of the DB (Veldhuijzen and Cuperus, 1976). Also, in young specimens of the stylommatophoran slug *D. reticulatum*, cauterization of the DB results in retarded growth of the female ASO (Wijdenes and Runham, 1976).

Thus, there is convincing evidence that the DBH has a stimulating effect exclusively on the female ASO, acting on cellular differentiation, growth, and synthetic activity. This may be true for all pulmonate snails.

An important question arising from the above discussion is whether the DBH action is exerted indirectly by stimulating the release of a female gonadal hormone. For basommatophorans this seems improbable, as castration does not seriously affect egg mass production (see above). The problem was solved in organ culture experiments. Goudsmit (1975, 1978) demonstrated that galactogen synthesis in albumin gland pieces of *H. pomatia* is stimulated by a factor of the cerebral ganglia. The origin of the factor is not yet clear. Absence of the effect in a calcium-free medium points to an active secretion by exocytosis. In *L. stagnalis* the cerebral ganglia also stimulate the polysaccharide synthesis in albumin glands kept *in vitro*. The effect could be attributed to the DBH and the caudodorsal cell hormone (CDCH; see below). Albumin glands of young, sexually mature snails cultured with isolated DB, or in the presence of DB homogenate or of an extract of intercerebral commissures (the storage–release organ of the CDCH) all showed a significant rise in polysaccharide synthesis (Fig. 21) (Wijdenes et al., 1983). The DBH is probably specific at the genus level (Fig. 21) (Wijdenes et al., 1981).

Thus, it is suggested that in Basommatophora both DBH and CDCH act directly on the female ASO. In Stylommatophora the female ASO are controlled

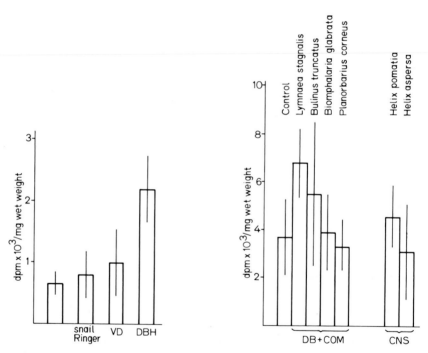

Fig. 21. Stimulation of incorporation of labeled glucose into galactogen in albumen glands of *L. stagnalis* in organ culture conditions. Left: Addition of extract of dorsal bodies (DBH; 2.5 an. equivalent) stimulates galactogen synthesis significantly, compared to an extract of vas deferens (VD) or snail Ringer solution. Right: Effects of brain homogenates (dorsal bodies and commissures; DB + COM or CNS) of different pulmonates on galactogen synthesis in the albumin gland of *L. stagnalis*. Only the response to the *Lymnaea* homogenate is statistically different from that of the controls. (Left, from Wijdenes et al., 1983; right, from Wijdenes et al., 1981.)

by a gonadal hormone and by DBH: Most probably DBH acts via a gonadal hormone on the cellular differentiation and growth of the ASO, whereas it has a direct effect on the synthetic activity of these organs. There are no indications of a neuroendocrine control of the gonadal hormone secretion in Stylommatophora.

An endocrine organ, analogous to the DB, for the control of male activity probably does not exist.

4. Control of Pheromone Production

A new dimension in the endocrinology of pulmonates has been introduced by Takeda (1980). He has called attention to a sex pheromone–secreting gland located between the optic tentacles of the stylommatophoran species *Euhadra peliomphala* (Takeda and Tsuruoka, 1979). This so-called head wart is most active just before courtship. Castration led to atrophy of the gland. In organ culture the immature head wart showed development only in a medium containing testosterone (Takeda, 1980). The author suggests that this gland is under the direct control of the gonad, probably by means of testosterone.

5. Endocrine Control of Ovulation and Oviposition

In several basommatophoran species, the occurrence of a specific neurohormone that controls ovulation and oviposition is well established. Geraerts and Bohlken (1976) were the first to demonstrate that in *L. stagnalis* cauterization of the caudodorsal cells (CDC) results in the accumulation of numerous ripe and degenerating oocytes in the gonad and in a halt to oviposition. Injection of homogenates of the intercerebral commissures (the neurohemal area of the CDC) induces oviposition not only in CDC-cauterized but also in control snails. The authors concluded that CDCH does not affect oocyte production, but it does induce ovulation and oviposition. Numerous studies on the CDC system of *L. stagnalis* have followed. A short survey will be presented; for more details, see Joosse et al. (1982).

The CDC are located in two clusters near the origins of the intercerebral commissure (see Fig. 1). The total number of cells is about 100. The maximum size of the cell bodies is about 90 μm. The CDC can be identified *in vivo* on the basis of their whitish color (Joosse, 1964). The hormone is synthesized in the RER and packed in the Golgi zones in elementary granules 150 nm in size. The secretory material stains with basophilic stains. The periphery of the intercerebral commissure is the neurohemal area of the CDC. Here, about 80,000 endings of CDC axon branches release the hormone by exocytosis (Wendelaar Bonga, 1971a). CDCH synthesis and release in snails kept under natural conditions show a basic diurnal rhythm (Wendelaar Bonga, 1971a), the synchronization of which is controlled by the eyes (Roubos, 1976). CDCH is a peptide with an isoelectric point of about 9.3 and a molecular weight of about 4700 (Geraerts et al., 1983a).

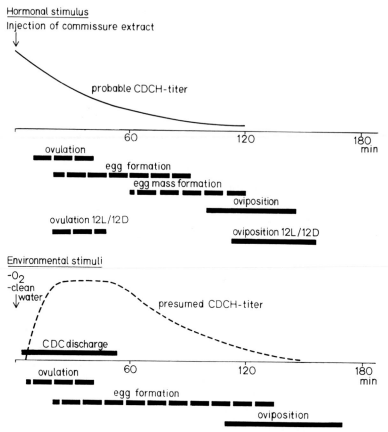

Fig. 22. Time schedules of egg mass formation in *L. stagnalis*. Egg-laying was induced either by injection of the ovulation hormone (CDCH) or after application of appropriate environmental stimuli. The schedules show a great similarity. Note the delayed reaction of the 12L/12D photoperiod-adapted snails compared to the 16L/8D snails (all other data) (cf. Fig. 15).

After a single injection of CDCH containing extract, all processes involved in the production of an egg mass are activated (Fig. 22). Ovulation of numerous oocytes (up to 200) occurs rapidly (within 5–10 min) (G. Dogterom et al., 1983a). The mechanism of ovulation in *Lymnaea* is not yet clear (de Jong-Brink et al., 1976). However, Saleuddin and Khan (1981) have presented evidence that in *Helisoma* oocytes may be liberated by activation of their own actin filaments. After transport of the oocytes to the carrefour region, where fertilization occurs, they are packaged individually as eggs and then assembled as the egg mass. This takes a considerable period of time. At 20°C, it may take 2 h (Fig. 22), but at low temperatures (4°C) it takes up to 7.5 h (G. Dogterom et al., 1983b). The packaging process is a highly complicated procedure. A complex program involving various secretory processes and ciliary and muscular activities is involved. The nerve supply to the female reproductive tract of freshwater pulmonates is well developed (Brisson and Collin, 1980; de Jong-Brink and Goldschmeding, 1983). Therefore, probably CDCH will control the packaging process indirectly by activation of the genital nervous plexus of the female tract.

In *Lymnaea* various environmental factors have long-term effects on the frequency of oviposition: an increase in the food supply (Scheerboom, 1978), high temperature (Joosse and Veld, 1972), and a long-day photoperiod (16L/8D) (Bohlken and Joosse, 1982). They all have a strong stimulating influence on the number of egg masses produced. At the short term, oviposition can rapidly be induced by a change from dirty to clean water and by the addition of oxygen to the water. The time sequence of the events resulting in the production of an egg mass after the application of these stimuli is presented in Fig. 22. The similarity to the effects of CDCH injection is striking. Two questions arise: Are these environmental factors acting via the CDC system, and if so, how does the external information reach the CDC?

Electrically, the CDC are usually silent. However, if a clean water stimulus is given to snails kept at long-day photoperiods, the CDC are active within 5 min (de Vlieger et al., 1980; Kits, 1980). As reported for the bag cells of *Aplysia,* the electrical activity of the CDC consists of a discharge, which is a series of action potentials, about 1 per 2 sec, lasting 30–60 min (Fig. 23). All CDC fire synchronously. The synchronization is achieved by electrotonic coupling of the cells in a loop area, where all axons of one cluster of CDC run closely parallel before turning to the periphery of the commissure for the release of the CDCH. The coupling of the activities of the two clusters is performed by ventral CDC. These cells (six in each cluster) have two axons: one directly releasing axon and a second one which runs to the contralateral loop area and subsequently to the release area (Fig. 24). Near the periphery of the commissure, the axons ramify frequently and end in beads of about 4 μm in size. These are the neurohemal endings, where release of hormone occurs. During a discharge the CDCH titer in the blood will increase rapidly, comparable to that of a CDCH injection. This has been demonstrated by quantitative studies of the exocytosis phenomena during an *afterdischarge,* which is a discharge elicited by electrical stimulation *in vitro* (Kits, 1981). It appeared that 5 min after the start of the afterdischarge, the number of exocytosis figures had increased ninefold, whereas after 15 min the increase was 36-fold (Roubos et al., 1981a). Not only ω-shaped exocytosis but also multiple exocytoses were observed (Roubos and Buma, 1982). Simultaneously, the shape of the action potentials changes: The initial and final action potentials are narrow, whereas during the main part of the firing period, slower, broad action potentials occur, consisting of a calcium-dependent part added to the initial sodium-dependent part (see Fig. 23) (Kits, 1981). The high release of

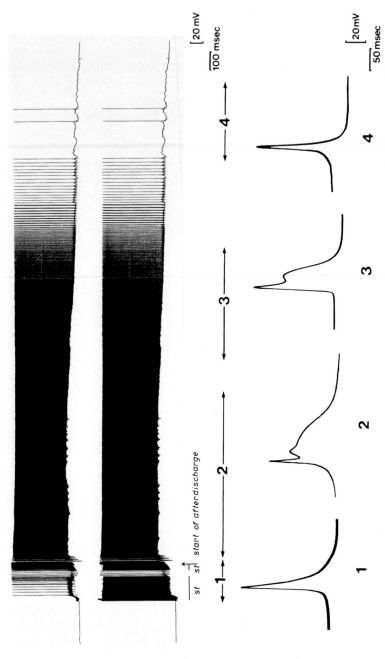

Fig. 23. Simultaneous recording of the afterdischarge activity of the CDC of *L. stagnalis*. After a number of electrical stimuli (lower cell), both cells start to fire synchronously for a period of about 50 min. During the afterdischarge, the shape of the action potentials changes: During phases 2 and 3, a calcium-dependent part is added to the sodium-dependent part. (After Kits, 1981.)

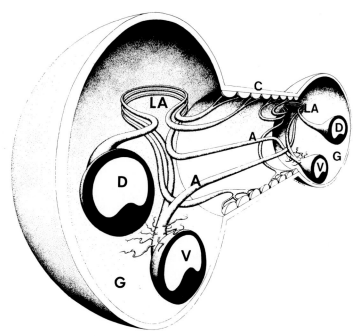

Fig. 24. Functional morphology of the CDC in the paired cerebral ganglia (G) of *L. stag-nalis.* Each dorsal (D) and ventral (V) CDC has an axon that runs via the ipsilateral loop area (LA). Ventral CDC have an additional axon (A) running via the contralateral loop area. All axons form neurohemal terminals in the periphery of the cerebral commissure (C). Ventral CDC have small dendritic branches near the cell body where input occurs. For further details, see text. (Courtesy of Dr. E. W. Roubos.)

the hormone occurs during the part of the afterdischarge with broad action potentials. Assays have demonstrated the presence of CDCH in saline, in which firing CDC had stayed for some time (Kits, 1980).

The discharge activity characterizes the *active state* of the CDC and accompanies the production of an egg mass. During the preparation of the egg mass, the CDC enter the *inhibited state,* during which it is impossible to induce an afterdischarge. The cells are able to respond only with action potentials of short duration. This state may last for 4–6 h. Then the CDC enter the *resting state.* In this state, the CDC can be activated to produce an afterdischarge (Kits, 1980).

The different states of the CDC probably play an important role in the reproductive physiology of the snail. Starvation (Joosse et al., 1968; Veldhuijzen and Van Beek, 1976) and stagnant water rapidly halt egg mass production. This is important in view of the high energy content of the egg mass in proportion to the total amount of reserve material present in the body. The CDC of starving snails have a more negative membrane potential. However, under these conditions, the cells can be easily activated by electrical stimulation. Thus, they are in the resting state. Apparently, the stop in oviposition is caused by a block in the transition from the resting to the active state due to a hyperpolarization of the membrane of the CDC (ter Maat et al., 1982). Starvation also results in a complex change in the ionic composition of the blood. In particular, the K^+ concentration is rapidly lowered after the onset of starvation and rises quickly after refeeding (de With, 1978; de With and Sminia, 1980). This change in ionic composition may be a signal to the CDC about the feeding condition of the snail. Starvation and refeeding also cause changes in blood glucose (Scheerboom et al., 1978) and amino acid levels. These components may also have a signaling effect.

After the production of an egg mass, the gonad has a refractory period of 6–20 h. This period tends to be shorter in snails kept at long-day photoperiods. It is clear that the duration of CDC inhibition is shorter than the refractory period of the reproductive system. A different situation is found in starved snails. Here, the CDC rapidly (within 6 d) stop spontaneous discharge activity. However, the gonad and female ASO are still able to react to CDCH injections by ovulation and egg mass production for a long time; the reactivity finally disappears completely after 25 d of starvation (G. Dogterom et al., 1983b). Upon refeeding, the reactivity is restored within a few days and precedes spontaneous oviposition, Thus, it is the CDC system that shows the most rapid reaction (by a change of state) to the adverse conditions.

The inhibitory action of food scarcity on oviposition can be overcome by the stimulating action of the long-day photoperiod. At this photoperiod, oviposition continues although food is absent. This unexpected behavior is part of the strategy of reproduction of animal as well as plant species, which have only one main reproductive season. During this period, the survival of the adults loses priority, because no second chance for reproduction might come. The absence of food is not necessarily a problem for the offspring, because these appear later and have a different food selection.

Probably all stimuli exerting short-term effects on ovulation act via the CDC system. Electrophysiological studies of ter Maat and Lodder (1980) have shown that the CDC react to electrical stimulation of all peripheral nerves by a biphasic response, of which the second (main) component is inhibitory. In isolated ganglia this stimulation results in an arrest of a running afterdischarge. Tactile stimulation of the body wall has the same effect, causing the CDC to change to the inhibited state. The presence of various types of synaptic contact on the CDC axons (Roubos and Moorer–van Delft, 1979) and of acetylcholine (ACh)-esterase activity and ACh receptors near and on fine lateral branches of the proximal part of the CDC axons demonstrates that the CDC receive diverse information (ter Maat et al., in press). This is in line with the great diversity of factors affecting CDC activity. The sense organs and pathways involved in the regulation and transduction of this information are as yet unknown.

Egg-laying behavior has been extensively studied in the pond snail (Goldschmeding et al., 1983). Four phases can be recognized: (a) stopping of locomotion, (b) turning, (c) oviposition, and (d) inspection of the egg mass (Fig. 25). Within 3–30 min after a clean water stimulus the snail ceases locomotion, the tentacles are dropped, and the shell is drawn down over the body. The second phase lasts about 90 min. Turns start by bending of the head and foot (usually clockwise). As a result, the foot is plied. Next, the shell is twisted (counterclockwise) until its apex points forward over the head. The number of radula movements increases during the last 30 min of the turning phase. These movements probably do not serve feeding but may have a surface-cleaning function (Fig. 26). Next, the egg mass is extruded and glued to the substrate (this phase lasts for about 10 min). The meaning of inspection of the egg mass is not clear.

This behavior, observed in the laboratory, fits the snail's behavior in the field. Egg masses have to be fixed on substrates in a microhabitat where conditions for embryonic development are optimal. For many freshwater snails, this habitat is the undersurface of floating leaves, e.g., of nympheids, or of submerged leaves near the water surface. Apparently, firing of the CDC is induced by the favorable environmental conditions in this area: the presence of oxygen, food, high temperature, and a solid substrate. Nearly all these factors are known to induce egg-laying. After the induction of CDCH release, the arrest of locomotion keeps the snails at this place. Turning and eating movements result in a clean substrate optimally suitable for the fixation of the egg mass. This has been elegantly demonstrated by showing the removal of a thin layer of starch from a glass plate by ovipositing snails: The animals cleared an area of the size and shape of the future egg mass (Goldschmeding et al., 1983; Fig. 26). As a result, in nature, the egg mass is prevented from sinking to the bottom, where often hardly any oxygen is present. It is to be expected that part of the egg-laying behavior is induced by direct effects of the CDC on centers in the brain.

This survey of the CDC system of *L. stagnalis* demonstrates that, like the bag cell system of *Aplysia,* it is a useful model for the study of basic characteristics of neuroendocrine cells. Undoubtedly, this system will play an important role in the ecology of this and related snail species.

Cells homologous to the CDC have been demonstrated in all species of freshwater pulmonates studied (Boer et al., 1977), but in only one terrestrial pulmonate (*H. aspersa;* Wijdenes et al., 1980). The CDCH of the Basommatophora is species specific at the genus level (G. Dogterom and van Loenhout, 1983). Saleuddin and Khan (1981) suggest that the effect of cerebral ganglion homogenates on oocyte motility and thus, possibly, on ovulation in *H. duryi,* might be attributed to the DBH present in the DB located on these ganglia. CDC are

Fig. 25. Characteristic postures during egg-laying behavior of *L. stagnalis*. From the upper left to the bottom right: resting phase, turning phase from dorsal, turning phase from ventral, deposition of the egg mass. (Courtesy of Dr. J. T. Goldschmeding and N. P. A. Bos.)

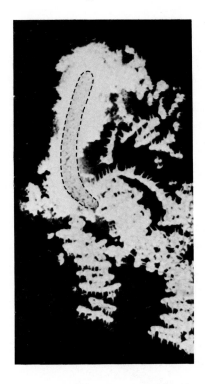

Fig. 26. Pattern of bites, made during egg-laying behavior of *L. stagnalis,* on a glass plate covered with starch. The area where the egg mass is deposited had been cleaned previously by the snail. The position of the egg mass is outlined. (Courtesy of Drs. J. T. Goldschmeding and M. Wilbrink.)

present in the cerebral ganglia of *Helisoma* (Saleuddin, personal communication), and it is therefore likely that CDCH exerts these effects.

Takeda (1979) stimulated egg-laying in *D. reticulatum* and *L. flavus* by injections of androgens and estrogens. The estrogens were particularly effective. However, this effect was primarily on oocyte development, not on ovulation.

In conclusion, it can be stated that our knowledge of the endocrine control of reproduction of the pulmonate gastropods is rapidly increasing. It already offers the most complete picture of this subject in the gastropods (Fig. 27). In the control of hermaphroditism, separate male and female control systems are operative. Of these, the female control system is the more complex. In terrestrial snails, as in vertebrates, neurohormones, a specific endocrine organ (DB), and a gonadal hormone are involved. In contrast to this third-order neuroendocrine mechanism, in freshwater pulmonates a second-order mechanism is present, due to the absence of a gonadal control of the ASO. In addition, the presence of a specific neurohormone for the control of the timing of ovulation, the preparation of the egg mass, and microhabitat selection have been demonstrated in the freshwater pulmonates. This detailed knowledge of the control of reproduction is

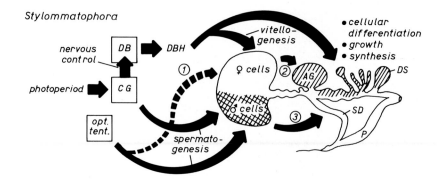

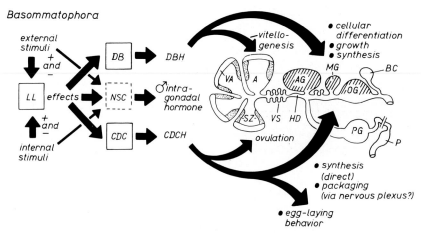

Fig. 27. Schematic representations summarizing the present data on the endocrine control of reproduction in pulmonate snails. The scheme for the Stylommatophora comprises data of various species: (1) inhibition of autodifferentiation of female cells; (2) female gonadal hormone controls female ASO; (3) male gonadal hormone controls cellular differentiation and growth of male ASO. The scheme for the Basommatophora is mainly based on the investigations of *L. stagnalis*. All actions of hormones concern the control of female activity, except for the stimulation of male maturation by center(s) in the Lateral Lobes (LL), which is probably effected via neurosecretory cells (NSC) in the central nervous system, and a male intragonadal hormone. For details, see text. A, Acinus; AG, albumin gland; BC, bursa copulatrix; CDC, caudodorsal cells; CDCH, caudodorsal cell hormone; CG, cerebral ganglia; DB, dorsal bodies; DBH, dorsal body hormone; DS, dart sac; HD, hermaphrodite duct; LL, lateral lobes; MG, muciparous gland; NSC, neurosecretory cells; OG, oothecal gland; opt. tent., optic tentacles; P, penis; PG, prostate gland; SD, sperm duct; SZ, spermatogenetic zones; VA, vitellogenetic areas; VS, vesiculae seminales.

available for only a few invertebrates and is unique with regard to hermaphroditic animals.

D. Cephalopoda: One Non-Sex-Specific, Gonadotropic Hormone

In contrast to the complex control patterns of reproduction in gastropods, the organization of the control of reproduction in cephalopods is strikingly simple. This may be due, first, to the terminal reproductive period in these gonochoristic molluscs. Probably all cephalopods normally breed only once and then die. Breeding of *O. vulgaris,* the most closely studied species, occurs in its second year. After spawning, the females hardly eat while caring for the eggs, and invariably die shortly after the eggs hatch. Various authors have shown that day length (Richard, 1970a; Wells and Wells, 1959, 1969, 1972), temperature (Mangold-Wirz, 1963), and feeding habits (e.g., Rowe and Mangold, 1975) have an influence on maturation. Mangold and Froesch (1977) could not find a relation between body size and start of reproduction in *O. vulgaris.* At the moment, the decisive factor which starts the reproductive activity is still unknown.

The second simplifying aspect is the control system itself. For the control of the reproductive organs, this seems to be restricted to one hormone, the production and release of which are under inhibitory nervous control. This scheme is unique among animals.

Various recent reviews of cephalopod reproductive endocrinology are available (Joosse, 1979a, 1979b; Wells, 1976; Wells and Wells, 1977a, 1977b). Research in this field started with the discovery of the precocious maturation of *O. vulgaris* after sectioning of the nerve supply to the optic glands (see Fig. 5) (Wells and Wells, 1959). As a result, the glands are activated, increase in size, and become bright orange. Upon removal of the glands, gonads that have already begun to develop regress (Wells et al., 1975). Transplanted glands or glands in organ culture release their hormone continuously (Richard, 1970b). All these observations point to an inhibitory nervous control of these endocrine glands. There is only one type of optic gland cell. As already mentioned (see Section II,D), opinions on the endocrine nature of these cells are conflicting: Mangold and Froesch (1977) could not find any histological evidence of hormone synthesis and release in the optic gland at any moment in the life cycle of an octopus. They even attributed a catabolic function to these glands (Froesch and Mangold, 1976). However, Froesch (1979) was able to induce the production and release of a particulated material, probably of a steroid nature, in the glands by using antigens (Froesch, 1979). However, there is no proof that this substance is the hormone. There is no doubt that in organ culture, cell divisions in the germinal epithelium of *Sepia* occur only in the presence of optic glands and thus of optic gland hormone (Durchon and Richard, 1967; Richard, 1976b).

In the male, the production of spermatophores and the growth of the male ducts are stimulated by the optic gland hormone. The hormone is not sex specific (Wells and Wells, 1975). In the female, it stimulates the production of proteinaceous yolk in the oocytes and the growth of the ASO. The oocytes of maturing octopods rapidly incorporate injected, labeled amino acids into protein. Isolated oocytes can be stimulated to do the same by the addition of optic gland extracts to the incubation medium (O'Dor and Wells, 1973, 1975). Wells et al. (1975) suggest that both uptake of amino acids and protein synthesis are independently controlled by the optic gland hormone.

The interesting and complicated sexual behavior of cephalopods is still performed by castrated animals or animals without optic glands. This suggests that this behavior is under nervous control (cf. Wells, 1976). There are no indications of an endocrine effect of the gonad on the accessory system (Wells and Wells, 1977b). However, the fact that the gonad of *Sepia officinalis* is able to synthesize steroids, and the presence of testosterone in the blood of these animals, point to an endocrine function of the gonad (Carreau and Drosdowsky, 1977).

There is an interesting relationship between maturation and metabolism in cephalopods (O'Dor and Wells, 1978). The optic gland hormone suppresses protein synthesis in the muscles of *Octopus*. This results in a considerable increase in the free amino acid pools in muscles and hemolymph. Normally, these amino acids are taken up by the ovary or the testis. However, after castration the amino acid levels are not much higher. Therefore, the authors suggest that a gonadal hormone will probably enhance the mobilization of amino acids. This supposed hormone would have no effect on the reproductive organs. In maturing animals, the regenerative capacity of cut arms is reduced. This also points to a breakdown of somatic structures in favor of reproduction. The optic gland hormone is the active factor coordinating these processes. This was convincingly demonstrated by Wodinsky (1977) in *O. hummelincki:* After removal of the optic glands from females brooding their eggs, the animals generally abandon the eggs, resume feeding and growth, and survive longer than intact animals.

At present, the chemical nature of the optic gland hormone is uncertain. The nervous origin of the glandular cells (Bonichon, 1967) points to a peptide nature. The hormone has no clear species specificity within the octopods (Wells and Wells, 1977b). It is important to note that no indications of a feedback action of the optic gland hormone have been observed.

The overall picture arising from these data is that in cephalopods the highly developed nervous system, together with the terminal ''self-destructive'' reproduction, has led to a reduction of the endocrine part of the control mechanisms. Only one hormone controls nearly all reproductive activities. The terminal reproductive period has eliminated the need to modify these activities, once they have started, in relation to changes in the external environment. The hypothetic hormone produced by the gonad has a role only in the control of secondary metabolic activities. For a comparative endocrinological view on this control scheme, see Joosse (1979a).

IV. Control of Body and Shell Growth and Shell Regeneration

In shell-bearing molluscs, growth of the soft body parts must be distinguished from growth of the shell. The mechanisms behind these two activities are basically different. Growth of the body is characterized by an increase in the dry weight of the organic material (mainly protein). A growth-stimulating or somatotropic hormone should stimulate the growth of all organs but should leave their proportional weight (percentage of total body weight) unchanged. In most bivalves and gastropods, shell growth implies an increase in its surface area, resulting in an increase in the shell cavity. Shell enlargement occurs at the shell edge by the secretion of organic material for periostracum, formation of which is followed by crystalline layers mainly consisting of (inorganic) calcium carbonate (cf. Wilbur, 1976). In most shell-bearing molluscs, body growth and shell growth have to occur simultaneously in order to ensure their correct functional interrelationship.

As will be shown below, during the past 10 years clear evidence has been presented for the occurrence of growth-controlling hormones in molluscs. It appears that both body and shell growth are stimulated by the same hormone. Particular attention will be given to the effects of the growth hormone of *L. stagnalis,* which are known in some detail.

Many molluscs are able to repair shell damage. During shell regeneration, body growth always stops. Consequently, shell repair may be stimulated by (endocrine) factors other than growth hormones. Therefore, shell regeneration will be discussed separately.

A. Control of Body and Shell Growth

The first suggestion of the presence in molluscs of an endocrine factor involved in the control of growth came from Lubet (1971). On the basis of extirpation and implantation experiments, he showed that the cerebral ganglia of the prosobranch *Crepidula fornicata* produce a factor that stimulates body growth. Recently, P. Le Gall (1980) has performed more detailed experiments on the same species, also using the techniques of cauterization of parts of the central nervous system (CNS) and sectioning of nerves. A temporary strong decrease in body growth was observed in animals after cauterization of anterior parts of the cerebral ganglia. This growth could be only partly restored by cerebral ganglia implantations. Moreover, sectioning of the cerebral nerves also resulted in a severe and persistent decrease in body growth. Le Gall concluded that the cerebral ganglia might stimulate growth in a hormonal as well as in a nervous way.

A further interesting observation on *C. fornicata* concerns the production of a mitogenic humoral factor by the cerebral ganglia (P. Le Gall, 1980). This substance, which is released at the periphery of cerebral nerves, stimulates the differentiation and growth of the organs and is identical to the factor that stimulates the mitoses of the male cells in the gonad (see Section III,A,1). The factor is present in various species of prosobranchs and bivalves. Its nature and chemistry are unknown.

In the basommatophoran *L. stagnalis,* Geraerts (1976a) was able to demonstrate that the neurosecretory LGC (see Fig. 1) produce a growth hormone. The LGC are located in the cerebral ganglia in two paired groups with a total number of about 200 cells. The LGC hormone is released at the peripheries of the median lip nerves (Joosse, 1964). Removal of the LGC from rapidly growing juvenile snails results in markedly retarded body growth, which can be restored by implantation of cerebral ganglia containing LGC (Fig. 28). The diminished body growth appears to be the result of retarded growth of all body parts. Because the LGC hormone stimulates the proportional growth of all body parts, it has the characteristics of a real growth hormone (see above). For growth of the female sex organs, the presence of the gonadotropin DBH (see Section III,C,3) in addition to this growth hormone is needed.

In the pulmonate slug *D. reticulatum,* the neurosecretory medial cells in the cerebral ganglia have the same effects on body growth as the LGC in *Lymnaea* (Wijdenes and Runham, 1977). After removal of the medial cells, body growth stops. This cessation is not due to a lack of nutrients, because the levels of glucose and glycogen in the nonreproductive tissues increase. The same was found in *Lymnaea* (see below).

In *L. stagnalis,* a second center is involved in the control of body growth. This is located in the LL of the cerebral ganglia (see Fig. 1) (Geraerts, 1976b). Cauterization of the LL results in a giant growth (Fig. 28), whereas simultaneous implantation of cerebral ganglia with LL restores normal growth. For this effect of the LL, the presence of the growth hormone producing LGC is necessary. Quantitative electron microscopic studies showed that the synthesis and release of the hormone in the LGC of LL-deprived snails are activated (Roubos et al., 1980). Therefore, the LL normally exert an inhibiting effect on the LGC and via these on body growth. The giant growth of LL-cauterized snails, however, is aberrant. The body dry weight of the large snails is not different from that of sham-operated controls, and their shell is extremely thin. The high water content of these snails is mainly due to a large blood volume (Scheerboom and Dogterom, 1978). However, the low glycogen content of LL-cauterized snails indicates that other metabolic effects of the increased growth hormone release are normal (Geraerts, 1976c).

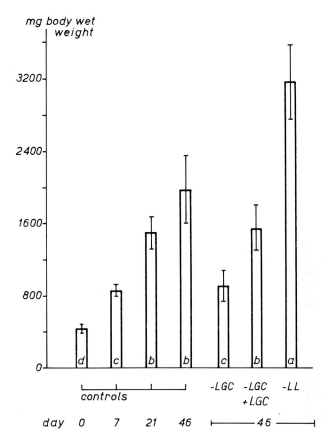

Fig. 28. Effects of the LGC and the LL on body growth of *L. stagnalis*. At day 46 after cauterization of the LGC (−LGC), body growth is considerably retarded. Growth is restored to nearly the normal rate after implantation of cerebral ganglia with complete LGC systems (+LGC). Removal of LL is followed by enormous growth. Groups sharing a common letter do not differ significantly. (Redrawn after Geraerts, 1976a.)

B. Antagonism of Growth and Reproduction

A second important effect of the LL in *Lymnaea* is the stimulation of reproduction (see Section II,C,2). This is again an indirect effect, as it is achieved by activation of the DB and CDC (Geraerts, 1976b; Roubos et al., 1980). Apparently, the LL are involved in the control of the antagonism between growth and reproduction. Probably the LL act by way of one or more hormone(s), although the involvement of neuronal pathways cannot be ruled out (Roubos et al., 1980).

With regard to *O. vulgaris*, good evidence has been presented for endocrine control of the antagonism of growth and reproduction. The start of reproduction coincides with cessation of body growth. Both are induced by the optic gland hormone. For gametogenesis, muscle protein is mobilized. For the release of amino acids from muscle cells into the blood, a second hormone from the gonad may be involved (O'Dor and Wells, 1978). It is not known whether body growth is under endocrine control prior to the onset of reproductive activity.

C. Effects of the Growth Hormone on the Tissues

In mammals, a single injection with the pituitary growth hormone very rapidly stimulates the activity of the enzyme ornithine decarboxylase (ODC) (cf. Dogterom et al., 1979b). Increased activity of this enzyme was also found in mantle edges of *L. stagnalis* after injection of a crude extract of the median lip nerves that contain the growth hormone. This response shows great variability, which hampers its application in a bioassay procedure (Dogterom and Robles, 1980). ODC activity is low in animals in which the LGC have been cauterized, and in starving snails, which also do not grow. Upon refeeding, ODC activity is restored within 4 h. The function of ODC in body growth is only poorly understood (Dogterom et al., 1979b).

The marked increase of the glycogen stores in LGC-deprived pond snails is considered to be a secondary effect of the cessation of growth in these animals (Dogterom, 1980; Geraerts, 1976a; Scheerboom and Dogterom, 1978). Glucose levels in the hemolymph tend to be higher and may cause a rise in glycogen storage (see Section V,C). Similar effects on glucose levels and glycogen stores were observed in *D. reticulatum* after cauterization of the medial cells (Wijdenes and Runham, 1977). After removal of the LGC, the food uptake of *Lymnaea* is proportional to the body weight, and assimilation of the ingested food is equal to that of controls (Geraerts, 1976a; Scheerboom and Dogterom, 1978).

These observations indicate that the growth hormones of gastropods, like that of the vertebrates, exert effects on all tissues.

D. Effects of the Growth Hormone(s) on Shell Formation

As mentioned above, shell-bearing molluscs have to coordinate the growth of soft body parts with that of the shell. It is therefore of great importance that Geraerts (1976a) has demonstrated in *L. stagnalis* that the growth hormone stimulates both processes. The effect on the shell concerns shell enlargement, i.e., formation of new shell by secretion of periostracal materials and of the outer crystalline layer. Detailed studies by A. A. Dogterom have elucidated the role of growth hormone in the control of these processes in *Lymnaea*.

The effect of the hormone on periostracum formation was studied by determining the rate of [^{3}H]tyrosine incorporation into the proteinaceous components of

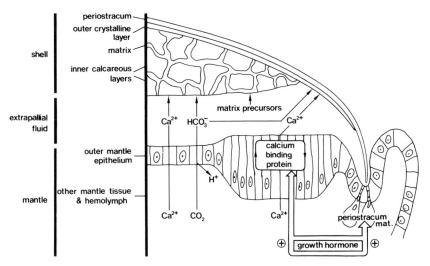

Fig. 29. Schematic transverse section of the edges of the mantle and the shell of *L. stagnalis*, showing the role of the growth hormone in shell enlargement. For details, see text. (Courtesy of Dr. A. A. Dogterom.)

the marginal shell edge (Fig. 29). This was compared with the labeling of the organic material in the remaining part of the shell (mainly shell matrix). It appeared that after LGC removal the periostracum formation decreases considerably, whereas shell matrix formation is not affected (Dogterom and Jentjens, 1980).

In experiments on the effects of growth hormone on the formation of the outer crystalline layer, attention was focused on the mantle edge. LGC removal had no effects on the calcium concentration of the soft body parts, with the exception of the mantle edge. In this tissue, the calcium concentration was 30% lower than normal. After the injection of ^{47}Ca in control snails, the shell edge only showed a continuous increase in specific activity. In the absence of growth hormone (after LGC removal) the incorporation of label in the shell edge was 70–90% lower (A. Dogterom et al., 1979a). The incorporation of injected [^{14}C]bicarbonate in the shell edge of LGC-deprived snails showed a similar decrease (with a minimum of 10% of the control value). In the mantle edge and the outer mantle epithelium of these animals, carbonic anhydrase was still present, which suggests that this enzyme is not the limiting factor in bicarbonate movements to the shell (Dogterom and van der Schors, 1980). The authors conclude that the effect of growth hormone on bicarbonate movements during shell formation is secondary, the primary effect being the calcium transport through the mantle edge.

In many cases, transport of calcium through cells is accompanied by the presence of calcium-binding proteins (CaBP) in these cells (for literature, see Dogterom and Doderer, 1981). By means of gel filtration and ion-exchange chromatography on calcium-saturated Chelex-100 columns, a CaBP fraction could be isolated from the mantle edge. This fraction was not present in other tissues. Removal of the LGC resulted in a strong decrease of the amount of CaBP (Fig. 30). Thus, *Lymnaea* has a hormone-dependent CaBP, which probably is responsible for the maintenance of the high calcium concentration in the part of the mantle that produces the outer crystalline layer of the shell (see Fig. 29). The details of the actual process are unknown (Dogterom and Doderer, 1981).

In conclusion, it can be stated that the growth hormone of *Lymnaea* stimulates shell growth by increasing the synthesis and release of periostracal materials and by maintaining a high concentration of CaBP in the mantle edge. The hormone is not involved in controlling the formation of the inner calcareous layers.

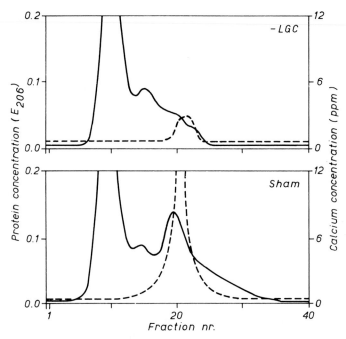

Fig. 30. Elution patterns of mantle edge homogenates of *L. stagnalis* processed through a calcium-saturated Chelex-100 column. After removal of the growth hormone–producing cells (−LGC, top), a calcium-binding protein, clearly present in the sham-operated animals (bottom), has nearly completely disappeared. For details, see text. Solid line, protein concentration; broken line, calcium concentration (ppm). (From Dogterom and Doderer, 1981.)

E. Chemical Nature of the Growth Hormone

Histological and ultrastructural characteristics of the LGC suggest that the growth hormone has a peptide nature (Wendelaar Bonga, 1970). Immunohistochemical studies have shown that the secretory material in the LGC and their neurohemal areas cross-reacts with antibodies against mammalian somatostatin (Schot et al., 1981). The reaction was also positive in homologous cells of *Physa* (Grimm-Jørgensen, in press). Moreover, immunoreactive somatostatin-like material was found in the hemolymph and in the gastrointestinal tract of *Physa*. The concentration in the hemolymph showed a positive relationship with the growth rate. With mammalian somatostatin, however, the synthesis of DOPA-rich protein (periostracin) in the mantle edge could not be stimulated. These results indicate that the gastropod growth hormone, although somatostatin-like, is distinctly different from mammalian somatostatin (Grimm-Jørgensen, in press).

F. Factors Controlling LGC Activity

Body and shell growth of freshwater pulmonate snails is known to change rapidly upon changes in the external (e.g., food supply, water composition, temperature, photoperiod) and internal (e.g., age, reproductive activity) conditions. Probably these changes in growth rate are effected by the growth hormone system. Little is known about the registration and integration centers of these parameters and how they are transduced to the LGC. As mentioned above, the LL will be involved in these pathways.

Recently, Roubos and van der Wal-Divendal (1982) discovered in *Lymnaea* an input system of the LGC (Fig. 31). Environmental stimuli are registered and integrated by two types of sensory cell located in the epidermis near the base of the tentacles. The axons of one of these cell types run through the tentacular nerves to the LGC, which are contacted synaptically. Furthermore, the LGC show large infoldings of the plasma membrane. This suggests that they can perceive extracelluar, nonneuronal stimuli. Nothing is as yet known of the factors registered by the sensory cells and the infoldings.

G. Control of Shell Regeneration

As mentioned in Section IV,A, the growth hormone of *Lymnaea,* and thus possibly also that of other freshwater and terrestrial pulmonates, stimulates shell enlargement only. This means that shell repair cannot be influenced by this hormone, except when the damage is at the shell edge. However, during shell repair shell growth stops. Therefore, probably local sensory information is important in the control of shell repair. In this respect, it is of interest that

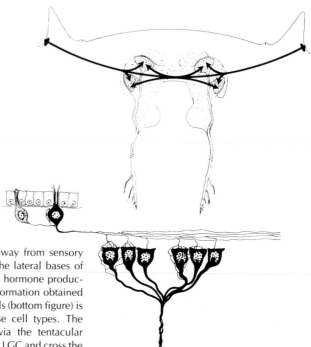

Fig. 31. Nervous pathway from sensory cells in the epidermis at the lateral bases of the tentacles to the growth hormone producing LGC of *L. stagnalis*. Information obtained by two types of sensory cells (bottom figure) is integrated by one of these cell types. The axons of the latter run via the tentacular nerves to the four groups of LGC and cross the cerebral commissure (top figure). The number of synapse-like contacts per LGC varies considerably. (Courtesy of Dr. E. W. Roubos.)

Saleuddin and Dillaman (1976) have shown that the mantle edge gland of the basommatophoran *H. duryi* is directly innervated by neurosecretory axons, which strongly suggests a local control mechanism. Dillaman et al. (1976), studying the same species, observed a correlation between changes in the activity of special neurosecretory cells in the visceral ganglion and shell regeneration. On the basis of the effects of injections of homogenates of the cerebral and subesophageal (including visceral) ganglia, Kunigelis and Saleuddin (1978) concluded that in *H. duryi* shell growth may be controlled by a factor from the cerebral ganglia and shell regeneration by a factor from the subesophageal ganglia. The first factor may be homologous to the growth hormone of *Lymnaea*.

For shell repair, large amounts of calcium have to be mobilized rapidly. Recently, Saleuddin and Jones (1978) have suggested that the type 1 cells in the visceral ganglion of *H. duryi* might affect shell regeneration by causing a rise of the blood calcium level, because hypercalcemia was observed after injection of homogenates of the subesophageal ganglia.

Whitehead (1977) and Whitehead and Saleuddin (1977) approached the problems in a completely different way. They injected ecdysterone and found that this depresses the incorporation of ^{45}Ca into intact shells but enhances the uptake by regenerating shells. These experiments were performed with starving snails. Cholesterol has the same effect. The authors suggest that these steroids mimic the action of a factor which, in normal circumstances, is released after shell damage.

V. Control of Energy Metabolism

In molluscs, glycogen is an important reserve material. It is present in all cells. However, these animals also have special cells for glycogen storage, which occur in large numbers in the mantle edge region: the so-called vesicular cells of bivalves (Lubet, 1959) and the vesicular connective tissue cells of gastropods (Sminia, 1972). For all these cells, the functional name *glycogen cells (GC)* is proposed, which stresses the analogy to the fat cells of vertebrates. In some families of bivalves (e.g., the Mytilidae), a second type of storage cell is found, the adipogranular cells. In addition to glycogen, these cells store mainly lipid and protein (Lubet, 1959).

Evidence for the occurrence of metabolic hormones in bivalves and gastropods is now becoming firm. For reviews of this field, see Gabbott, (1975), Joosse (1979b), and Plisetskaya and Joosse (in press).

A. Factors Affecting Hemolymph Glucose Concentrations

In molluscs the hemolymph glucose concentration shows considerable variation. In *Patella* seasonal variations occur, ranging from 30 to 60 μg/ml. In field specimens of *L. stagnalis* the concentrations range from 4 to 350 μg/ml, with extremely low mean values in winter when food uptake is negligible (Scheerboom and Van Elk, 1978).

In laboratory experiments, clear relationships between food uptake and hemolymph glucose levels have been demonstrated in bivalves (cf. Gabbott, 1975) and gastropods. The most detailed studies concerning gastropods have been performed by Veldhuijzen and van Beek (1976), Scheerboom (1978), and Scheerboom et al. (1978). They showed that in lettuce-fed specimens of *L. stagnalis* the blood glucose concentration is rather low (about 20 μg/ml). When higher amounts of lettuce are consumed and assimilated, it may rise to 55 μg/ml. A very rapid (within 4 h) and conspicuous increase, up to a mean value of about 760 μg/ml, occurs after a change to a carbohydrate- and protein-rich diet. Afterward, the concentration decreases gradually (within 24 h) to about 200–250 μg/ml. This decrease is caused by a lower food consumption and by the removal of glucose from the hemolymph. High hemolymph glucose levels repress locomotion and food consumption, probably by nervous pathways (Scheerboom and Doderer, 1978). Glucose itself may also activate endocrine centers (e.g., insulin cells; see below) to stimulate its removal from the hemolymph. The high tolerance of bivalves and gastropods to fluctuating blood glucose levels allows a low rate of reactivity of these control systems.

B. Hyperglycemia

In terrestrial pulmonate snails, hyperglycemia is known to occur as a "stress" reaction to mechanical trauma. For instance, in specimens of *Strophocheilus oblongus* the blood glucose concentration increases from 26 to 168 µg/ml within 40 min after the introduction of a heart catheter through the shell and after manipulation and injections (Fig. 32) (Marques and Falkmer, 1976). In the freshwater pulmonate *L. stagnalis* anoxia induces strong hyperglycemia: The glucose level increases from 40 to 775 µg/ml within 6 h (Fig. 32). Stressing

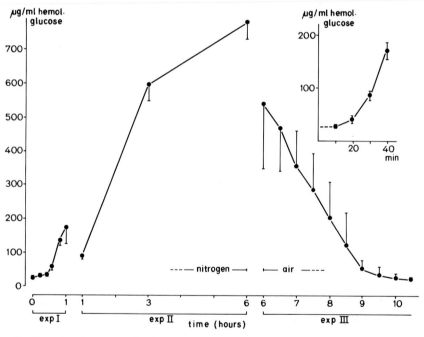

Fig. 32. Experiments I and II demonstrate the strong hyperglycemic reaction of *L. stagnalis* during anaerobiosis and the return to a normal glucose level in the hemolymph (experiment III) upon subsequent aeration of the water. Inset: Hyperglycemia in *Strophocheilus oblongus* evoked by the blood sampling procedure. For details, see text. (Experiments I–III courtesy of Dr. Th. C. M. Wijsman; inset redrawn after Marques and Falkmer, 1976.)

agents, such as NaCl (0.5%), $CuSO_4$ (5×10^{-6} M), and anesthetics (Nembutal and MS 222) also induce hyperglycemia, albeit less pronounced (Th. C. M. Wijsman, unpublished results). The way in which these responses are effected is not known. In view of the situation in insects and crustaceans, it is presumed that one or more hormones are involved.

C. Control of Glycogen Storage

In *Mytilus edulis* the glycogen stores show a clear annual periodicity which is related to reproductive activity (de Zwaan and Zandee, 1972; Gabbott, 1975; Gabbott and Bayne, 1973; Gabbott et al., 1979). The reserves start to increase after the reproductive period and are mobilized later during gametogenesis. In the mantle the key enzyme for glycogen synthesis, glycogen synthetase, shows simultaneous changes in activity. Both the local activity of the enzyme and the percentage of the enzyme which is in the active form increase two- to threefold in the mantle tissue during summer, thus coinciding with the glycogen accumulation in this period. Increased feeding activity and a concomitant increase in tissue glucose concentration probably are involved in changing the proportion of the enzyme which is in the active form.

However, hormones may also be involved. Lubet and his associates have studied, in *M. edulis,* the effects of various nervous centers on the storage cells (GC and adipogranular cells) and the gonad located in the mantle (Lubet, 1959; Lubet et al., 1976). In an anhormonal organ culture medium, pieces of mantle tissue are difficult to maintain. Addition of cerebral ganglia induces the start of lysis of the GC and adipogranular cells, which in normal animals accompanies gametogenesis. The sex cells seem to continue their development (Herlin-Houtteville and Lubet, 1974). Addition of visceral ganglia to cultures of mantle tissue, however, prevents lysis and induces an increase in reserves comparable to that observed in field specimens collected after the reproductive season. Gametogenesis stops under these conditions. Lubet et al. (1976) conclude that the visceral ganglia release a neurohormone which stimulates the storage activity of the GC and the adipogranular cells, whereas the cerebral ganglia produce a neurohormone with a mobilizing action on the reserves, which is needed for the survival and development of the gametes.

Thus, in bivalves, the storage activity in the mantle depends on the concentration of metabolites, such as glucose, in the hemolymph (i.e., on the feeding condition) and on the action of neurohormones. As will be shown below, the storage of glycogen in the muscles of bivalves is stimulated by an insulin-like hormone. Whether this hormone also affects the special storage cells is as yet unknown.

In *L. stagnalis* the feeding condition clearly influences the total amount of glycogen in the body. Starvation results in a gradual decrease of glycogen in the

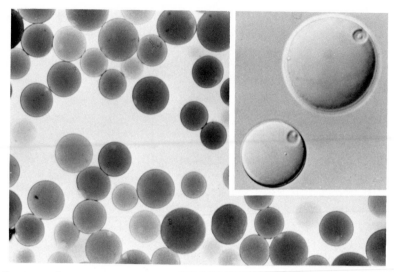

Fig. 33. Isolated glycogen cells of *L. stagnalis.* The diameter of these cells varies from 10 to 75 μm. Inset: Interference contrast light microscopy picture of isolated glycogen cells. Note the peripheral location of the small nuclei. The cells are nearly completely filled with glycogen. (Courtesy of Dr. M. Hemminga.)

mantle. In this organ, the largest numbers of GC are found. In the other body parts (e.g., muscles), glycogen starts to decrease only after 6 days of starvation. A carbohydrate-rich diet first leads to a rapid saturation of the muscles with glycogen, whereas the mantle and the other body parts show a more gradual increase (Veldhuijzen and van Beek, 1976). These changes in glycogen reserves are correlated with changes in the blood glucose levels. Recently, M. A. Hemminga (unpublished results) succeeded in isolating the GC of *L. stagnalis* (Fig. 33). This offers the possibility of studying directly the effects of the glucose content of the medium on glycogen synthesis in these cells. In the GC, the incorporation of labeled glucose into glycogen appears to be linearly and positively related to the glucose concentration in the medium. This suggests that *in vivo* hemolymph glucose levels play a major and direct role in the control of glycogen synthesis in the GC. However, glycogen mobilization cannot be explained in this way. The fact that during starvation glycogen stores are depleted, whereas the blood glucose levels are only slightly lower than in lettuce-fed snails, suggests that other factors are involved (Veldhuijzen and van Beek, 1976). In view of the general occurrence of hyperglycemic hormones in animals, mobilization of glucose in molluscs may be under endocrine control. Such a hormone may also be responsible for the hyperglycemia induced by anoxia, which coincides with a decrease of the glycogen stores in the mantle of *L. stagnalis* (Th. C. M. Wijsman, unpublished results).

In *L. stagnalis* the storage of glycogen can also be changed by a change of the photoperiod. A transfer of snails from long-day (16L/8D) to short-day (8L/16D) photoperiods induces an important increase in the amount of glycogen in the mantle (Fig. 34). At the same time, a decrease in ovipository activity occurs. Blood glucose levels of lettuce-fed animals are low. Therefore, this increase cannot be explained on the basis of a change in the concentration of glucose in the blood, as was the case during carbohydrate-rich food uptake (see below) (M. A. Hemminga and J. Joosse, unpublished results). In this influence of the photoperiod, hormones affecting the reproductive activity may be involved, as was shown in *Mytilus*.

In *L. stagnalis*, the changes in the glycogen stores in relation to the presence or absence of growth hormone have been discussed above (Dogterom, 1980; see

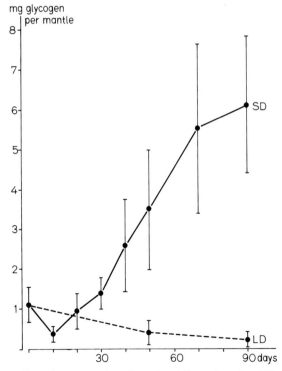

Fig. 34. The effect of transition from long-day (LD) to short-day (SD) photoperiods on glycogen storage in the mantle of *L. stagnalis*. (M. A. Hemminga and J. Joosse, unpublished results.)

Section IV,C). They are considered to be indirect effects. The same holds for the effects of the LL on glycogen metabolism (Geraerts, 1976c).

Thus, in gastropods, the increase in glycogen (in the mantle) may be dependent on glucose levels in the hemolymph, and therefore on the feeding condition. Mobilization of glycogen depends on other factors and may be controlled by hormones. As will be shown below, glycogen storage in muscles of some gastropods can be stimulated by insulin. Similar to the situation in bivalves, in gastropods the effect of an insulin-like hormone on the GC is not yet known.

D. Insulin-Like Factors in Molluscs

There is now good evidence for the occurrence of cells producing insulin-like substances (ILS) in molluscs. The demonstration of ILS in cells is based on the classical staining method with paraldehyde-fuchsin and/or on a positive reaction with antibodies against mammalian insulin. ILS cells have been identified in the gut epithelium in the region of the hepatopancreas of the marine bivalve *M. edulis* (Fritsch and Sprang, 1977; Fritsch et al. 1976) and the freshwater bivalves *Unio pictorum* and *Anodonta cygnea* (Plisetskaya et al., 1978b), in the intestine of the prosobranch *Buccinum undatum* (Boquist et al., 1971; Davidson et al., 1971), in the digestive gland and pylorus of the terrestrial pulmonate *Achatina fulica* (Gomih and Grillo, 1976), and in the CNS of *L. stagnalis*. In the last species also neurons reacting with antiglucagon have been observed (Schot et al., 1981).

The first physiological experiment demonstrating the presence of an ILS in molluscs was done by Collip as early as 1923. He extracted an ILS from the tissues of the marine bivalve *Mya arenaria* and injected it into a rabbit. This caused the typical convulsions of a hypoglycemic condition, which could be relieved by an injection of dextrose. Recently, detailed experiments have been performed by Plisetskaya and her associates in the freshwater bivalves *U. pictorum* and *A. cygnea*. They saw a loss of secretion product from the ILS cells in the small intestine after injections of glucose. At the same time, the insulin titer in the hemolymph, determined with a radioimmunoassay test for mammalian insulin, increased (Fig. 35). Intramuscular injections of mammalian insulin were followed by a decrease in the blood glucose level and an increase in the activity of the enzyme glycogen synthetase in the adductor muscles (Plisetskaya et al., 1978a; cf. Plisetskaya and Joosse, in press). Similar reactions were obtained in the marine bivalves (*M. galloprovincialis* and *Chlamys (Flexopecten) glaber ponticus* (Plisetskaya and Soltitskaya, 1979).

The ILS has been extracted from the freshwater bivalves *Unio* and *Anodonta*. The extract was more potent in these species than mammalian insulin. In the marine bivalve *M. galloprovincialis* the extract appeared to stimulate the activity of glycogen synthetase, whereas mammalian insulin had little effect (Plisetskaya

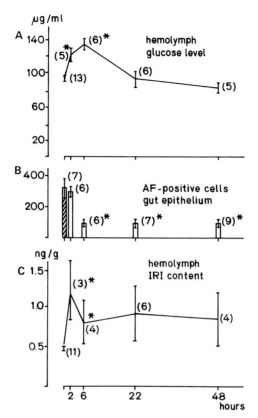

Fig. 35. Effects of a glucose injection (30 mg/100 g) on (**A**) the hemolymph glucose level, (**B**) the number of aldehyde-fuchsin (AF)-positive cells in the gut epithelium, and (**C**) the immunoreactive insulin (IRI) content of the hemolymph of *Anodonta cygnea*. An asterisk indicates the $p < .05$ compared to the controls (crosshatched column). (After Plisetskaya et al., 1978b.)

et al., 1979). Moreover, application of an antiserum against molluscan insulin revealed a 10× higher insulin titer in the radioimmunoassay test compared with that for mammalian insulin (Rusacov and Kazakov, 1979). The results clearly show that the molluscan insulin molecule differs from that of the mammals.

In gastropods, Marques and Falkmer (1976) have studied the effects of insulin. They saw a more rapid disappearance of a glucose load from the hemolymph of the terrestrial pulmonate *Strophocheilus oblongus* when the animals had been injected with a high dose of mammalian insulin. Normally, the glucose concentration in the blood of this snail is too low for the effect of insulin to be detected. In the slugs *Arion ater* and *Ariolimax columbianus* the hemolymph glucose concentration is high, up to 250 μg/ml. An injection of mammalian insulin in these species was followed by a significant decrease of the glucose levels in the blood (Plisetskaya and Joosse, in press). In contrast, mammalian insulin had no effect on normal or artificially raised hemolymph glucose levels of *L. stagnalis* (M. A. Hemminga, unpublished results).

To summarize these investigations, there is no doubt that bivalves and gastropods have specific types of insulin molecules. These remove glucose from the hemolymph by stimulating the synthesis of glycogen. It is important to note that this effect of insulin has been demonstrated in muscle cells. In view of the crucial function of the GC in the glycogen metabolism of molluscs, studies on the effects of insulin in these cells will be of great interest.

VI. Cardioactive Peptides

At present, there are a number of reasons to suggest that the heart rate in molluscs is very accurately controlled, and that neuropeptides probably play an important role in this control (cf. de With, 1980; Geraerts et al., 1983b; Koester et al., 1979; Greenberg and Price, 1980). First, evidence for the importance of the heart rate to various physiological functions will be presented. Second, evidence for the role of neuropeptides in cardiac control will be examined. Attention will be focused on the functioning of the peptides (neurohormonal, neuromodulator/neurotransmitter).

A. Importance of Cardiac Control

In molluscs the heart rate may change considerably in response to changes in external and/or internal environmental factors and behavior, e.g., temperature, availability of food, blood metabolite concentrations, and estivation (de With, 1978, 1980; de With and Sminia, 1980; Lloyd, 1978a, 1978b). Various studies have shown that the heart rate plays a crucial role in a number of physiological processes, e.g., energy metabolism, water and ion metabolism, acid–base balance, activation from dormancy, and flight and fight reactions (in cephalopods) (de With, 1977, 1978, 1980; de With and Sminia, 1980; de With et al., 1980; Geraerts et al., 1983b; Wells and Mangold, 1980). In many species the heart also acts as the primary filter in urine formation. Thus, changes in the heart rate may also affect the rate of ultrafiltration, resorption, and excretion (Geraerts et al., in press). An extensively studied species in this respect is the basommatophoran *L. stagnalis*. De With (1977, 1978, 1980; de With and Sminia, 1980; de With et al., 1980) has unraveled the complex interrelationships between the effects of starvation on the heart rate and on a number of physiological processes. He found that during starvation the heart rate decreases considerably (from 38 to 22 beats/min). According to this author, this decrease in heart rate is the direct cause of the reduced rates of renal filtration, water turnover, and energy metabolism. The lowered energy metabolism, in turn, causes changes in the blood pH and, therefore, in the ionic composition of the blood. This example clearly shows that the animal can manipulate many aspects of its physiology by simply changing its heart rate.

B. FMRFamide and Related Peptides

Molluscan cardiac activity is, in part, directly regulated by the cardiac nerves. Associated with this type of control are the usual neurotransmitters, such as 5-hydroxytryptamine (5-HT), acetylcholine (ACh), and catecholamines (cf. Cottrell and Osborne, 1969). During the last four to five decades, the actions of

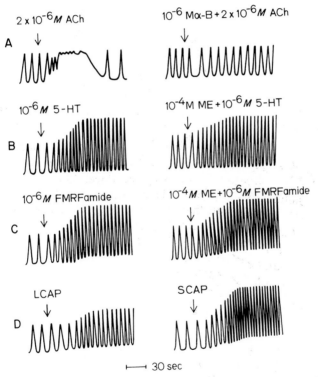

Fig. 36. Effects of neurotransmitters and peptides on the *in vitro* activity of a *Lymnaea* auricle preparation. (**A**) The effect of acetylcholine (ACh), which is blocked by the anti-acetylcholinergic agent α-bungarotoxin (α-B); (**B**) the effect of 5-hydroxytryptamine (5-HT), which is not blocked by the antiserotonergic agent methysergide (ME); (**C**) the effect of synthetic FMRFamide, which is not blocked by ME; (**D**) cardioexcitatory effects of a large cardioactive peptide (LCAP) and an FMRFamide-like small cardioactive peptide (SCAP) of *L. stagnalis*. (After Geraerts et al., 1981.)

these transmitters on the molluscan heart have been studied in great detail. This has mainly been done with *in vitro* heart preparations (Fig. 36). In fact, the hearts of some molluscan species, especially those of the bivalve family Veneridae, are very sensitive to neurotransmitters and, therefore, have.been used in physiology and pharmacology as standard objects for assaying neurotransmitters and their analogs.

In the 1960s cardioregulatory neuropeptides were isolated from molluscan brain extracts. Since then, cardioactive neuropeptides have been found in molluscan species belonging to all classes (cf. Greenberg and Price, 1980). Most peptides, however, have not been well characterized chemically and functionally.

Recently, one of these peptides has been purified, identified, and synthesized (Price and Greenberg, 1977a,b). It is the tetrapeptide amide Phe-Met-Arg-Phe-NH_2 (FMRFamide) of the bivalve *Macrocallista nimbosa*. Structure–activity relationship studies strongly suggest that the C-terminal with the free amide group is the operational portion of the FMRFamide molecule (Price and Greenberg, 1980). Initially, it was thought that the effects of FMRFamide on the molluscan heart and other molluscan muscles are exclusively excitatory, and that it is therefore a mimetic of 5-HT. However, subsequent (pharmacological) investigations have shown that the excitatory effects of FMRFamide are different from those of 5-HT (Greenberg and Price, 1980). In addition, FMRFamide has been tested on *in vitro* heart preparations of a large number of bivalve species. It was found that it can depress as well as excite heart activity, or that it does not elicit a response at all (Painter et al., 1979).

FMRFamide has also been shown to have excitatory effects on the heart of the basommatophoran *L. stagnalis* (Fig. 36) (Geraerts et al., 1981), the stylommatophoran *H. aspersa* (Greenberg and Price, 1980), and the cephalopod *O. vulgaris* (Kiehling et al., 1981). Thresholds are very low (e.g., about $10^{-9} M$ in *L. stagnalis*), suggesting that at least in these species, and perhaps in all molluscs, FMRFamide is present. This is, however, not the case. In a number of gastropods, e.g., *H. aspersa* (Greenberg and Price, 1980; Lloyd, 1978a), *L. stagnalis* (Geraerts et al., 1981) and *Tritonia* (Lloyd, 1979), peptides with activities and molecular weights similar to those of FMRFamide are often present. These peptides also do not show species specificity in regard to their functional activity. However, they must have structures different from that of FMRFamide, as indicated by differences in behavior during purification procedures.

Of interest is the demonstration of FMRFamide immunoreactivity in a large variety of nonmolluscan animals, e.g., the coelenterate *Hydra* (Grimmelikhuyzen et al., (1982), the CNS of crustaceans (J. H. U. van Deynen, personal communication) and insects (Boer et al., 1980), and the CNS, pituitary, gastrointestinal tract, and pancreas of various vertebrates (Boer et al., 1980; Dockray et al., 1981; Weber et al., 1981). This indicates that FMRFamide and related peptides have persisted for long periods in evolution. Research on the functions of this type of peptide in these animals should now be undertaken.

There are several arguments against a neurohormonal role of FMRFamide and in favor of a neurotransmitter/neuromodulator function. First, immunocytochemical studies have demonstrated that in the CNS of *L. stagnalis* and *H. aspersa* (C.-R. Marchand, personal communication), FMRFamide is present not only in the soma of neurons but also in axons terminating on other neurons. Second, FMRFamide-like positivity has been demonstrated not only in peripheral nerves running to the heart but also in nerves which innervate female ASO, the mantle, and other tissues (Schot and Boer, 1982). Third, FMRFamide also affects noncardiac muscles, e.g., the radula protractor muscle of the prosobranch *B. contrarium*, the anterior byssus retractor muscle of *M. edulis* (Greenberg and Price, 1980), and the penis retractor muscle of *L. stagnalis* (W. P. M. Geraerts, unpublished observation). Moreover, neurons in the CNS of *H. aspersa* (Cottrell, 1978, 1980) are affected.

C. Large Cardioactive Peptides

Evidence for the presence in gastropods of cardioactive peptides with a neurohormonal function has been presented for *H. aspersa*. The two cardioexcitatory neurohormones have identical molecular weights (about 7000); one has a neutral isoelectric point, and the other is basic. They occur in high concentrations in the subesophageal ganglia and their nerve trunks, as well as in the auricle, but not in the ventricle or in other tissues (Lloyd, 1978a, 1980). Lloyd (1978b), furthermore, has shown that these peptides are associated with granules with a diameter of about 122 nm that are present in the auricle as well as in the subesophageal ganglia. He also demonstrated transport of the peptides via the visceral nerve toward the auricle and the release from both the auricle and the ganglia after appropriate stimulation. In addition, Lloyd demonstrated that the ventricle isolated *in vitro* responds to concentrations of the peptides similar to those found in the hemolymph. In *L. stagnalis* similar high concentrations of (a) large cardioexcitatory peptide(s) have been found in the CNS (Fig. 36) and in the auricle (W. P. M. Geraerts, unpublished observation). These studies imply that (a) large cardioactive peptide(s) are (is) released at multiple sites, i.e., at the peripheries of several nerves originating from the subesophageal ganglia and in the auricle. This is supported by ultrastructural studies of a number of molluscs showing that the auriculoventricular junction, in particular, is a neurohemal area which releases this substance (cf. Greenberg and Price, 1980).

D. Cephalopod Cardioactive Neurohormone

In cephalopods, neurosecretory cells have been found which have neurohemal areas in the anterior vena cava (AVC; Alexandrowicz, 1964, 1965) and in the pharyngo-ophthalmic vein (POV; Boycott and Young 1956; Froesch, 1974). These neurohemal areas are very extensive and lie beneath the muscular coat adjacent to the inner surface of the vessels. Extracts of both vessel walls, when applied to the systemic heart *in vitro,* cause increases in the frequency and amplitude of the heartbeat (Berry and Cottrell, 1970; Blanchi, 1969; Blanchi and De Brisco, 1971; Froesch and Mangold, 1976). Blanchi and co-workers (1973) have partially purified and characterized a cardioactive peptide from AVC extracts of *O. vulgaris.* This peptide has a molecular weight of about 1300. The cardioexcitatory hormone is different in its effects from 5-HT, adrenaline, and noradrenaline (Berry and Cottrell, 1970). It is not species specific (Blanchi et al., 1973).

What is the physiological role of the products of the AVC and the POV in the intact animal? This question has been answered by Wells and Mangold (1980), who have tested the extracts of both neurohemal areas on the (branchial) hearts of intact, free-moving *O. vulgaris.* AVC and POV extracts each produce a different spectrum of effects which are unlike those of ACh, 5-HT, adrenaline, histamine, and tyramine. The product of the AVC is effective at very low doses (less than 2% of the material extractable from a single vein per kilogram of animal). It increases the force and amplitude of the heartbeat (cf. the *in vitro* effect: increase of amplitude and frequency; see above). Because it is released at a point just upstream of the branchial hearts, the AVC material must be relevant to the normal performance of the hearts. The hormone is probably released during stress. This supposition is based on the observation that blood from the vena cava of just-caught animals contains the cardioactive principle, but blood in quickly cut-off tentacles does not. POV extracts are effective only at doses equivalent to several veins per kilogram, suggesting that it does not have a role in cardiac regulation.

For other molluscs, similar *in vivo* studies are needed. However, these are hampered by the fact that, with the exception of *L. stagnalis,* adequate cannulation techniques are not available.

E. Vertebrate Peptides and Cardiac Control in Molluscs

With immunocytochemical techniques, the presence of substances reacting with the antibodies against several vertebrate peptide hormones has been demonstrated in the neurohemal areas of cardioactive peptides, e.g., α-MSH, arginine vasopressin, neurophysin, met- and leu-enkephalins (AVC of *O. vulgaris;* Martin et al., 1980), and vasotocin (heart of *L. stagnalis;* Schot and Wijdenes, unpublished observation). Experiments designed to elucidate the possible cardioactive effects of these peptides have not yet been conclusive (*L. stagnalis;* J. Wijdenes, unpublished observation) or have shown that they do not produce distinct effects in the *in vitro* system with the heart of *Octopus* (Kiehling et al., 1981).

VII. Endocrine Control of Body Volume and Ionic Composition of the Hemolymph

Evidence for the involvement of hormones in the control of body volume and the ionic composition of the hemolymph in molluscs is almost completely confined to gastropods. Among these, most attention has been given to freshwater pulmonate snails. (For a recent review, see Geraerts et al., 1981.)

A. Physiological Aspects of Hydromineral Regulation in Basommatophora

As in all animals living in freshwater habitats, the osmolarity of the hemolymph of freshwater pulmonates is much higher than that of the environment. Moreover, each of the main ions in the hemolymph (Na^+, K^+, Ca^{2+}, Mg^{2+}, Cl^-, and HCO_3^-) is present at a concentration far above that of the medium. The concentration in the hemolymph of most of these ions is controlled by independent mechanisms (de With, 1977). In this control, the main problem is to compensate for the continuous diffusional loss of ions. This is effected by diverse active uptake systems. This uptake occurs in the integument which is exposed to the environment, but also in the intestine. Moreover, in ion regulation, excretion and reabsorption of ions in the kidney are important factors.

Water enters the snails very easily: In *L. stagnalis* the turnover rate of water is five times the total amount of body water per hour (de With, 1980; Van Aardt 1968). On the other hand, the hemolymph volume has to be controlled carefully in view of its important function as a hydrostatic skeleton: The muscles in the body wall exert a pressure on the hemolymph, thus giving the necessary rigidity to the body. There must, therefore, be an equilibrium between the volume of the hemolymph and the tension of the muscles in the body wall. Consequently, water entering the snail by osmosis through the integument has to be removed by the production of urine. Volume regulation therefore has to be effected essentially via the renal excretion of excess water.

In molluscs the initial step in urine formation is pressure ultrafiltration of the hemolymph through the wall of the heart into the pericardial cavity. The heart rate will determine the rate of pro-urine production. Cardioactive peptides may control this activity (see Section VI). However, up to now, the role of these substances in urine formation has not been studied, Moreover, not in all species of freshwater pulmonate snails does the heart contain the classical type of podocyte, which is, in many animals, characteristic for ultrafiltration sites (Andrews, 1976; Boer et al., 1973; Boer and Sminia, 1976). In some of these snails, the pro-urine appears to be filtered directly into the kidney through the kidney sac epithelium (Khan and Saleuddin, 1979a, 1979b, 1981a). The renal blood circulation pattern makes it conceivable that in this case the heart rate also determines, at least to a large extent, the rates of ultrafiltration, secretion, and reabsorption in the kidney. As will be shown below, hormones affect the permeability of the kidney epithelium.

Recently, attention has been focused on the fact that in *L. stagnalis* the feeding condition has important effects on all aspects of hydromineral regulation (de With, 1977, 1980; de With and Sminia, 1980). In fact, the extent of this influence is often greater than that resulting from experimental changes in the environment. This means that in experiments on these problems, great attention has to be paid to the food uptake of the snails.

1. The Dark Green Cell System

The first studies demonstrating the involvement of neuroendocrine factors in hydromineral regulation are from Lever and co-workers. Hekstra and Lever (1960) found a marked swelling of the body of *L. stagnalis* after extirpation of the pleural ganglia. Injection of homogenates of the pleural ganglia into intact animals resulted in a decrease of the body weight during the first hours after the injection (Lever et al., 1961). The authors suggested that the pleural ganglia produce a hormonal factor which exerts a diuretic influence. This conclusion was sustained by experiments of Chaisemartin (1968) with *L. limosa*. He cauterized the pleural ganglia of this snail and found an increase in body weight and a reduced urine formation.

The presence of endocrine cells in the pleural ganglia of *L. stagnalis* was demonstrated by Wendelaar Bonga (1970). These ganglia contain the majority of the neurosecretory DGC. Other cells of this type are located in the parietal ganglia.

Wendelaar Bonga (1971b, 1972) investigated the reactions of the DGC in snails exposed to a saline solution or to deionized water. The cells appeared to be inactivated in the saline medium and activated in the deionized water condition. These results were in agreement with the hypothesis that the DGC may produce a factor with a diuretic effect.

DGC in ganglionic complexes of the pleural and parietal ganglia, when implanted in acceptor snails which were exposed to osmotic stimuli, reacted in the same way as the cells in the ganglia of the acceptor snails. DGC in isolated ganglia kept under *in vitro* conditions also reacted similarly to changes in the osmolarity of the culture medium (Roubos and Moorer-van Delft, 1976). Apparently, the changes in the osmolarity of the medium are registered by the neurosecretory cells or by sensory cells within the pleural ganglia. A synaptic input to the DGC, possibly from these sensory elements, was demonstrated by Benjamin et al. (1976).

Studies about the distribution of the release areas of the DGC have further elucidated the function of these cells. Swindale and Benjamin (1976a) demonstrated that the release of the DGC hormone is not restricted to the connectives of the pleural ganglia and the surrounding connective tissue, as had been described by Wendelaar Bonga (1970). The majority of the DGC axons project in nerves which innervate the ventral and anterior parts of the head–foot, neck region, and mantle. Release of the hormone may occur in these regions (Swindale and Benjamin, 1976b). This widespread peripheral secretion of the DGC hormone in the skin parts exposed to the environment might be an indication that the skin is the target organ of this hormone.

The latter suggestion is sustained by investigations of Grimm-Jørgensen. She attacked the problem in a new way. Using radioimmunoassay techniques to quantify immunoreactive TRH (IRTRH; TRH = thyroid-stimulating hormone-releasing hormone, a vertebrate hypothalamic neuropeptide), she was able to demonstrate the presence of IRTRH in the nervous system of various gastropods (Grimm-Jørgensen and Jackson, 1975). In *L. stagnalis* the distribution of this substance appears to be similar to that of the cell bodies and the neurohemal areas of the DGC. Moreover, the IRTRH content of the hemolymph changes with the salinity of the medium (Grimm-Jørgensen, 1978). Injections of synthetic TRH (a tripeptide) in *L. stagnalis* cause a loss of body weight. Furthermore, TRH appears to be preferentially taken up and degraded by the mantle, foot, and head regions of these snails (Grimm-Jørgensen, 1979). These are the very same regions where Swindale and Benjamin (1976a) had observed the axons of the DGC. Finally, the hypothesis has been tested that TRH might alter salt and water fluxes in the skin by changing the secretion of epidermal mucus (Grimm-Jørgensen, in press). This was done by measuring *in vitro* the effect of TRH on the synthesis and release of pronase-resistant sulfated polysaccharides (SP) by foot epidermal tissue of *Helisoma carabaceum*. TRH appeared to decrease the release of SP by the skin, but it did not alter SP synthesis. The author suggests that TRH affects mucus secretion. It is not unlikely that mucus will change the diffusion of ions and water. It may change the ionic composition of the unstirred aqueous layer beneath the mucous layer and in this way alter the ionic gradients between the epidermal cells and the external environment.

Thus, although it is still uncertain whether synthetic TRH is identical or nearly identical to the DGC hormone, these studies are helpful in further clarifying the role of the DGC. More direct measurements of the effect of TRH on water flow and sodium transport are now necessary to prove the suggestions of Grimm-Jørgensen.

2. Effects of Endocrine Centers in the Parietal and Visceral Ganglia

Wendelaar Bonga (1972) and Soffe et al. (1978) have shown that in *L. stagnalis* the yellow cells (YC) and yellow green cells (YGC) of the visceral and parietal ganglia, respectively, react to different osmotic environments in a way similar to that observed for the DGC. The axons of the YC and YGC show peripheral release activity in the heart-kidney area, where active uptake of ions

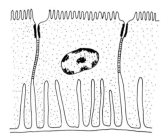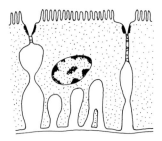

Fig. 37. Effects of *in vitro* treatment of the kidney sac epithelium of *Helisoma* with extract of the visceral ganglion. Left: Kidney sac epithelium of snails acclimated to isosmotic saline. Intercellular spaces and basal lacunae are narrow, zonulae adherentes are closed, and septate junctions are long, with compact septa. Right: Same kidney sac after *in vitro* treatment with visceral ganglion extract. Intercellular spaces and basal lacunae are widened; zonulae adherentes are open; and septate junctions are short, with few septa, or long, with misaligned septa having irregular periodicity. (After Khan and Saleuddin, 1979a, 1981b.)

from the pro-urine occurs. The latter observation fits in with the finding of Chaisemartin (1968) that cauterization of the parietal ganglia of *L. limosa* decreases the turnover of sodium.

The role of the neuroendocrine cells in the visceral and parietal ganglia has been studied in *H. duryi* and *H. trivolvis* in detail by Saleuddin and co-workers. Three cell types have been described in these ganglia (Dillaman et al., 1976). The position and stainability of these cell types cannot be clearly related to those described in *L. stagnalis* by Wendelaar Bonga (1970). All three cell types react to changes in the environmental osmotic pressure. In particular, cells of type 2 seem to be more active in synthesizing and releasing their secretion product when the snails are kept in distilled water. Therefore, Khan and Saleuddin (1979b) suggest that these cells may be involved in stimulating diuresis. Type 3 cells show a reverse reaction. There are no indications for the occurrence of DGC in *Helisoma*. Special attention was given to the kidney sac epithelium as a possible target of the hormone of type 2 cells. When snails are kept in a hypo-osmotic medium, the intercellular spaces in this epithelium are wide (Fig. 37), whereas they are minimal when the snails are kept in an isosmotic medium (Khan and Saleuddin, 1979a). During *in vitro* treatment of kidney sacs of saline-acclimated snails with crude extracts of visceral ganglia of snails kept in distilled water, the spaces rapidly expand. Extracts of visceral ganglia of snails kept in saline, and extracts of other ganglia, have no effect. This suggests that a factor of the visceral ganglion, probably a neurohormone, may act on the kidney sac epithelium to alter the rate of filtration and pro-urine transport (Khan and Saleuddin, 1979a).

Electron microscopic analysis has revealed that the kidney sac epithelium cells are interconnected by zonulae adherentes and septate junctions. *In vitro* treatment of the kidney sac with crude brain extract not only dramatically increased intercellular space width but affects also the junctional complexes (Fig. 37). Khan and Saleuddin (1981b) conclude that the septate junctions are involved in osmoregulation and that they are regulated by a brain factor.

B. Terrestrial Pulmonates, Opisthobranchs, and Bivalves

Subramanyam (1973) found a rise in hemolymph calcium within 30 min after injections of aquous extracts of ganglia in *Cryptozona semirugata*. Burton (1977) found the same effect on the hemolymph calcium concentration after injections of subesophageal ganglia homogenates in *H. pomatia*. In this snail, the hemolymph calcium concentration shows wide variations. Therefore, the putative hormone may be mainly involved in the prevention of hypocalcemia.

Sawyer et al. (in press) demonstrated the presence of an arginine-vasotocin-like substance (AVTLS) in the pleuropedal ganglia of the terrestrial pulmonate slugs *A. columbianus* and *L. maximus*. AVTLS resembles but clearly differs from vertebrate arginine-vasotocin. Both peptides enhance the water permeability of the body wall of slugs.

Kupferman and Weiss (1976) have shown that in *Aplysia californica* injections (into the hemocoel) of extracts of the large neurosecretory R15 neuron cause a rapid uptake of water. Thus, R15 is assumed to produce a neurohormone that is involved in hydromineral regulation. The presumed hormone is a peptide with a molecular weight of 1500 (Gainer et al., 1977). Electrically, R15 appears to be a steadily firing neuron, which suggests that the hormone is continuously released. Introduction of the animals in a medium with a lower salt content, which might occur in nature when the animals become isolated in a tide pool, inhibits the firing rate. Therefore, the hormone will exert an antidiuretic effect. The hypoosmotic stimuli are sensed by the osphradium (cf. Geraerts et al., 1981).

Graves and Dietz (1979, 1980) and Dietz and Graves (1981) have studied the effects of prostaglandin E_2 and serotonin on the sodium transport in gills of the freshwater mussel *Ligumia rostrata*. Prostaglandin E_2 inhibits sodium uptake. In the hemolymph, the presence of prostaglandin-like material has been demonstrated.

Acknowledgments

We are indebted to Professor J. Lever for critically reading the manuscript, to Dr. M. A. Hemminga and Dr. N. D. de With for their help in the preparation of Sections V and VII, respectively, and to Mr. R. van Elk for his assistance in the search of literature. We thank Drs. S. and P. Le

Gall for their cooperation in the preparation of Fig. 11. We are grateful to Mr. Bastiaanse, Mr. Van den Berg, and Mr. Van Groenigen for preparing the figures, and to Ms. T. Laan and Ms. D. Hoonhout for typing the manuscript.

References

Abeloos, M. (1943). Effets de la castration chez un mollusque, *Limax maximus* L. *C. R. Acad. Sci.* **216,** 90–91.

Alexandrowicz, J. Z. (1964). The neurosecretory system of the vena cava in Cephalopods. Vol. 1. *Eledona cirrosa. J. Mar. Biol. Assoc. U.K.* **44,** 111–132.

Alexandrowicz, J. S. (1965). The neurosecretory system of the vena cava in Cephalopods, Vol. 2. *Sepia officinalis* and *Octopus vulgaris. J. Mar. Biol. Assoc. U.K.* **45,** 209–228.

Andrews, E. B. (1976). The fine structure of the heart of some prosobranch and pulmonate gastropods in relation to filtration. *J. Moll. Stud.* **42,** 199–216.

Arch, S. (1976). Neuroendocrine regulation of egg laying in *Aplysia californica. Am. Zool.* **16,** 167–175.

Arch, S., Early, P., and Smock, T. (1976). Biochemical isolation and physiological identification of the egg-laying hormone in *Aplysia californica. J. Gen. Physiol.* **68,** 197–210.

Arch, S., Smock, T., Gurvis, R., and McCarthy, C. (1978). Atrial gland induction of the egg-laying response in *Aplysia californica. J. Comp. Physiol.* **128,** 67–70.

Arch, A., Lupatkin, J., Smock, T., and Beard, M. (1980). Evidence for an exocrine function of the *Aplysia californica* atrial gland. *J. Comp. Physiol. A* **141,** 131–138.

Bailey, T. G. (1973). The *in vitro* culture of reproductive organs in the slug *Agriolimax reticulatus* (Mull). *Neth. J. Zool.* **23,** 72–85.

Bardon, O., Lubet, P., and Drosdowsky, M. A. (1971). Biosynthèse des stéroides chez un mollusque gastéropode marin *Crepidula fornicata* (Phil.). *Steroidologia* **2,** 366–377.

Beeman, R. D. (1977). The anatomy and functional morphology of the reproductive system in the opisthobranch mollusc *Phyllaplysia taylori* Dall, 1900. *Veliger* **13,** 1–31.

Benjamin, P. R., Swindale, N. V., and Slade, C. T. (1976). Electrophysiology of identified neurosecretory neurones in the pond snail *Lymnaea stagnalis* (L.). *In* "Neurobiology of Invertebrates: Gastropoda Brain" (J. Salánki, ed.), pp. 85–100. Akadémiai Kiado, Budapest, Hungary.

Berry, C. F., and Cottrell, G. A. (1970). Neurosecretion in the vena cava of the cephalopod *Eledone cirrosa. Z. Zellforsch. Mikrosk. Anat.* **104,** 107–115.

Bierbauer, J., and Török, L. J. (1968). Histophysiological study of the optic tentacle in Pulmonates. I. Histological examination of the optic tentacle with special regard to the morphology of the collar and lateral cells. *Acta Biol. Acad. Sci. Hung.* **19,** 133–143.

Bierbauer, J., and Vigh-Teichmann, I. (1970). Histophysiological examination of the optic tentacle of Pulmonates. II. Cytochemistry of the special and secretory cells. *Acta Biol. Acad. Sci. Hung.* **21,** 11–24.

Björkman, N. (1963). On the ultrastructure of the optic gland in *Octopus. J. Ultrastruct. Res.* **8,** 195.

Blanchi, D. (1969). Experimenti sulla funzione del sistema neurosecretorio della vena cava nei cefalopodi. *Boll. Soc. Ital. Biol. Sper.* **45,** 1615–1619.

Blanchi, D., and De Brisco, R. (1971). Experimenti sulla funzioni neurosecretorio della vena cava nei cefalopodi. Purificazione e caratterizzazione parziale del principio attivo. *Boll. Soc. Ital. Biol. Sper.* **47,** 477–480.

Blanchi, D., Noviello, L., and Libonati, M. (1973). A neurohormone of cephalopods with cardioexcitatory activity. *Gen. Comp. Endocrinol.* **21,** 267–277.

Blankenship, J. (1980). Physiological properties of peptide-secreting neuroendocrine cells in the marine mollusc *Aplysia*. *In* "Role of Peptides in Neuronal Function" (J. Barker, ed.), pp. 159–187. Marcel Dekker, New York.

Boer, H. H., and Joosse, J. (1975). Endocrinology. *In* "Pulmonates: Functional Anatomy and Physiology" (V. Fretter and J. Peake, eds.), Vol. 1, pp. 245–307. Academic Press, London.

Boer, H. H., and Sminia, T. (1976). Sieve structure of slit diaphragms of podocytes and pore cells of gastropod molluscs. *Cell Tissue Res.* **170**, 221–229.

Boer, H. H., Slot, J. W., and van Andel, J. (1968). Electron microscopical and histochemical observations on the relation between medio-dorsal bodies and neurosecretory cells in the Basommatophoran snails *Lymnaea stagnalis, Ancylus fluviatilis, Australorbis glabratus* and *Planorbarius corneus*. *Z. Zellforsch.* **87**, 435–450.

Boer, H. H., Algera, N. H., and Lommerse, A. W. (1973). Ultrastructure of possible sites of ultrafiltration in some gastropods, with particular reference to the auricle of the freshwater prosobranch *Viviparus viviparus* L. *Z. Zellforsch.* **143**, 329–341.

Boer, H. H., Mohamed, A. M., van Minnen, J., and de Jong-Brink, M. (1976). Effects of castration on the activity of the endocrine dorsal bodies of the freshwater pulmonate snail *Bulinus truncatus*, intermediate host of *Schistosoma haematobium*. *Neth. J. Zool.* **26**, 94–105.

Boer, H. H., Roubos, E. W., van Dalen, H, and Groesbeek, J. R. F. Th. (1977). Neurosecretion in the basommatophoran snail *Bulinus truncatus* (Gastropoda, Pulmonata). *Cell Tissue Res.* **176**, 57–67.

Boer, H. H., Schot, L. P. C., Roubos, E. W., ter Maat, A., Lodder, J. C., and Reichelt, D. (1979). ACTH-like immunoreactivity in two electrotonically coupled giant neurons in the pond snail *Lymnaea stagnalis*. *Cell Tissue Res.* **202**, 231–240.

Boer, H. H., Schot, L. P. C., Veenstra, J. A., and Reichelt, D. (1980). Immunocytochemical identification of neural elements in the central nervous system of a snail, some insects, a fish and a mammal with an antiserum to the molluscan cardio-excitatory tetrapeptide FMRF-amide. *Cell Tissue Res.* **213**, 21–27.

Bohlken, S., and Joosse, J. (1982). The effect of photoperiod on female reproductive activity of the freshwater pulmonate snail *Lymnaea stagnalis* kept under laboratory breeding conditions. *Int. J. Invertebr. Reprod.* **4**, 213–222.

Bonichon, A. (1967). Contribution à l'étude de la neurosécrétion et de l'endocrinologie chez les Céphalopodes. I. *Octopus vulgaris*. *Vie Milieu* **18**, 228–252.

Boquist, L., Falkmer, S., and Mehrotra, B. K. (1971). Ultrastructural search for homologues of pancreatic β-cells in the intestinal mucosa of the mollusc *Buccinum undatum*. *Gen. Comp. Endocrinol.* **17**, 236–239.

Boycott, B. B., and Young, J. Z. (1956). The subpedunculate body and nerve and other organs associated with the optic tract of cephalopods. *In* "Bertil Hanström. Zoological Papers in Honour of his Sixty-fifth Birthday, November 20, 1956" (K. G. Wingstrand, ed.), pp. 76–105. Zoological Institute, Lund, Sweden.

Brisson, P. (1971). Castration chirurgicale et régénération gonadique chez quelques planorbides (Gastéropodes Pulmonés). *Ann. Embryol. Morphog.* **4**, 189–210.

Brisson, P., and Collin, J. P. (1980). Systèmes aminergiques des mollusques gastéropodes pulmonés, Vol. 4. Paraneurones et innervation catécholaminergiques de la région du carrefour des voies génitales; étude autoradiographique. *Biol. Cell.* **38**, 211–220.

Burton, R. F. (1977). Haemolymph calcium in *Helix pomatia*: Effects of EGTA, ganglion extracts, ecdysterone, cyclic AMP and ionophore A 23187. *Comp. Biochem. Physiol. C* **57**, 135.

Buijs, R. M., and Swaab, D. F. (1980). Immunoelectronmicroscopical demonstration of vasopressin and oxytocin synapses in the rat limbic system. *Cell Tissue Res.* **204**, 335–367.

Carreau, S., and Drosdowsky, M. (1977). The *in vitro* biosynthesis of steroids by the gonad of the cuttlefish *(Sepia officinalis* L.*)*. *Gen. Comp. Endocrinol.* **33**, 554–565.

Chaisemartin, C. (1968). Contrôle neuroendocrinien du renouvellement hydrosodique chez *Lymnaea limosa* L. *C.R. Séances Soc. Biol.* **162**, 1994–1998.

Chiu, A. Y., Hunkapiller, M. W., Heller, E., Stuart, D. K., Hood, L. E., and Strumwasser, F. (1979). Purification and primary structure of the neuropeptide egg-laying hormone of *Aplysia californica*. *Proc. Nat. Acad. Sci. U.S.A.* **76**, 6656–6660.

Choquet, M. (1971). Etude du cycle biologique et de l'inversion de sexe chez *Patella vulgata* L. (Mollusque Gastéropode Prosobranche). *Gen. Comp. Endocrinol.* **16**, 59–73.

Choquet, M., and Lemaire, J. (1969). Contribution à l'étude de la régénération tentaculaire chez *Patella vulgata* L. (Gastéropode prosobranche). *Arch. Zool. Exp. Gen.* **109**, 319–337.

Cobbs, J., and Pinsker, H. (1982) Role of bag cells in egg deposition of *Aplysia brasiliana:* I. Comparison of normal and elicited behaviors. *J. Comp. Physiol.* **147**, 523–536.

Coggeshall, R. E. (1967). A light and electron microscope study of the abdominal ganglion of *Aplysia californica*. *J. Neurophysiol.* **30**, 1263–1287.

Coggeshall, R. E. (1970). A cytologic analysis of the bag cell control of egg laying in *Aplysia*. *J. Morphol.* **132**, 461–485.

Collip, J. B. (1923). The demonstration of an insulin-like substance in the tissue of the clam. *J. Biol. Chem.* **55**, 39.

Cook, H. (1966). Morphology and histology of the central nervous system of *Succinea putris* (L.). *Arch. Neerl. Zool.* **17**, 1–72.

Cottrell, G. A. (1978). Actions of a "molluscan cardioexcitatory neuropeptide" on identified 5-hydroxytryptamine-containing neurons and their follower neurons in *Helix aspersa*. *J. Physiol.* **284**, 130–131P.

Cottrell, G. A. (1980). Voltage dependent and voltage independent actions of the molluscan neuropeptide Phe-Met-Arg-Phe-NH_2 on a snail neurone. *J. Physiol.* **300**, 42P.

Cottrell, G. A., and Osborne, N. N. (1969). Localization and amode of action of cardioexcitatory agents in molluscan hearts. *In* "Comparative Physiology of the Heart: Current Trends" (F. V. McCann, ed.), Experientia suppl. 15, pp. 220–231. Birkhäuser Verlag, Basel, Switzerland.

Davidson, I. K., Falkmer, S., Mehrotra, B. K., and Wilson, S. (1971). Insulin assay and light microscopical studies of digestive organs in protostomian and deuterostomian species and in coelenterates. *Gen. Comp. Endocrinol.* **17**, 388–401.

Davis, W. J., Mpitsos, G. J., and Pinneo, J. M. (1974). The behavioral hierarchy of the mollusc *Pleurobranchaea*, Vol. 2. Hormonal suppression of feeding associated with egg-laying. *J. Comp. Physiol.* **90**, 222–243.

Defretin, R., and Richard, A. (1967). Ultrastructure de la glande optique de *Sepia officinalis* (Mollusque, Céphalopode). Mise en évidence de la sécrétion et de son contrôle photopériodique. *C.R. Hebd. Séances Acad. Sci.* **265**, 1415–1418.

de Jong-Brink, M., and Geraerts, W. P. M. (1982). Oogenesis in gastropods. *Malacologia* **22**(1-2), 145–149.

de Jong-Brink, M., and Goldschmeding, J. T. (1983). Endocrine and nervous regulation of female reproductive activity in the gonad and albumen gland of *Lymnaea stagnalis*. *In* "Molluscan Neuro-endocrinology" (J. Lever and H. H. Boer, eds.), pp. 126–131. Mon. Kon. Ned. Akad. Wet., North-Holland, Amsterdam.

de Jong-Brink, M., and Lodder, J. C. (in press). Ultrastructure and innervation of the albumen gland of the pulmonate snail *Lymnaea stagnalis* (L). *Int. J. Invertebr. Reprod.*

de Jong-Brink, M., de Wit, Ankie, Kraal, G., and Boer, H. H. (1976). A light and electron microscope study on oogenesis in the freshwater pulmonate snail *Biomphalaria glabrata*. *Cell Tissue Res.* **171**, 195–219.

de Jong-Brink, M., ter Borg, J. P., Bergamin-Sassen, M. J. M., and Boer, H. H. (1979). Histology and histochemistry of the reproductive tract of the pulmonate snail *Bulinus truncatus*, with observations on the effects of castration on its growth and histology. *Int. J. Invertebr. Reprod.* **1**, 41–56.

de Jong-Brink, M., Schot, L. P. C., Schoenmaker, H. J. N., and Bergamin-Sassen, M. J. M. (1981). A biochemical and quantitative electron-microscope study on steroidogenesis in ovotestis and digestive gland of the pulmonate snail *Lymnaea stagnalis*. *Gen. Comp. Endocrinol.* **45**, 30–38.

de Jong-Brink, M., Boer, H. H., and Joosse, J. (1983). Oogenesis, oviposition and oosorption in Mollusca. *In* '':Reproductive Biology of Invertebrates'' (K. G. Adiyodi and G. Adiyodi, eds.). Vol. I, pp. 297–355. Wiley, Chichester, England.

de Lacaze Duthiers, H. (1872). Du système nerveux des mollusques gastéropodes pulmonés aquatiques et d'un nouvel organe d'innervation. *Arch. Zool. Exp. Gen.* **1**, 437–500.

de Longcamp, D., Lubet, P., and Drosdowsky, M. (1974). The *in vitro* biosynthesis of steroids by the gonad of the mussel *Mytilus edulis*. *Gen. Comp. Endocrinol.* **22**, 116–127.

de Vlieger, T. A., Kits, K. S., ter Maat, A., and Lodder, J. C. (1980). Morphology and electrophysiology of the ovulation hormone producing neuro-endocrine cells of the freshwater snail *Lymnaea stagnalis* (L.). *J. Exp. Biol.* **84**, 259–271.

de With, N. D. (1977). Evidence for the independent regulation of specific ions in the haemolymph of *Lymnaea stagnalis* (L.). *Proc. K. Ned. Akad. Wet. C* **80**, 144–157.

de With, N. D. (1978). The effects of starvation and feeding on the ionic composition of the haemolymph in the freshwater snail *Lymnaea stagnalis*. *Proc. K. Ned. Akad. Wet. C* **81**, 241–248.

de With, N. D. (1980). Water turn-over, ultrafiltration, renal water reabsorption and renal circulation in fed and starved specimens of *Lymnaea stagnalis*, adapted to different external osmolarities. *Proc. K. Ned. Akad. Wet. C* **83**, 109–120.

de With, N. D., and Sminia, T. (1980). The effects of the nutritional state and the external calcium concentration on the ionic composition of the haemolymph and the activity of the calcium cells in the pulmonate freshwater snail *Lymnaea stagnalis*. *Proc. K. Ned. Akad. Wet. C* **83**, 217–227.

de With, N. D., Witteveen, J., and van der Woude, H. A. (1980). Integumental Na^+/H^+ and Cl^-/HCO_3^- exchanges in the freshwater snail *Lymnaea stagnalis*. *Proc. K. Ned. Akad. Wet. C* **83**, 209–215.

de Wied, D., and Gispen, W. H. (1977). Behavioral effects of peptides. *In* ''Peptides in Neurobiology'' (H. Gainer, ed.), pp. 397–448. Plenum Press, New York, London.

de Zwaan, A., and Zandee, D. I. (1972). Body distribution and seasonal changes in the glycogen content of the common sea mussel *Mytilus edulis*. *Comp. Biochem. Physiol. A* **43**, 53–58.

Dietz, H., and Graves, S. Y. (1981). Sodium influx in isolated gills of the freshwater mussel, *Ligumia substrata*. *J. Comp. Physiol.* **143**, 185–190.

Dillaman, R. M., Saleuddin, A. S. M., and Jones, G. M. (1976). Neurosecretion and shell regeneration in *Helisoma duryi* (Mollusca: Pulmonata). *Can. J. Zool.* **54**, 1771–1778.

Dockray, G. J., Vaillant, C., and Williams, R. C. (1981). Neurovertebrate brain-gut peptide related to a molluscan neuropeptide and a opioid peptide. *Nature* **292**, 656–657.

Dogterom, A. A. (1980). The effect of the growth hormone of the freshwater snail *Lymnaea stagnalis* on biochemical composition and nitrogenous wastes. *Comp. Biochem. Physiol. B* **65**, 163–167.

Dogterom, A. A., and Doderer, A. (1981). A hormone dependent calcium-binding protein in the mantle edge of the freshwater snail *Lymnaea stagnalis*. *Calcif. Tissue Int.* **33**, 505–508.

Dogterom, A. A., and Jentjens, Th. (1980). The effect of the growth hormone of the pond snail *Lymnaea stagnalis* on periostracum formation. *Comp. Biochem. Physiol. A* **66**, 687–690.

Dogterom, A. A., and Robles, Bianca R. (1980). Stimulation of ornithine decarboxylase activity in *Lymnaea stagnalis* after a single injection with molluscan growth hormone. *Gen. Comp. Endocrinol.* **40**, 238–240.

Dogterom, A. A., and van der Schors, R. C. (1980). The effect of the growth hormone of *Lymnaea stagnalis* on (bi)carbonate movements, especially with regard to shell formation. *Gen. Comp. Endocrinol.* **41,** 334–339.

Dogterom, A. A., van Loenhout, H., and van der Schors, R. C. (1979a). The effect of the growth hormone of *Lymnaea stagnalis* on shell formation. *Gen. Comp. Endocrinol.* **39,** 63–68.

Dogterom, A. A., van Loenhout, H., and de Waal, R. (1979b). Ornithine decarboxylase in the freshwater snail *Lymnaea stagnalis* as related to growth and feeding. *Proc. K. Ned. Akad. Wet. Series C* **83,** 25–31.

Dogterom, G. E., and van Loenhout, H. (1983). Specificity of ovulation hormones in some basommatophoran species studied by means of iso- and heterospecific injections. *Gen. Comp. Endocrinol.* (in press).

Dogterom, G. E., Bohlken, S., and Joosse, J. (1983a). Effect of the photoperiod on the timeschedule of egg-mass production in *Lymnaea stagnalis,* induced by ovulation hormone injections. *Gen. Comp. Endocrinol.* **49,** 255–260.

Dogterom, G. E., Bohlken, S., and Geraerts, W. P. M. (in press a). A rapid *in vivo* bioassay of the ovulation hormone of *Lymnaea stagnalis. Gen. Comp. Endocrinol.*

Dogterom, G. E., Hofs, H. P., Wapenaar, P., Koomen, W., and van Loenhout, H. (1983b). Spontaneous oviposition in *Lymnaea stagnalis* kept under various experimental conditions. *In* "Molluscan neuroendocrinology" (J. Lever and H. H. Boer, eds.), pp. 111–118. Mon. Kon. Ned. Akad. Wet., North-Holland, Amsterdam.

Dudek, F. E., and Tobe, S. S. (1978). Bag cell peptide acts directly on ovotestis of *Aplysia californica:* Basis for an *in vitro* bioassay. *Gen. Comp. Endocrinol.* **36,** 618–627.

Dudek, F. E., Cobbs, J. S., and Pinsker, H. M. (1979). Bag cell electrical activity underlying spontaneous egg laying in freely behaving *Aplysia brasiliana. J. Comp. Neurophysiol.* **42,** 804–817.

Dudek, F. E., Weir, G., and Acosta-Urquidi, J. (1980). A secretion from neuroendocrine bag cells evokes egg release *in vitro* from ovotestis of *Aplysia californica. Gen. Comp. Endocrinol.* **40,** 241–244.

Durchon, M. (1967). "L'Endocrinologie des Vers et des Mollusques." Masson and Cie, Paris.

Durchon, M., and Joly, P. (1978). "L'Endocrinologie des Invertébrés." Presses Universitaires de France, Vendôme, France.

Durchon, M., and Richard, A. (1967). Etude, en culture organotypique, du rôle endocrine de la glande optique dans la maturation ovarienne chez *Sepia officinalis* L. (Mollusque Céphalopode). *C. R. Hebd. Séances Acad. Sci. Paris* **264,** 1497–1500.

Frazier, W. J., Kandel, E. R., Kupfermann, J., Waxiri, R., and Coggeshall, R. E. (1967). Morphological and functional properties of identified neurons in the abdominal ganglion of *Aplysia californica. J. Neurophysiol.* **30,** 1288–1351.

Fritsch, H. A. R., and Sprang, R. (1977). On the ultrastructure of polypeptide hormone-producing cells in the gut of the Ascidian, *Ciona intestinalis* L. and in the bivalve, *Mytilus edulis* L. *Cell Tissue Res.* **177,** 407–413.

Fritsch, H. A. R., van Noorden, S., and Pearse, A. G. E. (1976). Cytochemical and immunofluorescence investigations on insulin-like producing cells in the intestine of *Mytilus edulis* L. (Bivalvia). *Cell Tissue Res.* **165,** 365–369.

Froesch, D. (1974). The subpedunculate lobe of the octopus brain: Evidence for dual function. *Brain Res.* **75,** 277–285.

Froesch, D. (1979). Antigen-induced secretion in the optic gland of *Octopus vulgaris. Proc. R. Soc. London B* **205,** 379–384.

Froesch, D., and Mangold, K. (1976). On the structure and function of a neurohemal organ in the eye cavity of *Eledone cirrosa* (Cephalopods). *Brain Res.* **111,** 287–293.

Froesch, D., and Mangold, K., and Fritz, W. (1978). On the turnover of exogenous ferritin in the cephalopod optic gland. A microprobe study. *Experientia* **34,** 115–117.

Gabbott, P. A. (1975). Storage cycles in marine bivalve molluscs: A hypothesis concerning the relationship between glycogen metabolism and gametogenesis *Proc. Eur. Symp. Mar. Biol. 9th*, pp. 191–211. University Press, Aberdeen, Scotland.

Gabbott, P. A., and Bayne, P. L. (1972). Biochemical effects of temperature and nutritive stress on *Mytilus edulis* L. *J. Mar. Biol. Assoc. U.K.* **53**, 269–286.

Gabbott, P. A., Cook, P. A., and Whittle, M. A. (1979). Seasonal changes in glycogen synthase activity in the mantle tissue of the Mussel *Mytilus edulis* L.: Regulation by tissue glucose. *Biochem. Soc. Trans.* **7**, 895–898.

Gabe, M. (1966). "Neurosecretion." Pergamon Press, Oxford, England.

Gainer, H., Loh, Y. P., and Sarne Y. (1977). Biosynthesis of neuronal peptides. *In* "Peptides in Neurobiology" (H. Gainer, ed.), pp. 196–219. Plenum, New York.

Geraerts, W. P. M. (1976a). Control of growth by the neurosecretory hormone of the light green cells in the freshwater snail *Lymnaea stagnalis. Gen. Comp. Endocrinol.* **29**, 61–71.

Geraerts, W. P. M. (1976b). The role of the lateral lobes in the control of growth and reproduction in the hermaphrodite freshwater snail *Lymnaea stagnalis. Gen. Comp. Endocrinol.* **29**, 97–108.

Geraerts, W. P. M. (1976c). The effect of the lateral lobes on carbohydrate metabolism in the hermaphrodite freshwater snail *Lymnaea stagnalis. In* "Actualités sur les Hormones d'Invertébrés," pp. 125–137. Coll. internat. du C.N.R.S. No. 251.

Geraerts, W. P. M., and Algera, L. H. (1976). The stimulatory effect of the dorsal body hormone on cell differentiation in the female accessory sex organs of the hermaphrodite freshwater snail *Lymnaea stagnalis. Gen. Comp. Encocrinol.* **29**, 109–118.

Geraerts, W. P. M., and Bohlken, S. (1976). The control of ovulation in the hermaphrodite freshwater snail *Lymnaea stagnalis* by the neurohormone of the caudo-dorsal cells. *Gen. Comp. Endocrinol.* **28**, 350–357.

Geraerts, W. P. M., and Joosse, J. (1975). The control of vitellogenesis and of growth of female accessory sex organs by the dorsal body hormone (DBH) in the hermaphrodite freshwater snail *Lymnaea stagnalis. Gen. Comp. Endocrinol.* **27**, 450–467.

Geraerts, W. P. M., and Joosse, J. (in press). The reproductive biology of the Basommatophora. *In* "Biology of Molluscs" (K. Wilbur and A. S. M. Saleuddin, eds.). Academic Press, New York.

Geraerts, W. P. M., and Mohamed, A. M. (1981). Studies on the role of the lateral lobes and the ovotestis of the pulmonate snail *Bulinus truncatus* in the control of body growth and reproduction. *Int. J. Invertebr. Reprod.* **3**, 297–308.

Geraerts, W. P. M., van Leeuwen, J. P. Th., Nuyt, K., and de With, N. D. (1981). Cardioactive peptides of the CNS of the pulmonate snail *Lymnaea stagnalis. Experientia* **37**, 1168–1169.

Geraerts, W. P. M., de With, N. D., Roubos, E. W., and Joosse, J. (1981). Endocrine aspects of hydromineral regulation in molluscs. *In* "Neurosecretion, Molecules, Cells, Systems" (S. Farner and K. Lederis, eds.), pp. 337–347. Plenum, New York.

Geraerts, W. P. M., Cheeseman, P., Ebberink, R. H. M., Nuyt, K., and Hogenes, Th. M. (1983a). Partial purification and characterization of the ovulation hormone of the freshwater pulmonate snail *Lymnaea stagnalis. Gen. Comp. Endocrinol.* **51**, (in press).

Geraerts, W. P. M., de With, N. D., Ebberink, R. H. M., Casteleijn, E., and Hogenes, Th. M. (1983b). Cardioactive peptides in the freshwater pulmonate *Lymnaea stagnalis. In* "Molluscan neuro-endocrinology" (J. Lever and H. H. Boer, eds.), pp. 196–202. Mon. Kon. Ned. Akad. Wet., North-Holland, Amsterdam.

Golding, D. W. (1974). A survey of neuroendocrine phenomena in non-arthropod invertebrates. *Biol. Rev.* **49**, 161–224.

Goldschmeding, J. T., Wilbrink, M., and ter Maat, A. (1983). The role of the ovulation hormone in the control of egg-laying behaviour in *Lymnaea stagnalis. In* "Molluscan Neuro-endocrinology" (J. Lever and H. H. Boer, eds.), pp. 251–256. *Mon. Kon. Ned. Akad. Wet.*, North-Holland Publ. Co., Amsterdam.

Goldsworthy, G. J., Robinson, J., and Mordue, W. (1981), "Endocrinology." Blackie and Son, Glasgow, London.

Gomih, Y. K., and Grillo, T. A. I. (1976). Insulin-like activity of the extract of the digestive gland and the pyloris of the giant African snail, *Achatina fulica*; a preliminary report. *In* "The Evolution of Pancreatic Islets" (T. A. I. Grillo, L. Leibson, and S. Epple, eds.), pp. 153–162. Pergamon Press, Oxford, England.

Gomot, L. (1970). Analyse expérimentale du déterminisme du cycle de la gonade chez les Mollusques. *Bull. Soc. Zool. Fr.* **95,** 429–451.

Gottfried, H., and Dorfman, R. J. (1969). The occurrence of *in vivo* cholesterol biosynthesis in an invertebrate, *Ariolimax californicus*. *Gen. Comp. Endocrinol. Suppl.* **2,** 590–593.

Gottfried, H., and Dorfman, R. J. (1970a). Steroids of invertebrates. IV. On the optic tentacle-gonadal axis in the control of the male-phase ovotestis in the slug *(Ariolimax californicus)*. *Gen. Comp. Endocrinol.* **15,** 101–119.

Gottfried, H., and Dorfman, R. J. (1970b). Steroids of invertebrates, V. The *in vitro* biosynthesis of steroids by the male-phase ovotestis of the slug *(Ariolimax californicus)*. *Gen. Comp. Endocrinol.* **15,** 120–138.

Gottfried, H., and Dorfman, R. J. (1970c). Steroids of invertebrates, VI. Effect of tentacular homogenates *in vitro* upon post-androstenedione metabolism in the male phase of *Ariolimax californicus* ovotestis. *Gen. Comp. Endocrinol.* **15,** 139–142.

Goudsmit, E. M. (1975). Neurosecretory stimulation of galactogen synthesis within the *Helix pomatia* albumen gland during organ culture. *J. Exp. Zool.* **191,** 193–198.

Goudsmit, E. M. (1978). Calcium-dependent release of a neurochemical messenger from the brain of the land snail *Helix pomatia*. *Brain Res.* **141,** 418–423.

Graves, S. Y., and Dietz, T. H. (1979). Prostaglandin E_2 inhibition of sodium transport in the freshwater mussel. *J. Exp. Zool.* **210,** 195–201.

Graves, S. Y., and Dietz, T. H. (1980). Diurnal rhythms of sodium transport in the freshwater mussel. *Can. J. Zool.* **58,** 1626–1630.

Greenberg, M. J., and Price, D. A. (1980). Cardioregulatory peptides in molluscs. *In* "Peptides, Integrators of Cell and Tissue Function" (F. E. Bloom, ed.), pp. 107–126. Raven, New York.

Griffond, B. (1975). Analyse expérimentale en culture "*in vitro*" des facteurs de la gamétogénèse chez la paludine *Viviparus viviparus* (L.). *Ann. Sci. Univ. Besancon Zool.* **12,** 73–98.

Grimmelikhuyzen, C. J. P., Dockray, G. J., and Schot, L. P. C. (1982). FMRFamide-like immunoreactivity in the nervous system of *Hydra. Histochem.* **73,** 499–508.

Grimm-Jørgensen, Y. (1978). Immunoreactive thyrotropin-releasing factor in a gastropod: distribution in the central nervous system and haemolymph of *Lymnaea stagnalis. Gen. Comp. Endocrinol.* **35,** 387–390.

Grimm-Jørgensen, Y. (1979). Effect of thyrotropin releasing factor on body weight of the pond snail *Lymnaea stagnalis. J. Exp. Zool.* **208,** 169–275.

Grimm-Jørgensen, Y. (in press). Distribution and physiological roles of TRH and somatostatin in gastropods. *In* "Advances in Comparative Endocrinology" (B. Lofts, ed.). Hong Kong University Press, Hong Kong.

Grimm-Jørgensen, Y., and Jackson, I. M. D. (1975). Immunoreactive thyrotropin releasing factor in gastropod circumoesophageal ganglia. *Nature* **254,** 620.

Guyard, A. (1971a). Etude de la différenciation de l'ovotestis et des facteurs contrôlant l'orientation sexuelle des gonocytes de l'escargot *Helix aspersa* Müller. *Thèse de Sciences, Besançon No.* **56,** p. 187.

Guyard, A. (1971b). Nature endocrine des substances réglant la sexualisation de la gonade et son fonctionnement chez les Mollusques gonochoriques et hermaphrodites. *Haliotis* **1,** 167–183.

Harry, H. W. (1965). Evidence of a gonadal hormone controlling the development of the accessory reproductive organs in *Taphius glabratus* Say (Gastropoda, Basommatophora). *Trans. Am. Microsc. Soc.* **84,** 157.

Hekstra, G. P., and Lever, J. (1960). Some effects of ganglion extirpations in *Lymnaea stagnalis*. *Proc. K. Ned. Akad. Wet. C* **63**, 271–282.

Heller, E., Kaczmarek, K., Hunkapiller, M. W., Hood, L. E., and Strumwasser, F. (1980). Purification and primary structure of two neuroactive peptides that cause bag-cell afterdischarge and egg-laying in *Aplysia*. *Proc. Nat. Acad. Sci. U.S.A.* **77**, 2328–2332.

Herlin-Houtteville, P., and Lubet, P. E. (1974). Analyse expérimentale en culture organotypique, de l'action des ganglions cérébro-pleuraux et visceraux sur le manteau de la moule pâle. *C.R. Acad. Sci.* **278**, 2469–2472.

Highnam, K. C., and Hill, L. (1977). "The Comparative Endocrinology of the Invertebrates" 2nd edition. Edward Arnold, London.

Hogg, N. A. S., and Wijdenes, J. (1979). A study of gonadal organogenesis and the factor influencing regeneration following surgical castration in *Deroceras reticulatum* (Pulmonata: Limacidae). *Cell Tissue Res.* **198**, 295–307.

Idler, D. R., Sangalang, G. B., and Kanazawa, A. (1969). Steroid desmolase in gonads of a marine invertebrate, *Placopecten magellanicus* Gmelin. *Gen. Comp. Endocrinol.* **12**, 222–230.

Joosse, J. (1964). Dorsal bodies and dorsal neurosecretory cells of the cerebral ganglia of *Lymnaea stagnalis* L. *Arch. Neerl. Zool.* **15**, 1–103.

Joosse, J. (1972). Endocrinology of reproduction in molluscs. *Gen. Comp. Endocrinol. Suppl.* **3**, 591–601.

Joosse, J. (1975). Structural and endocrinological aspects of hermaproditism in pulmonate snails, with particular reference to *Lymnaea stagnalis* (L.). In "Intersexuality in the Animal Kingdom" (R. Reinboth, ed.), pp. 158–169. Springer-Verlag, Berlin.

Joosse, J. (1976). Endocrinology of molluscs. In "Actualités sur les Hormones d'Invertébrés," pp. 107–123. Coll. Internat. du C.N.R.S. No. 251.

Joosse, J. (1979a). Evolutionary aspects of the endocrine system and of the hormonal control of reproduction of molluscs. In "Hormones and Evolution" (E. J. W. Barrington, ed.), pp. 119–157. Academic Press, London.

Joosse, J. (1979b). Endocrinology of molluscs. In "Pathways in Malacology" (J. van der Spoel, A. C. van Bruggen and J. Lever, eds.), pp. 107–137. Bohn, Scheltema and Holkema, Utrecht, The Netherlands.

Joosse, J., and Reitz, D. (1969). Functional anatomical aspects of the ovotestis of *Lymnaea stagnalis*. *Malacologia* **9**, 101–109.

Joosse, J., and Veld, C. J. (1972). Endocrinology of reproduction in the hermaphrodite gastropod *Lymnaea stagnalis*. *Gen. Comp. Endocrinol.* **18**, 599–600.

Joosse, J., Boer, M. H., and Cornelisse, C. J. (1968). Gametogenesis and oviposition in *Lymnaea stagnalis* as influenced by γ-irradiation and hunger. *Symp. Zool. Soc. London* **22**, 213–234.

Joosse, J., de Vlieger, T. A., and Roubos, E. W. (1982). Nervous systems of lower animals as models, with particular reference to peptidergic neurons in gastropods. In "Chemical transmission in the brain, Progress in brain research" Vol. 55 (R. M. Buijs et al., eds.), pp.379–404. Elsevier, Amsterdam, New York.

Kaczmarek, L. K., Finbow, M., Revel, J. P., and Strumwasser, F. (1979). The morphology and coupling of *Aplysia* bag cells within the abdominal ganglion and in cell culture. *J. Neurobiol.* **10**, 535–550.

Khan, H. R., and A. S. M. Saleuddin (1979a). Effects of osmotic changes and neurosecretory extracts on kidney ultrastructure in the freshwater pulmonate *Helisoma*. *Can. J. Zool.* **57**, 1256.

Khan, H. R., and A. S. M. Saleuddin (1979b). Osmotic regulation and osmotically induced changes in the neurosecretory cells of the pulmonate snail *Helisoma*. *Can. J. Zool.* **57**, 1371.

Khan, H. R., and Saleuddin, A. S. M. (1981a). Involvement of actin and Na^+-K^+ATPase in urine formation of the freshwater Pulmonate *Helisoma*. *J. Morphol.* **269**, 243–251.

Khan, H. R., and Saleuddin, A. S. M. (1981b). Cell contacts in the kidney epithelium of *Helisoma*

(Mollusca: Gastropoda)—Effects of osmotic pressure and brain extracts: a freeze-fracture study. *J. Ultrastruct. Res.* **75**, 23–40.

Kiehling, C., Froesch, D., Martin, R., and Voight, K. H. (1981). Enkaphelin-related peptides: Structure activity relationship of cardioexcitatory substances in *Octopus vulgaris*. Conf. Eur. Soc. Comp. Endocrin., Jerusalem. (Abstract 11.)

Kits, K. S. (1980). States of excitability in ovulation hormone producing neuroendocrine cells of *Lymnaea stagnalis* (Gastropoda) and their relation to the egg-laying cycle. *J. Neurobiol.* **11**, 397–410.

Kits, K. S. (1981). Electrical activity and hormone output of ovulation hormone producing neuroendocrine cells in *Lymnaea stagnalis* (Gastropoda), Adv. Physiol. Sci., Vol. 23. *In* "Neurobiology of Invertebrates—Mechanisms of Integration" (J. Salanki, ed.), pp. 34–54. Adadémiai Kiadó, Budapest, Hungary.

Koester, J., Dieringer, N., and Mandelbaum, D. E. (1979). Cellular neuronal control of molluscan heart. *Am. Zool.* **19**, 103–116.

Krusch, B., Schoenmakers, H. J. N., Voogt, P. A., and Nolte, A. (1979). Steroid synthesizing capacity of the dorsal body of *Helix pomatia* L. (Gastropoda). An *in vitro* study. *Comp. Biochem. Physiol. B* **64**, 101–104.

Kuhlmann, D. (1966). Die Dorsalkörper der Stylommatophoren. *Z. Wiss. Zool.* **172**, 218–231.

Kunigelis, S. C., and Saleuddin, A. S. M. (1978). Regulation of shell growth in the pulmonate gastropod *Helisoma duryi*. *Can. J. Zool.* **56**, 1975–1980.

Kupfermann, I. (1967). Stimulation of egg-laying: possible neuroendocrine function of bag cells of abdominal ganglion of *Aplysia californica*. *Nature* **216**, 814–815.

Kupfermann, I. (1970). Stimulation of egg-laying by extracts of neuroendocrine cells (bag cells) of abdominal ganglion of *Aplysia*. *J. Neurophysiol.* **33**, 877–881.

Kupfermann, I., and Kandel, E. R. (1970). Electrophysiological properties and functional interconnections of two symmetrical neurosecretory clusters (bag cells) in abdominal ganglion of *Aplysia*. *J. Neurophysiol.* **33**, 865–876.

Kupfermann, I., and Weiss, K. (1976). Water regulation by a presumptive hormone contained in identified neurosecretory cell R15 of *Aplysia*. *J. Gen. Physiol.* **67**, 113–123.

Lane, N. J. (1962). Neurosecretory cells in the optic tentacles of certain pulmonates. *Q. J. Microsc. Sci.* **103**, 211–226.

Lane, N. J. (1964). The fine structure of certain secretory cells in the optic tentacles of the snail, *Helix aspersa*. *Q. J. Microsc. Sci.* **105**, 35–47.

Laviolette, P. (1954). Rôle de la gonade dans le déterminisme humoral de la maturité glandulaire du tractus génital chez quelques gastéropodes Arionidae et Limacidae. *Bull. Biol. Fr. Belg.* **88**, 310–332.

Le Gall, P. (1980). Etude expérimentale de l'association en chaîne et de son influence sur la croissance et la sexualité chez la crépidule *Crepidula fornicata* (Mollusque mésogastéropode). *Thesis*, University Caen, pp. 1–251.

Le Gall, S. (1974). Déterminisme de la morphogénèse et du cycle du tractus génital mâle externe chex *Crepidula fornicata* Phil (Mollusque hermaphrodite protandre). *Thesis*. University Caen, pp. 1–267.

Le Gall, S. (1978). Contrôle du facteur pédieux morphogénétique de pénis par les ganglions cérébropleuraux chez *Crepidula fornicata* Phil (Mollusque hermaphrodite protandre). *C. R. Acad. Sci.* **287**, 1305–1307.

Le Gall, S. (1981). Etude expérimentale du facteur morphogénitique contrôlant la différenciation du tractus génital mâle externe chez *Crepidula fornicata* L. (Mollusque hermaphrodite protandre). *Gen. Comp. Endocrinol.* **43**, 51–62.

Le Gall, S., and Streiff, W. (1975). Protandric hermaphroditism in prosobranch gastropods. *In*

"Intersexuality in the Animal Kingdom" (R. Reinboth, ed.), pp. 170–178. Springer-Verlag, Berlin, Heidelberg and New York.

Lehoux, J. G., and Williams, E. E. (1971). Metabolism of progesterone by gonadal tissue of *Littorina littorea* (L.). (Prosobranchia, Gastropoda). *J. Endocrinol.* **51**, 411–412.

Lever, J. (1957). Some remarks on neurosecretory phenomena in *Ferrissia sp.* (Gastropoda Pulmonata). *Proc. K. Ned. Akad. Wet. C* **61**, 235–242.

Lever, J. (1958). On the relation between the medio-dorsal bodies and the cerebral ganglia in some pulmonates. *Arch. Neerl. Zool.* **13**, 194–201.

Lever, J., and Joosse, J. (1961). On the influence of the salt content of the medium on some special neurosecretory cells in the lateral lobes of the cerebral ganglia of *Lymnaea stagnalis*. *Proc. K. Ned. Akad. Wet. C* **64**, 630–639.

Lever, J., Jansen, J., and de Vlieger, Th. A. (1961). Pleural ganglia and water balance in the freshwater pulmonate *Lymnaea stagnalis*. *Proc. K. Ned. Akad. Wet. C* **64**, 531–542.

Lloyd, P. E. (1978a). Distribution and molecular characteristics of cardioactive peptides in the snail, *Helix aspersa*. *J. Comp. Physiol.* **128**, 269–276.

Lloyd, P. E. (1978b). Neurohormonal control of cardiac activity in the snail *Helix aspersa*. *J. Comp. Physiol.* **128**, 277–283.

Lloyd, P. E. (1979). Central peptide-containing neurons modulate gut activity in *Tritonia*. *Neurosci. Abstr.* **5**, 252.

Lloyd, P. E. (1980). Biochemical and pharmacological analysis of endogenous cardioactive peptides in the snail, *Helix aspera*. *J. Comp. Physiol.* **138**, 265–270.

Lubet, P. (1959). Recherches sur le cycle sexuel et l'emission des gamètes chez les Mytilidés et les Pectinidés. *Rev. Trav. Inst. Peches Marit.* **23**, 395–548.

Lubet, P. (1971). Influence des ganglions cérébroides sur la croissance de *Crepidula fornicata* Phil. (Mollusque Mésogastéropode). *C. R. Acad. Sci. Paris* **273**, 2309–2311.

Lubet, P., and Silberzahn, N. (1971). Recherches sur les effets de l'ablation bilatérale des ganglions cérébroides chez la crépidule (Crepidula fornicata Phil., Mollusque gastéropode) effets somatotrope et gonadotrope. *C. R. Séances Soc. Biol.* **165**, 590–594.

Lubet, P., and Streiff, W. (1969). Etude expérimentale de l'action des ganglions nerveux sur la morphogénèse du pénis et l'activité génitale de *Crepidula fornicata* Phil. (Mollusque Gastéropode). *In* "Cours et Documents de Biologie" (Gordon and Breach, eds.), Vol. 1, pp. 141–159. Paris.

Lubet, P., Herlin, P., Mathieu, M., and Collin, F. (1976). Tissu de réserve et cycle sexuel chez les lamellibranches. *Haliotis* **7**, 59–62.

Lupo di Prisco, C., and Dessi Fulgheri, F. (1975). Alternative pathways of steroid biosynthesis in gonads and hepatopancreas of *Aplysia depilans*. *Comp. Biochem. Physiol. B* **50**, 191–195.

Maddrell, S. H. P. (1967). Neurosecretion in insects. *In* "Insects and Physiology" (J. W. L. Beament and J. E. Treherne, eds.), pp. 103–118. American Elsevier, New York.

Maddrell, S. H. P., and Nordmann, J. J. (1979). "Neurosecretion." Blackie and Son, Glasgow, London.

Mangold, K., and Froesch, D. (1977). A reconsideration of factors associated with sexual maturation. *Symp. Zool. Soc. London* **38**, 541–555.

Mangold-Wirz, K. (1963). Biologie des Céphalopodes benthiques et nectoniques de la Mer Catalane. *Vie Milieu 13, Suppl.*, 1–285.

Marques, M., and Falkmer, S. (1976). Effects of mammalian insulin on blood glucose level, glucose tolerance and glycogen content of musculature and hepatopancreas in a gastropod mollusc *Strophocheilus oblongus*. *Gen. Comp. Endocrinol.* **29**, 522–530.

Martin, R., Froesch, D., and Voigt, K. H. (1980). Immunocytochemical evidence for melanotropin- and vasopressin-like material in a cephalopod neurohemal organ. *Gen. Comp. Endocrinol.* **42**, 235–243.

Martoja, M. (1965). Existence d'un organe juxtaganglionnaire chez *Aplysia punctata* Cuv. (Gastéropode Opisthobranche). *C. R. Acad. Sci. Paris*, **260**, 4615–4617.

Martoja, M. (1968). Neurosecretion. *In* "Traité de Zoologie" (P. P. Grassé, ed.), pp. 927–986. Masson & Cie., Paris.

Mayeri, E., Brownell, P., and Branton, W. D. (1979a). Multiple, prolonged actions of neuroendocrine bag cells on neurons in *Aplysia*. II. Effects on beating pacemaker and silent neurons. *J. Neurophysiol.* **42**, 1185–1197.

Mayeri, E., Brownell, P., Branton, W. D., and Simon, S. B. (1979b). Multiple, prolonged actions of neuroendocrine bag cells on neurons in *Aplysia*. I. Effects of bursting pacemaker neurons. *J. Neurophysiol.* **42**, 1165–1184.

McCrone, E. J., and Sokolove, P. G. (1979). Brain-gonad axis and photoperiodically-stimulated sexual maturation in the slug, *Limax maximus*. *J. Comp. Physiol.* **133**, 117–123.

Morse, D. E., Duncan, H., Hooker, N., and Morse, A. (1977). Hydrogen peroxide induces spawning in molluscs, with activation of prostaglandin endoperoxide synthetase. *Science* **196**, 298–300.

Nishioka, R. S., Bern, H. A., and Golding, D. W. (1970). Innervation of the cephalopod optic gland. *In* "Aspects of Neuroendocrinology" (W. Bargman and B. Scharrer, eds.), pp. 47–54 (5th International Symposium on Neurosecretion). Springer Verlag, Berlin.

Nolte, A. (1965). Neurohämal-"Organe" bei Pulmonaten (Gastropoda). *Zool. Jahrb. Anat.* **82**, 365–380.

Nolte, A. (1978). Ultrastructure of the dorsal neurohemal area of the snail *Theba pisana* (Stylommatophora Gastropoda). *In* "Neurosecretion and Neuroendocrine Activity" (W. Bragman, A. Oksche, A. Polenov, and B. Scharrer, eds.), pp. 386–389. Springer-Verlag, Berlin.

Nolte, A., and Kuhlmann, D. (1964). Histologie und Sekretion der Cerebral-Drüse adulter Stylommatophoren (Gastropoda). *Z. Zellforsch.* **63**, 550–567.

O'Dor, R. K., and Wells, M. J. (1973). Yolk protein synthesis in the ovary of *Octopus vulgaris* and its control by the optic gland gonadotropin. *J. Exp. Biol.* **59**, 665–674.

O'Dor, R. K., and Wells, M. J. (1975). Control of yolk protein synthesis by Octopus gonadotropin *in vivo* and *in vitro*. *Gen. Comp. Endocrinol.* **27**, 129–135.

O'Dor, R. K., and Wells, M. J. (1978). Reproduction versus somatic growth: hormonal control in *Octopus vulgaris*. *J. Exp. Biol.* **77**, 15–31.

Painter, S. D., Price, D. A., and Greenberg, M. J. (1979). Responses of bivalve myocardia to 5-hydroxytryptamine (5HT) and the molluscan neuropeptide, FMRF-amide. *Am. Zool.* **19**, 959.

Pelluet, D. (1964). On the hormonal control of cell differentiation in the ovotestis of slugs (Gasteropoda, Pulmonata). *Can. J. Zool.* **42**, 195–199.

Pelluet, D., and Lane, N. J. (1961). The relation between neurosecretion and cell differentiation in the ovotestis of slugs (Gastropoda: Pulmonata) *Can. J. Zool* **39**, 691–805.

Pinsker, H., and Dudek, F. E. (1977). Bag cell control of egg-laying in freely behaving *Aplysia*. *Science* **197**, 490–493.

Plesch, B. E. C. (1977). An ultrastructural study of the innervation of the musculature of the pond snail *Lymnaea stagnalis* (L.) with reference to peripheral neurosecretion. *Cell Tiss. Res.* **183**, 353–369.

Plisetskaya, E., and Joosse, J. (in press). Endocrine control of metabolism in molluscs. *In* "Advances in Comparative Endocrinology" (B. Lofts, ed.). Hong Kong University Press, Hong Kong.

Plisetskaya, E. M., and Soltitskaya, L. P. (1979). Participation of insulin in metabolic regulation in marine bivalve molluscs. *Zh. Evol. Biokhim. Fiziol.* **15**, 288–294.

Plisetskaya, E., Soltitskaya, L., and Rusacov, Y. I. (1978a). Production and role of insulin in the regulation of carbohydrate metabolism in freshwater and marine bivalve molluscs. *In* "Comparative Endocrinology" (P. J. Gaillard and H. H. Boer, eds.), pp. 449–453. Elsevier/North Holland Biomedical Press, Amsterdam.

Plisetskaya, E., Kazakov, V. K., Soltitskaya, L., and Leibson, L. G. (1978b). Insulin-producing cells in the gut of freshwater bivalve molluscs *Anodonta cygnea* and *Unio pictorum* and the role of insulin in the regulation of their carbohydrate metabolism. *Gen. Comp. Endocrinol.* **35**, 133–145.

Price, D. A., and Greenberg, M. J. (1977a). Structure of a molluscan cardioexcitatory neuropeptide. *Science* **197**, 670–671.

Price, D. A., and Greenberg, M. J. (1977b). Purification and characterization of a cardioexcitatory neuropeptide from the central ganglia of a bivalve mollusc. *Prep. Biochem.* **7**, 261–281.

Price, D. A., and Greenberg, M. J. (1980). Pharmacology of the molluscan cardioexcitatory neuropeptide FMRFamide. *Gen. Pharmacol.* **11**, 237–241.

Ram, J. L. (1977). Hormonal control of reproduction in *Busicon:* Laying of egg capsules caused by nervous system extracts. *Biol. Bull.* **152**, 221–232.

Richard, A. (1967). Rôle de la photopériode dans le déterminisme de la maturation génitale femelle du Céphalopode *Sepia officinalis* L. *C. R. Acad. Sci., Paris* **164**, 1315–1318.

Richard, A. (1970a). Analyse du cycle sexuel chez les Céphalopodes: mise en évidence expérimentale d'une rythme conditionnée par les variations des facteurs externes et internes. *Bull. Soc. Zool. France* **95**, 461–469.

Richard, A. (1970b). Différenciation sexuelle des Céphalopodes en culture *in vitro*. *Annee Biol.* **9**, 409–415.

Rogers, D. G. (1969). The fine structure of the collar cells in the optic tentacles of *Helix aspersa*. *Z. Zellforsch.* **102**, 113–128.

Rohlack, S. (1959). Uber das Vorkommen von Sexualhormonen bei der Meeresschnecke *Littorina littorea*. *Z. Vgl. Physiol.* **42**, 164–180.

Röhlich, P., and Bierauer, J. (1966). Electron microscopic observations on the special cells of the optic tentacle of *Helicella obvia* (Pulmonata). *Acta Biol. Hung.* **17**, 359–373.

Röhnisch, S. (1964). Untersuchungen zur Neurosekretion bei *Planorbarius corneus* L. (Basommatophora). *Naturwissenschaften* **51**, 148.

Roubos, E. W. (1976). Neuronal and non-neuronal control of the neurosecretory caudo-dorsal cells of the freshwater snail *Lymnaea stagnalis* (L.). *Cell Tiss. Res.* **168**, 11–31.

Roubos, E. W., and Buma, P. (1982). Cellular mechanisms of neurohormone release in the snail *Lymnaea stagnalis*. *In* "Chemical transmission in the brain, Progress in brain research," Vol. 55, pp. 185–194. Elsevier, Amsterdam, New York.

Roubos, E. W., and Moorer-van Delft, Carry M. (1976). Morphometric *in vitro* analysis of the control of the activity of the neurosecretory Dark Green Cells in the freshwater snail *Lymnaea stagnalis* (L.). *Cell Tiss. Res.* **174**, 221–231.

Roubos, E. W., and Moorer-van Delft, Carry M. (1979). Synaptology of the central nervous system of the freshwater snail *Lymnaea stagnalis* (L.), with particular reference to neurosecretion. *Cell Tiss. Res.* **198**, 217–235.

Roubos, E. W., and van der Wal-Divendal, R. M. (1982). Sensory input to growth stimulating neuroendocrine cells of *Lymnaea stagnalis*. *Cell Tiss. Res.* **227**, 371–386.

Roubos, E. W., Geraerts, W. P. M., Boerrigter, G. H., and van Kampen, G. P. J. (1980). Control of the activities of the neurosecretory Light Green and Caudo-Dorsal Cells and of the endocrine Dorsal Bodies by the lateral lobes in the freshwater snail *Lymnaea stagnalis*. *Gen. Comp. Endocrinol.* **40**, 446–454.

Roubos, E. W., Boer, H. H., and Schot, L. P. C. (1981a). Peptidergic neurones and the control of neuroendocrine activity in the freshwater snail *Lymnaea stagnalis*. *In* "Neurosecretion, Molecules, Cells, Systems" (D. S. Farner and K. Lederis, eds.), pp. 119–127. Plenum, New York.

Roubos, E. W., Schmidt, E. D., and Moorer-van Delft, C. M. (1981b). Ultrastructural dynamics of exocytosis in the ovulation neurohormone producing Caudo-Dorsal Cells of the freshwater snail *Lymnaea stagnalis* (L.). *Cell Tiss. Res.* **215**, 63–73.

Rowe, V. L., and Mangold, K. (1975). The effect of starvation on sexual maturation in *Illex illecebrosus* (Cephalopoda: Teuthoidea). *J. Exp. Mar. Biol. Ecol.* **17**, 157–163.

Runham, N. W., and Laryea, A. A. (1968). Studies on the maturation of the reproductive system of *Agriolimax reticulutus* (Pulmonata, Limacidae). *Malacologia* **7**, 93–108.

Runham, N. W., Bailey, T. G., and Laryea, A. A. (1963). Studies on the endocrine control of the reproductive tract of the grey field slug *Agriolimax reticulatus*. *Malacologia* **14**, 135–142.

Rusacov, Y. I., and Kazakov, V. K. (1979). Preparation of insulin-like substance from molluscs and antiserum against it. *Zh. Evol. Biokhim. Fiziol.* **15**, 617–619 (in Russian).

Saleuddin, A. S. M., and Dillaman, R. M. (1976). Direct innervation of the mantle edge gland by neurosecretory axons in *Helisoma duryi* (Mollusca: Pulmonata). *Cell Tiss. Res.* **171**, 397–401.

Saleuddin, A. S. M., and Jones, G. M. (1978). Effects of shell regeneration on certain neurosecretory cells in *Helisoma* (Mollusca: Pulmonata). *Microsc. Soc. Can. Bull.* **6**, 24–26.

Saleuddin, A. S. M., and Khan, H. R. (1981). Motility of the oocyte of *Helisoma* (Mollusca). *Eur. J. Cell. Biol.* **26**, 5–10.

Saleuddin, A. S. M., Wilson, L. E., Khan, H. R., and Jones, G. M. (1980). Effects of brain extracts on oocyte maturation in *Helisoma* (Pulmonata: Mollusca). *Can. J. Zool.* **58**, 1109–1124.

Sawyer, W. H., Pang, P. K. T., Deyrup-Olsen, I., and Martin, A. W. (in press). Evolution of neurohypophysial peptides and their functions. *In* "Advances in Comparative Endocrinology" (B. Lofts, ed.). University Hong Kong Press, Hong Kong.

Saliot, A., and Barbier, M. (1971). Sur l'isolement de la progestérone et de quelques cétostéroides de la partie femelle des gonades de la coquille Saint-Jacques *Pecten maximus*. *Biochemie* **53**, 265–266.

Sanchez, S., and Bord, C. (1958). Origine cellules neurosécrétices chez *Helix aspersa* Mull. *C. R. Acad. Sci. Paris* **246**, 845–847.

Scharrer, B. (1935). Ueber das Hanströmsche Organ X bei Opisthobranchiern. *Pubbl. Stan. Zool. Napoli* **15**, 132–142.

Scheerboom, J. E. M. (1978). The influence of food quantity and food quality on assimilation, body growth and egg production in the pond snail *Lymnaea stagnalis* (L.) with particular reference to the haemolymph–glucose concentration. *Proc. K. Ned. Akad. Wet. C* **81**, 184–197.

Scheerboom, J. E. M., and Doderer, A. (1978). The effects of artificially raised haemolymph-glucose concentrations on feeding, locomotory activity, growth and egg production of the pond snail *Lymnaea stagnalis* (L.). *Proc. K. Ned. Akad. Wet. C* **81**, 377–386.

Scheerboom, J. E. M., and Dogterom, A. A. (1978). The effects of the light green cells and of the lateral lobes on growth and feeding in the pond snail *Lymnaea stagnalis* (L.). *Proc. K. Ned. Akad. Wet. C* **81**, 327–334.

Scheerboom, J. E. M., and van Elk, R. (1978). Field observations on the seasonal variations in the natural diet and the haemolymph–glucose concentration of the pond snail *Lymnaea stagnalis* (L.). *Proc. K. Ned. Akad. Wet. C* **81**, 365–376.

Scheerboom, J. E. M., Hemminga, M. A., and Doderer, A. (1978). The effects of a change of diet on consumption and assimilation and on the haemolymph–glucose concentration of the pond snail *Lymnaea stagnalis* (L.). *Proc. K. Ned. Akad. Wet. C* **81**, 335–346.

Schot, L. P. C., Boer, H. H., Swaab, D. F., and van Noorden, S. (1981). Immunocytochemical demonstration of peptidergic neurons in the central nervous system of the pond snail *Lymnaea stagnalis* with antisera raised to biologically active peptides of vertebrates. *Cell Tiss. Res.* **216**, 273–291.

Schot, L. P. C., and Boer, H. H. (1982). Immunocytochemical demonstration of peptidergic neurons in the pond snail *Lymnaea stagnalis* with antisera to vertebrate biologically active peptides and of a molluscan tetrapeptide in *Lymnaea stagnalis*, various insects, a fish, a mouse and a cat. *Gen. Comp. Endocrinol.* **46**, 372.

Simpson, L. (1969). Morphological studies of possible neuroendocrine structures in *Helisoma tenue* (Gastropoda: Pulmonata). *Z. Zellforsch.* **102**, 570–593.

Simpson, L., Bern, H. A., and Nishioka, R. S. (1966). Examination of the evidence for neurosecretion in the nervous system of *Helisoma tenue* (Gastropoda, Pulmonata). *Gen. Comp. Endocrinol.* **7**, 525–549.

Sminia, T. (1972). Structure and function of blood and connective tissue cells of the freshwater pulmonate *Lymnaea stagnalis*, studied by electron microscopy and enzyme histochemistry. *Z. Zellforsch.* **130**, 497–526.

Smith, B. J. (1966). The structure of the central nervous system of the slug *Arion ater* L., with notes on the cytoplasmic inclusions of the neurons. *J. Comp. Neur.* **126**, 437–452.

Soffe, S. R., Benjamin, P. R., and Slade, C. T. (1978). Effects of environmental osmolarity on blood composition and light microscope appearance of neurosecretory neurons in the snail *Lymnaea stagnalis* (L.). *Comp. Biochem. Physiol. A* **61**, 577.

Sokolove, P. G., and McCrone, E. J. (1978). Reproductive maturation in the slug, *Limax maximus,* and the effects of artificial photoperiod. *J. Comp. Physiol.* **125**, 317–325.

Streiff, W. (1966). Autodifférenciation ovarienne chez un mollusque hermaphrodite protancre *Calyptraea sinensis* L. *C. R. Acad. Sci. Paris* **263**, 539–542.

Streiff, W. (1967). Etude endocrinologique du déterminisme du cycle sexuel chez un mollusque hermaphrodite protandre, *Calyptraea sinensis* L. II. Mise en évidence par culture *in vitro* de facteurs hormonaux conditionnant l'évolution du tractus génital femelle. *Ann. Endocrinol.* **28**, 461–472.

Streiff, W., Lebreton, J., and Silberzahn, N. (1970). Non specificité des facteurs hormonaux responsables de la morphogénèse et du cycle du tractus génital mâle chez les Mollusques Prosobranches. *Ann. Endocrinol. Paris* **31**, 548–556.

Strumwasser, F., Jacklet, J. W., and Alvarez, R. B. (1969). A seasonal rhythm in the neural extract induction of behavioral egg-laying in *Aplysia. Comp. Biochem. Physiol.* **29**, 197–206.

Strumwasser, F., Kaczmarek, L. K., Chiu, A. Y., Heller, E., Jennings, K. R., and Viele, D. P. (1980). Peptides controlling behavior in *Aplysia. In* "Peptides: Integrators of Cell and Tissue Function" (F. E. Bloom, ed.), pp. 197–218. Raven Press, New York.

Stuart, D. K., and Strumwasser, F. (1980). Neuronal sites of action of a neurosecretory peptide, egg-laying hormone, in *Aplysia californica. J. Neurophysiol.* **43**, 499–519.

Stuart, D. K., Chiu, A. Y., and Strumwaser, F. (1980). Neurosecretion of egg-laying hormone and other peptides from electrically active bag cell neurons of *Aplysia. J. Neurophysiol.* **43**, 488–498.

Subramanyam, O. V. (1973). Neuroendocrine control of calcium levels in the blood of *Ariophanta semirugata,* a terrestrial pulmonate snail. *Endocrinol. Exp.* **7**, 315–317.

Swindale, N. V., and Benjamin, P. R. (1976a). The anatomy of neurosecretory neurones in the pond snail *Lymnaea stagnalis* (L.). *Phil. Trans. R. Soc. B* **274**, 169–202.

Swindale, N. V., and Benjamin, P. R. (1976b). Peripheral neurosecretion in the pond snail *Lymnaea stagnalis* (L.). *In* "Neurobiology of Invertebrates: Gastropoda Brain" (J. Salanki, ed.), pp. 75–84. Akadémiai Kiadó, Budapest, Hungary.

Takeda, N. (1979). Induction of egg-laying by steroid hormones in slugs. *Comp. Biochem. Physiol. A* **62**, 273–278.

Takeda, N. (1980). Hormonal control of head-wart development in the snail *Euhadra peliomphala. J. Embryol. Exp. Morphol.* **60**, 57–69.

Takeda, N., and Tsuruoka, H. (1979). A sex pheromone secreting gland in the terrestrial snail, *Euhadra peliomphala. J. Exp. Zool.* **207**, 17–26.

Teshima, S. I., and Kamazawa, A. (1971). Metabolism of progesterone in the hepatopancreas of whelk, *Buccinum undatum. Nippon Suisan Gakkaishi* **37**, 529–533.

ter Maat, A., and Lodder, J. C. (1980). A biphasic cholinergic input on the ovulation hormone producing Caudo-Dorsal Cells of the freshwater snail *Lymnaea stagnalis*. *Comp. Biochem. Physiol. C* **66**, 115–119.

ter Maat, A., Lodder, J. C., Veenstra, J., and Goldschmeding, J. T. (1982). Suppression of egg laying during starvation in the snail *Lymnaea stagnalis* by inhibition of the ovulation hormone producing caudo-dorsal cells. *Brain Res.* **239**, 535–542.

ter Maat, A., Roubos, E. W., Lodder, J. C., and Buma, P. (in press). Integration of synaptic input modulating pacemaker activity of electrotonically coupled neuroendocrine Caudo-Dorsal Cells in the pond snail. *J. Neurophysiol.*

Thompson, T. E., and Bebbington, A. (1969). Structure and function of the reproductive organs of three species of *Aplysia* (Gastropoda: Opisthobranchia). *Phil. Trans. R. Soc. B* **245**, 171–218.

Toevs, L. (1970). Identification and characterization of the egg-laying hormone from the neurosecretory bag cells of *Aplysia*. Ph.D. dissertation, California Institute of Technology.

Tuzet, O., Sanchez, S., and Pavans de Ceccatty, M. (1957). Données histologiques sur l'organisation neuroendocrine de quelques mollusques gastéropodes. *C. R. Acad. Sci. Paris* **244**, 2962–2964.

van Aardt, W. J. (1968). Quantitative aspects of the water balance in *Lymnaea stagnalis*. *Neth. J. Zool.* **18**, 253–312.

van Minnen, J., and Reichelt, D. (1980). Photoperiod dependent neural control of the activity of the neurosecretory canopy cells in the lateral lobes of the cerebral ganglia of the freshwater pulmonate snail *Lymnaea stagnalis* (L.). *Cell Tiss. Res.* **208**, 457–465.

van Mol, J. J. (1967). Etude morphologique et phylogénétique du ganglion cérébroïde des gastéropodes pulmonés (mollusques). *Mém. Acad. M. Belg. Cl. Sci.* **37**, 1–168.

van Noorden, S., Fritsch, H. A. R., Grillo, T. A. I., Polak, J. M., and Pearse, A. G. E. (1980). Immunocytochemical staining for vertebrate peptides in the nervous system of a gastropod mollusc. *Gen. Comp. Endocrinol.* **40**, 375–376.

Veldhuijzen, J. P., and Cuperus, R. (1976). Effects of starvation, low temperature and the dorsal body hormone on the *in vitro* synthesis of galactogen and glycogen in the albumen gland and the mantle of the pond snail *Lymnaea stagnalis*. *Neth. J. Zool.* **26**, 119–135.

Veldhuijzen, J. P., and van Beek, G. (1976). The influence of starvation and of increased carbohydrate intake on the polysaccharide content of various body parts of the pond snail *Lymnaea stagnalis*. *Neth. J. Zool.* **26**, 106–118.

Vianey-Liaud, M. (1979). Action de la castration partielle ou totale sur la croissance et le fonctionnement de l'appareil génital chez le Planorbe *Biomphalaria glabrata* (gastéropode pulmoné). *Experientia* **35**, 188–190.

Vicente, N. (1969). Contribution à l'étude des Gastéropodes Opisthobranches du Golfe de Marseille. Histophysiologie du système nerveux. II. Etude des phénomènes neurosécrétoires. *Rec. Trav. St. Mar. Endoume* **45**, 13–101.

Wattez, C. (1973). Effet de l'ablation des tentacules oculaires sur la gonade en croissance et en cours de régénération chez *Arion subfuscus* Draparnaud (Gastéropode Pulmoné). *Gen. Comp. Endocrinol.* **21**, 1–8.

Wattez, C. (1975). Effet d'injections répétées de broyats de tentacules oculaires sur la gonade en croissance et en cours de régénération d'Arionidés tentaculectomisés: étude chez *Arion subfuscus* Draparnaud (Gastéropode Pulmoné). *Gen. Comp. Endocrinol.* **27**, 479–487.

Wattez, C. (1978). Influence du complexe céphalique (tentacules-oculaires-cerveau) sur l'évolution, en culture *in vitro*, de gonades infantiles et juvéniles d'*Arion subfuscus* Drap. (Gastéropode Pulmoné). *Gen. Comp. Endocrinol.* **35**, 360–374.

Wattez, C. (1980). Recherches expérimentales sur le déterminisme de la différenciation sexuelle et du fonctionnement de la gonade chez le pulmoné stylommatophore hermaphrodite *Arion subfuscus* Draparnaud. Thesis, Univ. Lille, France.

Weber, E., Evans, J. C., Samuelson, S. J., and Barchas, J. D. (1981). Novel peptide neuronal system in rat brain and pituitary. *Science* **214**, 1248–1251.

Wells, M. J. (1976). Hormonal control of reproduction in cephalopods. *In* "Perspectives in Experimental Zoology" (P. Spencer Davies, ed.), Vol. 1, pp. 157–166. Pergamon Press, Oxford and New York.

Wells, M. J., and Mangold, K. (1980). The effects of extracts from neurosecretory cells in the anterior vena cava and pharyngo-ophthalmic vein upon the hearts of intact free-moving Octopuses. *J. Exp. Biol.* **84**, 319–334.

Wells, M. J., and Wells, J. (1959). Hormonal control of sexual maturity in *Octopus*. *J. Exp. Biol.* **36**, 1–33.

Wells, M. J., and Wells, J. (1969). Pituitary analogue in the *Octopus*. *Nature London* **222**, 293–294.

Wells, M. J., and Wells, J. (1972). Sexual displays and mating of *Octopus vulgaris* Cuvier and *O. cyanea* Gray and attempts to alter performance by manipulating the glandular condition of the animals. *Anim. Behav.* **20**, 293–308.

Wells, M. J., and Wells, J. (1975). Optic gland implants and their effects on the gonads of Octopus. *J. Exp. Biol.* **62**, 579–588.

Wells, M. J., and Wells, J. (1977a). Cephalopoda: Octopoda. *In* "Reproduction of Marine Invertebrates, Molluscs: Gastropods and Cephalopods" (A. C. Giese and J. S. Pearse, ed .), Vol. 4, pp. 291–336. Academic Press, London.

Wells, M. J., and Wells, J. (1977b). Optic glands and the endocrinology of reproduction. *Symp. Zool. Soc. London* **38**, 525–540.

Wells, M. J., O'Dor, R. K., and Buckley, S. K. L. (1975). An *in vitro* bioassay for a molluscan gonadotropin. *J. Exp. Biol.* **62**, 433–446.

Wendelaar Bonga, S. E. (1970). Ultrastructure and histochemistry of neurosecretory cells and neurohaemal areas in the pond snail *Lymnaea stagnalis* (L.). *Z. Zellforsch.* **108**, 190–224.

Wendelaar Bonga, S. E. (1971a). Formation, storage and release of neurosecretory material studied by quantitative electron microscopy in the freshwater snail *Lymnaea stagnalis* (L.). *Z. Zellforsch.* **113**, 490–517.

Wendelaar Bonga, S. E. (1971b). Osmotically induced changes in the activity of neurosecretory cells located in the pleural ganglia of the freshwater snail *Lymnaea stagnalis* (L.), studied by quantitative electron microscopy. *Neth. J. Zool.* **21**, 127–158.

Wendelaar Bonga, S. E. (1972). Neuroendocrine involvement in osmoregulation in a freshwater mollusc, *Lymnaea stagnalis*. *Gen. Comp. Endocrinol. Suppl.* **3**, 308–316.

Whitehead, D. L. (1977). Steroids enhance shell regeneration in an aquatic gastropod *(Biomphalaria glabrata)*. *Comp. Biochem. Physiol. C* **58**, 137–141.

Whitehead, D. L., and Saleuddin, A. S. M. (1978). Steroids promote shell regeneration in Helix aspersa (Mollusca; Pulmonata). *Comp. Biochem. Physiol. C* **59**, 5–10.

Wilbur, K. M. (1976). Recent studies of invertebrate mineralization. *In* "The Mechanisms of Mineralization in the Invertebrates and Plants" (N. Watabe and K. M. Wilbur, eds.), pp. 79–108. University of South Carolina Press, Columbia, South Carolina.

Wijdenes, J. (1981a). A comparative study on the neuroendocrine control of growth and reproduction in pulmonate snails and slug. Thesis, Free University, Amsterdam.

Wijdenes, J. (1981b). Experiments on the endocrine control of penial complex maturation in the slug *Agriolimax reticulatus* (Pulmonata: Limacidae). *Proc. K. Ned. Akad. Wet. C* **84**, 107–114.

Wijdenes, J., and Runham, N. W. (1976). Studies on the function of the dorsal bodies of *Agriolimax reticulatus* (Mollusca, Pulmonata). *Gen. Comp. Endocrinol.* **29**, 545–551.

Wijdenes, J., and Runham, N. W. (1977). Studies on the control of growth in *Agriolimax reticulatus* (Mollusca, Pulmonata). *Gen. Comp. Endocrinol.* **31**, 154–156.

Wijdenes, J., and Vincent, C. (1981). Ultrastructural evidence for the neuroendocrine innervation of the Dorsal Bodies of the Pulmonate snail *Helix aspersa*. *In* "Thesis J. Wijdenes." Free University, Amsterdam.

Wijdenes, J., van Minnen, J., and Boer, H. H. (1980). A comparative study on neurosecretion demonstrated by the alcian blue/alcian yellow technique in three terrestrial pulmonates (Stylommatophora). *Cell Tiss. Res.* **210,** 47–56.

Wijdenes, J., van Elk, R., Goudsmit, E. M., and Joosse, J. (1981). Species specificity of female gonadotropic hormones in Pulmonate snails. *In* "Thesis J. Wijdenes." Free University, Amsterdam.

Wijdenes, J., van Elk, R., and Joosse, J. (1983). Effects of two gonadotropic hormones on the polysaccharide synthesis in the albumen gland of *Lymnaea stagnalis,* studied with the organ culture technique. *Gen. Comp. Endocrinol.* (in press).

Wodinsky, J. (1977). Hormonal inhibition of feeding and death in Octopus: control by optic gland secretion. *Science* **198,** 948–951.

9

Physiological Energetics of Marine Molluscs

B. L. BAYNE

Natural Environment Research Council
Institute for Marine Environmental Research
Plymouth, England

R. C. NEWELL

Natural Environment Research Council
Institute for Marine Envionmental Research
Plymouth, England

I. Introduction

Physiological energetics is concerned with the study of gains and losses of energy, and the efficiencies of energy transformation, from the standpoint of the whole organism. This definition distinguishes physiological energetics from *bioenergetics,* which is the study of energy exchanges within the cell, and *ecological energetics,* which is concerned with the study of energy transfers between groups of organisms. Whereas these distinctions are convenient, all three topics are closely related and interdependent. The techniques of physiological energet-

THE MOLLUSCA, VOL. 4
Physiology, Part 1

ics have been widely applied to the study of the energetics of natural populations since the pioneer research of Odum and Smalley (1959) on the salt-marsh gastropod (*Littorina irrorata*), and of Odum and Odum (1955), Teal (1962), and Slobodkin (1962) on the relation between primary production and net energy utilization by heterotrophic communities. In this chapter we are mainly concerned with the physiological energetics of individual molluscs, but we also consider studies on the energetics of molluscan populations in order to bridge the gap between strictly physiological and ecological considerations.

Following the laws of thermodynamics (Kleiber, 1961; Wiegert 1968) and assuming steady-state conditions, the net energy exchange in the individual organism may be described by the familiar expression:

$$C = P + R + F + U \tag{1}$$

where ingestion (C; also called consumption) is the energy equivalent of food intake; F is the energy content of the ingested material that is voided as feces; U is the energy content of excreta (e.g., urine, mucus); R is the energy equivalent of metabolic heat losses; and P is the energy incorporated as growth (P_g) or reproductive products (P_r) (see Winberg, 1956; Ricker, 1968; Petrucewicz and MacFadyen, 1970; Crisp, 1971; Grodinski et al., 1975; Conover, 1978; Brett and Groves, 1979).

Figure 1 is a flow diagram to illustrate the main features of Eq. (1). Undigested material is voided as feces (F); digested material than represents the absorbed ration (A). The efficiency with which the ingested ration (C) is absorbed (i.e., A/C) is called the absorption efficiency (e). Two main sources of loss from the absorbed ration are those due to excretion (U; primarily the loss of end products of protein metabolism, which—unlike carbohydrate and lipid substrates—are not fully oxidized) and those due to the heat increment (R'), which is also called the specific dynamic action or specific dynamic effect (Kleiber, 1961). These two losses (excretion and the heat increment) are probably related, and the latter may be viewed as "non-utilised energy freed through deamination and other processes [Warren and Davis, 1967]" or energy losses "derived from the biochemical transformation of ingested food into a metabolisable, excretable form [Brett and Groves, 1979]."

The assimilated ration (the physiologically useful ration) can be partitioned into three components. Some is used to fuel the processes of body maintenance (maintenance metabolism) and the many other energy-demanding activities of the animal, including routine and active metabolism; together these comprise the metabolic energy losses R'', which are normally measured, with the heat increment R', as the total metabolic loss, R. The proportion of the assimilated ration that remains when the metabolic demands have been met is then available for either or both of somatic production (P_g, or growth) and the production of gametes (P_r). In addition to the absorption efficiency mentioned earlier, four

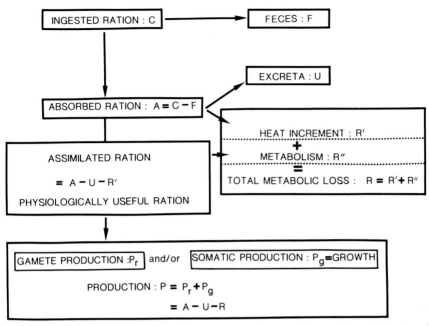

Fig. 1. Flow diagram to illustrate the components of energy balance. Three efficiencies of energy transfer are commonly considered: (1) absorption efficiency (A/C); (2) gross growth efficiency (P/C); (3) net growth efficiency (P/A).

other "efficiencies" are of general interest: (1) total gross growth efficiency (P/C), (2) total net growth efficiency (P/A), (3) gross somatic growth efficiency (P_g/C), and (4) net somatic growth efficiency (P_g/A).

All the processes that comprise the energy balance equation are capable of variation in response to changes in the environment. The animal must be able to balance its gains from the environment against its metabolic losses in order to allocate an optimal distribution of surplus energy to somatic growth and to reproduction. Variations in the balances between these processes are the substance of physiological adaptation. Inherent in any approach to understanding these adaptive responses is the view that they can only be fully understood when the entire energy budget, or energy balance equation, is considered.[1] Warren and Davis (1967) developed the concept of a "scope for growth," which they regarded as "the difference between the energy of the food an animal consumes and all other utilisations and losses." This term has descriptive validity for a

---

[1]Traditionally, Eq. (1) is considered in terms of energy; the logic applies equally to balances of carbon, of nitrogen, or of body weight, although detailed differences of criteria may then apply (Crisp, 1971).

wide range of environmental variables and has been widely used in the study of physiological adaptation in molluscs (Bayne 1976a; Newell and Branch, 1980).

II. The Energetics of Populations

The concept of a balanced system in which inflowing energy equals outflowing energy can be applied in whole populations at different trophic levels (see Lindeman, 1942), or even to communities of mixed populations (Odum and Odum, 1955). In this case, however, the production represents not only the growth of the individuals within the population but also the immigration of new individuals to the community, whereas losses include both the terms $(F + U + R)$ for the component individuals, and those from mortality, emigration, and the energetic cost of gametes from the population as a whole. Steady-state conditions that can be applied to the population are, however, rare because the biomass is commonly increasing or decreasing with time according to the relative rates of recruitment, mortality, and immigration or emigration. The values for the components of the balanced energy equation may also vary with the size class of individuals that comprise the population. For example, the amount of energy that is allocated to growth (P_g) or reproduction (P_r), may vary considerably with age class (Section V,C). The size structure of the population also has an effect on the net production efficiency (P/A), this ratio being greater in populations with a high proportion of young individuals than in stable or old-aged populations (Calow, 1977; Humphreys, 1979; McNeill and Lawton, 1970). In studying the energetics of populations, therefore, it is necessary not only to establish values for the components of the balanced energy equation for different-sized individuals within the population, but also to integrate these with information on the size class structure, production, and mortality of the population in the field (see Crisp, 1971; Grodinski et al., 1975).

Although most of the components of the equation can be determined experimentally, especially in filter-feeding bivalves, the application of the results of laboratory-based studies to the field situation presents many problems. It is almost certain, for example, that the consumption, and especially the absorption efficiency, of molluscs fed on their preferred diets is quite different from that obtained with cultures of algae such as *Phaeodactylum* or *Dunaliella*, which are commonly used to measure the clearance rate and ingestion in bivalves (see Bayne, 1976a; Carefoot, 1967a, 1967b, 1970; Griffiths, 1980b). Partly because of the problems involved in the extrapolation of experimental values of consumption (C), fecal losses (F), and resultant absorbed energy $(A = C - F)$ to field populations, many authors working on molluscs have estimated the value of A from the sum of field measurements of production (P) and laboratory estimates of respiration (R), that is, as $A = P + R$. Other components of the population energy balance, including the respiratory cost and production, can then be expressed as a proportion of the absorbed energy in order to obtain the net respiratory cost (R/A) and net production efficiency $(P_g + P_r)/A$.

Because organisms often tend to be quiescent in a respirometer, especially if they have been starved, values of R may need to be adjusted when applied to field populations. An example is the detailed study of the population energetics of the bivalve *Tellina tenuis* by Trevallion (1971). She showed that laboratory estimates of feeding yielded values for an average-sized individual of 73 mg C/year for consumption (C) and 65.6 mg C/year for feces (F), leaving only 7.3 mg C/year as the absorbed ration, a value that was too low to account for the measured energy expenditure by the population. This could have arisen because feeding was assumed to occur for only 50% of the time, or because the laboratory respiration rates were multiplied by 2 to give an estimate of population (R), or to a combination of these and other factors inherent in the extrapolation of the results of short-term experimental studies to whole populations under field conditions.

These difficulties are by no means confined to *T. tenuis*, and it is clear that the resolution that can be expected from energy budgets on whole populations is relatively low. But one useful feature of the energy equation is that because input (C) equals losses $(R + F + U)$ plus energy incorporated into the population as production (P), the value of any one term can be estimated provided the others are known. The net absorbed energy A, for example, may be obtained by subtraction of fecal losses (F) from consumption (C), and because A also equals $(R + P + U)$, the difference between $(C - F)$ and $(R + P)$ gives an estimate of U. In the case of metabolic energy losses from the absorbed ration A, it is possible to verify the measured value of respiratory losses (R_m) by rearranging the basic equation to obtain the calculated respiration $(R_c) = C - (P + F + U)$, which in turn equals the absorbed ration $(A) - (U + P)$.

Odum et al. (1962), van Hook (1971), Hagvar (1975), and, more recently Wightman (1977; see also Humphreys, 1979) have shown that when both R_m and R_c are estimated for a variety of arthropods, the value for R_c calculated from the completed energy budget is generally 2.5 times the experimentally measured respiration (R_m), and may be even higher in some instances. Partly for these reasons, an assumed factor of two has been used in many estimates of the population energy budgets of molluscs (see Carefoot, 1967a; Trevallion, 1971; Wright and Hartnoll, 1981), although Huebner and Edwards (1981), in a study of the population energy balance of the predatory marine gastropod *Polinices duplicatus*, found that approximately balanced budgets were obtained by extrapolating seasonal laboratory measurements of respiration to the field population without recourse to the two-to threefold adjustments used by some other authors. Unmeasured terms amounted to 6–10% of the annual carbon flow for the 1- to 2-year-old animals and 8–13% of annual carbon for older females, and probably

TABLE I

Basic Components of the Population Energy Budgets of Marine Molluscs Arranged in Trophic Categories[a]

Organism	Feces (F)	Absorbed ration A = (C − F)	Gross production efficiency		Pg + Pr = P/C	Respiration (R)	Net respiratory cost (R/A)×100	Urine (U)	Net production efficiency (Pg+Pr)/A ×100	Reference
			P_g	P_r	P/C					
Grazing gastropods										
Nerita tessellata	59.9	40.1	3.4	1.4	4.8	35.4	88.3	—	12.0	Hughes (1971b)
Nerita versicolor	61.1	38.9	4.5	0.6	5.1	33.8	86.9	—	13.1	Hughes (1971b)
Nerita peloronata (pop. 1)	58.9	41.1	7.2	0.7	7.9	33.1	80.5	—	19.2	Hughes (1971b)
Nerita peloronata (pop. 2)	56.6	43.4	4.3	1.1	5.4	38.0	87.6	—	12.4	Hughes (1971b)
Tegula funebralis	29.7	70.3	4.2	0.7	4.9	54.0	76.8	7.0	7.0	Paine (1971)
Patella vulgata	55.1	44.9	4.2	5.9	10.1	31.0	69.0	3.7	22.5	Wright and Hartnoll (1981)
Littorina littoralis	27.1	72.9	14.8	8.5	23.3	49.6	68.0	—	32.0	Wright (1977)
Fissurella barbadensis	66.4	33.6	8.1	0.9	9.0	24.6	73.2	—	26.8	Hughes (1971a)
Mean ± SD	51.8 ± 14.9	48.1 ± 14.9	6.3 ± 3.8	2.5 ± 3.0	8.8 ± 6.2	37.4 ± 9.7	78.8 ± 8.3	—	18.1 ± 8.5	
Suspension and deposit feeders[b]										
Mercenaria mercenaria	58.7	41.3	5.6	4.7	10.3	18.7	45.3	12.3	35.5	Hibbert (1977)
Ostrea edulis	30.7	69.3	6.4	5.5	11.9	29.0	28.4	—	17.2	Rodhouse (1979)
Scrobicularia plana	39.3	60.7	6.6	6.2	12.8	47.9	21.0	—	21.0	Hughes (1970)
Macoma balthica	40.0	60.0	—	—	18.0	42.0	70.0	—	30.0	Warwick et al. (1979)

412

Patinopecten yessoensis (yr 1)	21.1	79.9	39.7	1.6	41.3	39.2	49.1	—	24.9	Fuji and Hashizume (1974)
P. yessoensis (yr 2)	34.6	65.4	21.8	4.9	26.9	43.6	66.7	—	40.8	Fuji and Hashizume (1974)
P. yessoensis (yr 3)	32.5	67.5	19.6	8.2	27.8	47.9	70.9	—	41.2	Fuji and Hashizume (1974)
Tellina tenuis	90.0	10.0	1.0	0.8	1.8	6.7	67.0	—	18.0	Trevallion (1971)
Aulacomya ater	41.5	58.4	—	—	9.2	49.2	84.2	—	15.7	Griffiths and King (1979a, 1979b)
Hydrobia ventrosa	37.0	63.0	—	—	21.0	20.7	32.8	—	33.3	Kofoed (1975b)
Chlamys islandica (yr 5)	66.6	33.4	11.0	1.7	12.7	20.7	62.0	—	37.9	Vahl (1981b)
C. islandica (yr 10)	74.2	25.8	2.6	2.7	5.4	20.4	79.2	—	20.8	Vahl (1981b)
C. islandica (yr 15)	74.8	25.2	1.4	3.3	4.8	20.4	80.9	—	4.8	Vahl (1981b)
Mytilus edulis	54.0	45.9	8.9	4.8	13.7	25.8	56.1	6.4	29.8	Bayne (unpublished)
Mean ± SD	45.6 ± 14.0	54.4 ± 14.0	8.7 ± 6.8	4.8 ± 1.2	14.4 ± 6.6	33.0 ± 12.6	53.7 ± 23.0	—	27.1 ± 8.8	
Carnivores										
Navanax inermis	35.4	64.6	17.9	13.6	31.5	28.1	43.4	7.0	48.7	Paine (1965)
Polinices duplicatus	12.8	87.2	31.9	9.0	40.9	46.2	53.0	—	46.9	Huebner and Edwards (1981)
Archidoris sp.	48.0	52.0	37.0	—	37.0	—	—	—	71.1	Carefoot (1967b)
Mean ± SD	32.0 ± 17.8	67.9 ± 17.8	28.9 ± 9.9	—	36.5 ± 4.7	37.1 ± 12.8	48.2 ± 6.8	—	55.6 ± 13.5	
Wood-eating bivalves										
Lyrodus pedicellatus	66.7	33.3	9.0	4.3	13.3	19.8	59.5	—	39.9	Gallager et al. (1981)

[a] Data recalculated from the literature as a percentage of consumption (C). P_g, Growth; P_r, reproduction.

[b] Mean values for this category calculated ignoring the values for Tellina tenuis and taking values for only years 2 and 10 for Patinopecten and Chlamys, respectively.

413

represented mainly losses of carbon through mucus production (U), with a small proportion of about 1% in shell protein (see Vinogradov, 1953; Hughes, 1970). Gallagher et al. (1981), working on the shipworm *Lyrodus pedicellatus,* and Hughes (1970), on *Scrobicularia plana,* have also obtained budgets that approximately balance without having to assume correction values for population (R).

Partly because of these difficulties and because of the different methods and assumptions made by various authors, generalizations that can be made on the population energetics of marine molluscs are limited. The basic components of the population energy budgets of a variety of marine molluscs are shown in Table I, in which the organisms are arranged in trophic categories. The data have been assembled from a variety of sources and have been recalculated as a percentage of consumption (C). The values of A thus become the percentage absorption efficiencies, whereas those for P_g and P_r become the gross growth efficiency and gross reproductive efficiency, respectively. The net respiratory costs and net production efficiency are also shown.

It has been proposed for assemblages of organisms in general that because of the high nutritional content of their food, carnivores have a generally higher absorption efficiency than herbivores or detritivores (Welch, 1968). There is some evidence of such a trend within the gastropod molluscs, the mean absorption efficiency for the grazers being 48% and for the carnivores 68%; the filter and deposit feeders have a mean absorption efficiency of 54%. More strikingly, both the gross (P/C) and net (P/A) production efficiencies are high in the carnivores. The mean gross production efficiency, for example, is 36% for carnivores, 14% for deposit and suspension feeders, and only 9% for the grazing gastropods. All three groups expend a similar proportion of the ingested ration on respiration (R), but there is a suggestion from Table I that the grazing gastropods expend a higher proportion of the absorbed ration on respiration (79%) than do the carnivores (48%). These trends within the molluscs confirm those reported for a wide range of trophic groups from mammals and birds to insects (Huebner and Edwards, 1981; Humphreys, 1979), but unless the resolution obtained for each component of the population energy budget is improved, there is little likelihood of comparative studies on population energy budgets furthering our understanding of the basic strategies that are adopted to sustain a high production efficiency in the upper trophic levels.

An alternative approach is to study the energy flow and its allocation into P_g and P_r within the individual organisms. We then find that much of the inherent variability that enters the population energy budget as "noise" is in fact a reflection of a series of complex compensatory adjustments that many molluscs can make to manipulate the components of energy gain and expenditure, and thus sustain energy flow into growth and reproduction despite changes in a wide variety of environmental conditions.

III. Energy Acquisition

A. Feeding by Suspension and Deposit Feeders

The morphological adaptations for feeding by means of ciliary currents on the gills and labial palps of bivalves have been summarized by Yonge (1949; see also Pohlo, 1973) and reviewed by Purchon (1968). Some difficulties occur in attempts to categorize the various microphagous feeding types within the Mollusca (Pohlo, 1969), but the classification of Reid (1971) distinguishes suspension feeders (feeding primarily on suspended particulate matter) from fine deposit feeders (feeding on deposited material but ingesting only small particles, i.e., less than 50 μm diameter) and from sand grain feeders (ingesting larger particles including sand grains). For each of these feeding types, opportunities for control over the quantity and quality of the diet include variation (a) in the rate at which water (bearing food particles) is passed over the feeding surfaces, and in the rate and efficiency with which particles are removed from the feeding current; (b) in the degree to which some particles may be selected for ingestion while others are rejected; and (c) in the modification of processes of digestion and absorption of the food, including control over the rate of passage of material through the gut, the further selection of particles for digestion or rejection in the stomach, and changes in digestive enzymes.

In considering feeding by bivalves, it is necessary to distinguish between the overall rate of water transport across the feeding (and gas-exchange) surfaces and the rate of particle filtration; both processes are usually measured as volume units/time. Total water transport is referred to as the pumping or ventilation rate; particle filtration is referred to as the filtration or clearance rate (the volume of water cleared of particles of a certain size in a unit of time). When all particles presented to the filtering surface are removed from suspension (i.e., the filtration efficiency is 100%), the filtration rate is the same as the pumping rate. Particles filtered from suspension are concentrated by ciliary currents but all may not necessarily be ingested. Ciliary "rejection" tracts exist through which filtered material may be voided from the mantle cavity as pseudofeces. The rate of intake of particulate material into the mantle cavity of deposit (or sand grain) feeders may best be referred to as the processing rate (Taghon et al., 1978).

Many studies on the relationship between filtration (and processing) rate and body size have established that feeding is proportional to a power of the body weight as described by the familiar allometric equation:

$$Y = aX^b \qquad (2)$$

where Y represents the feeding rate, X the body weight, and a and b are fitted parameters. Table II presents some values for b taken from the literature, with

TABLE II

Values for the Slopes of Regression Lines Relating Various Measures of Feeding Rate to Body Weight by Means of the Allometric Equation, for a Variety of Molluscs

Organism	b	Reference
Grazing gastropods		
Patella oculus	0.381	Branch (1982)
Patella cochlear	0.363	Branch (1982)
Nerita tessellata	0.480	Hughes (1971b)
Nerita versicola	0.538	Hughes (1971b)
Nerita peloronata	0.532	Hughes (1971b)
Ancylus fluviatilis	0.670	Calow (1975)
Mean ± SD	0.494 ± 0.113	
Suspension feeders		
Mytilus edulis	0.555 ± 0.160	
Mytilus californianus	0.462 ± 0.171	
Pecten irradians	0.820	
Arctica islandica	0.660	
Modiolus modiolus	0.740	
Cerastoderma edule	0.580	From various studies reviewed
Mercenaria mercenaria	0.730	by Winter (1978)
Modiolus demissus	0.760	
Cerastoderma lamarcki	0.630 ± 0.095	
Didacna longipes	0.520	
Didacna trigonoides	0.750	
M. edulis	0.607 ± 0.130	From various studies ($n = 6$) not included by Winter (1978)
Argopecten irradians	0.580	Kirby-Smith (1972)
M. mercenaria	0.310	Hibbert (1977)
Ostrea edulis	0.480	Rodhouse (1978)
Aulacomya ater	0.770	Griffiths and King (1979a)
C. edule	0.560	Newell and Bayne (1980)
Choromytilus meridionalis	0.596	Griffiths (1980c)
Chlamys islandica	0.600	Vahl (unpublished data)
Mean ± SD	0.616 ± 0.127	
Deposit feeders		
Planorbis contortus	0.670	Calow (1975)
Scrobicularia plana	0.630	Hughes (1969)
S. plana	0.600	Worrall (unpublished data)
Hydrobia ulvae	0.484	Hylleberg (1975)
Abra longicallis	0.550	Wikander (1980)
Mean ± SD	0.587 ± 0.072	
Predators		
Thais lapillus	0.555	Bayne and Scullard (1978b)
Clione limacina	0.365	Conover and Lalli (1972)
Polinices duplicatus	0.783	Edwards and Huebner (1977)

the species arranged according to trophic categories, for comparison with Tables I and V. In spite of the many studies with suspension-feeding bivalves, some doubt still remains concerning the correct value of b for animals in their natural habitat. On theoretical grounds Jørgensen (1976b) argued for a b value of 0.75. Winter (1978) concluded that values lay between 0.66 and 0.82, in agreement with the exponent for rates of oxygen consumption (Table V). On the other hand, experiments with natural particulates offered as food (e.g., Bayne and Widdows, 1978; Hibbert, 1977; Kirby-Smith, 1972; Newell and Bayne, 1980) often record lower values: 0.445 ± 0.120 for $n = 10$. In these circumstances the various behavioral factors that contribute to reduced rates of water transport, particularly at high seston concentrations (see later), are often more evident in larger than in smaller individuals. Maximal pumping (or filtration) rates may be expressed to the two-thirds power of body weight whereas in natural conditions the exponent value may be less. This would be in agreement with Jørgensen's assertion (1975, 1976a) that larger individuals are more prone to disturbances than smaller individuals, but interpreted in a natural rather than strictly experimental context.

Other values for b in Table II, for the other trophic categories, also suggest a relative suppression of feeding activity in the larger individuals, relative to the exponent values for respiration rate (see also Table V). This would indicate an increase in the unit metabolic costs of feeding in larger individuals. This relationship is conveniently assessed for suspension feeders per unit time per unit of oxygen consumed (e.g., liters/h/ml O_2) as an index of the relative efficiency of pumping or filtration efficiency (Jørgensen, 1975). Vahl (1972, 1973a, 1973b) recorded the following regression equations for the pumping rate: oxygen consumption ratio (P/R) as a function of body size (dry flesh weight in grams) for three bivalve species:

Cardium (Cerastoderma) edule	$P/R = 5.4W^{-0.19}$
Mytilus edulis	$P/R = 10.5W^{-0.16}$
Chlamys opercularis	$P/R = 20.5W^{-0.17}$

For two populations of *M. edulis,* Bayne and Widdows (1978) calculated seasonally variable P/R ratios with annual mean values of $P/R = 4.7 \pm 1.6\ W^{-0.105}$ and $P/R = 3.6 \pm 1.3\ W^{-0.105}$, respectively. Other relationships can be calculated from data for b in Tables II and V. The results suggest variable relative efficiencies in different species and between different populations.

Correlations of rates of filtration with dry flesh weight are convenient for elucidating the energetics of a species but may disguise the features that ultimately limit the feeding abilities of the bivalves. Hughes (1969) related filtration rates in different species to gill area and so reduced some of the interspecific variability in observed feeding rates. Foster-Smith (1975, 1976) took this analysis

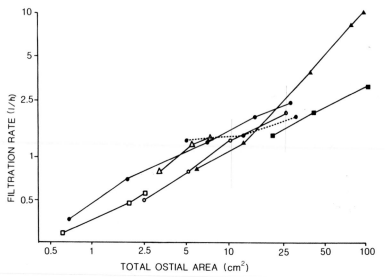

Fig. 2. The relationship between filtration rate and total ostial area for a number of bivalves. (Redrawn from Foster-Smith, 1976.) *Mytilus edulis* (●---●, Willemsen, 1952; ●——●, Theede, 1963), *Modiolus modiolus* (○——○, Winter, 1969), *Arctica islandica* (▲——▲, Winter, 1969), *Scrobicularia plana* (□——□, Foster-Smith, 1976), *Mya arenaria* (■——■, Foster-Smith, 1976); *Cerastoderma edule* (△——△, Foster-Smith, 1976).

further by measuring total gill ostial area (an estimate of the porosity of the gill surface) in 10 bivalve species and relating this to filtration rate using original and published data. Figure 2 shows this relationship for six species. The results suggest that porosity may set the upper limit on the rate of water transport; the data also indicate that there are no fundamental differences in the pumping performance of infaunal siphonate bivalves (e.g., *Mya arenaria* and *S. plana*) and epifaunal, nonsiphonate forms (e.g., *M. edulis* and *Modiolus modiolus*). Morphological and behavioral differences do exist, however, between deposit- and suspension-feeding species (Reid and Reid, 1969).

1. Behavioral Control of Pumping and Filtration Activity

In spite of the limits imposed by the size and the porosity of the gills, control of the rate of water transport through the mantle cavity may be effected by changes in the rate of beat of the lateral gill cilia, by alterations in the diameter and disposition of the gills within the mantle cavity (Dral, 1968), and possibly also by regulation of the diameter of the ostia themselves. In some cases rather slight postural changes may have large effects on ventilation rate. For example, *M. edulis* at full valve gape, but with the tentacles of the mantle edge converging across the inhalent siphon, shows a considerable reduction in rate of water transport relative to individuals with fully divergent tentacles.

Foster-Smith (1976) considered the various possibilities for control over pumping activity, confirming that closure of the exhalent siphon was the most common means of regulating pumping. In none of the species he studied did an individual continue to pump without interruption for long periods of time, even if the shell valves were open; this occasional cessation of pumping activity is a common feature in bivalves (Brand and Roberts, 1974; Earll, 1975). The periods of time for which bivalves do remain with their shell valves open seems to be a function largely of the ambient feeding conditions. Some authors have detected a diurnal rhythmicity (Brown et al., 1956; Morton, 1971; Salanki, 1966), but the presence of endogenous diurnal or tidal rhythms in pumping activity has been disputed (Higgins, 1980a; Winter, 1978). On the other hand, some studies suggest shorter term rhythms of pumping and feeding behavior in bivalves (e.g., Palmer, 1980; Davenport and Woolmington, 1982).

Higgens (1980a, 1980b) studied valve movements and feeding in *Crassostrea virginica* under different conditions of food availability. When food was present in suspension, the oysters were active (shell valves held open) for a greater proportion of the time than when food was absent, and activity could become entrained to artificial schedules of feeding. However, rates of removal of algal cells from suspension by oysters maintained in constant food-cell concentrations varied considerably from hour to hour even with the shell valves open (see also Palmer, 1980; Winter, 1973). Higgens (1980b) concluded that *C. virginica* does not regulate food intake to compensate for temporal variation in food availability (see also Palmer, 1980); rather, he suggests a regulation of feeding activity based on a postingestive detection of the overall nutritive value of the seston.

Considerable differences in filtration behavior exist between different species. Palmer (1980) compared filtration behavior in *C. virginica* and *Argopecten irradians*. Whereas the scallops had relatively constant rates of filtration over long periods (up to 26 h), the oysters showed alternating periods of high and low rates of filtration; the species differences were particularly clear when considered as a function of suspended algal concentration (discussed later). In both cases, however, patterns of feeding behavior could be interpreted as leading to the maintenance of relatively constant rates of ingestion of food (Palmer and Williams, 1980).

In addition to changes in the amount of time spent pumping (and filtering) and in the rates of water transport, the functional state of the gill and its efficiency as a filter will affect the rate of feeding; any capacity to alter the retention efficiency of the gill in response to changes in food quantity and quality would be of considerable value to the individual. Jørgensen (1976a) distinguishes three functional states in the gills of suspension-feeding bivalves: (1) a nonretentive state characteristic of disturbed animals; (2) a cleaning state characterized by copious mucus secretion in the gill surface, and apparently functional in conditions of very high particle concentrations; and (3) the normal feeding state characterized by high rates of water transport and highly retentive gills. Research is still needed to understand the functioning of the bivalve gill filter under natural conditions, and the morphological basis and physiological role of these three functional states.

2. Particle Selection and Retention Efficiency

The amount and the quality of food obtained by filter feeders depends not only on the volumes of water passed through the filtering surfaces but also on the retention efficiency of the gill filter for particles of different size and shape. The retention efficiencies (i.e., the proportion of particles of a given diameter retained by the gill) for many species of bivalves have been reviewed by Winter (1978) and Møhlenberg and Riisgard (1978). Figure 3 shows values for *C. edule* taken from Vahl (1973a). Retention efficiency was maximal for particles greater than 7 μm but was as high as 50% for particles as small as 2 μm diameter. From an energetics point of view, the volume (or the weight) of particles, relative to the total volume of particles available, is of significance. Haven and Morales-

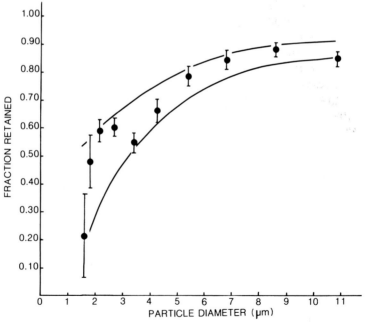

Fig. 3. Efficiency of particle retention by *Cerastoderma edule*. Values are means ± 2 SD. (Redrawn from Vahl, 1973a.)

Alamo (1970) found that for *C. virginica* feeding on natural particulates, although retention efficiency was reduced for particles 1–3 μm in diameter, such particles constituted the largest fraction removed from suspension, due to their greater abundance.

Vahl's data (1973a), shown in Fig. 3, illustrate a common feature in experiments of this kind, that is, increased variance in the retention efficiency estimates for smaller particles. In part this may be due to the resuspension of small particles from feces deposited in the experimental chamber (Hildreth, 1980); it may also represent some capacity to regulate filtration efficiency for such small particles. The diameter of the gill ostia may be varied by changes in the blood pressure in the gill filaments (Elsey, 1935) or by muscular action (Dral, 1968). The degree of synchrony between adjacent laterofrontal cirri (Owen 1974) may change with implications for the minimum mesh dimension on the gill (Hildreth and Mallet, 1980). Whatever the mechanism there is evidence that *C. virginica* (Haven and Morales-Alamo, 1970) and *M. edulis* (Bayne et al., 1977b) may regulate retention efficiency for smaller particles when feeding on natural suspended matter in their natural habitats (see also Palmer and Williams, 1980). Given the dominance of smaller particles in natural seston, this would be a significant adaptation toward control over energy acquisition.

3. The Production of Pseudofeces

At low particle concentrations, all particles filtered from suspension may be ingested; above a certain threshold concentration, a certain proportion of filtered material is rejected as pseudofeces (Foster-Smith, 1975; Owen and McCrae, 1976). The mechanisms for pseudofeces production are particularly well developed in deposit-feeding bivalves (Gilbert, 1977; Pohlo, 1973), where up to 97% by weight of processed material may be rejected from the mantle cavity (Hylleberg and Gallucci, 1975). The combined production of pseudofeces and true feces is referred to as *biodeposition* (Haven and Morales-Alamo, 1966; Tenore and Dunstan, 1973a, 1973b). Three aspects of pseudofeces production are of primary interest from an energetics point of view: (1) the threshold concentration below which no pseudofeces are produced; (2) the relationship between pseudofeces production and body size; and (3) the degree of particle selection, if any, that is made possible in the production of pseudofeces. Integration between particle concentrations, filtration rates, and pseudofeces production and ingestion are considered in the next section.

The threshold concentration for pseudofeces production varies with the type of particle. Loosanoff and Engle (1947; see also Winter, 1969, 1978) deduced an inverse relationship between the threshold and particle size for *C. virginica* fed with algal cultures. Foster-Smith (1975) recorded a similar trend for *M. edulis* fed inert particles of alumina and graphite but found no clear relationship between cell volume and the rejection threshold for three species of alga. From a

TABLE III

Threshold Concentrations for Pseudofeces Production in Some Suspension-Feeding Bivalves

Species	Food	Pseudofeces threshold (mg dry matter/liter)	Reference
Mytilus edulis			
4.5–5.2 cm	*Phaeodactylum*	≈1.0[a]	Foster-Smith (1975)
1.7–7.0 cm	Natural seston	2.6–5.0	Widdows et al. (1979a)
3.7 cm	*Phaeodactylum* and silt	≈1.0	Kiørboe et al. (1980)
Cerastoderma edule			
3.6–4.2 cm	*Phaeodactylum*	≈4.0[a]	Foster Smith (1975)
	Isochrysis	≈6.5	Foster Smith (1975)
	Dunaliella	≈6.0	Griffiths (1980a, 1980b)
Choromytilus meridionalis	Natural seston	3.0–5.0	Haven and Morales-Alamo (1966)
Crassostrea virginica	*Thalassiosira*	≈6.0	Palmer and Williams (1980)
Argopecten irradians			

[a] Weight estimated from published values of cells per milliliter.

variety of data (Table III), the threshold concentration seems to lie between 1 and 6 mg seston (dry weight) per liter. The threshold concentration for pseudofeces production also depends on the body size of the individual (Bayne and Worrall, 1980; Widdows et al., 1979a).

Pseudofeces production is accompanied by the secretion of increased amounts of mucus (Foster-Smith, 1975; Owen and McCrae, 1976). Palmer and Williams (1980) observed increased retention efficiencies for smaller particles by A. *irradians* feeding at concentrations above the pseudofeces threshold (6–11 mg/liter) and they suggest that smaller particles (2–4 μm) that would normally not be filtered are trapped on the mucus sheets. In similar experiments with C. *virginica*, however, retention efficiencies were not altered at high cell concentrations but transiently declined in some individuals, possibly due to increased gill porosity. Both mechanisms are seen as helping to regulate the amount of material ingested to relatively constant levels regardless of ambient particle concentrations.

Earlier studies with C. *virginica* (Haven and Morales-Alamo, 1966; Loosanoff and Engle, 1947) suggested that oysters were able to select in favor of carbon-rich particles from the material filtered, rejecting as pseudofeces the nutritionally poorer particles and ingesting the remainder. Foster-Smith (1975) found no evidence of this in *Mytilus* fed mixtures of alumina and algal cells, but Kiørboe et al. (1980) used naturally occurring silt and observed a relative enrichment of chlorophyll *a* in the ingested ration. Interestingly, whereas for mussels from the Oresund the ratio of chlorophyll in the suspended fraction to chlorophyll in the pseudofeces was 2.9 ± 0.4, for mussels from the Wadden Sea (north of the Netherlands) (which have larger labial palps and feed more efficiently at high suspended loads—Theisen, 1977) the equivalent ratio was 9.1 ± 1.5. Newell and Jordan (1983) have confirmed that C. *virginica* may selectively ingest particles with a high nitrogen, carbon, and energy content from natural supended particles retained on the gill.

4. Effects of Particle Concentration on Suspension Feeding

In reviewing the literature on this topic, Winter (1978) concluded that filtration rates decline at high particle loads but are held relatively constant at lower concentrations; there is also some evidence for a threshold concentration below which feeding does not occur. There is considerable variability between species, however, and between individuals from different populations. Although comparisons between different studies are difficult as a result of the use of different particles, units, and procedures, two points emerge: (1) It is essential to elucidate the complex relationships between filtration, pseudofeces production, and animal size in order to estimate the true ingestion rate; and (2) experiments with unialgal cultures in the laboratory must be interpreted with caution.

Foster-Smith (1975) compared filtration and pseudofeces production in three bivalves over a range of concentrations of the algal cells *Phaeodactylum* and *Isochrysis* (Fig. 4). *M. edulis* maintained a constant filtration rate over a wide range of *Phaeodactylum* concentrations (50 to ≈800 cells/μl), whereas in *C. edule* filtration rate declined above 100–200 cells/μl. The production of pseudofeces by *M. edulis* increased sharply above 10–20 cells/μl, but pseudofeces production by *C. edule* rose gradually from a threshold of 40–50 cells/μl. These two patterns (the third species, *Venerupis pullastra*, not shown in the figure, resembled *C. edule*) represent alternative means of regulating the ingestion rate in conditions of variable food supply.

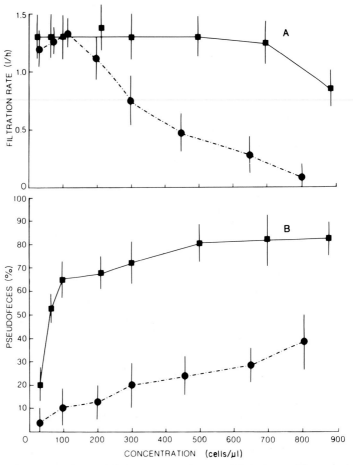

Fig. 4. Filtration rates (**A**) and the percentage rejection of filtered material as pseudofeces (**B**) by two bivalve species. ■, *Mytilus edulis;* ●, *Cerastoderma edule*. Values are means ± SD. (Redrawn from Foster-Smith, 1975.)

In conditions that more closely resembled the natural situation, Tenore and Dunstan (1973a) fed three bivalve species (*M. edulis, C. virginica,* and *Mercenaria mercenaria*) a mixture of algae (dominated by the diatom *Skeletonema*) and recorded constant feeding rates over a concentration range from 200 to 800 μg C/liter (= 150–500 cells/μl); at lower concentrations feeding rate was depressed. Tenore and Dunstan (1973a) interpreted these results in terms of natural levels of particulate carbon and suggested that feeding rates were maximal at food levels typical of the natural habitat; depressed feeding at lower concentrations was seen as an energy-saving adaptation.

The effects of natural particulate material on the various aspects of feeding have been best elucidated for oysters (*C. virginica*; Haven and Morales-Alamo, 1966) and for mussels (*M. edulis;* Widdows et al., 1979a, and Kiørboe et al., 1980). Figure 5 is modified from Widdows et al. (1979a). Filtration rate increases at low seston concentrations and is then held constant to high seston levels above which it declines to zero (at between 200 and 300 mg/liter for mussels from the population studied by Widdows et al., 1979a). The amount of material filtered from suspension increases to maximal values at ≈150 mg/liter, then declines.

From a threshold concentration of ≈5 mg/liter (a value dependent on the size of the mussel), pseudofeces production rises rapidly so that the ingested ration (weight of material ingested per hour) takes a constant value above a seston concentration slightly higher than the pseudofeces threshold. Following Foster-Smith (1975) and Bayne and Worrall (1980), the relationship between the seston concentration (*C*) and the proportion rejected as pseudofeces (*R*) may be described by an exponential function:

$$R = R_{\max} [1 - e^{K(C - T)}] \tag{3}$$

where T is the pseudofeces threshold, K is a weight-dependent constant operating on the seston concentration, and $R_{\max}$ is the maximum proportion rejected (approaches 80–95% at high but environmentally realistic seston concentrations). The results of Kiørboe et al. (1980) suggest that selection in favor of organic material occurs at the stage of pseudofeces production, so that the ingested ration is enriched relative to the available food. The role of processes occurring in the gut and digestive gland in further regulating the rate of energy acquisition are discussed later.

Whereas the general form of the relationships depicted in Fig. 5 may hold for this species, detailed differences between habitats are to be expected. For example, Theisen (1977) recorded very different filtration rates for *Mytilus* from the Wadden Sea compared with other sites in Danish waters. The extent to which such differences are genetically determined, or whether they represent a physiological plasticity in the species, remains a challenge for future research. The scheme suggested in Fig. 5 also does not necessarily accord with observations on

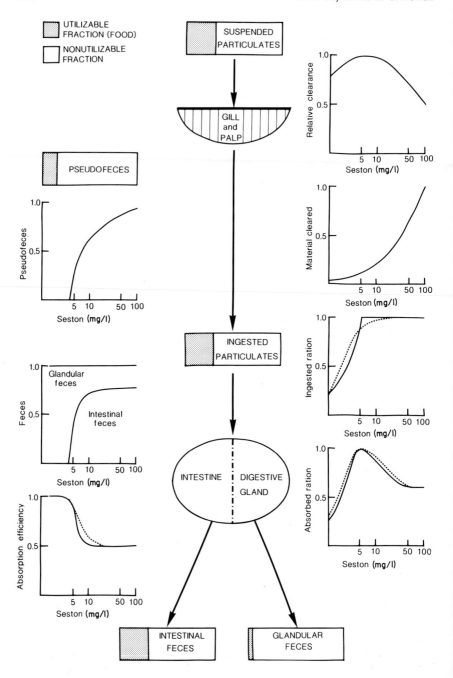

other species (e.g., *C. edule*), although *C. virginica* (Haven and Morales-Alamo, 1966; R. I. E. Newell, personal communication) appears to behave in a similar manner.

5. Particle Selection and Rates of Processing in Deposit Feeders

Cammen (1980) summarized values from the literature to relate rates of feeding to body weight in 19 species of deposit feeders, including 7 species of molluscs. Ingestion (or "processing") rate (I), when calculated as milligrams of total dry matter processed per day, correlated with both body size (W, in milligrams) and the fraction of organic matter in the sediment (OM) as:

$$I = 0.435 \ W^{0.771} \ OM^{-0.920}$$

The bivalve molluscs (three species only) were not included in this analysis, because their rates of processing seemed considerably lower than for the other species. Cammen (1980) suggested that the exponent -0.920, which was close to -1.0, implied that the ingestion rate of organic matter is independent of the concentration of the food. Therefore, ingestion rate (again excluding the bivalves) was analyzed as micrograms of organic matter per day and related to body size only; the relationship was described as:

$$C = 0.381 \ W^{0.742}$$

This exponent value (0.742) is higher than the mean calculated in Table II for deposit feeders (see also "Grazing Gastropods"). However, when values quoted by Cammen (1980) for molluscs only are supplemented by results from Hylleberg and Galluci (1975; on the bivalve *Macoma nasuta*) and Wikander (1980; on the bivalve *Abra nitida*), and analyzed by regression, the fitted exponent value ($b = 0.579$) is similar to values in Table II. As with so many studies of feeding, there is need for careful experiments on various aspects of behavior (including particle rejection), under natural conditions, in individuals of different weights.

Deposit-feeding bivalves select organically enriched particles from the available sediment for ingestion, rejecting large quantities of material as pseudofeces (Hylleberg and Galluci, 1975). Differences between species are to be expected,

---

Fig. 5. A schematic diagram of feeding and absorption in *Mytilus edulis*. All values are plotted on a relative scale (0–1) against total seston concentration (mg/liter; this scale is logarithmic). The dashed line in some figures represents the situation that might apply if selection in favor of the "utilizable fraction" occurs within the stomach. Shaded areas, utilizable fraction (food); clear area, nonutilizable fraction. For further discussion, see the text. (Based on Widdows et al., 1979a.)

however, depending on whether the animals are "fine deposit" or "sand grain" feeders (Reid and Reid, 1969). *M. nasuta* takes material from the sediment indiscriminately into the mantle cavity, but this material is then sorted and the organically rich fraction ingested (Hylleberg and Galluci, 1975). The degree of sorting can be assessed by comparing the organic content of the sediment with that of the feces; the assumption is made that absorption efficiency is so low that a sediment: feces ratio fairly reflects the sediment: ingested ration ratio (Cammen, 1980). Hughes (1970) recorded a ratio (in terms of energy content) of 1.77 for *S. plana*. In *M. nasuta* the ratio varied according to body size, between 1.34 and 2.28 (Hylleberg and Galluci, 1975). Some of this selection may be based on particle size; smaller particles, with a larger surface area:volume ratio, are likely to be relatively rich in surface organic matter (Fenchel, 1972). However, *M. nasuta* is also able to select particles on criteria other than size, rejecting as pseudofeces coarse organic and inorganic particles as well as smaller particles low in organic content (Hylleberg and Galluci, 1975).

Differences between species occur here as with other aspects of feeding behavior. Hughes (1973, 1975) studied the feeding behavior of three species of the deposit-feeding bivalve *Abra* (see also Wikander, 1980). The maximum size of particles taken into the mantle cavity depends on the diameter of the inhalent siphon. For *A. tenuis,* there was no evidence of selection for size within the mantle cavity, and little evidence for selection according to food value, although some larger silica particles that lacked on organic film were rejected as pseudofeces. In *A. abra,* however, material sampled from the stomach was significantly finer than material in the mantle cavity, suggesting an active selection against larger particles.

Feeding by the deposit-feeding gastropod *Hydrobia* has been analyzed in some detail (Fenchel et al., 1975; Hylleberg, 1975; Kofoed, 1975a, 1975b; Levinton and Bianchi, 1981; Lopez and Kofoed, 1980; Lopez and Levinton, 1978; Newell, 1965); see also Whitlack and Obrebski (1980) on two other deposit-feeding snails. *H. ulvae* employs three methods of feeding. For sediment particles <40 μm diameter, the snails deposit-feed (i.e., swallow all particles), although they may be capable of some selection according to size or the presence or absence of an organic film on the sediment surface. For particles >200 μm, the snails graze microorganisms (bacteria and diatoms) from the surface of the particle. For particles of intermediate size (40–160 μm diameter), Lopez and Kofoed (1980) describe feeding by epipsammic browsing; particles are taken into the buccal cavity, attached microorganisms are scraped off by the radula, and the particles are then ejected from the mouth. There is some evidence of differences in the distribution of different feeding behaviors between different hydrobid species. Hydrobid snails may also feed directly on free-living microorganisms in the sediment. However, with regard to the epiphytic species, Lopez and Levinton (1978) point out that a distinction is necessary between those microorganisms that are relatively safe from digestion and those that may be more readily digested. Attached microorganisms may be protected from digestion because the sediment particles absorb (and so neutralize) the digestive enzymes or because the particles are quickly egested via the intestine without being taken into the digestive gland. These processes have an important bearing on digestion and absorption efficiency.

6. *Absorption Efficiency and Gut Residence*

Table IV lists some values of absorption efficiency for various molluscs feeding on natural foods (see also Conover, 1978). Efficiencies may be very high (80%) for suspension feeders feeding on living algal diets (Bayne, 1976a; Winter, 1978), but values between 30 and 60% are more typical for natural seston. There is no evidence from the few data available that molluscan carnivores have higher absorption efficiencies than herbivores (but see p. 414), although deposit-feeding detritovores may have lower efficiencies than bacteria feeders (Calow, 1975; Hargrave, 1970). The structural carbohydrates of plant cells may not be digestible by some species (Hylleberg, 1975; Mathers, 1973) although the wood-boring bivalves may be a special case (Mann, 1982; Morton, 1978). The values quoted in Table IV for *Hydrobia ventrosa* (Kofoed, 1975a) may overestimate the absorption efficiencies in the natural habitat by a factor of two or three (Lopez and Levinton, 1978).

The factors affecting absorption efficiency in bivalves have been considered by Bayne (1976a), Winter (1978), Widdows et al. (1979a), and Vahl (1980). There appears to be little change in absorption efficiency with increase in body size. Bayne (1976a) concluded that increases in temperature slightly depressed absorption efficiencies in *Mytilus* sp. (0.7% decrease per degree Celsius), but Elvin and Gonor (1979) measured a higher efficiency by *M. californianus* at 15°C than at 9°C and concluded that exposure to elevated temperatures during low tide in spring and summer could enhance absorption efficiency. In laboratory experiments using pure cultures of algal cells, absorption efficiency is commonly observed to decline rapidly at high cell concentrations (Griffiths, 1980a; Thompson and Bayne, 1974; Widdows, 1978a), but with natural particles as food such a decline is not as evident (Bayne and Widdows, 1978). For example, Griffiths (1980b) observed a decline in absorption efficiency for *Choromytilus meridionalis* when fed on pure *Dunaliella primolecta* culture at 3 mg/liter, but when fed on natural particulates absorption efficiency averaged 40% at ration levels from 3 to 18 mg/liter. The presence of inorganic particles in suspension (at concentrations below the pseudofeces threshold) may, however, "dilute" the organic matter present (Widdows et al., 1979a) and bring about a reduction in absorption efficiency. Vahl (1980) described such an effect for *Chlamys islandica* where absorption efficiency declined from ≈50 to ≈20% with an increase in the fraction of inorganic particles from 0.5 to 0.6. Bayne et al. (1979)

TABLE IV

Values for Absorption Efficiency for a Variety of Molluscs Feeding on Natural Diets, Arranged According to Feeding Type[a]

Organism	Efficiency (range)	Reference
Grazing gastropods		
Aplysia punctata	0.63 (0.45–0.75)	Carefoot (1967a)
Aplysia dactylomela	0.58 (0.35–0.68)	Carefoot (1970)
Aplysia juliana	0.76 (0.69–0.84)	Carefoot (1970)
Littorina littorea	0.57 (0.03–0.83)	Grahame (1973)
Ancylus fluviatilis		
Diatoms	0.59 (0.56–0.60)	Calow (1975)
Blue-green algae	0.12 (0.10–0.14)	Calow (1975)
Suspension and deposit feeders		
Mytilus edulis	0.36 (0.18–0.50)	Bayne and Widdows (1978)
M. edulis	0.49 (0.42–0.59)	Widdows et al. (1979a)
M. edulis	0.60 (0.52–0.65)	Kiørboe et al. (1980)
M. edulis	0.47 (0.24–0.74)	Thompson (unpublished)
Cerastoderma edule	0.68 (0.58–0.74)	Newell and Bayne (1980)
Choromytilus meridionalis	0.40 (0.04–0.87)	Griffiths (1980b)
Bittium varium	0.47 (0.46–0.48)	Adams and Angelovic (1970)
Hydrobia ventrosa		
Bacteria	0.75	Kofoed (1975a)
Diatoms	0.65 (0.61–0.75)	Kofoed (1975a)
Blue-green algae	0.50 (0.49–0.51)	Kofoed (1975a)
Carnivores		
Navanax inermis	0.62 (0.50–0.70)	Paine (1965)
Archidoris pseudoargus	0.52	Carefoot (1967b)
Dendronotus frondosus	0.86	Carefoot (1967b)
Clione limacina	0.95 (0.82–0.98)	Conover and Lalli (1974)
Thais lapillus	0.66 (0.52–0.78)	Bayne and Scullard (1978b)
T. lapillus	0.38 (0.11–0.59)	Stickle and Bayne (unpublished)
Thais haemastoma	0.91 (0.81–0.97)	Stickle (unpublished)

summarized the relationship between absorption efficiency in *M. edulis* and the proportion of particulate organic matter present in the seston (% *POM*) in the form of a regression equation: Efficiency $= 0.50 \log_{10} \% POM - 0.32$ ($n = 26$; $r^2 = .76$).

Branch (1982) summarizes some values for absorption efficiency in limpets but also points out the difficulties of making these measurements for grazers. Values varied from 72 to 93% and, in general, species with an abundant food supply showed lower values than those with a more limited food supply, consistent with a hypothesis of "exploiter" species (high food abundance, low efficiency) compared with "conservers" (low food abundance, high efficiency; see Newell and Branch, 1980). For a carnivorous species, *Polinices alderi*, Ansell (1981) estimated absorption efficiencies between 28 and 50%; there was neither evidence of a dependence on body size nor any significant differences at different temperatures. W. B. Stickle and B. L. Bayne (unpublished observations) recorded absorption efficiencies for *Thais lapillus* (feeding on *M. edulis*) that were depressed at low temperature (5°C) and low salinity (17.5‰) but showed little consistent variation between temperatures of 10 and 20°C and salinities between 20 and 35‰.

The variability in absorption efficiencies, rather than the mean values, is probably of most physiological interest, for control over this efficiency would provide a powerful means of adapting to a variable food source. The absorption efficiency is functionally interrelated with gut capacity, the residence time for food in the gut, and the ingestion rate. For a sessile animal, therefore, unable to seek its own food and with a limited (albeit significant—see previous section) capacity to select food particles, changing the rate of ingestion and the gut residence time represent the main means for maximizing the rate of energy acquisition, at least as regards short-term responses to changes in food supply (rather than the time-averaged responses more apt to evolutionary adaptation; Levinton and Lopez, 1977).

If a steady state is assumed between ingestion and egestion rates, then the total number of particles in the gut (ΣN_i) will equal the product of the filtration or processing rate (P), the gut residence time (t), and the concentration of particles in the feeding current, weighted by a coefficient of selection (which includes selection following pseudofeces formation):

$$\Sigma N_i = Pt \, \Sigma \alpha D_i \tag{4}$$

This formulation follows Lehman (1976). Taghon et al. (1978) elaborated this expression (also following Lehman, 1976) in terms of maximum gut volume (V_g) and the volume of an ingested particle before (V_{o}) and after (V_v) digestion:

$$P = \frac{V_g}{\Sigma \alpha_i D_i [a(V_{o_i} - V_{f_i})(1 - e^{-t/a}) + V_{f_i} t]} \tag{5}$$

where a is a measure of the time-specific absorption efficiency. Three assumptions apply: The gut is assumed always to be full, feeding to be continuous, and absorption efficiency to increase asymptotically with increase in gut residence time.

Taghon et al. (1978) use this expression in a model to elucidate the optimal feeding behavior of deposit feeders, where net energy gain is calculated as the difference between absorbed energy and the sum of the energy costs of feeding and particle selection. The model illustrates the nonlinear relationships between the various components of feeding behavior and energy acquisition, and points to the "desirability of determining both the gut passage time and time-specific (absorption) efficiency of an organism" (Taghon et al., 1978, p. 755). For example, for an animal with a low absorption efficiency, gut retention time is not critical in determining the maximum net energy gain. Under certain circumstances, an animal that achieves a high absorption efficiency but with a long gut residence time may not gain as much energy per unit time as an animal with a lower absorption efficiency that is able to process more food items in the same amount of time.

Unfortunately, data on these processes in molluscs are very scarce, and no specific tests of these predictions have been made. The residence time of food in the gut of *S. plana* has been measured by Hughes (1969) and related to animal size and to temperature. [See also Calow (1975) on two freshwater gastropods; Hylleberg and Galluci (1975) on the bivalve *M. nasuta;* and Wikander (1980) on species of the bivalve *Abra.*] In *Scrobicularia,* residence time varied from ≈60 h at 4°C to ≈8 h at 24°C for animals of 35–40 mm shell length. Wikander (1980) recorded length-dependent residence time of similar magnitude in *Abra longicallus,* and Hylleberg and Galluci (1975) recorded 1–9 h at 12°C for *M. nasuta.* Newell (1977) found an allometric relationship for *C. edule* between gut residence time (hours) and shell height (h, in centimeters): $t = 2.19 \, h^{0.17}$, or 5–6 h for animals of 35–40 mm shell height. Calow (1975) determined the gut residence time for the grazing (freshwater) gastropod *Ancylus fluviatilis* as related to body size, temperature, and food quality; there was evidence of an inverse relationship between residence time and absorption efficiency for a diatom, a green alga, and a blue-green alga. Some values for the estimated gut capacity of four bivalve species are plotted in Fig. 6. There is, in general, a linear relationship between capacity and the square of the shell length.

However, relationships between gut capacity and absorption efficiency are complicated in the molluscs by the presence of the digestive gland (i.e., hepatopancreas), the site of intracellular digestion (see Purchon, 1968), and the possibilities for further sorting of particles in the stomach (or stomach–gizzard complex) for rapid egestion of one component of the food direct to the intestine (with a low absorption efficiency) compared with the retention for a longer period of another component within the digestive gland (with a higher absorption

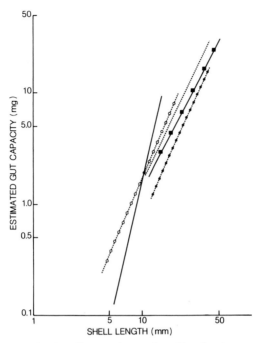

Fig. 6. The gut capacity, in milligrams dry weight, of four bivalve species, related to shell length. *Abra nitida* (———, Wikander, 1980); *Abra longicallus* (○--○, Wikander, 1980); *Scrobicularia plana* (■———■, Hughes, 1969; ————, C. Worrall, unpublished); *Cerastoderma edule* (●--●, Newell, 1977).

efficiency). These two components are often discernible in the feces (Calow, 1975; Van Weel, 1961) and have been termed "intestinal" and "glandular" feces, respectively, in bivalves (Widdows et al., 1979a). A tentative illustration of the balance between these two components in *Mytilus*, related to food concentration, is included in Fig. 5. Certainly the ability to vary the distribution and retention of ingested matter between these functionally separate parts of the gut provides the animal with a further suite of adaptive responses, to add to those of particle selection and ingestion rate.

The balance among gut residence time, compartmentalization within the gut, and absorption efficiency affects the energetic implications of any "decision" whether to feed more or less actively on food of reduced nutritive quality. The analysis by Cammen (1980) suggests that deposit feeders in general reduce their rates of feeding as the organic content of the sediment increases, resulting in increased residence time in the gut and the possibility for increased absorption. In a recent theoretical treatment by Taghon (1981), however, the energetically optimal behavior is predicted to be an increase in feeding rate with increase in food quality (see also Doyle, 1979); this may particularly apply when absorption efficiency is normally low and little is to be gained by increasing gut residence time (Calow, 1975). An additional complication is recognized by Lopez and Levinton (1978) for *Hydrobia,* and possibly for other deposit feeders, in which the proportion of available to unavailable food items might alter as total food concentration increases. Food quality then becomes a function of food concentration, and the proportions of microorganisms "safe from" and "available for" digestion must be considered in assessing the probable absorbed ration.

In addition to these mechanisms suggested for bringing about control over the absorbed ration, a further possibility exists, namely, regulation of the complement of digestive enzymes to match the requirements of the food (Van Weel, 1961; Kristensen, 1972). This topic is dealt with elsewhere in these volumes, but recent work on some bivalves from South Africa suggest certain implications for physiological energetics. For example, the "fitness" of the digestive enzymes to energetic demands imposed on the individual by the environment is illustrated by Seiderer and Newell (1979) who observed apparent thermal compensation in the activity of α-amylase extracted from the digestive style of *Choromytilus meridionalis;* the authors suggest that different iso-enzymes, with different thermal optima, may be available to cover the normal thermal range of the species. In a more far-reaching study, Seiderer et al. (1982) observed that both *C. meridionalis* and *Perna perna* possess a similar spectrum of carbohydrase enzymes in the crystalline style, but total carbohydrase activities differ by a factor of two. The authors used information on the carbon budgets of the animals to predict turnover times required for carbohydrase activities in the style in order to meet the carbon requirements, and found good agreement with measured turnover rates. In spite of similar enzyme complements, then, detailed differences in enzyme activities matched the energetic (or carbon) demands of the animals.

7. Phasic Activity in the Digestive Gland

Since the studies of J. E. Morton (1956) on the intertidal bivalve *Lasaea rubra,* there have been many descriptions of rhythms of intracellular digestion in bivalve molluscs, challenging the view that feeding in these animals is continuous (Purchon, 1971). Similar rhythms have also been described for some gastropods (Curtis, 1980; Robertson, 1979). The rhythmic, or cyclical, activity is often most apparent in intertidal individuals, and has been described in terms of changes in the morphological appearance of the tubules of the digestive gland, changes in the pH of the gut contents, and the presence or absence (and general appearance) of the crystalline style (Langton, 1975, 1977; Langton and Gabbott, 1976; Mathers, 1976; Mathers et al., 1979; Morton, 1971, 1973, 1977; Owen, 1974).

For an intertidal animal, with opportunities for feeding constrained to periods of immersion, a cyclicity in digestion is to be expected. Owen (1974) suggests that the digestive tubules adopt a "holding" phase during each low tide period, at which time a synchrony between tubules within the digestive gland is apparent. As feeding is initiated following reimmersion, food particles are delivered to the tubules over different periods of time, depending on their distance from the main ducts communicating with the stomach. Tubules enter the "absorptive" phase of the cycle at different times, therefore, and conditions within the gland become heterogeneous. As this phase gives way to a "disintegrating" phase, characterized by the exocytosis of residual bodies from the digestive cells, the tubules return to the holding phase (possibly via an intermediate or "reconstituting" phase) and the gland returns to synchrony. Morton (1977) has suggested that the residual bodies that are discharged into the stomach release proteolytic enzymes into the lumen that aid in the dissolution of the style and provide the means for some extracellular digestion, ready for the next influx of food with the incoming tide.

This cycle appears to be induced, in both tidal and subtidal animals, by an intermittent delivery of food to the digestive gland. Robinson et al. (1981; see also Wilson and La Touche, 1978) suggest that digestion, both extracellular and intracellular, is potentially continuous (Owen, 1972), but a cyclic or otherwise discontinuous supply of food induces the observed phasic processes in the digestive tubules. Many processes may lead to intermittent changes in food quantity or quality, even for subtidal animals (e.g., tidal variation in current pattern and velocity; Mathers, 1976). The bivalve is then seen as responding opportunistically to a variable food supply, but always able to respond rapidly by virtue of some digestive tubules being in a state of readiness for initiating intracellular digestion. Descriptions of these events require careful attention to the statistics of sampling (Robinson et al., 1981) but are clearly necessary for any further understanding of the range of adaptive responses available to the bivalves to enhance energy acquisition.

Hawkins et al (1983) have extended these studies by recording pronounced temporal fluctuations in fecal deposition by *Mytilus edulis* that relate, in terms of both amplitude and period, to a rhythmicity for ammonia excretion and changes in the morphology of digestive tubules. These coordinated rhythms of digestion, absorption, and excretion displayed different periodicities that varied with season. The results of this study imply discontinuous digestion by mussels feeding under environmentally invariant conditions and suggest a fundamental metabolic rhythmicity in this species, linked to the digestive rhythm.

B. Carnivorous Feeding

The rather scant data on mean feeding rates by carnivorous molluscs are summarized by Conover (1978). Values vary from 1.5 to 44.0% of body weight per day in various vermivorous species of *Conus* (Kohn, 1959, 1968), three mollusc-eating benthic species (Paine, 1956; Pearce and Thorson, 1967; Robilliard, 1971), and a petropod, *Clione limacina* (Conover and Lalli, 1971, 1974). Mean feeding rates vary with the size of the predator, however; Bayne and Scullard (1978b) recorded a seasonally variable relationship for *T. (Nucella) lapillus,* feeding on *M. edulis,* from 15 to 3% of body weight per day in October (over a size range for *Thais* of 20–400 mg dry flesh) to 6–1.5% in March, both at 15–16°C.

Mean feeding rates can be misleading, however; it is the variability in the feeding rate, caused by a complex of exogenous (temperature, salinity, prey size, and abundance) and endogenous (predator size, reproductive condition, past feeding history) factors, that is of real interest in assessing the energetics of the species. Such an assessment should start with an analysis of feeding behavior. MacArthur (1972) considered the feeding of a predator as comprising a period of search, during which the predator senses the prey; and pursuit, during which the prey item is pursued, captured, and eaten. Elements of pursuit and prey capture are usually considered together as the "handling time." Hughes (1980) distinguishes between "searchers," for which handling times per prey item are short relative to search time (e.g., vermivorous *Conus* species; Leviten, 1976); and "pursuers," for which pursuit or handling times are long relative to search time (e.g., the prosobranch *Acanthina punctulata* feeding on littorinid snails; Menge, 1974).

The amounts of time spent in the various stages of feeding behavior vary according to the size of the predator, the environmental conditions, and the season of the year. For example, feeding by *Acanthina* may comprise five phases (Menge, 1974): search, encounter and preliminary evaluation of the prey, pursuit, final evaluation, and drilling and feeding. In the laboratory, with suitable prey presented *ad libitum,* search occupied 1–2% of the total feeding time; drilling and eating, 98%; and the remaining phases all less than 1%. On the shore, however, more time was spent searching for prey (9–16%), only 17–33% of the time was spent drilling and eating, and 50–74% of the time (depending on season) was spent inactive in crevices. Constraints imposed by the environment (such as risks of predation, or of physical dislodgement from the substrate), which are sometimes difficult to reproduce in the laboratory, are clearly very important in regulating feeding behavior.

Laboratory experiments with *Thais* feeding on mussels have provided some data on seasonal and size-related variability in the feeding/time budget (Bayne and Scullard, 1978b; see also Connell, 1961). The length of time spent drilling and eating did not vary between individuals of different size, but decreased with increase in temperature and was disproportionately (considering the seasonal change in temperature) less in October than in March. The duration of the postfeeding phase, however (including a period for digestion, a postdigestive resting period, and subsequent search), did depend on body size, declining with increased weight of *Thais*. The result was that larger individuals acquired a larger ration than small individuals because they fed more frequently and on larger prey, although, because of relative size dependencies, the feeding rate as percentage of body weight per day declined with increase in size. The overall feeding rate was greater in October than in March.

Holling (1959) incorporated the concept of search and handling time in an analysis to elucidate the fundamental relationship between the density of prey (N_t) and the number of prey eaten (N_a):

$$N_a = a' \, T_s \, N_t \tag{6}$$

where T_s is the search time, and a' describes the rate of successful attack by the predator on the prey. However,

$$T_s = T - T_h \, N_a \tag{7}$$

where T is the total time and T_h is the handling time. Substituting in Eq. (6) for T_s gives:

$$N_a = \frac{a' \, N_t \, T}{1 + a' \, T_h \, N_t} \tag{8}$$

which is the "type 2 functional response" in which the number of prey eaten per predator increases with increasing prey density, but at a decreasing rate. The type 2 functional response, though common in experiments with invertebrate predators, refers to short-term measures of feeding rate (Royama, 1971). Murdoch (1971, 1973) described a "developmental" response to prey density, based on the type 2 response but incorporating the concept of the predator growing in size and therefore either consuming more prey or increasing the efficiency with which it handles and consumes each prey item. The number of prey eaten at each prey density now increases with time, and the functional response becomes sigmoidal, similar to Holling's (1959) "type 3 response." This is likely to be the common form of the functional response under natural conditions. For example, Morgan (1972) described how *T. lapillus* improved its efficiency of feeding on alternative prey following the disappearance of the barnacles that comprised its former diet.

Selection by the predator of prey of different sizes also affects the form of the functional response. A common observation is that larger individuals select larger prey. Leviten (1976) presents evidence for *Conus* species (see also Edwards and Huebner, 1977, on *P. duplicatus*) (i) that a minimum food size exists below which the predator does not feed, (ii) that this minimum prey size increases with increasing predator size, (iii) that there exists also a maximum prey size above which the predator cannot feed, and (iv) that this maximum size also increases with increasing predator body size. Where prey items of various sizes are available, therefore, the predators, eating and growing, can be expected to select larger prey. When a size range of prey is not available, larger predators will eat more prey per unit time.

This is well illustrated by data from Ansell (1982b) on *Polinices catena* preying on *T. tenuis* (Fig. 7). When the prey size offered in laboratory cultures remained constant (Fig. 7A), the predation rate (numbers of prey bored per week) increased due to predator growth. When the size of prey supplied to the culture was increased (Fig. 7B), the rate of predation declined as the predators acquired a greater ration per prey item eaten. Of course, the extent to which the predator can increase the rate of feeding is limited by the handling time; this limit determines the asymptotic feeding rate in the functional response.

The rate of feeding depends also on the feeding history of the individual. In the short term, periods of starvation may be compensated for by increased rates of feeding when food again becomes available. Here also, however, the constraints of search and handling times set limits on the capacity for compensation. After 16 days' starvation at 15°C in the summer, *T. lapillus* can only increase its ration intake by reducing the postingestion period, with implications for the time available for search and for activities not related to feeding (Bayne and Scullard, 1978b).

Temperature may have a dominating effect on feeding (Conver, 1978). Edwards and Huebner (1977) recorded a relationship between feeding rate (number of *Mya* consumed per snail per day, *Y*) and temperature (degrees Celsius, *X*) for

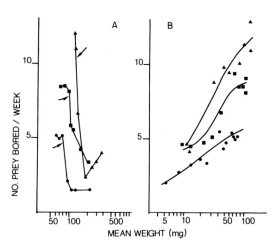

Fig. 7. Rates of predation by juvenile *Polinices catena* on *Tellina tenuis,* in laboratory cultures. Values graphed on the abscissa are per predator. (**A**) Decrease in predation rate when size of prey was increased (at the time of the arrow). (**B**) Increase in predation rate with increase in predator size, when prey size remained constant. Results from experiments at three temperatures: ●, 10°C; ■, 15°C; ▲, 20°C. (Redrawn from Ansell, 1982b.)

P. duplicatus that was described by the simple equation, $Y = 0.039X - 0.27$. Conover and Lalli (1972) combined the effects of predator's size (x_{wc}), size of prey (x_{ws}), and temperatures (x_t) on the feeding rate (y_{fc}) of *C. limacina* in the following multiple regression:

$$\log(y_{fc} \ 10^4) = 0.305 \log(x_{wc} \ 10^4) + 3.9 \log \log(x_{ws} \ 10^4) + 0.138x_t - 3.94$$

This equation described 78% of the variance in their feeding rate determinations. W. S. Stickle and B. L. Bayne (unpublished) determined the feeding rate of *T. lapillus* on mussels at different temperatures and salinities, and compared the results with *Thais haemastoma* feeding on oysters (W. S. Stickle, unpublished). The latter species fed at lower salinities than *T. lapillus* and showed a higher optimal temperature, consistent with observed ecological differences between the species.

Of all the endogenous factors likely to affect feeding rate, the reproductive condition of the predator is likely to be very important. Enhanced feeding rates by *T. lapillus* in October may be explained, in part, as a "preparation" for reproductive activity in the ensuing months. Ansell (1981) has reported the combined effects of temperature and reproductive activity (measured as an index of egg production) on feeding rate in *P. alderi* on *T. tenuis;* prey consumption increased with increase in both temperature and egg production, when prey of the preferred size range were provided.

Ultimately, the feeding behavior of a predator (and indeed of any animal) is a balance between the costs and risks associated with feeding and the energy (or other nutrient) gained (Emlen, 1966). Where risks are high (predation, physical factors, etc.), optimal behavior may require the animal to minimize the time spent on a particular activity such as feeding (the "time minimizer" option); in other circumstances the optimal solution may be to maximize the net energy gain (the "energy maximizer" option; see Schoener, 1971). The theories of optimal foraging predict how predators should seek and select their food under different environmental conditions (Pyke et al., 1977). A general form of the "optimal diet" theory, described by Hughes (1980) to include predators that hunt by chemical or tactile cues (as well as those that hunt visually), can be applied to molluscan predators. The "marginal value" model of Charnov (1976) and Parker and Stuart (1976) attempts to identify some of the rules governing optimal feeding in a patchy environment, and the model of Norberg (1977) is designed to identify optimal patterns of searching behavior.

The optimal diet model assumes that the animal optimizes its feeding by maximizing the term E/T, where E is the net gain from the diet during a foraging period of time T. The predator is assumed capable of "ranking" potential prey items in terms of values for E/T_h, where T_h is the handling time; prey items with a high net energy yield per unit handling times are to be preferred to those with a lower yield. Hughes (1980) summarizes the predictions of the model as follows:

1. The predator should eat only the prey yielding the highest net energy gain if it encounters such prey frequently and, whatever the encounter frequency, such prey should always be eaten when encountered.
2. As the highest energy-yielding prey become scarce, the predator should expand its diet to include prey items next in rank of net energy yield.
3. When the predator requires a long recognition time to identify the prey, lower energy-yielding prey should be eaten if they are frequently encountered. If the recognition time is short, lower yielding prey should not be eaten if better prey items are common.

In reviewing these ideas, Hughes (1980) summarizes relevant studies with molluscan (and other) predators and discusses the extension of the concepts of optimal foraging to suspension and deposit feeders by means of the models of Lam and Frost (1976), Lehman (1976), and Taghon et al. (1978). Bayne (1981) briefly reviews related topics such as switching between different prey items, problems of risk, and problems of resource depletion.

IV. Energy Expenditure

As had already been shown, an energetic gain can be achieved by adjustment of both consumption and absorption, and this may sustain energy flow into growth and reproduction even under conditions where energy resources are limited. Where food resources are abundant, a maximal gain may be derived by adoption of an "exploitative strategy" of a high consumption rate even though this may be associated with a relatively low absorption efficiency and high rate of metabolic energy expenditure (Newell, 1979, 1980; Newell and Branch, 1980). But reduction of metabolic costs, which may represent as much as 88% of the absorbed energy (see Table I), may also represent an important mechanism of energy conservation, especially under conditions of limited food availability where further optimization of consumption or absorption is not possible.

Because of the importance of measurement of energy losses from metabolism in estimating energy flow through both populations and the individual organism, the respiration of marine molluscs has attracted widespread attention over many years. Energy losses can be calculated from the oxygen consumed per unit time, from the carbon dioxide released, or from the liberation of heat. Difficulties can, however, arise under some experimental situations, when withdrawal or closure of the shell valves occurs (such as under conditions of low oxygen tension (P_{O_2}) or extremes of salinity or temperature), if oxygen consumption is assumed to equate to metabolic energy loss without simultaneous measurements of either heat production or anaerobic end-product accumulation (see Section IV,C). The energy equivalent of oxygen consumption also varies according to the substrate being used by the animal. Gnaiger (1983) gave a comprehensive guide to the calculation of energetic equivalents of respiratory oxygen consumption; estimates of energy flow based on oxygen consumption generally assume a mean oxycaloric equivalent of 20.08 J/ml O_2. Some of the respirometric methods available, and the practical aspects of their use, have been reviewed in Grodinski et al. (1975), Wightman (1977), and Gnaiger and Forstner (1983).

Part of the reason for the discrepancies that can occur between the measured rate of respiration (R_m) of molluscs in experimental situations and the calculated values of (R_c) from the other components of the energy budget (see Section 2) is that metabolic energy expenditure is affected by a wide variety of environmental factors including temperature, salinity, humidity, and P_{O_2}, as well as by endogenous factors including body size, activity, gametogenic stage, and sex, and by interactions between these and other time-dependent variables such as season (for reviews, see Bayne, 1976a; Newell, 1973, 1979). Where these have been ranked in order of their relative effect on the dependent variable oxygen consumption, the effects of body size and exposure temperature are often of main importance, with activity, gametogenic condition, and interactions between these and other independent variables, forming part of the complex multiple regression equations that are required to account for all the variations in metabolic energy expenditure observed (Newell and Roy, 1973; Widdows, 1978a).

A. Body Size and Metabolic Energy Expenditure

The relationship between body size and metabolic rate has been studied for many years since the fundamental allometric relationship established for mammals and birds was extended to include a wide range of invertebrates, including both adult molluscs and their larvae (Hemmingsen, 1960; Zeuthen, 1947, 1953; von Bertalanffy, 1957). Such studies have established that metabolism is proportional to a constant power of the body weight as described by the allometric equation:

$$Y = a X^b$$

where Y is the metabolic rate as oxygen consumption or energy units, X is the body size (in weight or equivalent units), b is the exponent, and a denotes the level of the metabolic rate of an organism of unit body weight. The value of a varies according to a wide variety of factors including, above all, activity and temperature. The value for the exponent b is less variable; when the data for all organisms are pooled, it is found that b approaches a value of between 0.67 and 0.75 (Zeuthen, 1953; Hemmingsen, 1960), signifying that metabolic rate becomes relatively slower with increasing body size. Because the metabolic rate is strongly dependent on body size, it is necessary to introduce a weight correction into comparisons between animals of different sizes. In this case the value of the weight exponent b can be used to correct the data to the rate of oxygen consumption of a standard-sized organism:

$$V_{O_2,s} = \left[\frac{W_s}{W_e} \right]^b \dot{V}_{O_2,e} \tag{9}$$

where $\dot{V}_{O_2,s}$ is the rate of oxygen consumption of a standard-sized animal, W_s is its weight, W_e is the weight of the experimental animal, $\dot{V}_{O_2,e}$ is the uncorrected rate of oxygen consumption of the experimental animal, and b is the weight exponent.

It is also often convenient to compare the metabolic rate of organisms per unit of body weight. This value, the weight-specific metabolic rate, is then described by:

$$\frac{Y}{X} = aX^{b-1}$$

The exponent of this relationship is usually negative from -0.33 to -0.25. That is, the weight-specific metabolic rate (ml O_2/g/h) is lower in large organisms than in small ones. This generalization applies in both interspecific comparisons between molluscs of different sizes and between individuals of any one species. It has the important implication that populations that are dominated by large or old individuals have a lower value for population (R) than populations of small individuals. It also shows that energy flow through small individuals or species may be much greater than might otherwise be supposed on the basis of their biomass alone.

The relation between body size and oxygen consumption for a variety of molluscs has been summarized in Table V. The organisms have been arranged in broad trophic categories for comparison with Table I, and the data for $\dot{V}_{O_2}$[2] are shown in the value of a (ml O_2/h) calculated for an individual of 1 g weight. The first and immediate feature of interest is that the allometric exponent b is approximately 0.7 in all the molluscs shown; the overall mean value is 0.7 ± 0.13 SD ($n = 50$). Obviously any factor that affects large and small individuals differently will alter the value of the slope of the regression. The data of Kuenzler (1961) for *Modiolus demissus,* for example, and of Newell and Roy (1973) for *Littorina littorea* suggest that the value of b varies with temperature. However, because of the scatter of the data about the regression lines in most experimental studies, the differences in the slopes are commonly found to be statistically insignificant, and for this reason a common mean weight exponent from the sum of all the experimental data is generally calculated.

---

[2]The symbol $\dot{V}$ signifies a volume change per unit time and is employed here in line with common usage. However, there are advantages to expressing oxygen consumption rates on a molar basis, $\dot{N}_{O_2}$ (μmol O_2/h) in order to align physiological energetics with usage in gas exchange physiology and biochemistry (Piiper et al., 1971).

TABLE V

Values for the Intercept (a) and Slope (b) of the Regression Lines Relating the Oxygen Consumption (V_{O_2}, ml/h) of a Variety of Molluscs to Body Weight (g)

Organism	a	b	Reference
Grazing gastropods			
Patella granularis	1.160	0.800	Branch and Newell (1978)
Patella granatina	3.450	0.660	Branch (1979)
Patella oculus	1.650	0.800	Branch and Newell (1978)
Patella cochlear	3.830	0.610	Branch and Newell (1978)
Nerita tessellata	1.431	0.720	Hughes (1971b)
Nerita versicolor	0.866	0.820	Hughes (1971b)
Nerita peloronata	1.117	0.750	Hughes (1971b)
Nerita articulata	0.810	0.531	Houlihan (1979)
Tegula funebralis	0.450	0.770	Paine (1971)
Littorina littorea			
10°C	1.327	0.690	Houlihan et al. (1981)
20°C	2.831	0.640	Houlihan et al. (1981)
L. littorea			
Active	4.508	0.670	Newell and Roy (1973)
Inactive	1.892	0.530	Newell and Roy (1973)
Gibbula cineraria			
10°C	1.986	0.560	Houlihan et al. (1981)
20°C	3.467	0.540	Houlihan et al. (1981)
Mean ± SD	2.052 ± 1.250	0.67 ± 0.10	
Suspension and deposit feeders			
Patinopecten yessoensis	0.341	0.810	Fuji and Hashizume (1974)
Scrobicularia plana	0.513	0.750	Hughes (1970)
Crassostrea virginica			
10°C	0.171	0.734	Dame (1972)
20°C	0.372	0.710	Dame (1972)
30°C	0.423	0.603	Dame (1972)
Ostrea edulis	0.455	0.659	Newell et al. (1977)
5°C	0.364	0.899	Rodhouse (1978)
15°C	0.962	0.753	Rodhouse (1978)
25°C	2.655	1.090	Rodhouse (1978)
Cerastoderma edule	0.410	0.530	Newell (1977)
	0.200	0.438	Boyden (1972a, 1972b)
Mytilus edulis (winter)			
Routine	0.549	0.744	Bayne et al. (1973)
Standard	0.263	0.724	
M. edulis (summer)			
Routine	0.339	0.702	Bayne et al. (1973)
Standard	0.164	0.670	

(continued)

TABLE V (Continued)

Organism	a	b	Reference
M. edulis (starved)	0.422	0.870	Famme (1980)
M. edulis	0.398	0.780	Gallager et al. (1981)
Mytilus californianus	0.630	0.840	Whedon and Sommer (1937)
M. californianus			
Fed	0.540	0.650	Bayne et al. (1976a)
Starved	0.230	0.650	
Modiolus demissus			
14°C	0.260	0.690	Kuenzler (1961)
22°C	0.629	0.798	Read (1962)
Aulacomya ater	0.170	0.660	Griffiths and King (1979a)
Crepidula fornicata	0.452	0.700	Newell and Kofoed (1977a)
Mean ± SD	0.496 ± 0.490	0.727 ± 0.130	
Carnivores			
Nassarius reticulatus			
Resting starved	0.200	0.950	Crisp et al. (1978)
Active	1.000	0.950	
Thais lapillus			
Summer	4.890	0.511	Bayne and Scullard (1978a)
Winter	1.970	0.511	
T. lapillus			
10°C	1.300	0.652	Houlihan et al. (1981)
15°C	1.945	0.666	
20°C	1.045	0.875	
Bullia digitalis	1.890	0.600	Brown and da Silva (1979)
Navanax inermis	0.420	0.885	Paine (1965)
Mean ± SD	1.63 ± 1.38	0.733 ± 0.180	
Wood borers			
Teredo navalis	1.200	0.680	Soldatova (1961a, 1961b)
Lyrodus pedicellatus	1.313	0.760	Gallager et al. (1981)
Mean	1.256	0.720	

Houlihan et al. (1981) have shown that the water content of the mantle cavity of intertidal gastropods affects the rate of oxygen consumption, a progressive decrease in $\dot{V}_{O_2}$ occurring with evaporative water loss. Because small individuals have a higher rate of evaporative water loss than large ones (for review, see Newell, 1979), the slopes of the regression lines relating aerial log $\dot{V}_{O_2}$ to log body weight in such molluscs may become steeper with time of exposure to air, especially under conditions where evaporative water loss is enhanced. It should be emphasized, therefore, that the value of b should at least be determined for the particular experimental organism, and if possible under conditions that are similar to those occurring in nature, before standard mean weight corrections are applied to experimental data. However, in the absence of such information, a common mean weight exponent of $b = 0.70$ is generally applicable.

A second feature of significance that emerges from inspection of Table V is that despite the variability of the oxygen consumption ($\dot{V}_{O_2}$) due to the inclusion of a wide variety of data for molluscs at different temperatures and activity levels, the values for mobile molluscs such as grazing gastropods and predators are much higher than those for the suspension and deposit feeders. The mean value of a for the grazing gastropods is 2.052 ± 1.250 SD ($n = 15$) and that for predators and scavengers is 1.63 ± 1.38 SD ($n = 9$), compared with only 0.496 ± 0.49 SD ($n = 24$) for the suspension- and deposit-feeding molluscs. This high value for $\dot{V}_{O_2}$ in the mobile gastropods corresponds well with the high proportion of the absorbed ration that is expended in respiration in population studies on such organisms (see Section I and Table I), and suggests that the energetic cost of crawling is much higher than that of filtration by sessile bivalves. Interestingly, the gastropod *Crepidula fornicata*, a sessile suspension-feeding organism, has a value for routine $\dot{V}_{O_2}$ of 0.452—recalculated from Newell and Kofoed (1977a) using a weight exponent of 0.70. This value is very close to that typical of the suspension-feeding bivalves rather than that of gastropods, and supports the view that the energetic cost of activity, rather than taxonomic grouping, accounts for the high energy expended in metabolism in the gastropods. The high costs of activity are also illustrated by data on the respiration rates of shipworms. For example, Gallager et al. (1981) recorded oxygen consumption rates by *Lyrodus pedicellatus* during boring activity that were four times greater than routine rates for non-wood-boring bivalves. It follows that quiescence constitutes an important compensatory adjustment that can be made, especially by gastropods, to sustain energy flow into growth and reproduction during periods of food deprivation.

B. Compensatory Adjustments in the Rate of Energy Expenditure

1. Activity and Ration

Where molluscs are subject to a reduction in food availability, or where environmental conditions are unsuitable for feeding, the routine energy requirements may exceed the energy captured from the environment. Under these conditions the ability to reduce the metabolic costs are of dominant importance in ensuring long-term survival, although short-term imbalances can be met by utilization of metabolic reserves. Gilfillan et al. (1977), for example, analyzed the flux of carbon in the bivalve *M. arenaria* and estimated that for 7 months of the year when environmental temperatures and food availability were low, respiratory losses exceeded gains from absorption. During this time the respiratory costs were kept low so that depletion of metabolic energy reserves was minimized. Only when temperatures rose and the phytoplankton became abundant did absorption increase sufficiently to exceed respiratory energy expenditure, resulting in a net gain of carbon. The interesting feature in these data is that the increase in absorption between May and June far exceeded the rise expected from the corresponding increase in temperature of 5°C, rising from 15 to 121 μg C/h in a 100-g animal. Conversely, the respiratory costs first increased but then decreased as the temperature rose. These responses suggest a compensatory adjustment of both energy gain and expenditure, which appear to be induced by seasonal variations in temperature and ration. Vahl (1978) measured the oxygen uptake of the scallop *C. islandica* during the year and found that a significant part of the seasonal variation was explained by the seasonal availability of food, whereas changes in temperature did not affect the seasonal metabolic pattern. Mackay and Shumway (1980) found no effect of feeding on the respiration rate of another scallop, *Chlamys delicatula,* and suggested that species that normally feed discontinuously may respond to changes in the food supply, in terms of their respiration rates, more noticeably than continuously feeding species.

The level of metabolic energy expenditure of intertidal limpets also appears to be adjusted to correspond with the availability of food. Branch (1979) and Newell and Branch (1980) have made a comparative study of metabolic energy

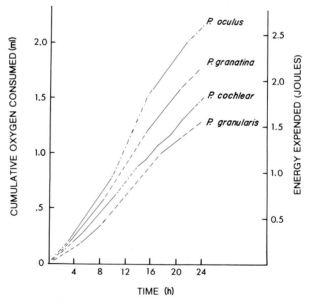

Fig. 8. Cumulative daily oxygen consumption (and expenditure of energy) for four *Patella* sp., standardized to 100 mg ash-free dry weight. Dotted lines indicate the aerial phase of the tidal cycle; solid lines, the aquatic phase. (After Branch, 1979; Newell and Branch, 1980.)

expenditure in a series of patellid limpets on the Cape Peninsula, South Africa. The midshore *Patella granatina* and *P. oculus* both have an abundant food supply, rapid growth, and high reproductive output, whereas the upper shore *P. granularis* experiences a shortage of food with a consequent density-dependent reduction in growth rate and reproductive output (Branch, 1974a, 1974b, 1975a, 1975b), even though it is covered for much of the tidal cycle. The oxygen consumption of the limpets was measured in air and in water over two tidal cycles at temperatures that were synchronized with the local tides and set to correspond with the sea and air temperatures on the shore. The cumulative metabolic energy expenditures for specimens of 100 mg ash-free dry weight of the four species over a period of 24 h is shown in Fig. 8. It is evident that the two midshore species have higher metabolic rates than either *Patella cochlear* or *P. granularis*, which, for different reasons, experience a reduction in availability of food compared with the midshore limpets. An abundant food supply is thus associated with the adoption of an "exploitative strategy" of high energy turnover in these limpets whereas a "conservationist strategy" of reduced metabolic energy expenditure occurs where food resources are limiting (see also Newell, 1980; Newell and Branch, 1980).

The amount of energy that can be conserved by a reduction in activity is related to the work performed by the organism. In fish, it has been common practice for many years to measure the oxygen consumption as a function of swimming speed in a water tunnel (see Bell and Terhune, 1970; Brett, 1971; Brett and Groves, 1979; Muir and Niimi, 1972). This then allows an important distinction to be made between the "active rate" of oxygen consumption at maximal sustained swimming speeds and the "standard rate" of the organism at rest. Between the limits set by these extremes, there is a variable "routine rate" that reflects different activity levels associated with intermediate motor responses.

Partly because of difficulties in quantifying levels of activity in a respirometer, there are rather few comparable studies for molluscs. Houlihan and Innes (1982), however, have related oxygen consumption to crawling speeds in four littoral gastropods and have calculated the relevant costs of transport. They observed linear increases in the rate of oxygen consumption with crawling speed, at two temperatures; increases in oxygen consumption of up to 300% of the inactive rate were recorded, in both air and in water. The costs of transport (the aerobic expenditure required to cover a unit of distance) declined with increasing speed (and they observed snails moving on the shore at speeds equivalent to minimal transport costs) and declined, per unit weight, with increase in body size:

$$E = 76.7 \ W^{-0.68} \ (r^2 = 0.78 \ \text{for} \ n = 5)$$

where E is measured in μl O_2/mg (dry flesh weight)/cm and W is the dry flesh weight in mg.

TABLE VI

Scope for Activity of Eight Mollusc Species

Species	Scope for activity	Activity	Reference
Mytilus edulis	×2–3	Pumping	Bayne et al. (1973)
Littorina littorea	×5–7	Crawling	Newell and Roy (1973)
Vivaparus contectoides	×2.3	Crawling	Fitch (1975)
Limaria fragilis	×4–8	Swimming	Baldwin and Lee (1979)
Bulla digitalis	×2	Burrowing	Brown (1979a)
Nassarius reticulatus	×1.2	Crawling	Crisp (1979)
Monodonta, Gibbula	×1.5–3	Crawling	Houlihan and Innes (1982)

Houlihan and Innes (1982) tabulated the scope for activity (or "aerobic expansibility," calculated as the increase in oxygen consumption associated with activity) in eight species of mollusc (Table VI). The costs of crawling for gastropods are high; for standard-sized animals of 1 g live body weight, locomotion costs for gastropods may be ~5 times higher than for running vertebrates and ants. At least part of this cost may be associated with the production of mucus.

In suspension-feeding organisms it is possible to quantify both oxygen consumption and filtration rate synchronously, and thus to arrive at an estimate of the energetic costs of feeding in such animals. Feeding is then found to be associated with a two- to threefold increase in oxygen consumption compared with the "standard rate" of the quiescent animal (Bayne et al., 1973; Newell, 1976, 1979; Newell and Branch, 1980; Newell and Kofoed, 1977b; Thompson and Bayne, 1972), a value that largely accounts for the discrepancy that has often been found between the measured rate of quiescent animals in a respirometer and that calculated from energy flow studies of active organisms in the field. It should be noted, however, that the metabolic costs of feeding in such organisms are not merely those associated with the demands of the gill cilia, but include a large component of postural activity as well as the energy dissipated in the processes of ingestion and absorption. The overall cost of feeding for the organism may therefore be high despite the fact that the cost of water transport itself may be rather low in filter-feeding organisms (see Jørgensen, 1976b).

The relationship between the rate of oxygen consumption and rate of filtration of the bivalve M. californianus and of the filter-feeding gastropod C. fornicata is shown in Fig. 9. The "standard rate" can be estimated, whether or not the

Fig. 9.   The relationship between filtration rate and oxygen consumption. A. Mytilus californianus, 1 g dry weight, at 13°C. (From Bayne et al., 1976a.) B. Crepidula fornicata, 124 mg dry weight, at 15°C. (From Newell and Kofoed, 1977a.)

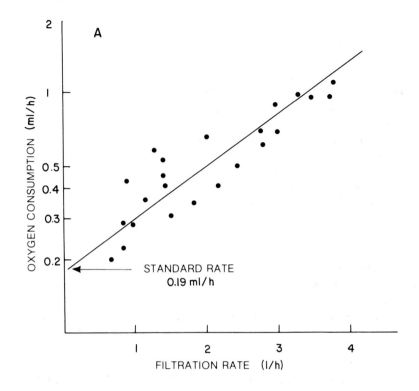

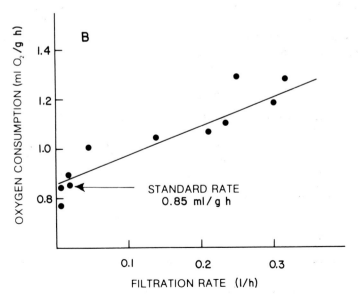

organism is at rest in the respirometer, by extrapolation of the regression line to a point where activity is zero. Conversely, the maximal sustained or "active rate" cannot be predicted and marks the extreme within which possible intermediate or "routine rates" of oxygen consumption occur. The energetic costs of feeding activity in *M. edulis* (Bayne et al., 1973; Thompson and Bayne, 1972; see also Bayne et al., 1976a) increase exponentially with filtration rate; a reduction in energy expenditure of two- to threefold from an active rate of approximately 0.65 ml/g dry flesh/h to a standard rate of 0.22 ml/g dry flesh/h can be achieved at 15°C. Assuming that 1 ml O_2 is equivalent to 20.08 J, this yields a conservation of 8.63 J/g/h. Similarly, in *Crepidula* at 15°C the maximum recorded rate is 0.640 ml O_2/g/h, which falls to 0.201 ml/g/h with cessation of filtration; the reduction of energy expenditure with quiescence leads to a conservation of energy of 8.81 J/g/h.

Bayne et al. (1976a) have shown that it is also possible to estimate the energetic costs of absorption and digestion of the food from the filtration and oxygen consumption rates of starved and fed individuals of *M. californianus*. The increment in filtration rate (liters per hour) and corresponding increase in oxygen consumption of these mussels held at 13°C is shown in Table VII. From this they showed that the relation between filtration and oxygen consumption is described by the expression: $\dot{V}_{O_2} = ae^{0.531} R_f$ ($n = 25$; $r^2 = .82$); where $\dot{V}_{O_2}$ is the rate of oxygen consumption (milliliters of O_2 per hour); R_f is the rate of filtration (liters per hour), and the intercept a is the standard metabolic rate. In fully fed mussels the latter value represents the sum of the "basal metabolic rate" plus the physiological cost of digestion and assimilation at a particular ration (Bayne et al., 1976a). If values for the rate of filtration and the oxygen consumption of starved animals of a similar size are inserted into this equation, the value of a becomes the basal metabolic rate whereas the difference between a

TABLE VII

The Energetic Cost of Feeding by *Mytilus californianus* at 13°C[a,b]

Increment in rate of filtration (liters/h)	Increment in rate of oxygen consumption (ml O_2/h)	Energy equivalents of increment in oxygen consumption (Joules/h)
0–1	0.117	2.35
1–2	0.198	3.97
2–3	0.332	6.66
3–4	0.556	11.16

[a] After Bayne et al. (1976a).
[b] Data expressed in terms of an individual of 1 g dry flesh weight.

for fed mussels and a for starved ones is a measure of the physiological cost of the particular ingested ration. For example, when specimens of *M. californianus* of 1 g dry flesh weight were fed at a ration level of 20.6 J/liter, they filtered an average 1.64 liters/h at 13°C, yielding an ingested ration of 33.8 J/h. In starved mussels of the same size, the basal rate a was 0.136 ml O_2/h. The physiological cost of feeding at this ration was thus $0.232 - 0.136 = 0.096$ ml O_2/h or 1.93 J/h. This represents a physiological cost of 5.7% of the ingested ration in *M. californianus* compared with an estimated value of 4% in *M. edulis* (Bayne and Scullard, 1977b).

From these and other experiments on *Mytilus* it becomes possible to identify two components of the metabolic rate: (1) a "basal rate," which is characteristic of starved organisms in which the physiological cost of digestion and assimilation is zero; and (2) the costs of water transport, filtration, digestion, and absorption (including the specific dynamic action of the ration). For a specimen of *M. californianus* of dry flesh weight 1 g held at 13°C at an average ration of 20.6 J/liter, the value for each of these components was as follows:

(1) Basal rate   = 0.136 ml O_2/h ($= 2.72$ J/h)
(2) Feeding costs   = 0.406 ml O_2/h ($= 8.15$ J/h)

The basal rate of a quiescent 1-g individual represents only 8.1% of the ingested ration of 33.8 J/h, whereas the costs of feeding add a further 24% to the metabolic costs in the presence of food. A value of 24% of the ingested ration of algal cells was also estimated for the cost of feeding in *M. edulis* (Bayne and Scullard, 1977b), and this, as in *M. californianus,* increased exponentially with filtration rate. There are thus sound energetic grounds for the very variable rates of filtration that occur in these and other bivalves, the majority of the energetic costs of activity being eliminated by a reduction of filtration when available ration is insufficient to meet the costs of feeding. In contrast, digestion and assimilation can be maintained after cessation of filtration with a cost of only 4–6% of the ingested ration.

The quantitative significance of a reduction in metabolic energy expenditure when food resources are scarce can be demonstrated when components of both energy gain and expenditure are compared. Table VIII compares the ingested ration, absorbed ration, respiratory losses, and consequent "scope for growth" (see Section V,B) in specimens of *M. californianus* held at a variety of temperatures between 13 and 26°C and presented with a ration of either 4–6% of the body weight per day or only 0.1–0.2%/day (Bayne et al., 1976a). At a high ration level the absorbed ration was high and increased with temperature from 561 to 804 J/day at 22°C before declining to 472 J/day at 26°C. Respiratory losses were less affected by temperature between 13 and 22°C, increasing from 258 J/day at 13°C to 307 J/day at 22°C and 412 J/day at 26°C. The scope for

TABLE VIII

Ingested and Absorbed Rations, Respiratory Losses, and Consequent Scope for Growth in
Specimens of *Mytilus californianus* Held at Various Temperatures in Seawater Containing
High and Reduced Available Rations

Temperature (°C)	Available ration (% of body weight)	Ingested ration (kJ/day)	Absorbed ration $(A)^a$ (kJ/day)	Respiratory losses (R) (kJ/day)	Scope for growth $(A - R)$ (kJ/day)
13	4–6	0.93	0.56	0.26	0.30
17.5	4–6	1.23	0.74	0.26	0.48
22	4–6	1.34	0.80	0.31	0.49
26	4–6	0.79	0.47	0.41	0.06
13	0.1–0.2	0.04	0.03	0.11	−0.08
17.5	0.1–0.2	0.05	0.04	0.11	−0.07
22	0.1–0.2	0.06	0.05	0.14	−0.10
26	0.1–0.2	0.03	0.03	0.18	−0.15

[a] The absorption efficiencies used to derive the absorbed ration were 0.6 for well-fed mussels and 0.8 for mussels at a low-ration level (see Bayne et al., 1976a; Thompson and Bayne, 1974).

growth in the well-fed animals thus achieved maximal values between 17.5 and 22°C, declining at both 13 and 26°C.

In contrast, the mussels held at a ration of only 0.1–0.2% of the body weight per day (Table VIII) reduced their metabolic energy expenditure to less than half of that in well-fed *M. californianus*, and because energetic losses do not increase much with temperature, the mussel is well adapted to tolerate periods of reduced food availabiilty coupled with high temperatures such as occur during low tide in the summer months. However, despite the reduction in activity and energetic losses, the ingested ration at minimal food availability was rather less than half of the energetic equivalent of losses through metabolism. The scope for growth was therefore negative in the mussels throughout the temperature range in these experiments.

In much the same way, many gastropods may achieve a significant reduction of energy expenditure with the onset of quiescence, and because of the high cost of activity in mobile organisms the reduction with quiescence is even more marked than in suspension-feeding molluscs. Crisp et al. (1978) have shown that following presentation of food, eversion of the proboscis occurs in *Nassarius reticulatus* and the oxygen consumption is increased to three or four times the rate of starved individuals. Coleman (1976) described how several species of neritid gastropods are active when first exposed to the air by the receding tide, but as the duration of exposure increases, they retreat under stones and into crevices and are quiescent for the remainder of the emersion period; the high rates of respiration characteristic of the active phase are then reduced by 20–50%. Similarly high values have been recorded for the ratio between the oxygen consumption of actively crawling *L. littorea* and that of organisms that are quiescent during the intertidal period (Newell, 1976; Newell and Roy, 1973).

Such phases of activity and quiescence may be linked not only to the tidal cycle but also to the local availability of food. Brown (1979b), for example, has shown that the gastropod *Bullia digitalis* remains buried in the sand below the low water mark for much of the time and consumes approximately 0.96 mg $O_2/g/h$ during this phase. The snails subsequently surf ashore with the foot expanded and feed on dead or moribund animal material cast up on the shore (Brown, 1971). But the cost of activity during transport in the surf zone is high and the animals then consume approximately 2.1 mg $O_2/g/h$. Thus, remaining inactive for most of the tidal cycle results in a considerable saving of energy compared with that expended in routine activity. Further, because the snails emerge and undertake active migratory movements in the surf zone only when they detect the presence of food in the water, periods of high energetic expenditure are linked with an increased chance of energy gain through feeding.

2. External Oxygen Availability

Many molluscs, especially those living in sands and muds, experience a decline in oxygen availability during intermittent pauses in irrigatory activity or during longer periods when the tide ebbs and the oxygen tension (P_{O2}) of the surrounding water falls to the low levels characteristic of the interstitial water. Other molluscs may be exposed to the air by the falling tide, and although many are able to carry out gas exchange with air and withstand the water loss that this entails, they must close the valves or withdraw into the shell at some stage and thereafter respire under conditions of declining oxygen tension, ultimately resorting to anaerobic pathways to supply an increasing proportion of their energy requirements. In the mussel *M. edulis,* which regularly experiences anoxia, there may be a temporary reduction in metabolism effected through operation of the anaerobic succinate pathway, to levels that are as low as 5% of the aerobic rate (De Zwaan and Wijsman, 1976).

In some species, such as the whelk *Buccinum undatum* (Lange et al., 1972), the mussel *M. demissus* (Booth and Mangum, 1978), and many other invertebrates (for review see Mangum, 1977; Mangum and van Winkle, 1973, 1977; Herreid, 1980), the rate of oxygen consumption is directly dependent on the ambient partial pressure of oxygen, and thus declines as the P_{O_2} in the water surrounding the gill surface is reduced. Such organisms are regarded as "conformers." In contrast, other species are able to maintain their rate of oxygen consumption under conditions of declining oxygen by increasing the ventilation rate, increasing the efficiency with which oxygen is removed from the water, or a combination of both responses. Such organisms, in which the V_{O2} is maintained at a relatively uniform level irrespective of a decline in ambient oxygen tension, are classed as "regulators." Bivalves, including *Mya* (Van Dam, 1935, 1938), *Pecten* (Van Dam, 1954), *Laevicardium* (Bayne 1971a), and several species of *Mytilus* (Moon and Pritchard, 1970; Bayne, 1967, 1971a, 1971b; Famme, 1980; Famme and Kofoed, 1980), as well as the gastropods *Busycon canaliculatum* (Mangum and Polites, 1980), *C. fornicata* (Newell et al., 1978), and many other invertebrates fall into this category (Mangum and van Winkle, 1973). The oxygen consumption of such regulators generally becomes increasingly dependent on the external oxygen tension as it declines below a certain critical value (Prosser and Brown, 1961); thereafter the metabolism may become increasingly anaerobic. The ability to maintain aerobic metabolism and thus compensate for reduced oxygen availability represents one way in which the energy yield from the absorbed ration is maximized over a wide range of external environmental conditions.

The validity of a sharp distinction into conformers and regulators has, however, been questioned (Mangum and van Winkle, 1973; Taylor and Brand, 1975b; Herreid, 1980), and it seems more likely that these represent merely the extremes of a continuous series of curves that may be affected by body size (Bayne, 1971a; Taylor and Brand, 1975a, 1975b), temperature (Newell et al., 1978), salinity (Bayne, 1973a; Polites and Mangum, 1980), and the conditions of oxygen availability that the molluscs experience in the natural environment (Bayne, 1973a; Mackay and Shumway, 1980). Such curves have been described by second-order polynomial equations (see Mangum and van Winkle, 1973), but an alternative way of comparing the degree of oxyconformity or regulation of the curves relating $\dot{V}_{O_2}$ to P_{O_2} is to use the transformation of Tang (1933) in which the ratio of P_{O_2}: $\dot{V}_{O_2}$ is plotted as a function of P_{O_2}. This gives a straight line whose slope and intercept can be used to compare the degree of oxyconformity of different organisms. This method has been used by Bayne (1971a, 1971b), Bayne and Livingstone (1977), Taylor and Brand (1975b), Newell et al. (1978; for review see Newell, 1979), and Mackay and Shumway (1980).

The ratio of the intercept a of the calculated regression lines obtained by this transformation, divided by the slope b, gives an index of respiratory dependence (Bayne, 1971a, 1971b). A low value for this ratio indicates a curve that is regulated independently of external P_{O_2}, whereas a high value indicates the dependent curve of an oxyconformer. Bayne (1973a) suggested that the capacity to regulate oxygen consumption in different bivalves may be correlated with the degree of hypoxia experienced in the natural environment. Subsequent work by Taylor and Brand (1975b) has tended to confirm this although they also point out that the bivalve *Arctica islandica* shows an increase in respiratory independence with increasing body size. Table IX, based on Mackay and Shumway (1980),

TABLE IX

Regression Equations for a/b (= Y) against Weight-Specific O_2 Consumption (X)
for Seven Species of Bivalves[a,b]

Species	Regression equation	Reference
Geloina ceylonica	$Y = 11.5\ X^{0.54}$	Bayne (1973a)[c]
Anadara granosa	$Y = 29.8\ X^{0.69}$	Bayne (1973a)[c]
Modiolus demissus	$Y = 62.7\ X^{0.93}$	Mackay and Shumway (1980)
Mytilus edulis	$Y = 75.5\ X^{0.82}$	Bayne (1971a)[c]
Chlamys delicatula	$Y = 115.7\ X^{0.77}$	Mackay and Shumway (1980)
Laevicardium crassum	$Y = 261.3\ X^{1.38}$	Bayne (1971a)[c]
Arctica islandica	$Y = 829.9\ X^{1.44}$	Taylor and Brand (1975a)[c]

[a] Based on Mackay and Shumway (1980).
[b] a and b are derived from the equation $V_{O_2} = P_{O_2}/a + b\ P_{O_2}$) (Bayne, 1971a; Tang, 1933).
[c] Equations recalculated from original data.

lists some regression equations for the ratio $a:b$ against weight-specific oxygen consumption for seven bivalve species.

The mechanisms for regulating oxygen consumption by bivalves during declining oxygen tension have been studied by Van Dam (1935, 1954), Bayne and Livingstone (1977), Taylor and Brand (1975b), Booth and Mangum (1978), Famme (1981), and Famme and Kofoed (1980). Species differ in their ability to maintain a constant rate of delivery of oxygen to the respiratory surfaces by means of increased pumping rate (Taylor and Brand, 1975b). In some species the efficiency of extraction of oxygen from the medium may increase at reduced P_{O_2}, although the role of the circulation of blood in enhancing extraction efficiency is controversial (Booth and Mangum, 1978; Famme, 1981). The functions, if any, of altered ventilation and perfusion rates and of diffusion distance (from the medium to the respiring tissue) in regulating gas exchange in bivalves remain to be elucidated.

In a review Herreid (1980) presents a simple model of gas exchange in which the rate of oxygen consumption depends on the "conductance" of oxygen through the animal and on the P_{O_2} gradient between the external and intracellular environments; various types of physiological compensation for environmental hypoxia are discussed and a framework for future research on these topics is suggested.

When viewed in terms of adaptation to a variable environmental parameter, conformity and regulation at low P_{O_2} have entirely different energetic implications, the former suggesting conservation, by reduction of metabolic costs (and possibly of feeding rate also), the latter implying the maintenance of both energy gain and expenditure. Species that normally experience transient hypoxia may be expected to respond by maintaining a relatively constant metabolic rate. However, further studies are needed, particularly with simultaneous aerobic, anaerobic, and heat loss (see Section IV,C) measurements, in order to elucidate the full energetic consequences of hypoxic exposures (Famme et al., 1981).

3. Aerial Oxygen Consumption

The ability to respire oxygen from the air is widespread among intertidal molluscs. Both gastropods (Baldwin, 1968; Branch and Newell, 1978; Houlihan, 1979; Houlihan and Newton, 1978; Houlihan et al., 1981; McMahon and Russell-Hunter, 1977; Micallef, 1967; Micallef and Bannister, 1967; Sandison, 1966) and many bivalves (Bayne et al., 1976a, 1976b; Boyden, 1972a, 1972b; Kuenzler, 1961; Widdows et al., 1979) can carry out gas exchange in air at a rate that, especially in gastropods, can approach or even exceed the minimum rate recorded in water.

The values for the weight-specific oxygen consumption (in milliliters per gram per hour) of a variety of bivalves and gastropods are summarized in Table X. Because the weight-specific rate is dependent on body size, all values are expressed in terms of a standard-sized individual of 1 g dry weight. The absolute levels of metabolic energy expenditure as well as the relative rates of energy expenditure in air and water can thus be compared, although it should be mentioned that in some cases a weight exponent of 0.7 has had to be assumed (see Table V). The examples have been arranged in groups representing the aspidobranch (bipectinate) ctenidia characteristic of bivalves and some archaeogastropods, the pectinibranch (monopectinate) gills of most intertidal gastropods (the mesogastropods and neogastropods), those such as the intertidal limpets that have lost the ctenidium in the mantle cavity and have instead evolved a series of secondary gills, and finally those that utilize a vascularized mantle cavity or lung. Within each group the species have been arranged in approximate order according to their normal intertidal position from low to high shore.

Despite the different experimental procedures used, and the variability this is likely to have introduced into the data, it is clear that the bivalves in general are less well adapted to carry out aerial gas exchange than the gastropods, which have tended to reduce the ctenidium and to replace this with secondary pallial gills or a lung. The overall rates of energy expenditure by the sessile suspension-feeding bivalves are lower than those of the browsing gastropods, a difference associated with the energetic cost of movement compared with that of filtration. Within each group the relative ability to sustain the metabolic energy requirements in air increases from lower shore molluscs, which can meet only some 4–9% of their oxygen requirements in air, to those that can sustain a high level of activity in air and in which aerial rates are considerably higher than in water.

TABLE X

Weight-specific O₂ Consumption for a Variety of Molluscs[a]

Species	O_2 Uptake (ml O_2/g/h) Air	Water	Aerial: aquatic ratio	Temperature (°C)	Reference
BIVALVIA					
Cardium glaucum	0.012	0.15	0.08	15	Boyden (1972b)
Mytilus edulis	0.017	0.42	0.04	10	Widdows et al. (1979b)
Mytilus galloprovincialis	0.044	0.33	0.13	25	Widdows et al. (1979b)
Cerastoderma edule	0.13	0.20	0.65	15	Boyden (1972a)
	0.14	0.50	0.28	10	Widdows et al. (1979b)
M. californianus	0.17	0.23	0.74	13	Bayne et al. (1976b)
Modiolus demissus	0.24	0.38	0.63	20	Kuenzler (1961)
	0.23	0.41	0.56	20	Widdows et al. (1979b)
GASTROPODA					
Aspidobranchs					
Calliostoma zizyphinum	0.79	1.34	0.59	20	Micallef (1967)
Gibbula cineraria	0.39	0.88	0.44	20	Micallef (1967)
Gibbula umbilicalis	0.49	0.54	0.91	20	Micallef (1967)
Monodonta lineata	0.40	0.37	1.08	20	Micallef (1967)
Monodonta turbinata	0.32	0.19	1.68	20	Micallef and Bannister (1967)
Nerita articulata	2.28	0.81	2.87	23	Houlihan (1979)
Pectinibranchs					
Acmaea testudinalis	0.76	3.30	0.23	20	McMahon and Russell-Hunter (1977)
Littorina littoralis	1.59	2.99	0.53	20	McMahon and Russell-Hunter (1977)
Littorina littorea	1.68	3.09	0.54	20	McMahon and Russell-Hunter (1977)
Crepidula fornicata	0.28	0.31	0.90	20	Newell and Kofoed (1977a)
Littorina saxatilis	1.20	1.66	0.72	20	McMahon and Russell-Hunter (1977)

(continued)

TABLE X (Continued)

Species	O₂ Uptake (ml O₂/g h) Air	Water	Aerial: aquatic ratio	Temperature (°C)	Reference
Secondary pallial gills					
Patella caerulea	0.47	1.31	0.35	20	Bannister (1974)
Patella cochlear	3.37	3.61	0.93	20	Branch and Newell (1978)
Patella granatina	2.82	4.52	0.62	20	Branch (1979)
Patella oculus	4.6	5.49	0.84	20	Branch and Newell (1978)
Patella granularis	1.22	1.22	1.00	20	Branch and Newell (1978)
Patella lusitanica	1.33	0.43	3.09	20	Bannister (1974)
Vascularized mantle cavity					
Cerithidea obtusa	2.35	0.38	6.18	23	Houlihan (1979)
Cassidula aurisfelis	2.80	0.51	5.49	23	Houlihan (1979)

[a] All values have been calculated for an organism of standard 1 g dry mass using either published values for the weight exponent (b) or the mean value of 0.7 (see Table V). The examples are arranged in major groups according to their respiratory organs. Within each group the species have been arranged in approximate order from aquatic to terrestrial conditions. The ratio of aerial to aquatic oxygen consumption is also shown. Note that within each group the relative importance of aerial gas exchange increases from aquatic to terrestrial conditions on the upper shore and that the rate of oxygen consumption in the mobile gastropods is greater than in the sessile filter-feeding bivalves.

Finally, among the browsing gastropods, the highest absolute rates of oxygen consumption are found among those lower to midshore species such as *P. cochlear*, *P. granatina*, and *P. oculus*, which exploit rich algal resources, whereas the upper shore gastropods such as *P. granularis* and the swamp-dwelling *Cerithidea* and *Cassidula* have reduced their metabolic energy expenditure as well as having an enhanced ability to carry out aerial gas exchange.

These trends reflect both behavioral responses and structural features such as the type of gill and degree of vascularization of the mantle cavity. Kuenzler (1961) and Widdows et al. (1979b) have shown, for example, that the ability of intertidal bivalves to sustain aerobic metabolism in air depends mainly on behavioral responses such as intermittent air-gaping. The aerial rate of *M. edulis* and *Mytilus galloprovincialis*, both of which close the valves on exposure to air (Coleman, 1973), is 4–17% of the minimum aquatic rate, and accumulation of anaerobic end products, signifying active anaerobiosis, occurs (Widdows et al.,

1979b). The mainly subtidal *Cardium glaucum,* which shows no air-gaping response, has a rate of oxygen consumption in air that is only 8% of the minimum rate in water (Boyden, 1972b). Conversely, *C. edule* and *M. demissus,* both of which gape on exposure to air, have a rate of oxygen uptake that is between 28 and 78% of the aquatic rate, and no accumulation of anaerobic end products occurs (Widdows et al., 1979b; see also Ahmad and Chaplin, 1977). Sandison (1966) has shown that the rates of oxygen consumption of a variety of intertidal gastropods in air is initially higher than in water, but that after 3 h in dry conditions the animals become quiescent and aerial oxygen consumption is then lower than in water.

Physiological factors, as well as behavioral ones, may also account for the different rates of oxygen consumption in air and in water. Houlihan et al. (1981) have shown that water loss from the mantle cavity may lead to the collapse of the gill in some intertidal gastropods with a subsequent suppression of the oxygen consumption in air. In *Gibbula cineraria* the aerial oxygen consumption of individuals in which the mantle water had been removed was only 36% of that in individuals with water in the mantle cavity. In *L. littorea* the rate without mantle fluid was 53% of that with mantle fluid, whereas in the stenoglossan *T. lapillus* the rate was suppressed to 63% of that with mantle fluid. The molluscan ctenidial gill is thus suited to only relatively short-term gas exchange in air and is dependent on mechanical support from the mantle fluid. It is not surprising, therefore, that most of the gill-bearing molluscs remain active only under moist conditions, such as immediately after exposure by the tide (Coleman, 1976), and subsequently withdraw until aquatic conditions return; on the other hand, those that have secondary gills around the vascularized mantle margins, or have developed a vascularized mantle margin or a vascularized mantle cavity, have become better adapted for aerial life and show conservation of metabolic energy expenditure in high-shore situations where environmental food resources are limiting.

These trends also occur intraspecifically in molluscs such as limpets, that occupy a range of zonational positions in the intertidal zone. The food-limited upper shore *Patella,* for example, migrates progressively up the shore, so that larger animals are found high on the shore whereas juveniles inhabit damp situations well below midtide level. Comparisons of the rate of oxygen consumption in air and in water show that small individuals respire more slowly in water than in air, so that their occurrence low on the shore ensures low respiration for most of the tidal cycle. The reverse is true of larger individuals, which occur high on the shore yet respire more slowly in air than in water. Thus they too have low respiratory rates, and hence low metabolic energy losses, under the conditions that prevail for much of the time in their local microhabitat (Branch, 1979; Branch and Newell, 1978; Newell and Branch, 1980). In this way the reduced food available at high-shore levels is offset by minimization of metabolic energy expenditure.

In contrast, the midshore limpets *P. oculus* and *P. granatina* have abundant food resources and have a high rate of feeding and defecation. These limpets are not limited by environmental food resources and have high metabolic rates under the normal conditions of temperature and aerial exposure that they experience on the shore, in comparison with *P. granularis* (Branch, 1979). *P. granatina* is also a migratory species, so that the larger individuals occur higher on the shore. In this case, however, the respiratory response in air is quite different from that of the "conservationist" *P. granularis*. Small limpets occur lower on the shore and respire faster in water whereas the larger high-shore individuals respire faster in air. Respiration in both size ranges is thus kept high. Similarly, in *P. oculus* the metabolic responses in air and water are such that a high level of metabolism is sustained under the local conditions encountered by the limpets. Thus the differing responses in air and water in the food-limited limpets such as *P. granularis* tend to minimize metabolic energy expenditure under the environmental conditions usually experienced by the limpets during their migration up the shore and also tend to maximize energy turnover in those species that have adopted an "exploitative" strategy of high growth and reproductive output in the presence of abundant food resources.

It is apparent that intertidal species, or those that regularly encounter conditions of declining P_{O_2} are able to sustain a significant proportion of the minimum oxygen requirements for aerobic metabolism both in air and in water. When environmental food resources are limiting, a decline in activity level, such as commonly occurs during the low-tide period, may allow much of the maintenance energy requirements of these molluscs to be met by aerobic pathways without resort to major utilization of metabolic reserves through anaerobiosis. However, where environmental P_{O_2} levels fail to below a critical value, or where the reduction of water loss in air by closure of the shell valves is necessary for survival, most molluscs are able to utilize anaerobic pathways and further to suppress their metabolic requirements so that depletion of energy reserves is minimized.

4. Temperature

The effects of temperature on the survival, activity, and metabolic energy expenditure of molluscs have been widely described in the literature (for reviews, see Kinne, 1970; Newell, 1979; Newell and Branch, 1980; Precht et al., 1973). In almost all instances the rate of metabolism varies with short-term exposure temperature, the increase or decrease over a 10°C interval being defined as the temperature coefficient (Q_{10}).

$$Q_{10} = \left[\frac{R_2}{R_1} \right]^{10/T_1 - T_2} \tag{10}$$

where R_2 and R_1 are the rates of reaction at temperatures T_2 and T_1, respectively. This short-term response to a change in exposure temperature may be accompanied by an overshoot, followed by a period of stabilization that takes place over a period of minutes or hours (see Precht et al., 1973). Following longer term exposure to changed temperature conditions lasting for days or weeks, the organism may show an adjustment of its rate of oxygen consumption to a level comparable with that which occurred before the temperature changed. When in response to temperature itself, this phenomenon is known as "thermal acclimation," whereas a compensatory adjustment in response to more complex seasonal changes in environmental conditions is distinguished as "seasonal acclimatization" (Prosser, 1973; Newell and Branch, 1980). Similar adjustments in response to other single environmental factors such as salinity may also be distinguished as acclimation effects.

Clearly, in terms of energy conservation, the organism would be expected to make compensatory adjustments to both components of energy gain and energy loss in such a way that energy gain is maximized and energy losses are minimized in the face of a change in environmental conditions. In some molluscs, as in the suspension-feeding gastropod *C. fornicata,* routine oxygen consumption increases with short-term exposure temperature, so that it might be inferred that metabolic costs would increase seasonally during the warm conditions of the summer months. However, as can be seen from Fig. 10, the acute rate: temperature curves are shifted to the right following acclimation to temperatures up to 25°C. Lateral translation of the rate: temperature curves in this case is in step with the increase in acclimation temperature, so that the response serves to maintain oxygen consumption at an almost constant level of between 73 and 75 μl/h, despite an increase in acclimation temperature from 10 to 25°C. The acclimated rate:temperature curve (in which exposure temperature T_e = acclimation temperature T_a) for routine oxygen consumption by *Crepidula* is shown in Fig. 10B, and suggests that the seasonal increase in metabolic energy expenditure that would have been anticipated from the acute response to short-term temperature change is minimized by thermal acclimation to the warm conditions that occur during the summer months (Newell and Kofoed, 1977b). This pattern of response may be widespread in molluscs and is commonly linked with changes in the filtration rate with thermal acclimation (Bayne et al., 1976a, 1976b; McLuskey, 1973; Read, 1962; Widdows, 1973a, 1973b, 1978a). In the case of *Crepidula,* the rate of filtration is increased with warm acclimation, so that by synchronous adjustment of components of both energy gain and expenditure, absorption is maximized during the warm conditions of the summer.

Other molluscs, however, show no such acclimatory adjustment of their metabolic rate in response to long-term changes in the environment, and thus their metabolic costs increase both with short-term and with seasonal changes in

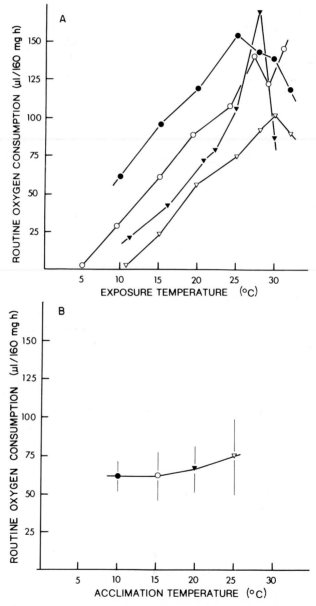

Fig. 10. The rate of routine oxygen consumption of *Crepidula fornicata* at a variety of exposure temperatures following acclimation to temperatures (T_a) of 10, 15, 20, and 25°C. **A.** Acute rate:temperature curves. **B.** Acclimated rate:temperature curves. ●——●, T_a 10; ○——○, T_a 15; ▼——▼, T_a 20; ▽——▽, T_a 25. (After Newell and Kofoed, 1977b.)

environmental temperature. Unless the increased cost of metabolism is offset by increased components of energy gain during the summer months, such organisms will clearly experience a metabolic deficit that may limit their distribution to low-tidal levels. Davies (1966, 1967), for example, showed that the upper shore *Patella vulgata* is capable of metabolic acclimation, both seasonally and in relation to position on the shore, whereas the low-shore *Patella aspersa* fails to acclimate and may thus be limited to a niche where environmental temperatures are more uniform.

The oyster *Ostrea edulis* also shows little evidence of adjustment of its metabolic energy expenditure to thermal acclimation (Buxton et al., 1981; Newell et al., 1977; see Fig. 11). In this case, however, the seasonal increase in metabolic losses is more than offset by compensatory adjustments of the filtration rate, which increases sharply with temperature and reaches a maximum following thermal acclimation to 20°C. As a result, the maximal scope for growth (see Section V,B), representing the difference between energetic gain and losses, is achieved during the warm conditions that prevail during the summer, though in this case it is achieved by increase of the components of gain, rather than a suppression of those of energy expenditure (see also Newell and Branch, 1980). A rather similar situation evidently occurs in the bivalve *Donax vittatus*. Ansell and Sivadas (1973) have shown that, as in *Ostrea,* metabolic costs increase rapidly with temperature, but in this case the increase in filtration rate is not so great as the increase in metabolism. The scope for growth may become negative at this stage, and heavy losses can occur as temperatures rise in the spring, for carbohydrate and lipid reserves are then at a low level, following depletion during the winter when food is limited.

Compensation for the effects of long-term changes in environmental temperature can also be achieved by rotation of the acute rate:temperature curve following warmth acclimation, or by a combination of translation and rotation. The rate:temperature curves for the routine oxygen consumption of the periwinkle *L. littorea* during acclimation to the warm conditions of the summer are shown in Fig. 12. It can be seen that translation of the curves to the right during the summer reflects the maintenance of a relatively uniform rate of routine oxygen consumption, which is probably linked with energy gained from the food. Rotation of the curves in addition shows that the oxygen consumption is much less dependent even on short-term temperature change than during the winter, and in this respect the winkle is well adapted for life in the intertidal zone where temperatures are likely to vary unpredictably.

A rather special case of rotation of the rate:temperature curve occurs in the case of many intertidal molluscs, and especially those of the upper shore, which are exposed not only to seasonal changes in environmental temperature but also to short-term changes such as occur with the ebb and flow of the tide. In the absence of conpensatory adjustment, metabolic energy losses would be very high

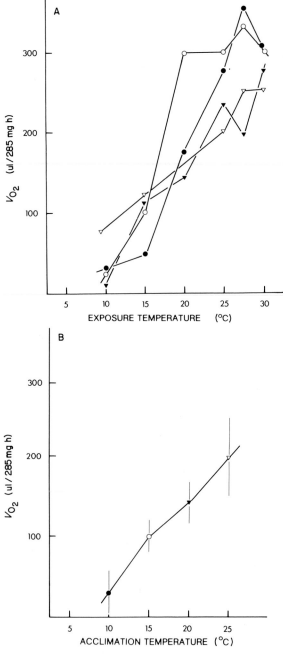

Fig. 11. The rate of routine oxygen consumption of *Ostrea edulis* at a variety of exposure temperatures following acclimation to temperatures (T_a) of 10, 15, 20, and 25°C. **A.** Acute rate: temperature curves. **B.** Acclimated rate:temperature curves. ●———●, T_a 10; ○———○, T_a 15; ▼———▼, T_a 20; ▽———▽, T_a 25. (After Newell et al., 1977.)

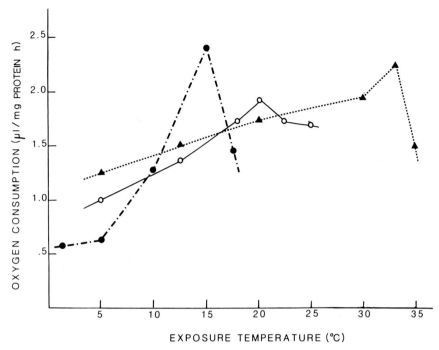

Fig. 12. The rate of routine oxygen consumption of *Littorina littorea* as a function of exposure temperature, following collection between February and June when the temperature at the collection site increased from T_a 12.5 to T_a 30°C. ●———●, T_a 12.5; ○———○, T_a 17; ▲---▲, T_a 30. (After Newell and Pye, 1971a.)

in these animals during the low-tide period in the summer, and moreover are likely to occur at stages in the tidal cycle when compensatory adjustments in the feeding rate are precluded by aerial exposure. Such organisms may show very low temperature coefficients for the standard rate of metabolism characteristic of the quiescent phase. *L. littorea* (Newell and Pye, 1970a, 1970b, 1971a, 1971b; McMahon and Russell-Hunter, 1977), *Macoma balthica* (Kennedy and Mihursky, 1972), and the mussels *M. edulis* (Widdows, 1973b) and *M. californianus* (Bayne et al., 1976b) each have rates of oxygen consumption that have very low temperature coefficients over the environmental temperature range normally experienced by the species, leading to a conservation of metabolic energy resources during periods of reduced food availability.

Brown et al. (1978) compared the oxygen consumption of the temperate South African *B. digitalis* with that of the tropical *Bullia (Dorsanum) melanoides* from India, and have shown that when measured at temperatures equivalent to those experienced in the field (15 and 30°C for the two species, respectively), the respiratory rate of *B. digitalis* is an order of magnitude lower than that of *B. melanoides*. If acclimation had occurred, it would be anticipated that the metabolic rates of the two species would have been similar under the local temperature conditions prevailing in their habitats. Brown et al. (1978) suggested that these differences may reflect the fact that *B. melanoides* breeds twice a year as opposed to once, and that *B. melanoides* has a shorter life span of 2 years compared with a longevity of at least 6 years in *B. digitalis*. *B. digitalis* experiences lengthy periods of starvation, which makes energy conservation important, and in this species the oxygen consumption of even active animals is almost independent of temperature between 10 and 25°C, so that metabolic energy losses are not elevated while the snails are feeding in the intertidal zone. The tropical *Donax incarnatus* also respired much faster at its environmental temperature of 30°C than the temperate *D. vittatus* at 15°C, suggesting that this is a further example of nonacclimation. *D. incarnatus,* like *B. melanoides,* also has a very high filtration rate, high growth rate, and shorter life than its temperate counterpart (McLusky and Stirling, 1975).

Acclimatory adjustments in metabolic energy expenditure thus appear to be linked primarily with conservation of energy resources. As discussed earlier, in situations where food resources are plentiful a maximal energetic gain may be achieved by the adoption of an ''exploitative'' strategy of high energy turnover where ingestion is maximized despite the increased fecal and metabolic losses that are incurred (Newell, 1980; Newell and Branch, 1980). Such organisms have a high growth and reproductive rate and short life span, and may show strongly temperature-dependent rates of oxygen consumption and feeding. Conversely, where food resources are limiting, especially where they are linked with short-term increases in environmental temperature, the ability to balance the metabolic losses with the energy gained from the environment is essential if energy is to be available for growth and reproduction. It is thus not surprising to find that there is a relative reduction in metabolic energy expenditure in gastropods at high-shore levels, nor that it is among this group of molluscs that acclimatory adjustments of the metabolic rate have been generally described.

The ability to sustain an energetic gain from the environment and to channel energy into growth and reproduction is clearly of central importance in controlling the ability of a species to compete under conditions where environmental energy resources are limited. As has been shown already, compensatory adjustments of both components of energy gain and expenditure are involved in this process. When the effects of both components are combined we can gain some insight into the endogenous and environmental factors that interact to control the conditions under which an energetic gain, and therefore growth and reproduction, can occur.

C. Direct Calorimetry

The use of rates of oxygen consumption as measures of metabolic heat loss, using standard oxycaloric coefficients (i.e., indirect calorimetry) has been validated for circumstances of aerobic metabolism (Kleiber, 1961), although the implicit assumptions (Crisp, 1971) should always be borne in mind. For example, Famme et al. (1981) established, by simultaneous direct and indirect calorimetry, that processes of anaerobic metabolism may account for less than 5% of total heat loss by *Mytilus edulis* under normoxic conditions (see also Shick et al., 1983). Nevertheless, it has become increasingly evident in recent years that many molluscs that are capable of efficient anaerobic metabolism may resort to anaerobiosis frequently in the natural habitat (due to periodic exposures to hypoxic or anoxic conditions) and may even invoke both aerobic and anaerobic metabolic processes simultaneously in different tissues. This subject is reviewed by DeZwaan and by Livingstone and DeZwaan (Volume 1, Chapters 4, 5). Here we concentrate on the implications of this research for understanding the physiological energetics of individuals.

In situations of partial anaerobiosis indirect calorimetry is an inadequate measure of heat loss and only direct calorimetry will provide a realistic measure of energy expenditure in such circumstances. Furthermore, as Pamatmat (1980) has pointed out, even if the energy demands of anaerobiosis (or the energy yields of anaerobic pathways) may be negligible in terms of a population energy budget, the adaptive value of anaerobiosis to the individual may be considerable. For example, Shick et al. (1983), working with *M. edulis,* have calculated that for a 12-h cycle, in which the animals are exposed to air for 5 h, the anaerobic contribution to total heat loss may amount to less than 5% However, this result is not to deny that the capacity for anaerobic metabolism during periods of tidal air exposure amounts to a major adaptive feature in this species.

Studies by direct calorimetry of total heat production by molluscs have been made by Pamamat (1979, 1980), Hammen (1979), Shick and Widdows (1981), Shick et al. (1983) and discussed by Gnaiger (1979, 1980) and by De Zwaan et al. (1981). Pamatmat (1979) calculated the following relationship between body size (*W,* in grams dry flesh weight) and heat production (*Y,* in watts, where 1 joule/h $= 0.28$ mW):

$$Y = 0.089 \times 10^{-4} + 1.58 \times 10^{-4}\ W$$

based on six species of bivalve and one polychaete worm, all at 20°C. Pamatmat argues that an isometric relstionship between total heat loss and body size in aquatic invertebrates is to be expected and that an allometric relationship (Section IV,A) between the rate of oxygen consumption and body size is the outcome of limitations on gas exchange as the animals grow in size. More work is needed to elucidate this fundamental problem, which has far-reaching consequences for the relationships between growth, growth efficiency, reproductive effort and body size (Section V,B).

Pamatmat (1979) determined a Q_{10} of 1.56 for anaerobic heat production by two bivalves (*Polymesoda caroliniana* and *Guekensia demissus*) between 20 and 30°C. Variability in anaerobic heat output was due largely to variable activity levels, but was generally less than during the measurement of total (largely aerobic) heat loss. As with aerobic metabolism, quiescence is an effective means of energy conservation during anaerobiosis; for example, for *Arctica islandica* under long-term anoxia, periods of activity raised heat output by an average of 31%. Individual bivalves fully adapted to regular hypoxic–anoxic excursions, for example during tidal air exposure, may show less frequent bursts of activity than individuals less well adapted (M. Schick, personal communication). A distinction between active and inactive rates may be just as important in studies of anaerobic heat loss as it is in aerobic measurements.

Measurements by Pamatmat (1979, 1980), Famme et al. (1981) and Shick and Widdows (1981) all support the contention that, for various bivalves, anaerobic heat loss during normoxic conditions in water represents a small percentage (5–7.5%) of total heat loss, giving confidence in the use of oxygen-consumption determinations in studies of energy balance under normal circumstances. During some situations, however, this discrepancy may be greater. For example, during starvation (Pamatmat, 1980) anaerobic heat production may increase to greater than 20% of total; during aerial exposure at low tide anaerobic heat production accounted for 72% of the total aerial rate of heat loss by *M. edulis* (Shick et al., 1983). Calculations of this kind suggest that the full advantages of direct calorimetry to studies of physiological energetics will accrue when simultaneous flow-through direct and indirect calorimetric techniques become generally available (Gnaiger and Forstner, 1983). Using such a technique Famme et al. (1981) concluded that there exists a dependence, in *M. edulis* between the levels of anaerobic and aerobic metabolism. Under anoxia, anaerobic heat loss represented 5–10% of total heat loss at normoxia. With increasing P_{O_2} up to 40–80 mm Hg, anaerobic heat production increased, then decreased at higher partial pressures; the maximum anaerobic heat production was 3–7 times greater than at anoxia.

Using a batch, rather than flow-through, calorimeter, Shick and Widdows (1981; see De Zwaan et al., 1981) measured total heat output and rates of oxygen uptake by *Mytilus* exposed to air and following reimmersion. These mussels demonstrated a typical "oxygen debt" on reimmersion (Widdows et al., 1979b) and over this period there was a good agreement between the results of direct and indirect calorimetry (Fig. 13). In a related study Shick et al. (1983) measured simultaneously the biochemical end products of anaerobiosis (and calculated the associated heat equivalents) and total heat output during prolonged anoxia. For periods up to 12 h of anoxia there was a reasonable agreement between the two methods (biochemical and direct)—at least, the 35% difference between means (calorimetry determinations exceeding the biochemical calculations) was not statistically significant. During 48 h of anoxia, however, this difference increased to 63%, and was significant. Processes that are so far little understood (e.g., shell dissolution, tissue autolysis) may be the explanation for these high

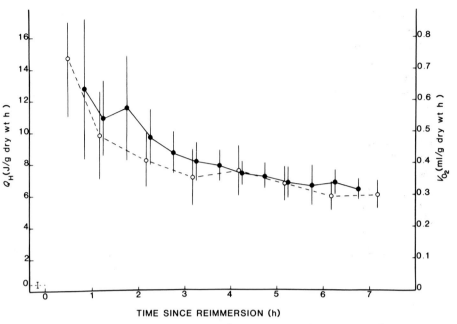

Fig. 13. The rate of oxygen consumption ($\dot{V}_{O_2}$, ●——●) and total heat production (Q_H, ○--○) by *Mytilus edulis* during reimmersion in water after 24 h exposure to air (12°C). Values are means ± SD. (From Shick and Widdows after De Zwaan et al., 1981.)

levels of heat production. The use of direct calorimetry in future studies promises to elucidate many such fundamental aspects of metabolic adaptation.

D. Energy Loss Due to Excretion

The losses due to excretion (U; see Fig. 1) are often ignored in energy balance studies, or estimated by difference. Where measurements have been made, however, there is evidence that U may comprise a significant component of total energy loss. This is particularly the case where the animal produces mucus for use in locomotion, feeding, or in the production of feces; in *A. fluviatilis,* mucus production amounted to ≈16% of the energy intake (Calow, 1974, 1977). Carefoot (1967b) found that in three opisthobranches 15% of absorbed energy was lost as excretion, and Paine (1971) reported a 10% loss in *Tegula funebralis;* in both cases much of this loss may have been due to mucus production. In *H. ventrosa* (Kofoed, 1975b) 9% of absorbed ration was lost as mucus and as much as 30% was lost as total excreta when measured as assimilated ^{14}C subsequently excreted. Branch (1982; see Fig. 23) estimated energy loss as mucus amounting to ≈40% of total production in a limpet.

The production of mucus does not represent a total loss to the animal, however, because it is produced in the first place to make locomotion or feeding more efficient. In bivalves, much of the mucus produced in feeding may be reingested, but the production of pseudofeces involves considerable production of mucus, which must be considered a net loss to the individual. Measuring mucus production poses many difficulties but the energy losses entailed should not be ignored in careful budgeting studies.

The other major component of excretory loss, and one estimated more readily than mucus, is nitrogenous excreta. The composition of this component varies between species and as a result of environmental conditions, but in most marine molluscs ammonia is assumed to be the dominant end product of protein catabolism. Among a variety of bivalves (reviewed by Bayne, 1976a), ammonia comprised between 60 and 90% of total measured nitrogen excretion; in certain circumstances, however, primary amines may comprise a significant component (Bayne and Scullard, 1977a), and in *T. lapillus* at 15°C and high salinity, primary amine loss was 23% of the measured ammonia excretion (Stickle and Bayne, 1982). Most attention has nevertheless focused on ammonia excretion.

In bivalve molluscs the relationship between ammonia excretion rates and body size can be very variable, due to a disproportionate reliance on protein catabolism for energy production by smaller individuals. Bayne and Scullard (1977a) found the exponent in the allometric equation with body size (for *M. edulis*) to vary between 0.482 and 1.480. During the winter and spring, small individuals had relatively depressed rates of excretion of ammonia, due presumably to a primary reliance on carbohydrates for energy metabolism; larger individuals were more reliant on protein catabolism at this time of year with higher ammonia excretion rates as a result. In gastropods the exponent value for ammonia excretion may be less variable. Maće and Ansell (1982) recorded exponents of 0.711 and 0.872 for *P. alderi,* and Stickle and Bayne (1982) found a common exponent of 0.610 ± 0.05 for *T. lapillus.*

Changes in the rates of nitrogen excretion are best understood, in the context of physiological energetics and nitrogen balance, when related to overall metabolic rate by means of the oxygen:nitrogen (or O:N) ratio. This ratio, when calculated by atomic equivalents, may be used to indicate the proportion of protein catabolized relative to carbohydrate and lipid; see Conover and Corner (1968). A low value of O:N (≈ 10) signifies considerable protein catabolism. In both *Polinices* (Maće and Ansell, 1982) and *T. lapillus* (Stickle and Bayne, 1982), the O:N ratio does not alter with size; that is, the exponents for rates of oxygen consumption and ammonia excretion against body weight are similar. With *Mytilus,* however, the O:N ratio varies considerably with size in a complex interaction with season, temperature, and ration (Bayne and Scullard, 1977a).

Langton et al. (1977) observed a reciprocal relationship between the rate of ammonia excretion and the weight of protein–nitrogen consumed by the bivalve

Tapes japonica; for small clams the excretion rate halved for an increase of 3.5-fold in ingested protein–nitrogen. An increase in ammonia excretion during starvation is commonly observed, as the animals are forced to catabolize protein for maintenance metabolism. In *M. edulis* in the winter, this increase was more evident in small than in large individuals. In contrast, excretion rates in the summer declined during starvation (Bayne and Scullard, 1977a). Seasonal differences are related to changes in the gametogenic condition of the animals and in the levels of endogenous energy reserves available for utilization during starvation. Crisp et al. (1981) also observed a decline in the rate of ammonia excretion by *Nassarius* during starvation, but the O:N ratio was not measured. In *T. lapillus* starvation resulted in a decline in O:N as proteins were utilized. Both *Nassarius* and *Buccinium* increased ammonia excretion 16–48 h after a meal, confirming similar observations on *Mytilus* (Bayne and Scullard, 1977b).

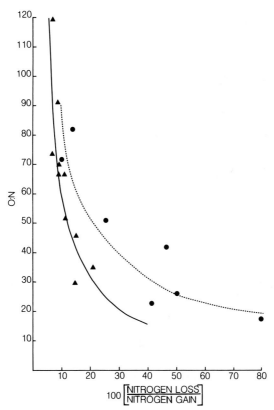

Fig. 14. The oxygen:nitrogen ratio (by atomic equivalents) as a function of proportional nitrogen balance in *Mytilus edulis* feeding on natural particulates. ▲, New Foundland (from R. J. Thompson, unpublished data). ●, Southwestern England (from Bayne and Widdows, 1978).

Bayne and Widdows (1978) and R. J. Thompson (unpublished data) related the O:N ratio to the proportion of nitrogen gained from the diet that was subsequently lost as ammonia (Fig. 14) in mussels feeding on natural particulates. In both cases the result is a negative exponential. During the main periods of tissue growth, nitrogen loss is a small proportion of nitrogen gained from the diet and O:N ratios >50 are to be expected. During periods of minimal, or even negative growth the O:N ratios are reduced, and approximate to theoretically minimal values (Mayzaud, 1973).

In spite of high and varied rates of ammonia excretion by both bivalves and gastropods, the contribution of this to the total energy losses in the energy budget may normally be rather small but nevertheless significant: 1–10% for *M. edulis*, 9–16% for *T. lapillus*, and <10% for *P. alderi*. But knowledge of these losses may be important in another way in the accurate construction of energy budgets. As Kersting (1972) and Beukema and de Bruin (1979) point out, proteins in the diet are not completely oxidized, not only as regards nitrogen, but also in terms of the carbon skeletons of amino acids. Physiologically realistic calorific values for dietary proteins should therefore be less than the commonly employed theoretical values (Gnaiger, 1983). In order to assess the magnitude of this potential error in the energy budget, the nature and the amounts of the nitrogenous end products must be known. However, as Ansell (1982a) argues, when dietary protein is used for growth rather than as an energy substrate, different constraints apply, about which much less is known at present.

V. Growth and Reproduction

Growth results when energy acquisition is in excess of energy expenditure. If, on the other hand, energy intake is less than expenditure, "negative growth" occurs and endogenous reserves of energy must be utilized to maintain the body in a viable condition (maintenance metabolism). Viewed in terms of the balanced energy equation,

$$A = (R + U) + \Delta P/\Delta t \qquad (11)$$

where $\Delta P/\Delta t$ is production (P) per unit time, A is the absorbed ration, and U is excreta. During complete starvation, the metabolic losses ($R + U$) must equal the (negative) rate of growth ($-\Delta P/\Delta t$). When the absorbed ration (A) exactly balances the metabolic rate (R) plus excreta (U), growth will be zero and the value for A then represents the "maintenance ration." At higher values of A, growth, as either somatic (P_g) or reproductive (P_r) production, or both, becomes possible.

The techniques of physiological energetics allow the analysis of these processes in relation to environmental and endogenous changes. We have earlier suggested some of the ways in which molluscs may regulate individual components of energy balance in order to maximize (or optimize) net energy gain. In this section we discuss some aspects of growth from the point of view of integrated studies of the energy budget. We do not attempt to review growth studies in general but rather to suggest some insights into the growth process that are possible through an understanding of physiological energetics.

Certain aspects of growth can best be viewed in the relationship between growth rate (as percentage of body weight per day) and the ingested (or absorbed) ration, also calculated as percentage of body weight per day. Figure 15 is based on Brett (1979); Fig. 15A illustrates the curve of growth rate rising from negative values to the maintenance ration (at zero growth), the optimum ration (coinciding with the tangent to the growth curve), and the values for maximum ration at the maximum rate of growth. Values for growth efficiency may be derived from this relationship either as the gross efficiency, P/C or as the net efficiency, P/A. Efficiency (Fig. 15B) rises through zero at the maintenance ration, to take maximal values at the optimal ration, and then to decline with further increase in ration. The form of the efficiency:ration curve (the K line; Paloheimo and and Dickie, 1966) beyond the optimal ration may vary, from no change with further increase in ration (e.g., Conover and Lalli, 1974) to a logarthmic decline. This is discussed in more detail later, and we consider the important distinction between growth efficiency considered in terms of total production ($P_g + P_r$) and in terms of P_g and P_r separately.

A. Negative Growth and Maintenance Ration

In the absence of food the metabolic rate indicates the rate of weight loss as energy reserves are utilized to maintain viable cell condition. For example, Conover and Lalli (1974) measured the respiration rate of C. limacina, and its weight losses during starvation, at different temperatures. If weight losses are designated as T (milligrams of dry tissue lost per day) and the respiration rates (R, in microliters of O_2 per day) are converted to units of milligrams dry weight (1 μl O_2 = 1.19 μg dry weight; Conover and Lalli, 1974), the following equations apply:

$$3°C: \quad T = 0.014 \ W^{1.128}$$
$$R = 0.018 \ W^{0.931}$$
$$12°C: \quad T = 0.020 \ W^{1.066}$$
$$R = 0.038 \ W^{0.931}$$

The agreement between both exponent and intercept values is fair, given that the respiration measurements were made on freshly caught material and probably overestimate true "standard" metabolism.

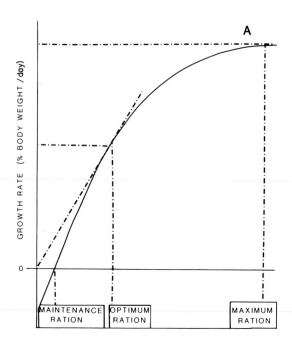

A

GROWTH RATE (% BODY WEIGHT /day)

0

| MAINTENANCE RATION | OPTIMUM RATION | MAXIMUM RATION |

RATION (% BODY WEIGHT /day)

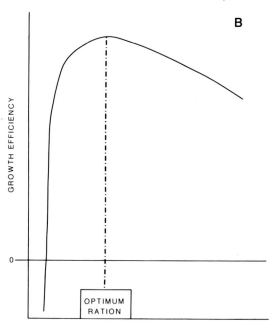

B

GROWTH EFFICIENCY

0

| OPTIMUM RATION |

RATION (% BODY WEIGHT /day)

In natural situations, however, animals will seldom be faced with the complete and prolonged absence of food. Rather, situations of food shortage may arise, leading to partial starvation $[A < (R + U)]$. Such conditions occur commonly for bivalves during the winter months in temperate environments. Figure 16 shows observed and predicted weight losses (the latter derived from oxygen consumption measurements), calculated as percentages of body weight per day, for *C. edule* (Newell and Bayne, 1980) and for *M. edulis* (Bayne and Widdows, 1978; Bayne and Worrall, 1980), from southwestern England, from November to January. Individuals of both species lost weight, depending on body size, from ≈1.0 to ≈0.5% of body weight per day; the predicted weight losses were much greater. The differences between predicted and observed values indicate the

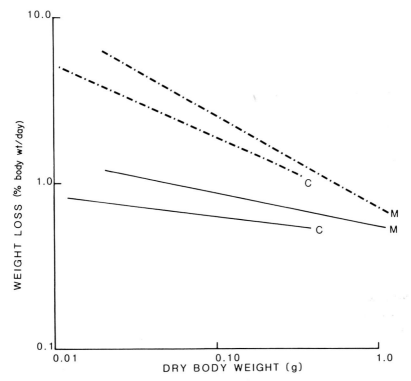

Fig. 16. Predicted (dashed lines) and observed (solid lines) weight loss by *Cerastoderma edule* (C) and *Mytilus edulis* (M) as a function of body weight during winter months in southwestern England. Predicted values are from measured rates of oxygen consumption.

Fig. 15. Growth rate (**A**) and growth efficiency (**B**) as a function of ration, to illustrate the concepts of maintenance, optimum, and maximum rations. (Redrawn from Brett, 1979.)

rations required to compensate for the metabolic energy losses. Smaller individuals were better able to effect this compensation than larger individuals due, at least in part, to the expenditure of less energy per unit volume of water cleared of food. Vahl (1981a) has recorded a similar phenomenon in the different age classes of the Icelandic scallop C. *islandica*, where, because of seasonal differences in the amount of food available, the larger individuals were forced into negative growth for a greater part of the year than smaller animals.

A common response during periods of reduced food availability is a reduction in the rate of oxygen consumption (Bayne, 1973b; Calow, 1977). This may follow from a depression in the level of activity, a depletion of certain endogenous stores of energy, or a simple loss of material from the gut. Whatever the cause (and it may be equally due to an active regulation of metabolic rate), the result is a relative conservation of energy (Section IV,B). Bayne and Scullard (1978a) recorded an exponential decline in the respiration rate of *T. lapillus* during starvation, with a halving of the rate over 18 days. However, in those species, such as *Thais*, that must seek out their food, a decline in the standard (or resting) rate of metabolism may be offset against higher activity as the search for food is increased. Calow (1974) starved the pulmonate *Planorbis contortus* over 4 days either constrained in cages or allowed to move freely. Constrained snails showed a marked decline in respiration rate, whereas mobile snails had higher rates, at least over 72 h. This amounts to an increase in the proportion of total energy expenditure available for activity (i.e., an increased "scope for activity") as the demands for search activity increase.

During periods of starvation, metabolic energy demands must be met from endogenous reserves. Where food shortages occur with some regularity, such as for suspension feeders over the temperate winter, adaptive responses are to be expected. A common response is to lay down reserves (carbohydrate, protein, or lipid) during periods when food is abundant for later utilization when food resources are low. In some species these reserves serve only the needs of maintenance metabolism; in others they are used also to fuel the demands of gametogenesis (see Volume 2, Chapter 5).

Ansell and Sivadas (1973) studied the clam *D. vittatus* in Scotland. Individuals lost weight in the winter due to the depletion of energy reserves. Of the total energy demand, estimated from measurements of oxygen consumption, ≈50% was met by the loss of the sum of carbohydrate (15%), lipid (14%), and protein (22%), the remainder being due to the absorbed ration. *C. edule*, like *Donax*, has a quiescent gametogenic period in the winter (Newell and Bayne, 1980). Metabolic and feeding activity are suppressed for 3–4 months. In spite of this, body weight declines by between 0.8 and 0.5%/day (Fig. 16). In smaller individuals ≈15% of this weight loss is due to loss of carbohydrate; in larger individuals up to 100% of the weight loss is accountable as carbohydrate. Measurement of the energy content of the tissues suggested that the noncarbohydrate substrates lost in smaller individuals were largely protein.

The role of protein as an energy substrate for molluscs during starvation requires further research, particularly into the source of the protein and the mechanism of its catabolism. In *C. islandica,* immature individuals lose about 56% of their body protein between August and February, whereas mature individuals lose only 16%. Sundet and Vahl (1981) suggest that the immature animals invest in rapid growth but at the risk of not possessing a large carbohydrate store for energy metabolism during food shortage in the winter. Mature individuals, in contrast, store more glycogen in order to maintain a high output of gametes, and this is done at the expense of more rapid growth.

In all these species there are intricate relationships among energy storage, maintenance metabolism, and gametogenesis (Gabbott, 1975). In *M. edulis,* for example, the energy demands of starvation in the winter are met largely by the catabolism of protein, at least in larger individuals (e.g., 75% due to protein, 10% to carbohydrate, and 15% to lipid; Gabbott and Bayne, 1973). Special connective tissue cells provide a means for sequestering glycogen and protein for use in gametogenesis (Bayne et al., 1982). In the summer, however, during gametogenic quiescence, starvation is countered by the utilization of carbohydrate reserves alone. In considering the physiological aspects of growth in these molluscs, seasonal factors are paramount, and the different demands of somatic and gametogenic production, set against environmental changes in food availability, must be understood.

When the absorbed ration exactly balances the sum of the metabolic demand and the energy losses due to excretion, growth rate and growth efficiency are zero and the ration level is called the maintenance ration (Fig. 15). The maintenance ration is a function of body size and is affected in different ways by various other endogenous and exogenous factors. Figure 17 shows the maintenance ration for *M. edulis,* taken from Winter and Langton (1976), Widdows (1978b), and some unpublished data. The relationship with body size is described by the allometric equation:

$$C_{\text{maint}} = 1.0 \ W^{-0.325}$$

This equation agrees well with the equation describing the routine respiration rate of *M. edulis* (Bayne, 1976a), which, converted to units of percentage of body weight per day, may be written:

$$R_{\text{routine}} = 1.1 \ W^{-0.305}$$

A higher maintenance ration in smaller individuals reflects their higher metabolic rate. It is pertinent to establish, however, whether smaller individuals are more efficient at utilizing the available ration for maintenance than are larger

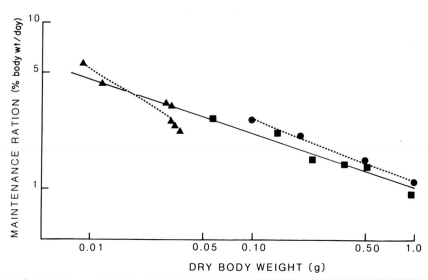

Fig. 17.   The maintenance ration of *Mytilus edulis* related to dry flesh weight. The solid line is drawn according to the equation: Maintenance ration = 1.0 $W^{-0.325}$ (see text). (▲, Winter and Langton, 1976; ●, Widdows, 1978b; ■, B. L. Bayne, unpublished data.)

individuals (Kleiber, 1961). Given two ration levels, both below maintenance, a differential energy loss may be expected, reflecting the efficiency with which the ration difference is utilized for maintenance metabolism. Data from Thompson and Bayne (1974) suggest a maintenance efficiency that declines in mussels from ≈60% in small individuals to ≈30% at a body size of 580 mg dry flesh weight, and <15% in the largest individuals. These values are maximum estimates, however, because they do not consider energy losses other than those due to respiration. Data for the same species published by Winter and Langton (1976) suggest a maintenance efficiency of 20–40% for mussels of 35 mg dry weight. If smaller individuals are indeed more efficient at converting absorbed energy for maintenance, this might help to explain the size-related differences between observed and predicted weight losses during partial starvation discussed earlier (Fig. 16).

   The maintenance ration increases with increase in temperature. For mussels feeding on pure cultures of algae in the laboratory, Widdows (1978b) recorded an increase (for individuals of 1 g dry flesh weight) from 0.16 mg/liter at 5°C to 1.40 mg/liter at 25°C. The temperature coefficient (Q_{10}) was 1.6 between 5 and 20°C, but increased to >6.0 for 15–25°C, consistent with metabolic regulation over much of the temperature range and considerable stress at 25°C. Temperature is here acting as a controlling factor (Brett, 1979; Fry, 1971). Similarly, endogenous factors such as gametogenesis may also affect the maintenance ration. A rise in metabolic rate resulting from the increased energy demands of vitello-genesis, for example, may increase the maintenance ration and lower the max-imum growth efficiency (Thompson and Bayne, 1974). Nothing is known of any size-related differences in the efficiency with which stored energy reserves are utilized for gametogenesis.

One potential "cost" of reproduction (Section V,C) may therefore be a rise in the maintenance ration and a reduction in the energy available for growth. Related to this, there may be a reduction in the maximum size to which the individual may grow, as a result of a greater proportion of available energy, in excess of maintenance, being directed toward the production of gametes. Ansell (1982a) has assessed this relationship in *P. alderi* feeding on the bivalve *T. tenuis*, by identifying the ration levels that are just sufficient to support the requirements of maintenance and somatic growth, and those sufficient to support maintenance, somatic growth, and certain levels of gamete production. By vary-ing these relationships in circumstances of potential food limitation, the animal may direct more energy to reproduction at the cost of reducing its maximum size. Ansell (1982a) interprets these relationships in *Polinices* as an opportunistic adaptation toward maximizing reproduction when the food resource diminishes.

B. Scope for Growth and Growth Efficiency

Warren and Davis (1967), drawing on Fry's concept (1947) of the "scope for activity" as the difference between active and standard metabolic rates, defined the "scope for growth" as "the difference between the energy of the food an animal consumes and all other energy utilisations and losses." Brett (1976, 1979) demonstrated that the difference between maximum and maintenance ra-tions for sockeye salmon gave a reliable estimate of growth at different tempera-tures, and he termed the ration difference the "growth scope." The scope for growth, used in the sense of Warren and Davis (1967), has been widely applied to the analysis of the components of growth and responses to environmental change in molluscs (Bayne, 1976a; Newell and Branch, 1980). When routine respiration rate is a reliable predictor of the maintenance ration, the equivalence between the "scope for growth" and "growth scope" is apparent. The impor-tant point, as discussed by Brett (1979), is that the environment acts, not directly on growth, but on the mechanisms of energy supply and demand that influence the scope for growth.

The extend to which the scope for growth, when estimated by a detailed analysis of the energy budget, provides a reasonable predictor of growth itself, was evaluated by Dame (1972) for the American oyster (*C. virginica*), and by Bayne et al. (1979) and Bayne and Worrall (1980) for *M. edulis*. Figure 18A shows monthly estimates of assimilation, respiration, and growth for oysters; Fig. 18B shows weight changes during 2 years for mussels, estimated as the

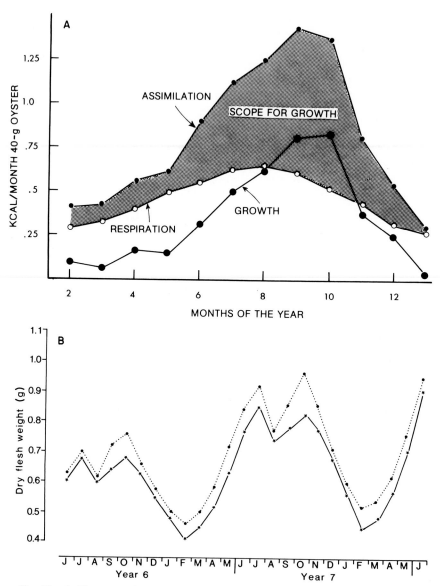

Fig. 18. **A.** The scope for growth (stippled area) and measured growth of *Crassostrea virginica.* (From Dame, 1972.) **B.** Growth predicted from physiological measurement (dotted line) and as measured (solid line) in *Mytilus edulis* over 2 years. (From Bayne and Worrall, 1980.)

scope for growth and from length/weight regressions. In both cases the agreement between the physiological estimations and the observed growth is reasonable. These results also illustrate the highly seasonal nature of growth in both species. This seasonality is the result of complex interactions among temperature, food, reproductive activity, and energy balance.

1. The Scope for Growth and Temperature

Buxton et al. (1981) examined the scope for growth in juvenile oysters (*O. edulis*) as a function of acclimation and exposure temperatures. The results demonstrated a maximum scope for growth between 15 and 20°C, although brief exposure to higher temperatures (25°C) further enhanced the growth potential. Adaptive adjustments of respiration and feeding rates resulted in the maintenance of high values for the scope for growth over a wide temperature range; the authors concluded that gradual warm acclimation during the summer months, together with brief exposures to higher temperatures (such as might occur in shallow water during low tide) represent optimal conditions for production in this species.

The regulation of the scope for growth over a large proportion of a species' normal temperature range is a common feature in shallow coastal or intertidal species, and has been described in detail for *M. edulis* (Bayne et al., 1973; Widdows and Bayne, 1971). Widdows (1978b) measured the scope for growth and its derivative, gross growth efficiency, at different temperatures and ration levels. The scope for growth was virtually independent of temperature over the range 5–20°C, but declined sharply at 25°C, indicative of thermal stress. Growth efficiency declined from 0.35 at 5°C to 0.20 at 20°C. R. Thompson (unpublished data), working on a population of mussels in Newfoundland, recorded net growth efficiencies of between 0.50 and 0.75, with no significant correlation with temperature (0.5–15°C). In an earlier study, Widdows (1976) compared mussels from two populations in experiments with fluctuating temperatures. Individuals that experienced elevated temperatures in their natural habitat were better able to survive (Bayne et al., 1977a) and to grow at high temperatures (25°C) than individuals from a population experiencing lower temperatures in the field. Exposure to high temperatures (>20°C) in a fluctuating regime enhanced the scope for growth in both sets of animals; these results were due in large part to the maintenance of high filtration rates, without concomitant increases in respiration, as a result of adaptation to the temperature cycles.

These responses (and see also reviews by Newell, 1979; Newell and Branch, 1980) may have the effect, in the natural environment, of maintaining the scope for growth and growth efficiency much less dependent on temperature than on food availability. Elvin and Gonor (1979) concluded that food level explained 96% of the seasonal variance in scope for growth in *M. californianus*, and average tissue temperature explained only 3%. Conover and Lalli (1974) examined the effects of temperature and ration on the gross growth efficiency (K_1) of *C. limacina*. There was a slight positive effect of temperature on log K_1 that, although not statistically significant, served to counter a negative correlation between log K_1 and ration. Increasing temperatures, at a time of increasing size of prey, would result in a maintained high growth efficiency over a wide range of prey availability. Ansell (1982a) concluded, from laboratory experiments with *Polinices,* that whereas different temperatures within the normal ecological range for the species affected the rates of various processes of energy acquisition and expenditure, and also tended to set upper limits to growth, growth efficiency was relatively insensitive to temperature. For two populations of *M. edulis* in south-western England, multiple regression analysis (B. L. Bayne, unpublished data) indicates that ration alone explains as much of the seasonal variance in growth efficiency as a model in which temperature changes were also included.

   The range of temperatures over which the scope for growth and growth efficiency is held relatively constant will, nevertheless, differ considerably between species and between different populations of the same species. Some species may be adversely affected by both high and low temperatures in the natural habitat. For example, *M. balthica* in the Wadden Sea may enter a period of negative growth in the summer, after spawning, when rising temperatures inflate the metabolic energy demand beyond the capacity of the feeding processes to compensate (De Wilde, 1975; Lammens, 1967). However, in comparing growth increments among 15 intertidal stations over 8 years, Beukema et al. (1977) found significant correlations with food availability and with the time available for feeding, but not with mean temperature. *Chlamys varia* in the Bay of Brest (north of France) also experiences periods of negative growth efficiency in the summer, due to high temperature. However, food is abundant at this time and the effect on growth efficiency is slight compared with a period in the winter, when low temperatures, combined with low ration, cause marked depression of the scope for growth and growth efficiency (Shafee, 1980). In contrast, Kirby-Smith and Barber (1974) concluded that natural variations in the levels of phyto-plankton in Beaufort Channel (North Carolina, United States) were not likely to influence growth of the scallop *A. irradians,* which was instead regulated by temperature. The overriding conclusion to emerge from studies of this kind is, nevertheless, that it is the combined effects of ration and temperature that have the most profound influences on physiological energetics in the natural habitat.

2. The Scope for Growth and Salinity

   There has not been much work reported on the effects of salinity on the physiological energetics of molluscs. J. Widdows (unpublished data) acclimated *M. edulis* to 15 and 30‰ and also exposed the mussels to a 12-h cycle of fluctuating salinities between 15 and 30‰ (see Livingstone et al., 1979). At 15‰ (constant salinity), both food consumption and respiration were slightly reduced; absorption efficiency was significantly depressed, and the scope for growth, though positive, was lower than at 30‰. In the fluctuating-salinity cycle, all components of the energy budget were held constant between 20 and 30‰, but depressed between 15 and 20‰; the scope for growth was therefore maintained over a wide salinity range, but depressed at 15‰. In experiments at low and fluctuating salinity, excretory losses due to amines must be considered a potentially significant energy drain.

Salinity is best thought of as a limiting factor (Fry, 1947) for marine molluscs acting, when reduced, to restrict the potential for energy acquisition and so setting limits on overall energy balance. This is well illustrated in experiments by Stickle and Bayne (1982) in *T. lapillus*. At 15°C, feeding and absorption were depressed at salinities below 25‰ although metabolic losses were not considerably affected. The result was negative values for the scope for growth below 20‰, increasing to maximum (positive) values at the highest salinity tested (35‰). These results were consistent with the known salinity limits on the distribution of the species (Boyden et al., 1977).

3. The Scope for Growth and Ration

Figure 19 illustrates the scope for growth in two species of mussels, *M. edulis* and *Aulacomya ater,* respectively, as a function of algal cell concentration

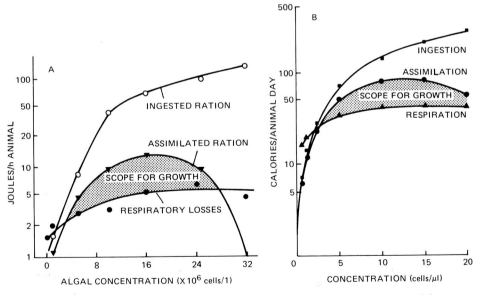

Fig. 19. The scope for growth (stippled areas) in (**A**) *Aulacomya ater* and (**B**) *Mytilus edulis,* related to ration of unialgal cultures in laboratory experiments. (**A** from Griffiths and King, 1979b; **B** from Thompson and Bayne, 1974.)

(*Tetraselmis suecica* and *D. primolecta*), determined in laboratory experiments. In both cases the scope for growth increased with increased food concentration, from negative values at <2 cells/μl to maximum values at between 10 and 15 cells/μl, declining again at higher cell concentrations as the absorption efficiency was depressed. In experiments such as these, with pure cultures of algae, the growth efficiency increases rapidly at low ration levels and declines at higher concentration. Results from four studies with *M. edulis* (Fig. 20) show the general form of the relationships commonly observed and illustrate the decline in maximum growth efficiency (K_{max}) and increase in optimal ration, with increase in body size.

In the natural environment, however, suspension feeders seldom experience a pure phytoplankton diet; the ration more often comprises a mixture of algal cells, detritus, and inorganic silt. Winter (1976), Murken (1976), Kiørboe et al. (1980, 1981), and Møhlenberg and Kiørboe (1981) have examined the effects of silt on the growth and energetics of *M. edulis* and of *Spisula subtruncata*. Winter (1976)

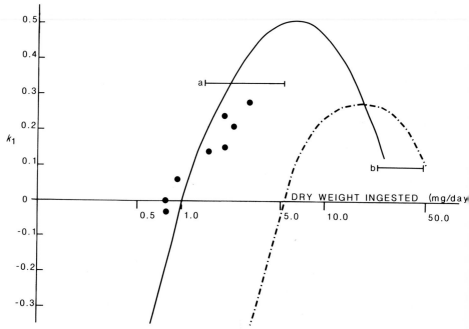

Fig. 20. Gross growth efficiency of *Mytilus edulis* as a function of ingested ration. ●, Individuals of 20–35 mg dry flesh weight. (From Winter and Langton, 1976.) a, Individuals of 20–27 mg dry flesh weight. (From Kiørboe et al., 1981.) b, Individuals of 260 mg dry flesh weight, recalculated from original data, which were in units of carbon. (From Tenore et al., 1973.) The solid and dashed lines are drawn from Thompson and Bayne (1974) for individuals of 100 and 500 mg dry weight, respectively.

observed increased feeding activity, a higher ingested ration, and increased growth in *Mytilus* exposed to optimal concentrations of algal cells with 12.5 mg (dry weight) of silt added per liter of seawater. Kiørboe et al. (1981) confirmed a stimulation of growth by *Mytilus* in the presence of silt and suggested that the silt induced more rapid feeding (see also Thompson and Bayne, 1972), with the further possibility that the mussels utilize organic matter present (presumably as surface aggregates) on the silt. Similarly, in *S. subtruncata* growth was increased by 10–110% by the addition of silt to a ration of algal cells (Mohlenberg and Kiørboe, 1981). In this case, clearance and respiration rates were independent of suspended silt but there was a suggestion of increased absorption efficiency in the presence of silt in the stomach.

Even the addition of inert particles to algal cultures does not fully mimic the complex suspensions that comprise the natural ration of bivalves. A fundamental problem, still largely unresolved, concerns the true nature and energy value of the suspended particulates in coastal and estuarine environments. Methods for estimating the food value of natural particulates usually include a measure of "organic matter" as weight lost on igniting the filtered particles, followed by the application of a calorific conversion value (e.g., Vahl, 1980). Widdows et al. (1979a) analyzed particulate nitrogen, carbohydrate, and lipid, and derived a seasonally variable calorific value. Bayne and Widdows (1978) and Bayne et al. (1979) measured the nitrogen content (by Kjeldahl digestion) and derived a regression relating particulate organic nitrogen to the calorific value of the particulate organic matter (POM).

The relationships between the scope for growth, growth efficiency, and level of natural ration available vary, depending on the absolute concentration of particles (e.g., whether above or below the pseudofeces threshold; see Section III,A), on the ratio of POM to particulate inorganic matter (PIM), and on the true food value of the organic matter. When concentrations are below the pseudofeces threshold, the presence of PIM reduces the quality of the diet by "dilution" of the POM (Haven and Morates-Alamo, 1966; Widdows et al., 1979a). For example, Vahl (1980) found that seasonal changes in the relative concentrations of POM and PIM explained the seasonal pattern of growth of *C. islandica* in Norway. When particle concentrations are high enough to lead to pseudofeces production, ingestion rate may be dependent on the pseudofeces threshold concentration (which varies among species and with animal size), on the clearance rate, and on the seasonally varying energy value of the seston (Bayne and Worrall, 1980). In all cases, particular care is necessary in arriving at a reliable estimate of the true absorption efficiency, which may be markedly dependent on the proportion of POM in the total particulate suspension.

Figure 21 illustrates the gross (K_1) and net (K_2) growth efficiency of *M. edulis* (1 g dry body weight) as a function of absorbed ration, derived from experiments with natural particulates. Unlike the results from experiments with pure algal

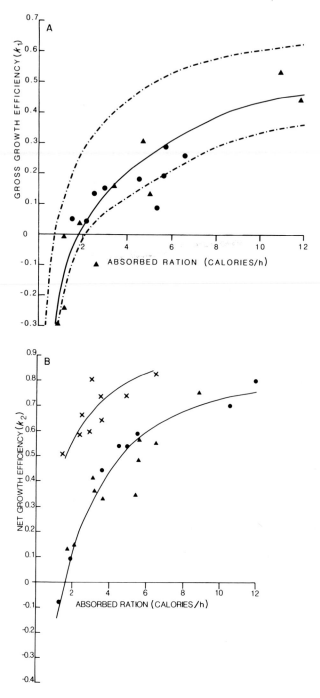

cultures, these K lines do not show a region of negative slope at high ration, but rather increase monotonically toward asymptotic values of ≈ 0.40 (K_1) and ≈ 0.80 (K_2). In Fig. 21A are also plotted curves for individuals of different weights showing a reduction in gross growth efficiency with increase in size (see also p. 490).

It appears that suspension-feeding bivalves in the natural habitat may normally function at food concentrations for which growth efficiency is a rapidly increasing function of food availability. At the lower end of the normal range of ration levels, growth efficiency may decline to negative values and the animals then lose weight, as observed for *M. arenaria* (Gillfillan et al., 1977), *M. edulis* (Bayne and Widdows, 1978), *C. edule* (Newell, 1977), *C. islandica* (Vahl, 1980), and *C. varia* (Shafee, 1980). Incze et al. (1980) recorded increased mortality in natural populations of *M. edulis* when conditions of reduced food availability coincided with increased temperatures, resulting in a negative scope for growth and rapid depletion of endogenous energy reserves. It is probably a commonplace for many bivalve species to be limited in their distribution by seasonally varying conditions that lead to periods of negative growth efficiency.

There have been few studies of growth efficiency measured in terms of nitrogen. Langton et al. (1977), however, did this for the bivalve *T. japonica* feeding on diatoms, and recorded nitrogen growth efficiencies of 0.37–0.48, with the highest values recorded at the lowest ration tested. Their results suggested higher efficiencies for nitrogen than for carbon (or energy values) at equivalent levels of ration (Winter, 1978). Bayne and Widdows (1978) measured nitrogen losses as a proportion of total nitrogen absorbed in a field population of *M. edulis;* values varied from 0.10 to 0.82 (mean $\pm$ SEM $= 0.38 \pm 0.09$), but were generally lower than the ratio for calories lost over calories gained (0.51 ± 0.09). For a mussel population in Newfoundland, R. J. Thompson (unpublished) recorded an equivalent nitrogen ratio of 0.11 ± 0.02. An interesting and possibly extreme case of nitrogen conservation concerns the shipworm, *L. pedicellatus,* in which no ammonia excretion was detectable (Gallager et al., 1981). These bivalves may be capable of recycling the end products of protein catabolism via symbiotic relationships with bacteria, in order to augment the nitrogen-poor diet of wood (see also Mann, 1982).

The problem of natural food quality as affecting energy balance is particularly relevant to carnivorous feeding, where the energy costs of search and prey

---

Fig. 21. **A.** Gross growth efficiency of *Mytilus edulis* related to absorbed ration. Data points refer to animals of 1 g dry flesh weight; dashed lines refer to individuals of 0.1 g (top) and 2 g (bottom) dry weight. (After Bayne and Widdows, 1978.) **B.** Net growth efficiency of *M. edulis* of 1 g dry weight, related to absorbed ration. ×, a population in Newfoundland. (From R. J. Thompson, unpublished data.) ▲ and ●, Two populations in southwestern England. (From Bayne and Widdows, 1978.)

handling may be high relative to the potential energy gain from the meal (Section III,B). It is the balance between these gains and losses (or, possibly, other dietary components) that is presumably maximized and then set against the risks involved in feeding. In experiments with *P. duplicatus,* Edwards and Huebner (1977) demonstrated a lack of growth when no molluscan prey were available; even with an abundant supply of the preferred sizes of some molluscan prey (*Nassarius, Mytilus*), growth was much reduced compared with the preferred prey species, *M. arenaria.* The energy value of all these potential prey tissues is similar and the growth efficiency is presumably governed by the energy expended in prey handling (i.e., the energy gain per unit prey handling time) (p. 437). When the time taken to recognize a prey item (Hughes, 1980) is a large component of the handling time, the predator may be constrained to feed in accordance with the availability, rather than the potential net energy gain, of the prey (Bayne, 1981), with consequent reduction in growth efficiency. As Edwards and Huebner (1977) point out, a predator in its natural habitat must often exercise its choice of prey in the context of suboptimal prey availability.

4. The Scope for Growth and Tidal Height

Distribution on the shore imposes special problems of food supply on both gastropods and bivalves. Grazers and carnivorous feeders, although in the presence of abundant food, may have to balance the risks of dessication at low tide, or of being washed away at high tide, against the requirements of feeding. Menge's study (1974) of the feeding behavior of the prosobranch *A. punctulata* illustrates the complex nature of the interactions among risk, prey selection, and net energy gain. In a recent review, Branch (1982) considers the impact of dessication, temperature, and other factors on the physiology and energetics of limpets on the shore. Branch (1982) discusses (and see also Section IV,B), the possibility that some species may be adapted (as "exploiters") to an abundant food supply by means of inefficient use, but high turnover, of the available energy, whereas others (the "conservers") are adapted to conditions of less abundant food by means of high efficiencies and low rates of turnover. Both types, as well as those intermediate between the two extremes, may be found at different levels of the same shore.

For suspension feeders, aerial exposure at low tide may impose, not only periods without food, but also periods at high temperatures. Griffiths and Bufenstein (1981) and Griffiths (1981a, 1981b) examined in detail the effects of a littoral distribution on the scope for growth of *C. meridionalis.* In this species there is no compensation for periods of tidal exposure in terms of increased feeding rate during immersion, or of increased absorption efficiency; energy acquisition is reduced in strict proportion to height on the shore. On the other hand, *Choromytilus* reduces its respiration rate during aerial exposure (Section IV,B), so that there is some energy conservation, which results in a positive scope for growth up to a limit at 50% aerial exposure over 14 days. In addition, mussels in the littoral zone allocated more of the available energy to reproduction than did subtidal individuals; this reduces somatic growth and maximum size, but maintains to a high level of reproductive input by these individuals to the population as a whole. Suchanek (1981) discusses a different situation among mussels on the West Coast of the United States.

Elvin and Gonor (1979) analyzed the tissue temperatures of *M. californianus* at different levels of the shore, and the relationships between shore height and the scope for growth. Due to temperature effects on absorption efficiency, mussels exposed to increased temperatures during low tides increased food absorption in the winter; this was accompanied by a reduction in metabolic rate during exposure, although the net energy gain that resulted was offset in the summer by reduced feeding rates as the animals were reimmersed in colder water. Overall, tidal exposure reduced the scope for growth, but the effects during the winter were not as marked as might be expected, and scope for growth could be maintained for up to 8 h exposure per day (Fig. 22).

More recently, Gillmore (1982) has assessed the energetics of adaptation to littoral conditions in six bivalve species. He resolved growth curves into compo-

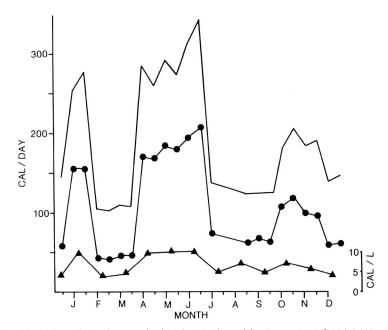

Fig. 22. A hypothetical energy budget for *Mytilus californianus* at +3 ft M.L.L.W. on the central Oregon coast, United States. ——, Absorbed ration; ●——●, the scope for growth; ▲——▲, food concentration as calories per liter. (From Elvin and Gonor, 1979.)

nents reflecting energy intake and energy loss, and found that the rates of energy loss due to intertidal exposure were lower in species that normally occur high on the shore (e.g., *Geukensia (Modiolus) demissa, C. virginica*). Also, *M. edulis* and *C. virginica* were able to supplement energy input so that growth per unit time of immersion was faster at certain intertidal levels than subtidally. This study demonstrates how some bivalves have adapted to a littoral existence by modifications involving both energy conservation and improved energy acquisition.

5. The Scope for Growth and Growth Efficiency Related to Age

Figure 23, taken from Rodhouse (1978) and Branch (1982), illustrates age-related changes in the energy budget that are typical of many studies. The scope for growth is equivalent to total production (P) and is the sum of reproductive output and growth of somatic tissue and shell. As the individual ages, a greater proportion of P is due to the production of gametes; eventually, somatic production declines to zero, the animal achieves its asymptotic size, and P_r/P (a measure of reproductive effort; Section V,C) reaches its maximum value at 1.0. It is clear that, in studies of the energetics of growth, the distinction between P_r and P_g should be made. With increasing age, somatic growth efficiency (P_g/A) declines to zero, whereas total growth efficiency ($P_g + P_r)/A$) may be held constant once the animals have reached maturity (Fig. 24).

Edwards and Huebner (1977; *P. duplicatus*), Broom (1981; *Natica maculosa*), and Ansell (1982a; *P. alderi*) have described these relationships for three carnivorous gastropods. Growth efficiency declines with increasing age when estimated on snails not producing gametes; for example, in *P. duplicata* feeding on *M. arenaria* (Edwards and Huebner, 1977), gross growth efficiency declined from 0.48 in 2-year-old snails to 0.16 2 years later. Ansell (1982a) was able to distinguish somatic from gamete production in *P. alderi* feeding on *T. tenuis;* overall growth efficiency did not decline with age, remaining at 0.38–0.42.

In considering the factors controlling maximum (asymptotic) body size in *M. edulis,* Jørgensen (1976b) considered that above a certain mean body size the rate of water transport per unit body weight may decline faster than the reduction in metabolic rate (see Section III). The negative relationship between body size and the ratio of clearance rate to respiration rate (or $V_w:V_{O_2}$) in bivalves supports this contention as one factor in reducing the growth efficiency of large individuals. However, this factor alone is not sufficient to regulate body size. Rather it is the gradual allocation of all available surplus energy to the production of gametes that ultimately sets the limit on size (Fig. 25).

Not only is growth efficiency related to age; it is also related to the rate of growth itself. Winter (1978) and Møhlenberg and Kiørboe (1981) present values for two species of bivalves that suggest a similar monotonic rise to K_{max} of ≈0.40 at high rates of growth.

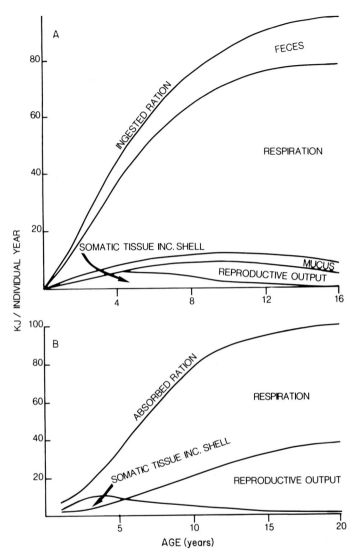

Fig. 23. Energy balance in (**A**) *Patella longicosta* (from Branch, 1982) and (**B**) *Ostrea edulis* (from Rodhouse, 1978), as a function of age.

C. Fecundity and Reproductive Effort

Life-cycle patterns and some of the evolutionary constraints that determine reproductive processes in different molluscs are considered elsewhere (Volume 6, Chapter 15). In this section, some aspects of reproductive energetics are discussed with particular reference to differences between species and between

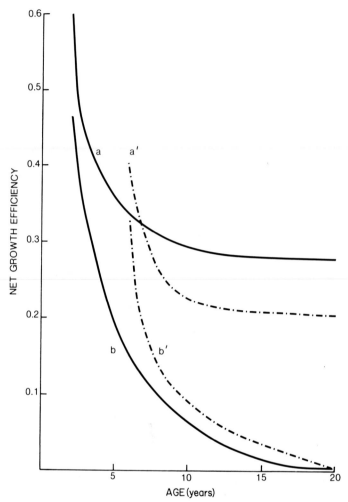

Fig. 24. Net growth efficiency in *Ostrea edulis* (solid lines; from Rodhouse, 1978) and *Chlamys islandica* (dashed lines; from Vahl, 1981a), related to age. a and a', Total efficiency, that is, $(P_g + P_r)/A$; b and b', somatic efficiency, that is, P_g/A.

populations of the same species, with regard to the allocation of available energy to reproduction and to growth. In some mollusc species, reproduction is not initiated until somatic growth ceases; in others, growth continues after the age of first maturity, but an increasing proportion of surplus energy is allocated to gametogenesis. As we have seen, overall growth efficiency may remain relatively constant under these conditions, whereas somatic growth efficiency declines (Fig. 24). In experiments with *Polinices,* in which production was not limited by food, Ansell (1982a) recorded a hyperbolic increase in net overall growth efficiency with increase in gametogenic activity.

Not only does reproductive activity tend to limit body size by directing energy away from somatic growth, but reproductive capacity (or fecundity) is itself a function of size. The form of the relationship between size and fecundity varies between species. Where egg numbers are related to shell length (or height) by a power function of 3.00 (e.g., see Hughes and Roberts, 1980), a constant fraction of the body weight is being invested in the production of eggs by females of all sizes; where the function is >3.00, parental involvement of body mass in gamete production is an increasing function of body weight. Relationships within the molluscs vary; among the *Thais* species investigated by Spight and Emlen (1976), and some bivalve species (e.g., Broussean, 1978; Rodhouse, 1978;

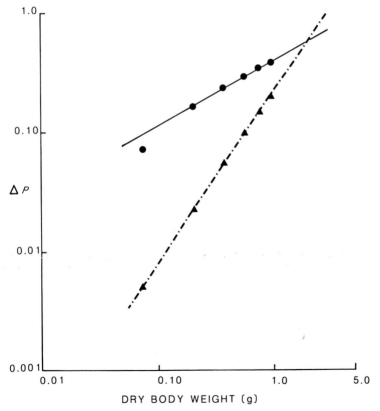

Fig. 25. Cumulative total production (●) and gametogenic production (▲) in *Mytilus edulis* as a function of body size. (Data for the "Lynher population" from Bayne and Widdows, 1978.)

Griffiths and King, 1979b) an increasing investment in P_r is made as the individuals age.

The term *reproductive effort* is used to refer to the level of energy allocation to reproduction. The term has many working definitions (Browne and Russell-Hunter, 1978; Calow, 1979; Hirshfield and Tinkle, 1975; Hughes and Roberts, 1980; Vahl, 1981c). A common approximation is to express reproductive effort as the proportion of total production (or "nonrespired assimilation"; Browne and Russell-Hunter, 1978) that is due to reproduction ($P_r/[P_r + P_g]$). Hirshfield and Tinkle (1975) recommend the proportion of absorbed energy that is devoted to reproduction. Calow (1979) argued that an index of reproductive effort should define the extent to which reproduction detracts from the nonreproductive energy demands of the parent:

$$C = 1 - \left[\frac{A - P_r}{R^*} \right] \tag{13}$$

where R^* is the metabolic energy demand of the somatic tissue. These three indices of reproductive effort are increasingly difficult to estimate with accuracy, due to the problems associated with measuring the absorbed ration A, and the "nonreproductive" metabolic rate R^*. But as Hughes and Roberts (1980) point out, because interpretations are likely to differ depending on the index used, different methods should be employed, particularly in comparative studies.

Figure 26 illustrates P_r as a proportion of total production in six bivalve species, related to age (see also Bayne, 1976b; Bayne et al., 1983). In all cases reproductive effort increases with increasing age, although the shapes of the curves differ. In making similar comparisons between prosobranch species, Hughes and Roberts (1980) could detect no general relationships between the magnitude of current (i.e., age-related) reproductive effort and either the type of reproduction or the population ecology of the different species. These authors also observed large differences within a single species (*Littorina rudis*) from different types of shores, and they concluded that environmental rather than demographic factors were responsible. A similar conclusion follows from studies with *M. edulis*.

Although more information is needed on a variety of species and ecotypes before curves of current reproductive effort can be interpreted in general terms (but see Browne and Russell-Hunter, 1978) careful analysis of reproductive effort within a population can provide important ecological insight. Such a study has been carried out by Vahl (1981a, 1981b, 1981c; Sundet and Vahl, 1981) on the Icelandic scallop *C. islandica*. In this species reproductive effort increases with age, but only in animals over 13 years old does this increase affect the expectation of future reproductive success—that is, there is no energetic trade-off between current reproduction and future growth and survival in individuals younger than 13 years. This is consistent with scallops older than 13 years being

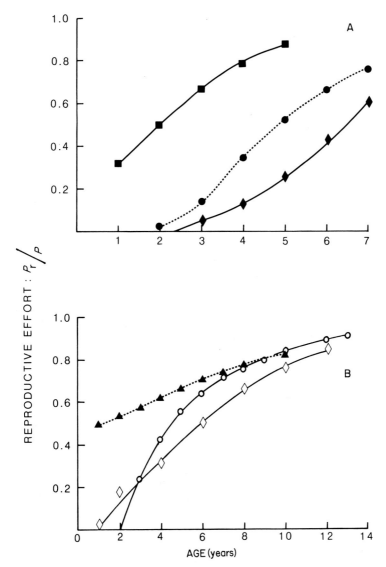

Fig. 26. Reproductive effort, calculated as P_r/P, in six bivalve species, related to age. **A.** ■, *Crassostrea virginica.* (From Dame, 1972, 1976.) ●, *Mytilus edulis.* (From Bayne and Widdows, 1978.) ◆, *Lissarca miliaris.* (From Richardson, 1979.) **B.** ○, *Placopectin magellanicus.* (From R. J. Thompson, unpublished data.) ◇, *Ostrea edulis.* (From Rodhouse, 1978.) ▲, *Choromytilus meridionalis.* (From Griffiths, 1980b.)

forced into partial starvation outside the spring period of food abundance, by a combination of seasonal change in metabolic rate and in food utilization, as well as by a difference in the size dependence of metabolic and feeding rates. A peak in reproductive value some years after first maturity may be a common feature in bivalves (e.g., Brosseau, 1978, on *M. arenaria*).

Comparisons of reproduction effort per unit time as a function of age are best made following standardization of the age axis with respect to growth rate (Hughes and Roberts, 1980). Figure 27 shows values for P_r/P for four populations of *M. edulis* plotted in this way; there is considerable variability among populations. Some of this variability is due to environmental stresses acting to reduce the allocation of available energy to reproduction, in favor of somatic growth. Experiments in the laboratory (Bayne et al., 1981) have shown that whereas fecundity declined linearly with a reduction in the scope for growth, reproductive effort was maintained at high values until the experimental stress was severe; with further decline in the scope for growth there was a disproportionate reduction in reproductive effort effected by resorption of ripe gametes and the utilization of endogenous energy stores for maintenance metabolism, rather than for gametogenesis. In a population of mussels exposed to unseasonable thermal stress the production of gametes was suppressed disproportionately. Calow's (1979) index C serves the useful purpose of ascribing an estimate of

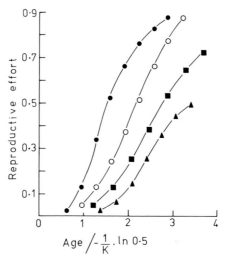

Fig. 27. Reproductive effort, calculated as P_r/P, of *Mytilus edulis* from four populations, related to age standardized for growth rate (see text). ●, Bellevue, Newfoundland. (From R. J. Thompson, unpublished data.) ○, Lynher; ■, Mothecombe; ▲, Cattewater. (All southwestern England; from Bayne and Widdows, 1978 and Bayne et al., 1983.)

energetic cost to particular levels of reproductive effort. Bayne et al. (1983) estimated C for individuals from four populations of *Mytilus edulis*, related to flesh weight. In all cases reproduction was basically "restrained," in not proceeding at the expense of somatic maintenance metabolism (i.e., $C < 0$), but became increasingly "reckless" with age ($C > 0$). This analysis suggested that mussels at two sites subjected to high levels of environmental stress reduced their reproductive effort, but incurred greater costs, than individuals from less stressed environments.

Environmental factors, acting on the components of the energy budget, may therefore have considerable impact on the energetics of reproduction. As with most of the physiological processes considered in this chapter, it is the interacting effects of temperature and food availability, set in a seasonal context, that conditions both the reproductive effort and the timing of reproductive events (Bayne, 1976b; Giese and Pearse, 1974; Newell et al., 1982; Sastry, 1979). Studies on *C. varia* by Shafee (1980) allow a detailed analysis of the reproductive effort by one species in different seasons of the year. In the Bay of Brest this species spawns twice in the year (June and September in 1977 and 1978). Gametes spawned in June were not produced at the expense of endogenous food reserves in the somatic tissue, because somatic and gonadal growth occurred together. For the September spawning, however, food reserves were utilized in the production of gametes, and somatic tissue declined in weight. Reproductive effort, calculated as P_r/P, was greater for the later than for the earlier spawning. When calculated as the index C, the September spawning is seen to have been more "reckless," because it proceeded at the expense of body maintenance. If the observed weight loss of somatic tissue (Shafee, 1980) is included in the "absorbed ration" term of Eq. (13), the cost of the September spawning is reduced, though it still exceeds that of the June spawning, which occurs immediately following a period of maximal food abundance.

Variability within and between species in the proportional allocation of available energy to somatic growth and to reproduction may therefore be seen in the context of a diversity of environmental factors, the presence or absence of an energy store available for gametogenesis, and the timing of gamete production relative to seasonal cycles of food availability and temperature. In addition, age-related effects, which follow from different allometric relationships between feeding and metabolism, and from differences in growth efficiency, add a further level of complexity.

It appears (Bayne et al., 1983) that certain fundamental properties of the reproductive strategy of a species may be maintained across a wide range of environmental qualities; for example, such properties as mean egg size, the unit energetic costs of gamete production and the relationship between maximum rate of increase of age-related reproductive effort and the weight-related change in energetic surplus. Other properties of reproduction are more vulnerable to environmental changes, including the timing of gametogenesis and spawning, the fecundity of the individual, and the maximum reproductive effort attainable in the individual life span. In trying to understand these processes further, it is essential that the approaches of physiological energetics be employed, to elucidate the various constraints that act on energy acquisition and/or energy utilization for maintenance, growth, and reproduction.

VI. Synopsis

Measurement of the components of the energy balance equation for individual organisms—namely, ingestion, absorption, excretion, and respiration—allows the derivation of the energy available to the animal for growth and for reproduction. In this way it is possible to analyze the relationships between growth and various endogenous and exogenous variables in terms of underlying physiological processes. If, in addition, energy allocation to the production of gametes and to reproductive processes in general are measured, a comprehensive understanding of animal adaptation becomes possible. In this chapter the various physiological components of growth in marine molluscs (excluding the cephalopods) are discussed in relation to the major variables of body size, seasonal effects, temperature, oxygen concentration, and ration level. A review of studies on energy flow at the population level indicates the average rates and efficiencies that have been reported in the major trophic categories (grazing gastropods, suspension and deposit feeders, carnivores, and wood-eating bivalves). However, for more fundamental insights into the causes of variability in the physiological processes of growth and production, research on individual animals is required and most of this chapter considers such studies.

It emerges that considerable progress has been made in understanding the physiology of growth in these animals and in formulating this understanding so that quantitative hypotheses of adaptation are possible. The majority of these studies, however, are based on laboratory experiments, and the need is recognized for more experimental work under natural field conditions. Results with controlled, and often monospecific, diets in the laboratory can be quite misleading when applied to the natural, inherently more diverse rations experienced in the field. This applies with equal force, though possibly for very different reasons, to all trophic categories considered. Results from experiments at controlled temperatures may also mislead when related to the very varied temperatures (over time spans from hours to months) normally experienced. The need to apply current techniques and understanding to real ecological situations is emphasized.

In applying the results of studies into physiological energetics to the real ecological domain, it is essential that the energies available for somatic growth and for reproduction be distinguished, analyzed separately, and used distinctly in assessing the efficiencies of energy transfer and production. Particularly when discussing the adaptive responses of marine molluscs to environmental variables, it is the relative allocation to growth and to the production of gametes, considered in terms of animal size, age, or both, that is fundamental. The procedures of physiological energetics are very useful here, but the integrated nature of adaptation, considering events at the biochemical, physiological, and population levels, must be borne in mind. In particular, recent advances in knowledge of the population genetics of marine molluscs pose a challenge to the "whole-animal" physiologist, and future research in either discipline must clearly take account of recent understanding that derives from the other. The integration, within the context of meaningful ecological questions, of studies by biochemists, physiologists, and geneticists, and the enhancement of physiological studies employing direct calorimetry for estimating true metabolic losses, are seen as major areas for development in the immediate future.

References

Adams, S. M., and Angelovic, J. W. (1970). Assimilation of detritus and its associated bacteria by three species of estuarine animals. *Ches. Sci.* **11**, 249–254.

Ahmad, T. A., and Chaplin, A. E. (1977). The intermediary metabolism of *Mytilus edulis* (L.) and *Cerastoderma edule* (L.) during exposure to the atmosphere. *Biochem. Soc. Trans.* **5**, 1320–1323.

Ansell, A. D. (1981). Experimental studies of a benthic predator-prey relationship. Feeding, growth and egg-collar production in long-term cultures of the gastropod drill *Polinices alderi* (Forbes) feeding on the bivalve *Tellina tenuis* (da Costa). *J. Exp. Mar. Biol. Ecol.* **56**, 1–21.

Ansell, A. D. (1982a). Experimental studies of a benthic predator-prey relationship. II. Energetics of growth and reproduction, and food-conversion efficiencies, in long-term cultures of the gastropod drill *Polinices alderi* (Forbes) feeding on the bivalve *Tellina tenuis* (da Costa). *J. Exp. Mar. Biol. Ecol.* **61**, 1–29.

Ansell, A. D. (1982b). Experimental studies of a benthic predator-prey relationship. Factors affecting rate of predation and growth in juveniles of the gastropod drill *Polinices catena* (da Costa) in laboratory cultures. *Malacologia.* **22**, 367–375.

Ansell, A. D., and Sivadas, P. (1973). Some effects of temperature and starvation on the bivalve *Donax vittatus* (da Costa) in experimental laboratory populations. *J. Exp. Mar. Biol. Ecol.* **13**, 229–262.

Baldwin, S. (1968). Manometric measurements of respiratory activity in *Acmaea digitalis* and *Acmaea scabra*. *Veliger* **11**, 79–82.

Baldwin, J. and Lee, A. K. (1979). Contributions of aerobic and anaerobic energy production during swimming in the bivalve mollusc *Limaria fragilis* (Family Limidae). *J. Comp. Physiol.* **129**, 361–364.

Bannister, J. V. (1974). The respiration in air and in water of the limpets *Patella caerulea* (L.) and *Patella lusitanica* (Gmelin). *Comp. Biochem. Physiol. A* **49**, 407–411.

Bayne, B. L. (1967). The respiratory response of *Mytilus perna* L. (Mollusca, Lamellibranchia) to reduced environmental oxygen. *Physiol. Zool.* **40**, 307–313.

Bayne, B. L. (1971a). Oxygen consumption by three species of lamellibranch mollusc in declining ambient oxygen tension. *Comp. Biochem. Physiol. A* **40**, 955–970.

Bayne, B. L. (1971b). Ventilation, the heart beat and oxygen uptake by *Mytilus edulis* L. in declining oxygen tension. *Comp. Biochem. Physiol. A* **40**, 1065–1085.

Bayne, B. L. (1973a). The responses of three species of bivalve mollusc to declining oxygen tension at reduced salinity. *Comp. Biochem. Physiol. A* **45**, 793–806.

Bayne, B. L. (1973b). Aspects of the metabolism of *Mytilus edulis* during starvation. *Neth J. Sea Res.* **7**, 399–410.

Bayne, B. L. (editor). (1976a), "Marine Mussels, their Ecology and Physiology." Cambridge University Press, London.

Bayne, B. L. (1976b). Aspects of reproduction in bivalve molluscs. *In* "Estuarine Processes" (M. Wiley, ed.), Vol. 1, pp. 432–448. Academic Press, New York.

Bayne, B. L. (1981). Theory and observation: Benthic predator-prey relationships. *In* "Analysis of Marine Ecosystems" (A. Longhurst, ed.), pp. 573–606. Academic Press, New York.

Bayne, B. L., and Livingstone, D. R. (1977). Responses of *Mytilus edulis* L. to low oxygen tension: Acclimation of the rate of oxygen consumption. *J. Comp. Physiol.* **114**, 129–142.

Bayne, B. L., and Scullard, C. (1977a). Rates of nitrogen excretion by species of *Mytilus* (Bivalvia: Mollusca). *J. Mar. Biol. Assoc. U. K.* **57**, 355–369.

Bayne, B. L., and Scullard, C. (1977b). An apparent specific dynamic action in *Mytilus edulis* L. *J. Mar. Biol. Assoc. U. K.* **57**, 371–378.

Bayne, B. L., and Scullard, C. (1978a). Rates of oxygen consumption by *Thais (Nucella) lapillus* (L.). *J. Exp. Mar. Biol. Ecol.* **32**, 97–111.

Bayne, B. L., and Scullard, C. (1978b). Rates of feeding by *Thais (Nucella) lapillus* (L.) *J. Exp. Mar. Biol. Ecol.* **32**, 113–129.

Bayne, B. L., and Widdows, J. (1978). The physiological ecology of two populations of *Mytilus edulis* L. *Oecologia* **37**, 137–162.

Bayne, B. L., and Worrall, C. M. (1980). Growth and production of mussels, *Mytilus edulis* from two populations. *Mar. Ecol. Prog. Ser.* **3**, 317–328.

Bayne, B. L., Thompson, R. J., and Widdows, J. (1973). Some effects of temperature and food on the rate of oxygen consumption by *Mytilus edulis* L. *In* "Effects of Temperature on Ectothermic Organisms" (W. Wieser, ed.), pp. 181–193. Springer-Verlag, Berlin.

Bayne, B. L., Bayne, C. J., Carefoot, T. C., and Thompson, R. J. (1976a). The physiological ecology of *Mytilus californianus* Conrad. Metabolism and energy balance, Vol. 1. *Oecologia* **22**, 211–228.

Bayne, B. L., Bayne, C. J., Carefoot, T. C., and Thompson, R. J. (1976b). The physiological ecology of *Mytilus californianus* Conrad. 2. Adaptations to low oxygen tension and air exposure. *Oecologia* **22**, 229–250.

Bayne, B. L., Widdows, J., and Worrall, C. W. (1977a). Some temperature relationships in the physiology of two ecologically distinct bivalve populations. *In* "Physiological Responses of Marine Biota to Pollutants" (F. J. Vernberg, A. Calabrese, F. P. Thurberg, and W. Vernberg, eds.), pp. 379–400. Academic Press, New York.

Bayne, B. L., Widdows, J., and Newell, R. I. E. (1977b). Physiological measurements on estuarine bivalve molluscs in the field. *In* "Biology of Benthic Organisms" (B. F. Keegan, P. O'Ceidigh, and P. J. S. Boaden eds.), pp. 57–68. Pergamon Press, Oxford, England.

Bayne, B. L., Moore, M. N., Widdows, J., Livingstone, D. R., and Salkeld, P. (1979). Measurements of the responses of individuals to environmental stress and pollution: Studies with bivalve molluscs. *Phil Trans. R. Soc. (B).* **286**, 563–581.

Bayne, B. L., Clarke, K. R., and Moore, M. N. (1981). Same practical considerations in the measurement of pollution effects on bivalve molluscs, and some possible ecological consequences. *Aquat. Toxicol.* **1**, 159–174.

Bayne, B. L., Bubel, A., Gabbott, P. A., Livingstone, D. R., Lowe, D. M., and Moore, M. N.

(1982). Glycogen utilisation and gametogenesis in *Mytilus edulis* L. *Mar. Biol. Lett.* **3,** 89–105.

Bayne, B. L., Salkeld, P. N. and Worrall, C. M. (1983). Reproductive effort and reproductive value in different populations of *Mytilus edulis* L. (Bivalvia; Mollusca). *Oecologia (Berl.)* (in press).

Bell, W. H., and Terhune, L. D. B. (1970). Water tunnel design for fisheries research. *Fish. Res. Board. Can. Tech. Rep.* **195,** 69.

Beukema, J. J., and de Bruin, W. (1979). Calorific values of the soft parts of the tellinid bivalve *Macoma balthica* (L.) as determined by two methods. *J. Exp. Mar. Biol. Ecol.* **37,** 19–30.

Beukema, J. J., Cadee, G. C., and Jansen, J. J. M. (1977). Variability in the growth rate of *Macoma balthica* (L.) in the Wadden Sea in relation to the availability of food. *In* "Biology of Benthic Organisms" (B. F. Keegan, P. O'Ceidigh, and P. J. S. Boaden, eds.), pp. 69–77. Pergamon Press, Oxford, England.

Booth, C. E., and Mangum, C. P. (1978). Oxygen uptake and transport in the lamellibranch mollusc, *Modiolus demissus. Physiol. Zool.* **51,** 17–32.

Boyden, C. R. (1972a). The behaviour, survival and respiration of the cockles *Cerastoderma edule* and *C. glaucum* in air *J. Mar. Biol. Assoc. U. K.* **52,** 661–680.

Boyden, C. R. (1972b). Aerial respiration of the cockle *Cerastoderma edule* in relation to temperature. *Comp. Biochem. Physiol. A* **43,** 697–712.

Boyden, C. R., Crothers, J. H., Little, C., and Mettam, C. (1977). The intertidal invertebrate fauna of the Severn Estuary. *Field Study* **4,** 477–554.

Branch, G. M. (1974a). The ecology of *Patella* Linaeus from the Cape Peninsula, South Africa, Reproductive cycles. *Trans. R. Soc. S. Afr.* **41,** 111–160.

Branch, G. M. (1974b). The ecology of *Patella* Linaeus from the Cape Peninsula, South Africa. Growth-rates. *Trans. R. Soc. S. Afr.* **41,** 161–193.

Branch, G. M. (1975a). Intraspecific competition in *Patella cochlear* Born. *J. Anim. Ecol.* **44,** 263–282.

Branch, G. M. (1975b). Mechanisms reducing intraspecific competition in *Patella* spp.: Migration differentiation and territorial behavior. *J. Anim. Ecol.* **44,** 575–600.

Branch, G. M. (1979). Respiratory adaptations in the limpet *Patella granatina:* A comparison with other limpets. *Comp. Biochem. Physiol. A* **62,** 641–647.

Branch, G. M. (1982). The biology of limpets: Physical factors, energy flow and ecological interactions. *Oceanogr. Mar. Biol. Ann. Rev.* **19,** 235–380.

Branch, G. M., and Newell, R. C. (1978). A comparative study of metabolic energy expenditure in the limpets *Patella cochlear, P. oculus* and *P. granularis. Mar. Biol.* **49,** 351–361.

Brand, A. R. and Roberts, D. (1973). The cardiac responses of the scallop *Pecten maximus* (L.) to respiratory stress. *J. Exp. Mar. Biol. Ecol.* **13,** 29–43.

Brett, J. R. (1971). Energetic responses of salmon to temperature. A study of some thermal relations in the physiology and freshwater ecology of sockeye salmon *(Onchorhynchus nerka). Am. Zool.* **11,** 99–113.

Brett, J. R. (1976). Scope for metabolism and growth of sockeye salmon, *Onchorhynchus nerka,* and some related energetics. *J. Fish. Res. Board. Can.* **33,** 307–313.

Brett, J. R. (1979). Environmental factors and growth. *In* "Fish Phisiology" (W. S. Hoar, D. R. Randall, and J. R. Brett, eds.), Vol. 8, pp. 599–675. Academic Press, New York.

Brett, J. R., and Groves, T. D. D. (1979). Physiological energetics. *In* "Fish Physiology" (W. S. Hoar, D. R. Randall, and J. R. Brett, eds.), Vol. 8, pp. 279–352. Academic Press, New York.

Broom, M. J. (1981). Size-selection, consumption rates and growth of the gastropods *Natica maculosa* (Lamarck) and *Thais carinifera* (Lamarck) preying on the bivalve *Anadara granosa* (L.). *J. Exp. Mar. Biol. Ecol.* **56,** 213–233.

Brousseau, D. J. (1978). Population dynamics of the soft shell clam, *Mya arenaria*. *Mar. Biol.* **50**, 63–71.

Brown, A. C. (1971). The ecology of the sandy beaches of the Cape Peninsula, South Africa, Part 2. The mode of life of *Bullia* (Gastropoda: Prosobranchia). *Trans. R. Soc. S. Afr.* **39**, 281–319.

Brown, A. C. (1979a). Respiration and activity in the sandy beach whelk, *Bullia digitalis* (Dillwyn) (Nassariidae). *S. Afr. J. Sci.* **75**, 451–452.

Brown, A. C. (1979b). Oxygen consumption of the sandy-beach whelk *Bullia digitalis* Meuschen at different levels of activity. *Comp. Biochem. Physiol. A* **62**, 673–675.

Brown, A. C., and da Silva F. M. (1979). The effects of temperature on oxygen consumption in *Bullia digitalis* Meuschen (Gastropoda, Nassaridae). *Comp. Biochem. Physiol. A* **62**, 573–576.

Brown, A. C., Ansell, A. D., and Trevallion, A. (1978). Oxygen consumption by *Bullia (Dorsanum) melanoide* (Deshayes) and *Bullia digitalis* Meushcen (Gastropoda, Nassaridae)—an example of non-acclimation. *Comp. Biochem. Physiol. A* **61**, 123–125.

Brown, F. A., Bennett, M. F., Webb, H. M., and Ralph, C. L. (1956). Persistent daily, monthly and 27 day cycles of activity in the oyster and quahog. *J. Exp. Zool.* **131**, 235–262.

Browne, R. A., and Russell-Hunter, W. D. (1978). Reproductive effort in molluscs. *Oecologia* **37**, 23–27.

Buxton, C. D., Newell, R. C., and Field, J. G. (1981). Response-surface analysis of the combined effects of exposure and acclimation temperatures on filtration, oxygen consumption and scope for growth in the oyster *Ostrea edulis*. *Mar. Ecol. Prog. Ser.* **6**, 73–82.

Calow, P. (1974). Some observations on locomotory strategies and their metabolic effects in two species of freshwater gastropods, *Ancylus fluvuatilis* Mull. and *Planorbis contortus* Linn. *Oecologia* **16**, 149–161.

Calow, P. (1975). Defaecation strategies of two freshwater gastropods, *Ancylus fluviatilis* Mull. and *Planorbis contortus* Linn., with a comparison of field and laboratory estimates of food absorption rate. *Oecologia* **20**, 51–63.

Calow, P. (1976). "Biological machines." Arnold, London.

Calow, P. (1977). Ecology, evolution and energetics: A study in metabolic adaptation. *Adv. Ecol. Res.* **10**, 1–62.

Calow, P. (1979). The cost of reproduction—A physiological approach. *Biol. Rev.* **54**, 23–40.

Cammen, L. M. (1980). Ingestion rate: An empirical model for aquatic deposit feeders and detritovores. *Oecologia* **44**, 303–310.

Carefoot, T. H. (1967a). Studies on a sublittoral population of *Aplysia punctata*. *J. Mar. Biol. Assoc. U. K.* **47**, 335–350.

Carefoot, T. H. (1967b). Growth and nutrition of three species of Opisthobranch molluscs. *Comp. Biochem. Physiol.* **21**, 627–652.

Carefoot, T. H. (1970). A comparison of absorption and utilisation of food energy in two species of tropical *Aplysia*. *J. Exp. Mar. Biol. Ecol.* **5**, 47–62.

Charnov, E. L. (1976). Optimal foraging: The marginal value theorem. *Theor. Popul. Biol.* **9**, 129–136.

Coleman, N. (1973). The oxygen consumption of *Mytilus edulis* in air. *Comp. Biochem. Physiol.* **45**, 393–402.

Coleman, N. (1976). Aerial respiration of nerites from the north-east coast of Australia. *Aust. J. Mar. Freshw. Res.* **27**, 455–466.

Connell, J. H. (1961). Effects of competition, predation by *Thais lapillus,* and other factors on natural populations of the barnacle *Balanus balancoides*. *Ecol. Monogr.* **31**, 61–104.

Conover, R. J. (1978). Transformation of organic matter. *In* "Marine Ecology" (O. Kinne, ed.), Vol. 4, pp. 221–500. Wiley, Chichester, England.

Conover, R. J., and Corner, E. D. S. (1968) Respiration and nitrogen excretion by some marine zooplankton in relation to their life cycles. *J. Mar. Biol. Assoc. U. K.* **48,** 49–75.

Conover, R. J., and Lalli, C. M. (1972). Feeding and growth in *Clione limacina* (Phipps), a pteropod mollusc. *J. Exp. Mar. Biol. Ecol.* **9,** 279–302.

Conover, R. J., and Lalli, C. M. (1974). Feeding and growth in *Clione limacina* (Phipps), a pteropod mollusc. Assimilation metabolism and growth efficiency, Vol. 2. *J. Exp. Mar. Biol. Ecol.* **16,** 131–154.

Crisp, D. J. (1971). Energy flow measurements. *In* "Methods for the Study of Marine Benthos" (N. A. Holme and A. D. McIntyre, eds.), pp. 197–323. (Int. Biol. Prog. 16.) Blackwell, Oxford, England.

Crisp, M. (1979). The effect of activity on the oxygen uptake of *Nassarius reticulatus* (Gastropoda, Prosobranchia). *Malacologia* **18,** 445–447.

Crisp, M., Davenport, J., and Shumway, S. E. (1978). Effects of feeding and of chemical stimulation on the oxygen uptake of *Nassarius reticulatus* (Gastropoda: Prosobranchia). *J. Mar. Biol. Assoc. U. K.* **58,** 387–399.

Crisp, M., Gill, C. W., and Thompson, M. C. (1981). Ammonia excretion by *Nassarius reticulatus* and *Buccinum undatum* (Gastropoda: Prosobranchia) during starvation and after feeding. *J. Mar. Biol. Assoc. U. K.* **61,** 381–390.

Curtis, C. A. (1980). Daily cycling of the crystalline style in the omnivorous, deposit-feeding estuarine snail, *Illyanassa obsoleta. Mar. Biol.* **59,** 137–140.

Dame, R. F. (1972). The ecological energetics of growth, respiration and assimilation in the intertidal American oyster, *Crassostrea virginica. Mar. Biol.* **17,** 243–250.

Dame, R. F. (1976). Energy flow in an intertidal oyster population. *Estuarine Coastal Mar. Sci.* **4,** 243–253.

Davenport, J. and Woolmington, A. D. (1982). A new method of monitoring ventilatory activity in mussels and its use in a study of the ventilatory patterns of *Mytilus edulis* L. *J. Exp. Mar. Biol. Ecol.* **62,** 55–67.

Davies, P. S. (1966). Physiological ecology of *Patella.* The effect of body size and temperature on metabolic rate, Vol. 1. *J. Mar. Biol. Assoc. U. K.* **46,** 647–658.

Davies, P. S. (1967). Physiological ecology of *Patella.* Effect of environmental acclimation on metabolic rate, Vol. 2. *J. Mar. Biol. Assoc. U.K.* **47,** 61–74.

De Wilde, P. A. W. J. (1975). Influence of temperature on behaviour, energy metabolism and growth of *Macoma balthica* (L.). *In* "Proc. 9th Europ. Mar. Biol. Symp." (H. Barnes, ed.), pp. 239–256. Aberdeen University Press, Aberdeen, Scotland.

De Zwaan, A. (1977). Anaerobic energy metabolism in bivalve molluscs. *Oceanogr. Mar. Biol.* **15,** 103–187.

De Zwaan, A., and Wijsman, T. C. M. (1976). Anaerobic metabolism in Bivalvia (Mollusca). Characteristics of anaerobic metabolism. *Comp. Biochem. Physiol. B* **54,** 313–324.

De Zwaan, A., Widdows, J., Bayne, B. L., de Bont, A. M. T., and Shick, J. M. (1981). Estimation of metabolic rates of sea mussels by direct calorimetry, biochemical analysis and oxygen uptake. *In* "Proc. 3rd. Europ. Congr. Comp. Biochem. Physiol." pp. 123–124.

Doyle, R. W. (1979). Ingestion rate of a selective deposit feeder in a complex mixture of particles: Testing the energy-optimization hypothesis. *Limnol. Oceanogr.* **24,** 867–874.

Dral, A. D. G. (1968). On the feeding of mussels (*Mytilus edulis* L.) in concentrated food suspensions. *Neth. J. Zool.* **18,** 440–441.

Edwards, D. C., and Huebner, J. D. (1977). Feeding and growth rates of *Polinices duplicatus* preying on *Mya arenaria* at Barnstaple Harbor, Massachusetts. *Ecology* **58,** 1218–1236.

Elsey, C. R. (1935). On the structure and function of the mantle and gill of *Ostrea gigas* Thunberg and *Ostrea lurida* Carpenter. *Trans. R. Soc. Can.* **5,** 131–158.

Elvin, O. W., and Gonor, J. J. (1979). The thermal regime of an intertidal *Mytilus californianus* Conrad population on the central Oregon coast. *J. Exp. Mar. Biol. Ecol.* **39,** 265–279.

Emlen, J. M. (1966). The role of time and energy in food preference. *Am. Nature* **100,** 611–617.

Famme, P. (1980). Oxygen-dependence of the respiration by the mussel *Mytilus edulis* L. as a function of size. *Comp. Biochem. Physiol. A* **67,** 171–174.

Famme, P. (1981). Haemolymph circulation as a respiratory parameter in the mussel *Mytilus edulis* L. *Comp. Biochem. Physiol. A* **69,** 243–247.

Famme, P., Knudsen, J., and Hansen, E. S. (1981). The effect of oxygen on the aerobic-anaerobic metabolism of the marine bivalve *Mytilus edulis* L. *Mar. Biol. Lett.* **2,** 345–351.

Famme, P., and Kofoed, L. H. (1980). The ventilatory current and otenidial function related to oxygen uptake in declining oxygen tension by the mussel *Mytilus edulis* L. *Comp. Biochem. Physiol. A* **66,** 161–171.

Fenchel, T. (1972). Aspects of decomposer food chains in marine benthos. *Verh. zool. Bot. Ges. Wien* **65,** 14–22.

Fenchel, T., Kofoed, L. H., and Lappalainen, A. (1975). Particle size selection of two deposit feeders, the amphibod *Corophium volutator* and the prosobranch *Hydrobia ulvae. Mar. Biol.* **30,** 119–128.

Fitch, D. D. (1975). Oxygen consumption in the prosobranch snail *Viviparus contectoides* (Mollusca: Gastropoda) I. Effects of weight and activity. *Comp. Biochem. Physiol.* **51,** 815–820.

Foster-Smith, R. L. (1975). The effect of concentration of suspension on the filtration rates and pseudofaecal production for *Mytilus edulis* L., *Cerastoderma edule* (L.) and *Venerupis pullastra* (Montagu). *J. Exp. Mar. Biol. Ecol.* **17,** 1–22.

Foster-Smith, R. L. (1976). Some mechanisms for the control of pumping activity in bivalves. *Mar. Behav. Physiol.* **4,** 41–60.

Fry, F. E. J. (1947). Effects of the environment on animal activity. *Univ. Toronto Stud. Biol. Ser.* **55,** 1–62.

Fry, F. E. J. (1971). The effect of environmental factors on the physiology of fish. *In* "Fish Physiology" (W. S. Hoar, and D. J. Randall, eds.), pp. 1–98. Academic Press, New York.

Fuji, A., and Hashizume, M. (1974). Energy budget for a Japanese common scallop, *Patinopecten yessoensis* (Jay), in Mutsu Bay. *Bull. Fac. Fish. Hokkaido Univ.* **25,** 7–19.

Gabbott, P. A. (1975). Storage cycles in marine bivalve molluscs: an hypothesis concerning the relationship between glycogen and gametogenesis. *In* "Proc. 9th Europ. Mar. Biol. Symp." (H. Barnes ed.), pp. 191–211. Aberdeen Univ. Press, Aberdeen, Scotland.

Gabbott, P. A., and Bayne, B. L. (1973). Biochemical effects of temperature and nutritive stress on *Mytilus edulis* L. *J. Mar. Biol. Assoc. U. K.* **53,** 269–286.

Gallager, S. M., Turner, R. D., and Berg, C. J. (1981). Physiological aspects of wood consumption, growth and reproduction in the shipworm *Lyrodus pedicellatus* Quatrefages (Bivalvia: Teredinidaw). *J. Exp. Mar. Biol. Ecol.* **52,** 63–77.

Giese, A. C., and Pearse, J. S. (1974). Introduction: General Principles. *In* "Reproduction of Marine Invertebrates. Acoelomate and Pseudocoelomate Metazoons" (A. C. Giese, and J. S. Pearse, eds.), pp. 1–49. Academic Press, New York.

Gilbert, M. A. (1977). The behaviour and functional morphology of deposit feeding in *Macoma balthica* (L.) in New England. *J. Moll. Stud.* **43,** 18–27.

Gilfillan, E. S., Mayo, D., Hanson, S., Donovan, D., and Jiang, L. C. (1977). Reduction in carbon flux in *Mya arenaria* caused by a spill of No. 6 fuel oil. *Mar. Biol.* **37,** 115–123.

Gillmore, R. B. (1982). Assessment of intertidal growth and capacity adaptations in suspension feeding bivalves. *Mar. Biol.* (In press.)

Gnaiger, E. (1979). Direct calorimetry in ecological energetics: long term monitoring of aquatic animals. *Experientia Suppl.* **37,** 155–165.

Gnaiger, E. (1980). Direct and indirect calorimetry in the study of animal anoxybiosis; a review and the concept of ATP-turnover. *In* "Thermal Analysis" (W. Hemminger ed.), Vol. II, pp. 547–552. Birkhäuser, Basel.

Gnaiger, E., and Forstner, H. (1983). Polarographic oxygen sensors: Aquatic and physiological applications. Springer-Verlag, Berlin.

Grahame, J. (1973). Assimilation efficiency of *Littorina littorea* (L.) (Gastropoda: Prosobranchiata). *J. Anim. Ecol.* **42**, 383–389.

Griffiths, C. L., and King, J. A. (1979a). Some relationships between size, food availability and energy balance in the ribbed mussel *Aulacomya ater*. *Mar. Biol.* **51**, 141–149.

Griffiths, C. L., and King, J. A. (1979b). Energy expended on growth and gonad output in the ribbed mussel *Aulacomya ater*. *Mar. Biol.* **53**, 217–222.

Griffiths, R. J. (1980a). Ecophysiology of the black mussel *Choromytilus meridionalis* (Krauss). Ph.D. thesis, Dept. Zoology, Univ. Cape Town, South Africa.

Griffiths, R. J. (1980b). Natural food availability and assimilation in the bivalve *Choromytilus meridionalis*. *Mar. Ecol. Prog. Ser.* **3**, 151–156.

Griffiths, R. J. (1980c). Filtration, respiration and assimilation in the black mussel *Choromytilus meridionalis*. *Mar. Ecol. Prog. Ser.* **3**, 63–70.

Griffiths, R. J. (1981a). Aerial exposure and energy balance in littoral and sub-littoral *Choromytilus meridionalis* (Kr.) (Bivalvia). *J. Exp. Mar. Biol. Ecol.* **52**, 231–241.

Griffiths, R. J. (1981b). Production and energy flow in relation to age and shore level in the bivalve *Choromytilus meridionalis* (Kr.). *Estuarine Coast. Shelf. Sci.* **13**, 477–493.

Griffiths, R. J., and Buffenstein, R. (1981). Aerial exposure and energy input in the bivalve *Choromytilus meridionalis* (Kr.). *J. Exp. Mar. Biol. Ecol.* **52**, 219–229.

Grodinski, W., Klekowski, R. Z., and Duncan, A. (1975). "Methods for ecological bioenergetics." (Int. Biol. Prog. 24.) Blackwell, Oxford, 367 pp.

Hagvar, S. (1975). Energy budget and growth during the development of *Molesoma colleris* (Coleoptera). *Oikos* **26**, 140–146.

Hammen, C. (1979). Metabolic rates of marine bivalve mollusc determined by calorimetry. *Comp. Biochem. Physiol. A* **62**, 955–959.

Hargrave, B. T. (1970). The ultization of benthic microflora by *Hyalella azteca* (Amphipoda). *J. Anim. Ecol.* **39**, 427–437.

Haven, D. A., and Morales-Alamo, R (1966). Aspects of biodeposition by oysters and other invertebrate filter feeders. *Limnol. Oceanogr.* **11**, 489–498.

Haven, D. S., and Morales-Alamo, R. (1970). Filtration of particles from suspension by the American oyster *Crassostrea virginica*. *Biol. Bull.* **139**, 248–264.

Hawkins, A. J. S., Bayne, B. L., and Clark, K. R. (1983). Co-ordinated rhythms of digestion, absorption and excretion in *Mytilus edulis* (Bivalvia: Mollusca). *Mar. Biol.* (in press).

Hemmingsen, A. M. (1960). Energy metabolism as related to body size and respiratory surfaces and its evolution. *Rep. Steno. Mem. Hosp. Copenh.* **9**, 7–110.

Herreid, C. F. (1980). Hypoxia in invertebrates. *Comp. Biochem. Physiol.* **67A**, 311–320.

Hibbert, C. J. (1977). Energy relations of the bivalve *Mercenaria mercenaria* on an intertidal mudflat. *Mar. Biol.* **44**, 77–84.

Higgens, P. J. (1980a). Effects of food availability on the valve movements and feeding behaviour of juvenile *Crassostrea virginica* (Gmelin). Valve movements and periodic activity, Vol. 1. *J. Exp. Mar. Biol. Ecol.* **45**, 229–244.

Higgens, P. J. (1980b). Effects of food availability on the valve movements and feeding behaviour of juvenile *Crassostrea virginica* (Gmelin). Feeding rates and behaviour, Vol. 2. *J. Exp. Mar. Biol. Ecol.* **46**, 17–27.

Hildreth, D. I. (1980). Bioseston production by *Mytilus edulis* and its effect in experimental systems. *Mar. Biol.* **55**, 309–315.

Hildreth, D. I., and Mallet, A. (1980). The effect of suspension density on the retention of 5 μm diatons by the *Mytilus edulis* gill. *Biol. Bull.* **158**, 316–323.

Hirshfield, M. E., and Tinkle, D. W. (1975). Natural selection and the evolution of reproductive effort. *Proc. Natl. Acad. Sci. U.S.A.* **72**, 2227–2231.

Holling, C. S. (1959). Some characteristics of the simple types of predation and parasitism. *Can. Entomol.* **91**, 385–398.

Houlihan, D. F. (1979). Respiration in air and water of three mangrove snails. *J. Exp. Mar. Biol. Ecol.* **41**, 143–161.

Houlihan, D. F. and Innes, A. J. (1982). Oxygen consumption, crawling speeds and cost of transport in four Meditarranean intertidal gastropods. *J. Comp. Physiol.* **147**, 113–122.

Houlihan, D. F., and Newton, J. R. L. (1978). Respiration of *Patella vulgata* on the shore. *In* "Physiology and behaviour of marine organisms" (D. S. McLusky, and A. S. Berry, eds.), pp. 39–46. Pergamon Press, Oxford, England.

Houlihan, D. F., Innes, A. J., and Dey, D. G. (1981). The influence of mantle cavity fluid on the aerial oxygen consumption of some intertidal gastropods. *J. Exp. Mar. Biol. Ecol.* **49**, 57–68.

Huebner, J. D., and Edwards, D. C. (1981). Energy budget of the predatory marine gastropod *Polinices duplicatus*. *Mar. Biol.* **61**, 221–226.

Hughes, R. N. (1969). A study of feeding in *Scrobicularia plana*. *J. Mar. Biol. Assoc. U. K.* **49**, 805–823.

Hughes, R. N. (1970). An energy budget for a tidal flat population of the bivalve *Scrobicularia plana* (Da Costa). *J. Anim. Ecol.* **39**, 357–381.

Hughes, R. N. (1971a). Ecological energetics of the keyhole limpet, *Fissurella barbadensis* Gmelin. *J. Exp. Mar. Biol. Ecol.* **6**, 167–178.

Hughes, R. N. (1971b). Ecological energetics of *Nerita* (Archaeogastropoda, Neritacea) populations on Barbados, West Indies. *Mar. Biol.* **11**, 12–22.

Hughes, R. N. (1980). Optimal foraging theory in the marine context. *Oceanogr. Mar. Biol.* **18**, 423–481.

Hughes, R. N., and Roberts, D. J. (1980). Reproductive effort of winkles (*Littorina* spp.) with contrasted methods of reproduction. *Oecologia* **47**, 130–136.

Hughes, T. G. (1973). Deposit feeding in *Abra tenuis* (Bivalvia: Tellinacea). *J. Zool. London* **171**, 499–512.

Hughes, T. G. (1975). The sorting of food particles by *Abra* sp. (Bivalvia: Tellinacea). *J. Exp. Mar. Biol. Ecol.* **20**, 137–156.

Humphreys, W. F. (1979). Production and respiration in animal populations. *J. Anim. Ecol.* **48**, 427–453.

Hylleberg, J. (1975). The effect of salinity and temperature on egestion in mud snails (Gastropoda: Hydrobiidae). *Oecologia* **21**, 279–289.

Hylleberg, J., and Galluci, V. F. (1975). Selectivity in feeding by the deposit-feeding bivalve *Macoma nasuta*. *Mar. Biol.* **32**, 167–178.

Incze, L. S., Lutz, R. A., and Watling, L. (1980). Relationships between effects of environmental temperature and seston on growth and mortality of *Mytilus edulis* in a temperate Northern estuary. *Mar. Biol.* **57**, 147–156.

Jørgensen, C. B. (1975). Comparative physiology of suspension feeding. *Ann. Rev. Physiol.* **37**, 57–79.

Jørgensen, C. B. (1976a). Comparative studies on the function of gills in suspension feeding bivalves, with special reference to effects of serotonin. *Biol. Bull.* **151**, 331–343.

Jørgensen, C. B. (1976b). Growth efficiencies and factors controlling size in some mytilid bivalves, especially *Mytilus edulis* L.: A review and interpretation. *Ophelia* **15**, 175–192.

Kennedy, V. S., and Mihursky, J. A. (1972). Effects of temperature on the respiratory metabolism of three Chesapeake Bay bivalves. *Chesapeake Sci.* **13**, 1–22.

Kersting, K. (1972). A nitrogen correction for caloric values. *Limnol. Oceanogr.* **17**, 643–644.

Kinne, O. (1970). Temperature—invertebrates. *In* "Marine Ecology" (O. Kinne, ed.), Vol. 1, Part 1, pp. 407–514. Wiley (Interscience), London.

Kiørboe, T., Møhlenberg, F., and Nøhr. O. (1980). Feeding, particle selection and carbon absorption in *Mytilus edulis* in different mixtures of algae and resuspended bottom material. *Ophelia* **19**, 193–205.

Kiørboe, T., Møhlenberg, F., and Nøhr, O. (1981). Effect of suspended bottom material on growth and energetics in *Mytilus edulis. Mar. Biol.* **61**, 283–288.

Kirby-Smith, W. W. (1972). Growth of the bay scallop: The influence of experimental water-currents. *J. Exp. Mar. Biol. Ecol.* **8**, 7–18.

Kirby-Smith, W. W., and Barber, R. T. (1974). Suspension feeding acquaculture systems: effects of phytoplankton concentration and temperature on growth of the bay scallop. *Aquaculture* **3**, 135–145.

Kleiber, M. (1961). "The fire of life: An introduction to animal energetics." Wiley (Interscience), New York/London.

Kofoed, L. H. (1975a). The feeding biology of *Hydrobia ventrosa* (Montagu). The assimilation of different components of the food. *J. Exp. Mar. Biol. Ecol.* **19**, 233–241.

Kofoed, L. H. (1975b). The feeding biology of *Hydrobia ventrosa* (Montagu). Allocation of the components of the carbon-budget and the significance of the secretion of dissolved organic material. *J. Exp. Mar. Biol. Ecol.* **19**, 243–256.

Kohn, A. J. (1959). The ecology of *Conus* in Hawaii. *Ecol. Monogr.* **25**, 47–90.

Kohn, A. J. (1968). Microhabitats, abundance and food of *Conus* on atoll reefs in the Maldive and Chagos Islands. *Ecology* **49**, 1046–1062.

Kristensen, J. H. (1972). Carbohydrases of some marine invertebrates with notes on their food and on natural occurrence of the carbohydrates studied. *Mar. Biol.* **14**, 130–142.

Kuenzler, E. J. (1961). Structure and energy flow of a mussel population in a Georgia salt marsh. *Limnol. Oceanogr.* **6**, 191–204.

Lam, R. K., and Frost, B. R. (1976). Model of copepod filtering responses to changes in size and concentration of food. *Limnol. Oceanogr.* **21**, 490–500.

Lammens, J. J. (1967). Growth and reproduction of a tidal flat population of *Macoma bathica* (C.). *Neth. J. Sea Res.* **3**, 315–382.

Lange, R., Staaland, H., and Mostad, A. (1972). The effect of salinity and temperature on solubility of oxygen and respiratory rate in oxygen-dependent marine invertebrates. *J. Exp. Mar. Biol. Ecol.* **9**, 217–229.

Langton, R. W. (1975). Synchrony in the digestive diverticula of *Mytilus edulis* L. *J. Mar. Biol. Assoc. U. K.* **55**, 221–229.

Langton, R. W. (1977). Digestive rhythms in the mussel, *Mytilus edulis. Mar. Biol.* **41**, 53–58.

Langton, R. W., and Gabbott, P. A. (1974). The tidal rhythm of extracellular digestion and the response to feeding in *Ostrea edulis. Mar. Biol.* **24**, 181–187.

Langton, R. W., Winter, J. E., and Roels, O. A. (1977). The effect of ration size on the growth and growth efficiency of the bivalve mollusc, *Tapes japonica. Aquaculture* **12**, 283–292.

Lehman, J. T. (1976). The filter-feeder as an optimal forager, and the predicted shapes of feeding curves. *Limnol. Oceanogr.* **21**, 501–516.

Leviten, P. J. (1976). The foraging strategy of vermivorous conid gastropods. *Ecol. Monogr.* **46**, 157–178.

Levinton, J. S., and Bianchi, T. S. (1981). Nutrition and food limitation of deposit-feeders. The role of microbes in the growth of mud-snails (Hydrobiidae), Vol. 1. *J. Mar. Res.* **39**, 531–545.

Levinton, J. S., and Lopez, G. R. (1977). A model of renewable resources and limitation of deposit-feeding benthic populations. *Oecologia* **31**, 177–190.

Lindeman, R. L. (1942). The trophic-dynamic aspect of ecology, *Ecology* **23**, 399–418.

Livingstone, D. R., and Bayne, B. L. (1977). Responses of *Mytilus edulis* L. to low oxygen tension; anaerobic metabolism of the posterior adductor muscle and mantle tissues. *J. Comp. Physiol.* **114**, 143–155.

Livingstone, D. R., and Widdows, J., and Fieth, P. (1979). Aspects of nitrogen metabolism of the common mussel, *Mytilus edulis:* adaptation to abrupt and fluctuating changes in salinity. *Mar. Biol.* **53**, 41–55.

Loosanoff, V. L., and Engle, J. B. (1947). Effect of different concentrations of micro-organisms on the feeding of oysters *(O. virginica). Fish. Bull. U.S.* **51**, 31–57.

Lopez, G. R., and Kofoed, L. H. (1980). Epipsammic browsing and deposit-feeding in mud snails (Hydrobiidae). *J. Mar. Res.* **38**, 585–599.

Lopez, G. R., and Levinton, J. S. (1978). The availability of microorganisms attached to sediment particles as food for *Hydrobia ventrosa* Montagu (Gastropod: Prosobranchia). *Oecologia* **32**, 263–275.

MacArthur, R. H. (1972), "Geographical ecology." Harper and Row, New York.

Macé, A-M. and Ansell, A. D. (1982). Respiration and nitrogen excretion of *Polinices elderi* (Forbes) and *Polinices catena* (da Costa) (Gastropoda: Naticidae). *J. Exp. Mar. Biol. Ecol.* **60**, 275–292.

Mackay, J., and Shumway, S. E. (1980). Factors affecting oxygen consumption in the scallop *Chlamys delicatula. Ophelia* **19**, 19–26.

Mangum, C. P. (1977). The analysis of oxygen uptake and transport in different kinds of animals. *J. Exp. Mar. Biol. Ecol.* **27**, 125–140.

Mangum, C. P., and Polites, G. (1980). Oxygen uptake and transport in the prosobranch mollusc *Busycon canaliculatum.* Gas exchange and the response to hypoxia, Vol. 1. *Biol. Bull.* **158**, 77–90.

Mangum, C. P., and van Winkle, W. (1973). Responses of aquatic invertebrates to declining oxygen conditions. *Am. Zool.* **13**, 529–541.

Mann, R. (1982). Nutrition in the Teredinidae. *In* "Proc. Symp. Marine Biodeterioration" (J. Costlow, ed.). (In press.)

Mathers, N. F. (1973). Carbohydrate digestion in *Ostrea edulis. Proc. Malac. Soc. London* **40** 359–367.

Mathers, N. F. (1976). The effects of tidal currents on the rhythm of feeding and digestion in *Pecten maximus* L. *J. Exp. Mar. Biol. Ecol.* **24**, 271–283.

Mathers, N. F., Smith, T., and Colins, N. (1979). Monophasic and diphasic digestive cycles in *Venerupis decussata* and *Chlamys varia. J. Moll. Stud.* **45**, 68–81.

Mayzaud, P. (1973). Respiration and nitrogen excretion of zooplankton. Studies of the metabolic characteristics of starved animals. *Mar. Biol.* **21**, 19–28.

McLusky, D. S. (1973). The effect of temperature on the oxygen consumption and filtration rate of *Chlamys (Aequipectan) opercularis* (L.) (Bivalvia). *Ophelia* **10**, 141–154.

McLusky, D. W., and Stirling, A. (1975). The oxygen consumption and feeding of *Donax incarnatus* and *D. spiculum* from tropical beaches. *Comp. Biochem. Physiol.* **51**, 942–947.

McMahon, R. F., and Russell-Hunger, W. D. (1977). Temperature relations of aerial and aquatic respiration in six littoral snails in relation to their vertical zaration. *Biol. Bull.* **152**, 182–198.

McNeill, S., and Lawton, J. H. (1970). Annual production and respiration in animal populations. *Nature. London* **225**, 472–474.

Menge, J. L. (1974). Prey selection and foraging period of the predaceous rocky intertidal snail, *Acanthina punctulata. Oecologia* **17**, 293–316.

Micallef, H. (1967). Aerial and aquatic respiration of certain trochids. *Experimenta* **23**, 52.

Micallef, H., and Bannister, W. H. (1967). Aerial and aquatic oxygen consumption of *Monodonta turbinata* (Mollusca: Gastropoda). *J. Zool. London* **151**, 479–482.

Møhlenberg, F., and Kiørboe, T. (1981). Growth and energetics in *Spisula subtruncata* (da Costa) and the effect of suspended bottom material. *Ophelia* **20**, 79–90.

Møhlenberg, F. and Riisgard, H. U. (1978). Efficiency of particle retention in 13 species of suspension feeding bivalves. *Ophelia* **17**, 239–246.

Moon, T. W., and Pritchard, A. W. (1970). Metabolic adaptations in vertically-separated populations of *Mytilus californianus* Conrad. *J. Exp. Mar. Biol. Ecol.* **5**, 35–46.

Morgan, P. R. (1972). The influence of prey availability on the distribution and predatory behaviour of *Nucella lapillus* (L.). *J. Anim. Ecol.* **41**, 257–274.

Morton, B. S. (1971). The daily rhythm and the tital rhythm of feeding and digestion in *Ostrea edulis*. *Biol. J. Linn. Soc.* **3**, 329–342.

Morton, B. S. (1973). A new theory of feeding and digestion in the filter-feeding Lamellibranchia. *Malacologia* **14**, 63–79.

Morton, B. S. (1977). The tidal rhythm of feeding and digestion in the Pacific oyster *Crassostrea gigas* (Thunberg). *J. Exp. Mar. Biol. Ecol.* **26**, 135–151.

Morton, B. S. (1978). Feeding and digestion in shipworms. *Oceanogr. Mar. Biol.* **16**, 107–144.

Morton, J. E. (1956). The tidal rhythm and action of the digestive system of the lamellibranch *Lasaea rubra*. *J. Mar. Biol. Assoc. U. K.* **35**, 563–586.

Muir, B. S., and Niimi, A. J. (1972). Oxygen consumption of the euryhaline fish aholehole (*Kuhlia sandvicensis*) with reference to salinity, swimming and food consumption. *J. Fish Res. Board Can.* **29**, 66–77.

Murdoch, W. W. (1971). The developmental response of predators to changes in prey density. *Ecology* **52**, 132–137.

Murdoch, W. W. (1973). The functional response of predators. *J. Appl. Ecol.* **10**, 335–342.

Murken, J. (1976). Feeding experiments with *Mytilus edulis* L. Feeding of waste organic products from the fish industry of Bremerhaven as a means of recycling biodegradable wastes, Vol. 3. *In* "Proc. 10th Europ. Mar. Biol. Symp." (G. Persoone and E. Jaspers, eds.), pp. 273–284. Universa Press, Wetteren.

Newell, R. C. (1965). The role of detritus in the nutrition of two marine deposit-feeders, the prosobranch *Hydrobia ulvae* and the bivalve *Macoma balthica*. *Proc. Zool. Soc. Lonson* **144**, 24–45.

Newell, R. C. (1973). Factors affecting the respiration of intertidal invertebrates. *Am. Zool.* **13**, 513–528.

Newell, R. C. (1976). Adaptations to intertidal life. *In* "Adaptation to Environment: Essays on the Physiology of Marine Animals" (R. C. Newell, ed.), pp. 1–88. Butterworths, Lond.

Newell, R. C. (1979). "Biology of intertidal animals." Marine Ecological Surveys, Faversham, Kent, England.

Newell, R. C. (1980). The maintenance of energy balance in marine invertebrates exposed to changes in environmental temperature. *In* "Animals and Environmental Fitness" (R. Gilles, ed.), pp. 561–582. Pergamon Press, Oxford, England.

Newell, R. C., and Branch, G. M. (1980). The effects of temperature on the maintenance of metabolic energy balance in marine invertebrates. *Adv. Mar. Biol.* **17**, 329–396.

Newell, R. C., and Kofoed, L. H. (1977a). The energetics of suspension-feeding in the gastropod *Crepidula fornicata* L. *J. Mar. Biol. Assoc. U. K.,* **51**, 161–180.

Newell, R. C., and Kofoed, L. H. (1977b). Adjustment of the components of energy balance in the gastropod *Crepidula fornicata* in response to thermal acclimation. *Mar. Biol.* **44**, 275–286.

Newell, R. C., and Pye, V. I. (1970a). Seasonal changes in the effect of temperature on the oxygen consumption of the winkle *Littorina littora* L. and the mussel *Mytilus edulis*. *Comp. Biochem. Physiol.* **34**, 367–383.

Newell, R. C., and Pye, V. I. (1970b). The influence of thermal acclimation on the relation between oxygen consumption and temperature in *Littorina littorea* (L.) and *Mytilus edulis* L. *Comp. Biochem. Physiol.* **34**, 385–397.

Newell, R. C., and Pye, V. I. (1971a). Quantitative aspects of the relationship between metabolism and temperature in the winkle *Littorina littorea* (L.). *Comp. Biochem. Physiol. B* **38**, 635–650.

Newell, R. C., and Pye, V. I. (1971b). Temperature-induced variations in the respiration of mitochondria from the windle *Littorina littorea* (L.). *Comp. Biochem. Physiol. B* **40**, 249–261.

Newell, R. C., and Roy, A. (1973). A statistical model relating the oxygen consumption of a mollusk (*Littorina littorea*) to activity, body size, and environmental conditions. *Physiol. Zool.* **46**(4), 252–275.

Newell, R. C., Johnson, L. G., and Kofoed, L. H. (1977). Adjustment of the components of energy balance in response to temperature change in *Ostrea edulis*. *Oecologia* **30**, 97–110.

Newell, R. C., Johnson, L. G., and Kofoed, L. H. (1978). Effects of environmental temperature and hypoxia on the oxygen consumption of the suspension-feeding gastropod *Crepidula fornicata* L. *Comp. Biochem. Physiol. A* **59**, 175–182.

Newell, R. I. E. (1977). The eco-physiology of *Cardium edule* (Linne). Ph.D. thesis, University of London, London, England.

Newell, R. I. E., and Bayne, B. L. (1980). Seasonal changes in the physiology, reproductive condition and carbohydrate content of the cockle *Cardium (Cerastoderma) edule* (Bivalvia: Cardiidae). *Mar. Biol.* **56**, 11–19.

Newell, R. I. E. and Jordan, S. J. (1983). Preferential ingestion of organic material by oysters. *Science,* (in press).

Newell, R. I. E., Hilbish, T. J., Koehn, R. K. and Newell, C. J. (1982). Temporal variation in the reproductive cycle of *Mytilus edulis* L. (Bivalvia, Mytilidae) from localities on the east coast of the United States. *Biol. Bull.* **162**, 299–310.

Norberg, R. A. (1977). An ecological theory on foraging time and energetics and choice of optimal food-searching method. *J. Anim. Ecol.* **46**, 511–529.

Odum, E. P., and Odum, H. T. (1955). Trophic structure and productivity of a windward coral reef community on Eniwetok Atoll. *Ecol. Monogr.* **25**, 291–320.

Odum, E. P., and Smalley, A. E. (1959). Comparisons of population energy flow of a harbivorous and a deposit-feeding invertebrate in a salt marsh ecosystem. *Proc. Natl. Acad. Sci. U.S.A.* **45**, 617–622.

Odum, E. P., Connell, C. E., and Davenport, L. B. (1962). Population energy flow of three primary consumer components of old-field ecosystems. *Ecology* **43**, 88–96,

Owen, G. (1972). Lysosomes, peroxisomes and bivalves. *Sci. Prog.* **60**, 299–318.

Owen, G. (1974). Feeding and digestion in the bivalvia. *Adv. Comp. Physiol. Biochem.* **5**, 1–35.

Owen, G., and McCrae, J. M. (1976). Further studies on latero-frontal tracts of bivalves. *Proc. R. Soc. Ser. B.* **194**, 527–544.

Paine, R. T. (1965). Natural history, limiting factors and energetics of the opisthobranch *Navanax inermis. Ecology* **46**, 603–619.

Paine, R. T. (1971). Energy flow in a natural population of the herbivorous gastropod *Tegula funebralis. Limnol. Oceanogr.* **16**, 86–98.

Palmer, R. E. (1980). Behavioural and rhythmic aspects of filtration in the bay scallop *Argopecten irradians concentricus* (Say) and the oyster *Crassostrea virginica* (Gmelin) *J. Exp. Mar. Biol. Ecol.* **45**, 273–295.

Palmer, R. E., and Williams, L. G. (1980). Effect of particle concentrations on filtration efficiency of the bay scallop *Argopecten irradians* and the oyster *Crassostrea virginica. Ophelia* **19**, 163–174.

Paloheimo, J. E., and Dickie, L. M. (1966). Food and growth of fishes. Relations among food, body size and growth efficiency, Vol. 3. *J. Fish. Res. Board Can.* **23,** 1209–1248.

Pamatmat, M. M. (1979). Anaerobic heat production of bivalves *(Polymesoda caroliniana* and *Modiolus demissus)* in relation to temperature, body size and duration of anoxia. *Mar. Biol.* **53,** 223–229.

Pamatmat, M. M. (1980). Faculative anaerobiosis of benthos. *In* "Marine Benthic Dynamics" (E. R. Tenore and B. C. Coull, eds.), pp. 69–90. University of South Carolina Press, Columbia, South Carolina.

Parker, G. A., and Stuart, R. A. (1976). Animal behaviour as a strategy optimizer: Evolution of resource assessment strategies and optimal emigration thresholds. *Am. Nature* **110,** 1055–1076.

Pearce, J. B., and Thorson, G. (1967). The feeding and reproductive biology of the red whelk *Neptunea antiqua* L. (Gastropoda, Prosobranchia). *Ophelia* **4,** 277–314.

Petrusewicz, K., and MacFadyen, A. (1970). "Productivity of terrestrial animals: Principles and methods" (Int. Biol. Progm. Handbook 13.) Blackwell, Oxford, England.

Phillipson, J. (1962). Respirometry and the study of energy turnover in natural systems with particular reference to harvest spiders (Phalangiida). *Oikos* **13,** 311–322.

Piiper, J., Dejours, P., Haab, P. and Rahn, H. (1971). Concepts and basic quantities in gas exchange physiology. *Respir. Physiol.* **13,** 292–304.

Pohlo, R. H. (1969). Confusion concerning deposit feeding in the *Tellinacea. Proc. Malcol. Soc. London* **38,** 361–364.

Pohlo, R. W. (1973). Feeding and associated functional morphology in *Tagelus californianus* and *Florimetis obesa* (Bivalvia: Tellinacea). *Malacologia* **12,** 1–11.

Polites, G., and Mangum, C. P. (1980). Oxygen uptake and transport in the prosobranch mollusc *Busycon canaliculatum* (L.) Influence of acclimation salinity and temperature, Vol. 2. *Biol. Bull.* **157,** 118–128.

Precht, H. Christophersen, J., Hensel, H., and Larcher, W. (1973). "Temperature and Life," Springer-Verlag, Berlin.

Prosser, C. L. (1973). "Comparative Animal Physiology." Saunders, Philadelphia. (3rd edition.)

Prosser, C. L., and Brown, F. A. (1961), "Comparative animal physiology." Saunders, Philadelphia, Pennsylvania. (2nd edition.)

Purchon, R. D. (1968), "The biology of the Mollusca." Pergamon Press, Oxford, England.

Purchon, R. D. (1971). Digestion in filter-feeding bivalves—a new concept. *Proc. Malacol. Soc. London* **39,** 253–262.

Pyke, G. H., Pulliam, H. R., and Charnov, E. L. (1977). Optimal foraging: A selective review of theory and tests. *Q. Rev. Biol.* **52,** 137–154.

Read, K. R. H. (1962). Respiration of the bivalved molluscs *Mytilus edulis* L. and *Brachidontes demissus plicatulus* Lamark, as a function of size and temperature. *Comp. Biochem. Physiol.* **7,** 89–101.

Reid, R. G. B. (1971). Criteria for categorizing feeding types in bivalves. *Veliger* **13,** 358–359.

Reid, R. G. B., and Reid, A. (1969). Feeding processes of the members of the genus *Macoma* (Mollusca: Bivalvia). *Can. J. Zool.* **47,** 649–657.

Richardson, M. G. (1979). The ecology and reproduction of the brooding antarctic bivalve *Lissarca miliaris. Br. Antarct. Surv. Bull.* **49,** 91–115.

Ricker, W. E. (1968), "Methods for assessment of fish production in fresh waters." (Int. Biol. Prog. Handbook 3.) Blackwell, Oxford, England.

Robertson, J. R. (1979). Evidence for tidally correlated feeding rhythms in the eastern mud snail, *Illyanassa obsoleta. Nautilus* **93,** 38–40.

Robilliard, G. A. (1971). Predation by the nudibranch *Dirona albolineata* on three species of prosobranchs. *Pac. Sci.* **25,** 429–435.

Robinson, W. E., Pennington, M. R., and Langton, R. W. (1981). Variability of tubule types within the digestive glands of *Mercenaria mercenaria* (L.)., *Ostrea edulis* L., and *Mytilus edulis* L. *J. Exp. Mar. Biol. Ecol.* **54**, 265–276.

Rodhouse, P. G. (1978). Energy transformations by the oyster *Ostrea edulis* L. in a temperate estuary. *J. Exp. Mar. Biol. Ecol.* **34**, 1–22.

Rodhouse, P. G. (1979). A note on the energy budget for an oyster population in temperate estuary. *J. Exp. Mar. Biol. Ecol.* **37**, 205–212.

Royama, T. (1971). A comparative study of models for predation and parasitism. *Res. Popul. Ecol. Suppl.* **1**, 1–91.

Salanki, J. C. (1966). Daily activity rhythm of two Mediterranean Lamellibranchia. *Ann. Inst. Biol. Tihany* **33**, 135–142.

Sandison, E. C. (1966). The oxygen consumption of some intertidal gastropods in relation to zonation. *J. Zool. London* **149**, 163–173.

Sastry, A. N. (1979). Pelecypoda (excluding Ostreidae). *In* "Reproduction of Marine Invertebrates. Molluscs: Pelecypods and Lesser Classes" (A. C. Giese and J. S. Pearse, eds.), pp. 113–292. Academic Press, New York.

Schoener, T. W. (1971). Theory of feeding strategies. *Ann. Rev. Ecol. Syst.* **2**, 369–404.

Seiderer, L. J. and Newell, R. C. (1979). Adjustment of the activity of α-amylase extracted from the style of the black mussel *Choromytilus meridionalis* (Krauss) in response to thermal acclimation. *J. Exp. Mar. Biol. Ecol.* **39**, 79–86.

Seiderer, L. J., Newell, R. C. and Cook, P. A. (1982). Quantitative significance of style enzymes from two marine mussels (*Choromytilus meridionalis* Krauss and *Perna perna* Linnaeus) in relation to diet. *Mar. Biol. Lett.* **3**, 257–271.

Shafee, M. S. (1980), "Ecophysiological studies on a temperate bivalve *Chlamys vaira* (L.) from Lanevoc (Bay of Brest)." Thèse de Dr. es-Sci. Nat., Université de Bretagne Occidentale, France.

Shick, J. M. and Widdows, J. (1981). Direct and indirect calorimetric measurements of metabolic rate in bivalve molluscs during aerial exposure. *Am. Zool.* **21**, 983.

Shick, J. M., De Zwaan, A. and De Bont, A. M. Th. (1983). Anoxic metabolic rate in the mussel *Mytilus edulis* L. estimated by simultaneous direct calorimetry and biochemical analysis. *Physiol. Zool.* **56**, 56–63.

Slobodkin, L. B. (1962). Energy in animal ecology. *Adv. Ecol. Res.* **1**, 69–101.

Soldatova, I. N. (1961a). Oxygen consumption rate and size of the bivalve mollusk, *Teredo navalis* L. *In* "Marine Fouling and Borers" (I. V. Sterestin, ed.), pp. 161–166. *Trans. Inst. Oceanol.* **59**, Acad. Sci. USSR.

Soldatova, I. N. (1961b). Effects of various salinities on the bivalve mollusk, *Teredo navalia* L. *In* "Marine fouling and borers" (I. V. Starostin, ed.), pp. 167–183. *Trans Inst. Oceanol.* **49**, Acad. Sci. USSR.

Spight, T. M., and Emlen, J. (1976). Clutch sizes of two marine snails with a changing food supply. *Ecology* **57**, 1162–1179.

Stickle, W. B., and Bayne, B. L. (1982). Effects of temperature and salinity on oxygen consumption and nitrogen excretion in *Thais (Nucella) lapillus*. *J. Exp. Mar. Biol. Ecol.* **58**, 1–17.

Suchanek, T. H. (1981). The role of disturbance in the evolution of life history strategies in the intertidal mussels *Mytilus edulis* and *Mytilus californianus*. *Oecologia* **50**, 143–152.

Sundet, J. H., and Vahl, O. (1981). Seasonal changes in dry weight and biochemical composition of the tissues of sexually mature and immature Iceland scallops, *Chlamys islandica*. *J. Mar. Biol. Assoc. U. K.* **61**, 1001–1010.

Taghon, G. L. (1981). Beyond selection: Optimal ingestion rate as a function of food value. *Am. Nature* **118**, 202–214.

Taghon, G. L., Self, R. F. L., and Jumars, P. A. (1978). Predicting particle selection by deposit feeders: A model and its implications. *Limnol. Oceanogr.* **23**, 752–759.

Tang, P. S. (1933). On the rate of oxygen consumption by tissues and lower organisms as a function of oxygen tension. *Q. Rev. Biol.* **8**, 260–274.

Taylor, A. C., and Brand, A. R. (1975a). Effects of hypoxia and body size on the oxygen consumption of the bivalve *Arctica islandica* (L). *J. Exp. Mar. Biol. Ecol.* **19**, 187–196.

Taylor, A. C., and Brand, A. R. (1975b). A comparative study of the respiratory responses of the bivalves *Arctica islandica* (L.) and *Mytilus edulis* L. to declining oxygen tension. *Proc. R. Soc. London B,* **190**, 443–456.

Teal, J. M. (1962). Energy flow in the salt marsh ecosystem of Georgia. *Ecology* **43**, 614–624.

Tenore, K. R., and Dunstan, W. M. (1973a). Comparison of feeding and biodeposition of three bivalves at different food levels. *Mar. Biol.* **21**, 190–195.

Tenore, K. R., and Dunstan, W. M. (1973b). Comparison of rates of feeding and biodeposition of the American oyster, *Crassostrea virginica* Gmelin, fed different species of phytoplankton. *J. Exp. Mar. Biol. Ecol.* **12**, 19–26.

Tenore, K. R., Goldman, J. C., and Clarner, J. P. (1973c). The food chain dynamics of the oyster, clam and mussel in an aquaculture food chain. *J. Exp. Mar. Biol. Ecol.* **12**, 157–165.

Theede, H. (1963). Experimnetelle Unterssuchimgen über die Filtrationsleistung der Miesmuschel *Mytilus edulis* L. *Kiel. Meereoforsch.* **19**, 20–41.

Theisen, B. F. (1977). Feeding rate of *Mytilus edulis* L. (Bivalvia) from different parts of Danish waters in water of different turbidity. *Ophelia* **16**, 221–232.

Thomson, R. J., and Bayne, B. L. (1972). Active metabolism associated with feeding in the mussel *Mytilus edulis* L. *J. Exp. Mar. Biol. Ecol.* **8**, 191–212.

Thompson, R. J., and Bayne, B. L. (1974). Some relationships between growth, metabolsim and food in the mussel, *Mytilus edulis*. *Mar. Biol.* **27**, 317–326.

Trevallion, A. (1971). Studies on *Tellina tenuis* Da Costa. III. Aspects of general biology and energy flow. *J. Exp. Mar. Biol. Ecol.* **7**, 95–122.

Vahl, O. (1972). Particle retention and relation between water transport and oxygen uptake in *Chlamys opercularis* (L.) (Bivalvia). *Ophelia* **10**, 67–74.

Vahl, O. (1973a). Porosity of the gill, oxygen consumption and pumping rate in *Cardium edule* (L.) (Bivalvia). *Ophelia* **10**, 109–118.

Vahl, O. (1973b). Pumping and oxygen consumption rates of *Mytilus edulis* L. of different sizes. *Ophelia* **12**, 45–52.

Vahl, O. (1978). Seasonal changes in oxygen consumption of the Iceland scallop (*Chlamys islandica* (O. F. Fuller) from 70°N. *Ophelia* **17**, 143–154.

Vahl, O. (1980). Seasonal variations in seston and in the growth rate of the Iceland scallop, *Chlamys islandica* (O. F. Muller) from Balsfjord 70°N. *J. Exp. Mar. Biol. Ecol.* **48**, 195–204.

Vahl, O. (1981a). Energy transformations by the Iceland scallop, *Chlamys islandica* (O. F. Muller) from 70°N.I. The age-specific energy budget and net growth efficiency. *J. Exp. Mar. Biol. Ecol.* **53**, 281–296.

Vahl, O. (1981b). Energy transformations by the Iceland scallop, *Chlamys islandica* (O. F. Muller) from 70°N. II. The population energy budget. *J. Exp. Mar. Biol. Ecol.* **53**, 297–303.

Vahl, O. (1981c). Age-specific residual reproductive value and reproductive effort in the Iceland scallop, *Chalmys islandica* (O. F. Muller). *Oecologia* **51**, 53–56.

van Dam, L. (1935). On the utilisation of oxygen by *Mya arenaria*. *J. Exp. Biol.* **12**, 86–94.

van Dam, L. (1938). On the utilisation of oxygen and the regulation of breathing in some aquatic animals. Ph.D. dissertation, Drukkerij 'Volharding' Groningen.

van Dam, L. (1954). On the respiration in scallops (Lamellibranchia) *Biol. Bull.* **107**, 194–202.

Van Hook, R. I. (1971). Energy and nutrient dynamics of spider and orthopteran populations in a grassland ecosystem. *Ecol. Mnongr.* **41**, 1–26.

Van Weel, P. B. (1961). The comparative physiology of digestion in molluscs. *Am. Zool.* **1**, 245–252.

Vinogradov, A. P.(1953). The elementary chemical composition of marine organisms. *Mem. Sears Fdn. Mar. Res.* **2**, 647.

von Bertalanff, L. (1957). Quantitative laws in metabolism and growth. *Q. Rev. Biol.* **32**, 217–231.

Warren, C. E., and Davis, G. E. (1967). Laboratory studies on the feeding bioenergetics and growth of fish. *In* "The Biological Basis of Freshwater Fish Production" (S. D. Gerking, ed.), pp. 175–214. Blackwell Scientific Publications, Oxford, England.

Warwick, R. M., Joint, I. R., and Radford, P. J. (1979). Secondary production of the benthos in estuarine environment *In* "Ecological Processes in Coastal Environments" (R. L. Jefferies and A. J. Davy, eds.), pp. 429–450. Blackwell Scientific Publications, Oxford, England.

Welch, M. E. (1968). Relationships between assimilation efficiencies and growth efficiencies for aquatic consumers. *Ecology* **49**, 755–759.

Whedon, W. F., and Sommer, H. (1937). Respiratory exchange of *Mytilus californianus*. *Zeit. vgl. Physiol.* **25**, 523–528.

Whitlach, R. B., and Obrebski, S. (1980). Feeding selectivity and coexistence in two deposit-feeding gastropods. *Mar. Biol.* **58**, 219–225.

Widdows, J. (1973a). The effect of temperature on the heart beat, ventilation rate and oxygen uptake of *Mytilus edulis*. *Mar. Biol.* **20**, 269–276.

Widdows, J. (1973b). The effect of temperature on the metabolism and activity of *Mytilus edulis*. *Neth. J. Sea Res.* **7**, 387–398.

Widdows, J. (1976). Physiological adaptation of *Mytilus edulis* to cyclic temperatures. *J. Comp. Physiol.* **105**, 115–128.

Widdows, J. (1978a). Combined effects of body size, food concentration and season on the physiology of *Mytilus edulis*. *J. Mar. Biol. Assoc. U. K.* **58**, 109–124.

Widdows, J. (1978b). Physiological indices of stress in *Mytilus edulis*. *J. Mar. Biol. Assoc. U. K.* **58**, 125–142.

Widdows, J., and Bayne, B. L. (1971). Temperature acclimation of *Mytilus edulis* with reference to its energy budget. *J. Mar. Biol. Assoc. U. K.* **51**, 827–843.

Widdows, J., Fieth, P., and Worrall, C. M. (1979a). Relationship between seston, available food and feeding activity in the common mussel *Mytilus edulis*. *Mar. Biol.* **50**, 195–207.

Widdows, J., Bayne, B. L., Livingstone, D. R., Newell, R. I. E., and Donkin, P. (1979b). Physiological and biochemical responses of bivalve molluscs to exposure to air. *Comp. Biochem. Physiol.* A **62**, 301–308.

Wiegert, R. G. (1968). Thermodynamic considerations in animal nutrition *Am. Zool.* **8**, 71–81.

Wightman, J. A. (1977). Respirometry techniques for terrestrial invertebrates and their application to energetics studies. *N. Z. J. Zool.* **4**, 453–469.

Wightman, J. A. (1981). Why insect energy budgets do not balance. *Oecologia* **50**, 166–169.

Wikander, P. B. (1980). Quantitative aspects of deposit feeding in *Abra nitida* (Muller) and *A. longicallus* (Scacchi) (Bivalvia, Tellinaces). *Sarsia* **66**, 35–48.

Willemsen, J. (1952). Quantities of water pumped by mussels *(Mytilus edulis)* and cockles *(Cardium edule)*. *Arch. Neerl. Zool.* **10**, 153–160.

Wilson, J. H., and La Touche, R. W. (1978). Intracellular digestion in two sublittoral populations of *Ostrea edulis (Lamellibranchia)*. *Mar. Biol.* **47**, 71–77.

Winberg, G. C. (1956). "Rate of metabolism and food requirements of fishes (Russian)." Belorussian State Univ., Minsk. *(Trans. Ser. Fish. Res. Board Can. 194)*.

Winter, J. E. (1969). Uber den Einfluss der Nahrungskonzentration und anderer Faktoren auf

Filtrierleistung und Nahrungsausnutzung der Musheln *Arctica islandica* und *Modiolus modiolus*. *Mar. Biol.* **4,** 87–135.

Winter, J. E. (1973). The filtration rate of *Mytilus edulis* and its dependence on algal concentration, measured by a continuous automatic recording apparatus. *Mar. Biol.* **22,** 317–328.

Winter, J. E. (1976). Feeding experiments with *Mytilus edulis* L. at small laboratory scale. The influence of suspended silt in addition to algal suspensions on growth, Vol. 2. *In* "Proc. 10th Europ. Mar. Biol. Symp." (G. Persone and E. Jaspers, eds.), pp. 583–600. Universa Press, Wetteren.

Winter, J. E. (1978). A review of the knowledge of suspension-feeding in lamellibranchiate bivalves, with special reference to artificial aquaculture systems. *Aquaculture* **13,** 1–33.

Winter, J. E., and Langton, R. W. (1976). Feeding experiments with *Mytilus edulis* L. at small laboratory scale. The influence of the total amount of food ingested and food concentration on growth, Vol. 1. *In* "Proc. 10th Europ. Mar. Biol. Symp." (G. Persone and E. Jaspers, eds.), pp. 565–581. Universa Press, Wetteren.

Wright, J. R. (1977), The construction of energy budgets for three intertidal rocky shore gastropods, *Patella vulgata, Littorina littoralis* and *Nucella lapillus.* Ph.D. Thesis, University of Liverpool, England.

Wright, J. R., and Hartnoll, R. G. (1981). An energy budget for a population of the limpet *Patella vulgata. J. Mar. Biol. Assoc. U. K.* **61,** 627–646.

Yonge, C. M. (1949). On the structure and adaptations of the Tellinacea, deposit-feeding Eulamellibranchia. *Phil. Trans R. Soc, London B* **23,** 29–76.

Zeuthen, E. (1947). Body size and metabolic rate in the animal kingdom. *C. R. Lab. Carlsberg. Ser. Chim.* **26,** 15–161.

Zeuthen, E. (1953). Oxygen uptake as related to body size in organisms. *Q. Rev. Biol.* **28,** 1–21.

Index